STRUCTURAL PHASE TRANSITIONS

STRUCTURAL PHASE TRANSITIONS

A. D. BRUCE and R. A. COWLEY
Department of Physics, University of Edinburgh

TAYLOR & FRANCIS LTD
London
1981

First published in ADVANCES IN PHYSICS, *Volume 29, No. 1, January/February 1980.*
This edition published 1981 by Taylor & Francis Ltd,
4 John Street, London WC1N 2ET.

Printed and bound in Great Britain by Taylor & Francis (Printers) Ltd,
Rankine Road, Basingstoke, Hampshire.

ISBN 0 85066 206 0

Preface

The last ten years have witnessed considerable activity in the field of phase transitions; much progress has been made in the matching of theory and experiment. The most clear-cut advances have undoubtedly occurred in the context of critical-point phenomena in magnetic systems and in fluids. This book is intended to provide a comprehensive survey of the physics of the other major class of systems exhibiting phase transformations—crystalline solids undergoing structural phase transitions. Many of the problems posed by such systems have magnetic analogues, and may be treated with the theoretical techniques recently developed in the magnetic context. In addition, however, there are a number of problems unique to structural phase transitions, as a result of which the agreement between theory and experiment is less satisfactory in this area. We hope that this review succeeds in identifying the principal outstanding challenges posed to theorist and experimentalist.

The first of the three parts into which the book is divided is introductory in character: the fundamental concepts of order parameter and susceptibility are introduced, and the nature of the information accessible to the basic experimental probes is reviewed. The rich variety of phenomena associated with structural phase transitions are then surveyed, within the framework of Landau–Devonshire theory. Particular attention is devoted to topics of recent interest, including acoustic instabilities, improper ferroelectrics, competing interactions, incommensurably-modulated phases and lock-in transitions. Finally, the problems arising in the calculation of the parameters of the Landau theory from first principles are elaborated, and the inadequacies of the understanding of this aspect of phase-transitions are chronicled.

The second major part of the book is concerned with the statistical mechanical problems encountered in formulating a more satisfactory theory of static critical behaviour at structural phase transitions: the requisite techniques (the renormalization group and the ε-expansion) are introduced and developed in detail. The theory's predictions for a wide variety of structural transitions are reviewed, and compared with experiment.

The final section addresses the problems posed by the dynamical behaviour of crystals near structural phase transitions, with particular emphasis on quasi-elastic scattering phenomena ("central peaks"). Experimental studies of quasi-elastic scattering in a number of systems are surveyed in detail. The possible origins of the effect (phonon-density fluctuations, intrinsic clusters of precursor order, and defect-induced clusters) are presented and examined critically in the light of the experimental results.

The book owes much to the interactions we have had with many colleagues; to them we extend our thanks. We are also grateful for the patience of Eleanor Rose, who typed much of the script.

Contents

I. Landau theory

R. A. Cowley

II. Static critical behaviour

A. D. Bruce

III. Critical dynamics and quasi-elastic scattering

A. D. Bruce and R. A. Cowley

I

Landau Theory

By

R. A. COWLEY

Summary

Structural phase transitions can be studied by a variety of techniques: macroscopic fields, X-ray, light and neutron diffraction, and resonance methods. The information which can be obtained by these different techniques about the order parameter and susceptibility at structural phase transitions is described and compared. Landau theory coupled with the 'soft mode' concept provides a simple picture of many structural phase transitions in terms of relatively few phenomenological constants. A review is given of the application of this type of theory to the wide variety of different phenomena which can occur at structural phase transitions. In particular, emphasis is placed on the nature of the fluctuations within Landau–Ginzburg theory and on the role of secondary order parameters. The nature of incommensurate phases and of lock-in phase transitions are discussed using both the conventional lattice dynamical approach and the soft soliton theories. The parameters of the Landau theory can, in principle, be obtained from a microscopic model. We critically discuss the attempts that have been made to develop models for anharmonic crystals, order–disorder systems and various electronic instabilities.

§I.1. Introduction

Structural phase transitions occur when a material changes its crystallographic structure. The first studies of these transitions were measurements of macroscopic properties; specific heat, dielectric susceptibility or thermal expansion, and it was through these measurements that the transitions were discovered and initially characterized. Because the macroscopic properties change at the transition, their study is of importance for the various different applications of these materials: an example is the use of ferroelectrics as piezoelectric components and as pyroelectric detectors. There has also been a great deal of effort devoted to elucidating the nature of structural phase transitions at a more microscopic level. This work began with the careful determination of the crystal structure of the phases on both sides of the transition. Some of the earliest measurements were performed on $BaTiO_3$ by Megaw (1947), Kanzig (1951) and by Shirane *et al.* (1955) and most recently by Harada *et al.* (1970). The results of their investigations are illustrated in fig. I.1, which shows the low-temperature phase of $BaTiO_3$, $T < 393$ K, in which the atoms are displaced away from the centro-symmetric positions they have in the high-temperature cubic phase. In ferroelectric $BaTiO_3$ the static displacements of the atoms in every unit cell of the distorted phase are the same.

Fig. I.1

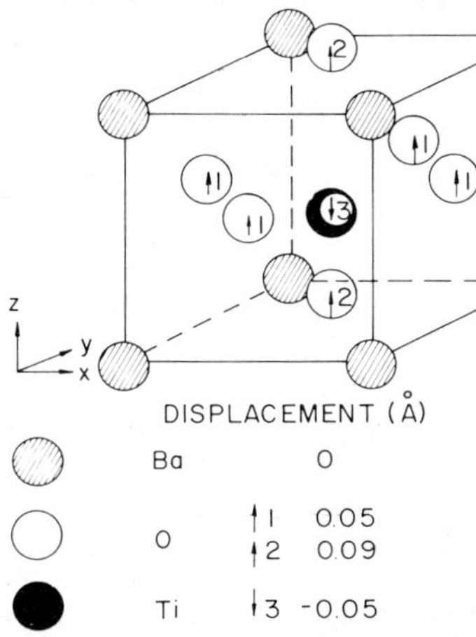

The distorted perovskite structure of ferroelectric $BaTiO_3$.

At other structural phase transitions the unit cell in the distorted phase consists of two, four or maybe more unit cells of the high-temperature phase. As an example of this type of phase transition we show, in fig. I.2, the structure of $SrTiO_3$ below the transition at 110 K. In this case the high-temperature perovskite phase is distorted by an anti-phase rotation of neighbouring oxygen octahedra (Unoki and Sakudo 1967).

Fig. I.2

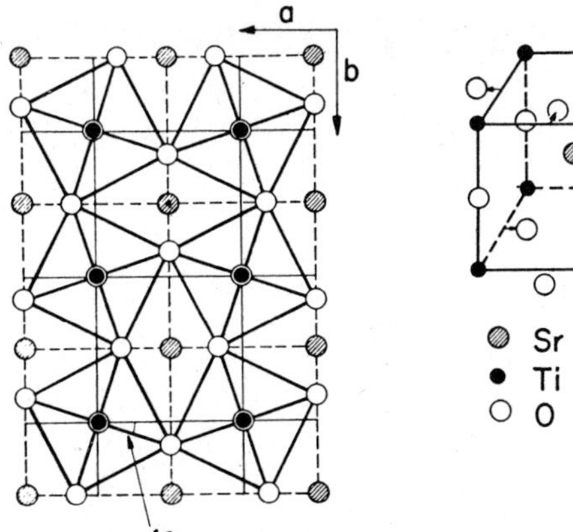

A projection down a cube axis of the distorted phase of $SrTiO_3$ as deduced by Unoki and Sakudo (1967), after Shirane and Yamada (1969).

These measurements showed that at least at some structural phase transitions the atoms in the distorted phase are slightly displaced away from the equilibrium positions of the high-temperature phase. This then led to the suggestion by Cochran (1960) and Anderson (1960), and earlier but largely unnoticed comments by others (Raman and Nedungadi 1940 and Saksena 1940, for example) that these phase transitions might be the result of an instability of the crystal against a particular normal mode of vibration of the high-temperature phase. Since this suggestion, there have been numerous measurements of the so-called 'soft modes' associated with structural phase transitions as reviewed for example by Scott (1974) and by Shirane (1974), and at least for these 'displacive' phase transitions their existence is well established.

There are other structural phase transitions where the phase transition is associated with an ordering of some variable which is disordered in the high-temperature phase. One example of this 'order–disorder' type of behaviour is the ferroelectric phase transition in $NaNO_2$. In the high-temperature (paraelectric) phase the triangular NO_2 groups have the N ions along either the positive or negative b axis with equal probability, but below 436 K the N ions tend to align in the same direction, as shown in fig. I.3 (Hoshino and Motegi 1967). Since the reorientation of the NO_2 groups involves very large displacements, particularly of the N ions, these motions cannot be described in terms of small displacements about the equilibrium positions. Consequently in these order–disorder systems 'soft mode' ideas are less useful than in the displacive materials. Finally, we mention another type of order–disorder phase transition, namely that occurring in alloys. In fig. I.4 we show the structure of ordered CuZn (β-brass); above the ordering temperature, 741 K, the Cu and Zn atoms are randomly arranged on the body-centred cubic lattice, but below the transition they take up the regular arrangement shown.

Fig. I.3

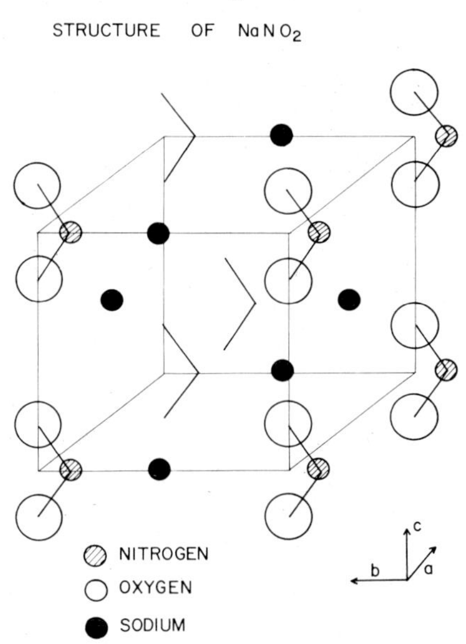

The structure of ferroelectric $NaNO_2$ (Hishano and Motegi 1967). Diagram taken from Sakurai *et al.* (1970).

Fig. I.4

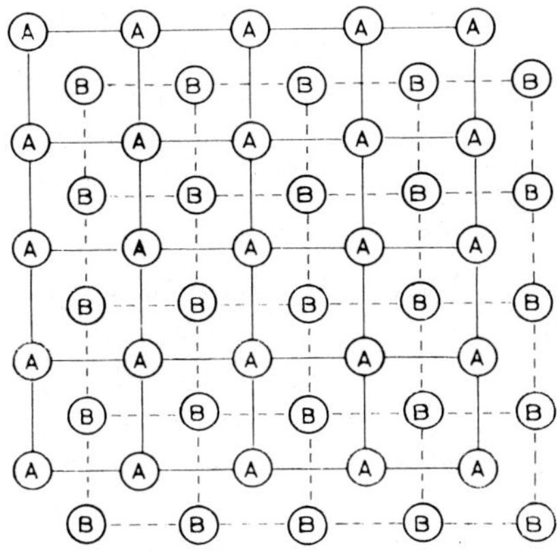

The structure of ordered β-brass.

During the 1960s and 1970s our understanding of structural phase transitions progressed rapidly—in a large part owing to the development of experimental techniques which enabled the microscopic properties to be probed more accurately. In particular, the measurements by neutron and Raman spectroscopy and by magnetic resonance techniques have proved to be especially powerful in exposing the richness and variety of phenomena displayed by systems undergoing structural phase transitions.

It is our intention in this, the first of three articles, to describe these developments within the unified framework afforded by the phenomenological theory of Landau (1937); we defer to parts II and III a discussion of those areas where Landau theory has proved to be inadequate. In this review we shall use as examples the phase transitions in $BaTiO_3$, $SrTiO_3$, $NaNO_2$ and β-brass to illustrate the wide range of the phenomena occurring at structural phase transitions. We shall concentrate on providing a unified framework and approach rather than to provide an exhaustive account of the experimental data and theoretical treatments of any particular material. Excellent books using this latter approach have been written by Jona and Shirane (1962) and more recently by Lines and Glass (1977).

Section I.2 of this article is a review of the notation and the basic lattice dynamical results for the susceptibility and scattering cross-sections of a crystal. In § I.3 we describe and compare the different experimental techniques which can be used to measure the order parameter and susceptibilities close to a phase transition. We also discuss the advantages of different techniques and illustrate the results with particular regard to the phase transitions discussed above.

The simplest theory of structural phase transitions is Landau's (1937) phenomenological theory, which was developed independently in great detail for $BaTiO_3$ by Devonshire (1949). We believe that the first stage in the understanding of a structural phase transition is the development of the appropriate Landau theory in terms of various phenomenological parameters. In § I.4, we describe the application of Landau theory to a very wide range of different structural phase transitions. In view of the recent interest in incommensurate phases, we discuss these in considerable detail.

Landau theory involves a considerable number of parameters which, at the phenomenological level, must be obtained from experimental measurements. In principle they can also be obtained from a microscopic theory which will then answer such questions as why the undistorted phase is unstable, and what is the transition temperature, T_c. In § I.5 we review the different microscopic models which have been developed to describe structural phase transitions and to obtain the parameters of a Landau expansion.

§ I.2. Preliminaries

I.2.1. *The hamiltonian*

Below the critical temperature for a structural phase transition the mean position of an atom is displaced from its position in the high-symmetry phase. Consequently the theory of structural phase transitions makes use of the results and notation developed to describe displacements of the atoms from their equilibrium positions. We adopt the notation which is now commonly used in lattice dynamics (Cochran 1973, Cowley 1968). In the high-temperature, high-symmetry phase the mean equilibrium positions of the atoms are $\mathbf{R}(l\kappa)$, where l denotes the unit cell and κ the κth type of s different atoms within the unit cell, so that $\mathbf{R}(l\kappa)=\mathbf{R}(l)+\mathbf{R}(\kappa)$. The displacements of the atoms from these equilibrium positions are written $\mathbf{u}(l\kappa)$, so that the position of the atoms at any instant is $\mathbf{r}(l\kappa)=\mathbf{R}(l\kappa)+\mathbf{u}(l\kappa)$.

The potential energy of the crystal, V, may then be expanded, formally, as a power series in the displacements, $\mathbf{u}(l\kappa)$. In the harmonic approximation all of the terms of higher order than second in this expansion are neglected. Within the harmonic approximation the equations of motion for an atom can be diagonalized by introducing a plane-wave representation, when the equations decouple for different wave-vectors, and the resultant dynamical matrices can be diagonalized, to give the frequencies $\omega(\mathbf{q}j)$ and normalized eigenvectors $\mathbf{e}(\kappa,\mathbf{q}j)$ for the normal modes. In terms of these normal modes the displacements are written

$$\mathbf{u}(l\kappa)=\sum_{\mathbf{q}j}(NM_\kappa)^{-1/2}\mathbf{e}(\kappa,\mathbf{q}j)\exp(i\mathbf{q}\,.\,\mathbf{R}(l\kappa))Q(\mathbf{q}j), \tag{I.2.1}$$

where M_κ is the mass of atoms of type κ, N is the number of unit cells in the crystal and $Q(\mathbf{q}j)$ is the amplitude of the normal mode, which, since $\mathbf{u}(l\kappa)$ is real, must satisfy the constraint,

$$Q(\mathbf{q}j)=Q^*(-\mathbf{q}j). \tag{I.2.2}$$

In terms of these normal modes the harmonic part of the hamiltonian becomes

$$H_2=\frac{1}{2}\sum_{\mathbf{q}j}[\dot{Q}(\mathbf{q}j)\dot{Q}(-\mathbf{q}j)+\omega(\mathbf{q}j)^2Q(\mathbf{q}j)Q(-\mathbf{q}j)], \tag{I.2.3}$$

where

$$\dot{Q}(\mathbf{q}j)=\frac{dQ(\mathbf{q}j)}{dt}=P(\mathbf{q}j),$$

the momentum conjugate to $Q(\mathbf{q}j)$.

In the usual development of lattice dynamics the normal mode coordinates are written in terms of phonon creation and destruction operators;

$$Q(\mathbf{q}j)=\left(\frac{\hbar}{2\omega(\mathbf{q}j)}\right)^{1/2}(a(\mathbf{q}j)+a^\dagger(-\mathbf{q}j)) \tag{I.2.4}$$

from which we can obtain the thermodynamic average

$$\langle Q(\mathbf{q}j)Q(-\mathbf{q}j)\rangle=\left(\frac{\hbar}{2\omega(\mathbf{q}j)}\right)(2n(\mathbf{q}j)+1), \tag{I.2.5}$$

where $n(\mathbf{q}j)$ is the Bose occupation number, $n(\mathbf{q}j)=1/(\exp(\beta\hbar\omega(\mathbf{q}j))-1)$, and $\beta=(k_{\mathrm{B}}T)^{-1}$. In most applications classical statistics are adequate when eqn. (I.2.5) becomes

$$\langle Q(\mathbf{q}j)Q(-\mathbf{q}j)\rangle=(\beta\omega(\mathbf{q}j)^2)^{-1}. \tag{I.2.6}$$

Structural phase transitions occur only in crystals that are significantly anharmonic and for which the strictly harmonic frequencies, $\omega(\mathbf{q}j)$, may be imaginary. The results of eqns. (I.2.4)–(I.2.6) can, however, be used provided that $\omega(\mathbf{q}j)$ is interpreted as a suitably renormalized frequency as we shall discuss in § I.5.1 (Cowley 1964). We shall need to consider the terms in the expansion of the potential energy beyond second order which may be written as

$$H_{\mathrm{A}}=\sum_{\substack{\mathbf{q}_1\mathbf{q}_2\mathbf{q}_3\\ j_1j_2j_3}} V\begin{pmatrix}\mathbf{q}_1\mathbf{q}_2\mathbf{q}_3\\ j_1j_2j_3\end{pmatrix} Q\begin{pmatrix}\mathbf{q}_1\\ j_1\end{pmatrix} Q\begin{pmatrix}\mathbf{q}_2\\ j_2\end{pmatrix} Q\begin{pmatrix}\mathbf{q}_3\\ j_3\end{pmatrix}+\ldots, \tag{I.2.7}$$

where the detailed form of the coefficients is given in terms of the atomic potential functions and the normal mode eigenvectors by Born and Huang (1954). Since in a pure system wave-vector is conserved $\mathbf{q}_3=-\mathbf{q}_1-\mathbf{q}_2+\boldsymbol{\tau}$, where $\boldsymbol{\tau}$ is any reciprocal lattice vector of the high-symmetry phase.

At many structural phase transitions the size and shape of the crystal alters. These changes are described by introducing infinitesimal strain parameters, $\eta_{\alpha\beta}$. The displacement of an atom due to an infinitesimal strain is

$$u_\alpha(l\kappa)=\sum_\beta \eta_{\alpha\beta}R_\beta(l\kappa). \tag{I.2.8}$$

A comparison of eqns. (I.2.8) and (I.2.1) then suggests that

$$\eta_{\alpha\beta}=\lim_{q_\beta\to 0}\frac{iq_\beta}{\sqrt{(N\sum_\kappa M_\kappa)}}Q(q_\beta,\alpha), \tag{I.2.9}$$

where $Q(q_\beta,\alpha)$ is the acoustic mode amplitude with eigenvector polarized in the α direction and wave-vector in the β direction. Alas, however, it is not possible to represent any particular combination of strain parameters in terms of the acoustic mode amplitudes: consequently, the terms involving the strains (or long wavelength acoustic modes) must be excluded from the summation over $(\mathbf{q}j)$ in eqns. (I.2.3) and (I.2.7) (Born and Huang 1954) and written as further contributions to the hamiltonian, with the form

$$H_{\mathrm{D}}=\sum_{\alpha\beta}\eta_{\alpha\beta}\left[\sum_j V_{\alpha\beta}\begin{pmatrix}0\\ j\end{pmatrix}Q\begin{pmatrix}0\\ j\end{pmatrix}+\sum_{\mathbf{q}jj'}V_{\alpha\beta}\begin{pmatrix}\mathbf{q} & -\mathbf{q}\\ j & j'\end{pmatrix}Q\begin{pmatrix}\mathbf{q}\\ j\end{pmatrix}Q\begin{pmatrix}-\mathbf{q}\\ j'\end{pmatrix}+\ldots\right]$$
$$+\frac{v}{2}\sum_{\alpha\beta\gamma\delta}\eta_{\alpha\beta}\eta_{\gamma\delta}[C_{\alpha\beta\gamma\delta}+\ldots], \tag{I.2.10}$$

where we have written only the three most important terms in the double power series expansion about the equilibrium positions of the high-symmetry phase. The first term in eqn. (I.2.10) describes the coupling of a strain to a long wavelength optic mode of vibration, while in the second term use has been made of wave-vector conservation. The third term contains the usual elastic constant multiplied by the volume of a unit cell.

At order–disorder phase transitions, as discussed in the introduction (figs. I.3 and I.4), the different possible positions of the atoms may be a considerable distance apart. Although it is always possible to express the hamiltonian in a power series in the displacements, the series will be, at best, only very slowly convergent and consequently very difficult to employ in practice. An alternative approach is to introduce a new Ising spin variable which describes the large-scale motions. In the case of $NaNO_2$ the spin variable locates the N atom of each NO_2 group in either the positive, $S(l\kappa)=\frac{1}{2}$, or negative, $S(l\kappa)=-\frac{1}{2}$, b direction. In the case of $NaNO_2$ there is only one NO_2 group in each unit cell so that we may then drop the suffix, κ, and have a single spin variable $S(l)$ describing the orientation of each NO_2 group.

A similar notation may be developed for a binary, AB, order–disorder alloy. If the crystal has only one type of site (i.e., one value of κ) in the high-symmetry phase then we may introduce a spin variable $S(l)=\frac{1}{2}$ if the site l is occupied by an A atom and $-\frac{1}{2}$ if it is occupied by a B atom. This model may also be extended to include more complex alloys and more complex high-temperature phases.

In the case of the AB alloy the potential energy between atoms at $\mathbf{R}(l)$ and $\mathbf{R}(l')$ may have different functional forms according to whether both atoms are of type A, both are of type B, or if there is one of each. Denoting these interaction energies by ϕ_{AA}, ϕ_{BB} and $\phi_{AB}=\phi_{BA}$ the energy of interaction between the two sites may then be written as

$$\phi(ll')=\phi_0(ll')+2\phi_1(ll')(S(l)+S(l'))+4\phi_2(ll')S(l)S(l'), \tag{I.2.11}$$

where

$$\phi_0=\frac{1}{4}(\phi_{AA}+\phi_{BB}+2\phi_{AB}),$$

$$\phi_1=\frac{1}{4}(\phi_{AA}-\phi_{BB}),$$

$$\phi_2=\frac{1}{4}(\phi_{AA}+\phi_{BB}-2\phi_{AB}).$$

This is then a general representation of the potential between l and l' irrespective of the occupation of the sites.

The third term of eqn. (I.2.11) gives rise to an interaction between the spin variables at different sites. It is frequently useful to diagonalize this term with the atoms at their mean equilibrium positions to obtain eigenvectors $E(\kappa,\mathbf{q}j)$ and eigenvalues $J(\mathbf{q}j)$ such that

$$S(l\kappa)=\frac{1}{\sqrt{N}}\sum_{\mathbf{q}j}E(\kappa,\mathbf{q}j)\exp(i\mathbf{q}\,.\,\mathbf{R}(l\kappa))S(\mathbf{q}j), \tag{I.2.12}$$

where $S(\mathbf{q}j)$ is the amplitude of the spin normal mode.

I.2.2. *Susceptibility*

The change in the property of a system when an external field is applied is described by the susceptibility: for example, the dielectric susceptibility gives the change in the dielectric polarization on application of an electric field. More generally, we consider a property of the system described by an operator, $B(\mathbf{r})$, which is dependent upon position, and apply to the system some time-dependent 'field' $E(\mathbf{r})$ whose interaction with the system we write as $H' = -AE$ with $A(\mathbf{r})$ some other operator. We may then introduce a position-dependent susceptibility

$$\chi_{BA}(\mathbf{r}-\mathbf{r}') = \frac{\partial}{\partial E(\mathbf{r}')}\langle B(\mathbf{r})\rangle|_{E(\mathbf{r}')=0}. \tag{I.2.13}$$

Writing the response in the form

$$\langle B(\mathbf{r})\rangle = \mathrm{Tr}\,\{\exp(-\beta(H+H'))B\}/\mathrm{Tr}\,\{\exp(-\beta(H+H'))\}, \tag{I.2.14}$$

where H is the hamiltonian of the unperturbed system, the susceptibility follows from eqn. (I.2.13) in the classical limit, as

$$\chi_{BA}(\mathbf{r}-\mathbf{r}') = \beta\langle B(\mathbf{r})A(\mathbf{r}')\rangle. \tag{I.2.15}$$

Since most of our interest is directed towards pure materials (where wave-vector conservation plays an important role), it proves useful to introduce the Fourier transform of this susceptibility,

$$\chi_{BA}(\mathbf{K}) = \beta\langle B(\mathbf{K})A(-\mathbf{K})\rangle, \tag{I.2.16}$$

where the operators $B(\mathbf{K})$ are the Fourier transforms of $B(\mathbf{r})$

$$B(\mathbf{K}) = \int B(\mathbf{r})\exp(-i\mathbf{K}\,.\,\mathbf{r})\,d^3r. \tag{I.2.17}$$

The operators $B(\mathbf{K})$ may be expanded formally in powers of normal mode coordinates $Q(\mathbf{q}j)$ (or, in the appropriate cases, the 'spin' coordinates $S(\mathbf{q}j)$): thus we write

$$B(\mathbf{K}) = B^0(\mathbf{K})\Delta(\mathbf{K}) + \left(\frac{1}{N}\right)^{1/2}\sum_{\mathbf{q}j} B(\mathbf{K},\mathbf{q}j)Q(\mathbf{q}j)\Delta(\mathbf{K}+\mathbf{q}), \tag{I.2.18}$$

where $B^0(\mathbf{K})$ is the operator for the undistorted crystal. Wave-vector conservation is ensured by the delta function which is defined by

$$\Delta(\mathbf{K}) = \sum_l \exp(i\mathbf{K}\,.\,\mathbf{R}(l)) = \frac{(2\pi)^3}{v}\sum_{\boldsymbol{\tau}} \delta^{(3)}(\mathbf{K}-\boldsymbol{\tau}), \tag{I.2.19}$$

where the summation is over the reciprocal lattice vectors, $\boldsymbol{\tau}$, of the undistorted lattice, while $\delta^{(3)}(\mathbf{K})$ is the normal three-dimensional delta function.

The expressions for the coefficients in the expansion, eqn. (I.2.18), may be found once an expression is known for the operator $B(\mathbf{r})$. For example, if $B(\mathbf{r})$ is the dipole moment operator

$$B(\mathbf{r}) = \mathbf{M}(\mathbf{r}) = \sum_{l\kappa} Z_\kappa \mathbf{u}(l\kappa)\delta(\mathbf{r}-\mathbf{R}(l\kappa)),$$

where Z_κ is the apparent charge, then

$$\mathbf{M}(\mathbf{K},\mathbf{q}j) = \sum_\kappa Z_\kappa M_\kappa^{-1/2}\mathbf{e}(\kappa,\mathbf{q}j)\exp(i(\mathbf{K}+\mathbf{q})\,.\,\mathbf{R}(\kappa)), \tag{I.2.20}$$

while $B^0(\mathbf{K}) = 0$.

Now, returning to the susceptibility function, eqn. (I.2.16), we make use of the expansion (I.2.18) for $B(\mathbf{K})$ and $A(\mathbf{K})$, picking out, in each case, the term in the expansion that is linear in the normal coordinate. The resulting contribution to the susceptibility is

$$\chi_{BA}(\mathbf{K})=\sum_{\mathbf{q}jj'}\Delta(\mathbf{K}+\mathbf{q})B(\mathbf{K},\mathbf{q}j)A(-\mathbf{K},\mathbf{q}j')\chi(\mathbf{q}jj'). \tag{I.2.21}$$

The importance of this result is that it shows that any one-phonon susceptibility can be obtained in terms of the susceptibility for the normal mode operators,

$$\chi(\mathbf{q}jj')=\beta\langle Q(\mathbf{q}j)Q(-\mathbf{q}j')\rangle. \tag{I.2.22}$$

Using the results quoted above, eqn. (I.2.6), for the harmonic crystal with classical statistics, this becomes,

$$\chi_{\mathrm{H}}(\mathbf{q}jj')=\omega(\mathbf{q}j)^{-2}\delta_{jj'}. \tag{I.2.23}$$

A similar analysis may be carried through for an order–disorder system. The operator, $B(\mathbf{r})$, is expanded in a power series in the spin operators (or, if necessary a double power series in the spin and displacement operators) with

$$B(\mathbf{K})=B^{0}(\mathbf{K})\Delta(\mathbf{K})+\left(\frac{1}{N}\right)^{1/2}\sum_{\mathbf{q}j}B_{s}(\mathbf{K},\mathbf{q}j)S(\mathbf{q}j)\Delta(\mathbf{K}+\mathbf{q})+\ldots, \tag{I.2.24}$$

where the coefficients $B_s(\mathbf{K},\mathbf{q}j)$ are given in terms of the spin eigenvectors. If the dipole moment associated with the κth type of molecular group is $\mathbf{p}_\kappa$, then the coefficients in the expansion of the dipole moment operator are given by

$$\mathbf{M}_{s}(\mathbf{K},\mathbf{q}j)=2\sum_{\kappa}\mathbf{p}_{\kappa}E(\kappa,\mathbf{q}j)\exp\left(i(\mathbf{K}+\mathbf{q})\,.\,\mathbf{R}(\kappa)\right). \tag{I.2.25}$$

The susceptibility is then given by eqn. (I.2.21) but with the normal mode susceptibility replaced by

$$\chi_{s}(\mathbf{q}jj')=\beta\langle S(\mathbf{q}j)S(-\mathbf{q}j')\rangle. \tag{I.2.26}$$

The susceptibility can be generalized to describe the response of a system to a time-dependent field, $E=\frac{1}{2}E_0\ (\exp(i\omega\tau)+\exp(-i\omega\tau))$. The susceptibility is then dependent upon the frequency ω as well as the wave-vector $\mathbf{K}$ and written $\chi_{BA}(\mathbf{K},\omega)$. The calculation of this frequency-dependent susceptibility is more difficult than that of the static susceptibility described above but is detailed in many texts such as Landau and Lifshitz (1959). Since a time-dependent field may lose energy to the system, the frequency-dependent susceptibility has both real and imaginary parts which are connected by a Kramers–Kronig relation

$$\chi^{\mathrm{R}}_{BA}(\mathbf{K},\omega)=\frac{1}{\pi}\int\frac{\chi^{\mathrm{I}}_{BA}(\mathbf{K},\omega')}{(\omega'-\omega)_{p}}\,d\omega', \tag{I.2.27}$$

and which satisfy the symmetry properties

$$\chi^{\mathrm{R}}_{BA}(\mathbf{K},\omega)=\chi^{\mathrm{R}}_{AB}(-\mathbf{K},-\omega)\quad\text{and}\quad\chi^{\mathrm{I}}_{BA}(\mathbf{K},\omega)=-\chi^{\mathrm{I}}_{AB}(-\mathbf{K},-\omega).$$

Since energy must be lost to the system, the imaginary part of the diagonal part of the susceptibility, $\chi^{\mathrm{I}}_{AA}(\mathbf{K},\omega)$, is positive for $\omega>0$.

Finally the dynamic susceptibility can be obtained for one-phonon processes as described above for the static susceptibility. The dynamic one-phonon susceptibility $\chi(\mathbf{q}jj',\omega)$ can then be calculated explicitly for harmonic normal modes as

$$\chi^{\mathrm{R}}(\mathbf{q}jj',\omega)=(\omega(\mathbf{q}j)^2-\omega^2)^{-1}\delta_{jj'}, \tag{I.2.28 a}$$

and

$$\chi^{\mathrm{I}}(\mathbf{q}jj',\omega)=\frac{\pi}{2\omega(\mathbf{q}j)}[\delta(\omega-\omega(\mathbf{q}j))-\delta(\omega+\omega(\mathbf{q}j))]\delta_{jj'}. \tag{I.2.28 b}$$

I.2.3. *Scattering cross-sections*

The calculation of the cross-section for the scattering of an incident plane-wave beam of particles with wave-vector $\mathbf{k}_1$ and energy E_0 into a plane-wave beam with wave-vector $\mathbf{k}_2$ and energy E is described by Van Hove (1954). Within the first Born approximation the differential cross-section can be written as

$$\frac{d^2\sigma}{d\Omega d\omega}=\frac{k_2}{k_1}S(\mathbf{K},\omega), \tag{I.2.29}$$

where $\mathbf{K}$ is the wave-vector transfer $\mathbf{K}=\mathbf{k}_1-\mathbf{k}_2$, and $\hbar\omega=E_0-E$ is the energy transfer. The Van Hove scattering function, $S(\mathbf{K},\omega)$, is given by

$$S(\mathbf{K},\omega)=\sum_{\mathrm{if}}p_{\mathrm{i}}F(\mathbf{K})_{\mathrm{if}}F(-\mathbf{K})_{\mathrm{fi}}\delta(\omega-\omega_{\mathrm{fi}}), \tag{I.2.30}$$

where $F(\mathbf{K})_{\mathrm{if}}$ is the expectation value of an operator describing the interaction between the initial and final states of the system. It is sometimes more convenient to rewrite eqn. (I.2.30), making use of the Heisenberg representation of the operators:

$$S(\mathbf{K},\omega)=\frac{1}{2\pi}\int_{-\infty}^{\infty}\langle F(\mathbf{K},\tau)F(-\mathbf{K},0)\rangle\exp(i\omega\tau)\,d\tau. \tag{I.2.31}$$

In the case of X-ray or neutron scattering, the interaction between the particles and the specimen can be expressed in terms of a scattering length $b_{l\kappa}$ as

$$F(\mathbf{K},\tau)=\sum_{l\kappa}b_{l\kappa}\exp(i\mathbf{K}\,.\,\mathbf{r}(l\kappa,\tau)), \tag{I.2.32}$$

where $\mathbf{r}(l\kappa,\tau)$ is the instantaneous position of the atoms $(l\kappa)$. For the scattering of X-rays by rigid atoms of a particular type κ, the scattering length $b_{l\kappa}$ is the same for each site l, being equal to the scattering power of each single electron multiplied by the atomic form factor, which is dependent upon the wave-vector, $\mathbf{K}$. For neutrons the scattering length may vary from site to site, owing to the randomness in the isotopic distribution and the spin alignment of the nuclei. This randomness results in incoherent scattering which usually gives rise to a troublesome background: we shall largely neglect it. The coherent scattering reflects the average scattering length $b_\kappa=1/N\sum_l b_{l\kappa}$ and hence includes the interference effects between the different ions. The X-ray or coherent neutron scattering is then given by

$$F(\mathbf{K},\tau)=\sum_{\kappa}b_\kappa\sum_{l}\exp(i\mathbf{K}\,.\,\mathbf{r}(l\kappa,\tau)). \tag{I.2.33}$$

The operator $F(\mathbf{K},\tau)$ can then be obtained by expanding the position coordinate $\mathbf{r}(l\kappa,\tau)$ in powers of the displacements. When these displacements are transformed into the normal mode coordinates one finds

$$F(\mathbf{K},\tau)=F^0(\mathbf{K})\Delta(\mathbf{K})+\left(\frac{1}{N}\right)^{1/2}\sum_{\mathbf{q}j}F(\mathbf{K},\mathbf{q}j)\Delta(\mathbf{K}+\mathbf{q})Q(\mathbf{q}j,\tau)+\ldots, \quad (\text{I}.2.34)$$

where $Q(\mathbf{q}j,\tau)$ is the instantaneous amplitude of the normal mode coordinate, and the coefficients are given by

$$F^0(\mathbf{K})=\sum_{\kappa}b_\kappa \exp(i\mathbf{K}\,.\,\mathbf{R}(\kappa))\exp(-W_\kappa), \quad (\text{I}.2.35)$$

while

$$F(\mathbf{K},\mathbf{q}j)=\sum_{\kappa}b_\kappa M_\kappa^{-1/2}(\mathbf{K}\,.\,\mathbf{e}(\kappa,\mathbf{q}j))\exp(i(\mathbf{K}+\mathbf{q})\,.\,\mathbf{R}(\kappa))\exp(-W_\kappa), \quad (\text{I}.2.36)$$

where W_κ is the Debye–Waller factor. Using these results $S(\mathbf{K},\omega)$ can be divided into a Bragg part, $S^0(\mathbf{K},\omega)$, and a one-phonon part, $S^1(\mathbf{K},\omega)$, where

$$S^0(\mathbf{K},\omega)=N|F^0(\mathbf{K})|^2\delta(\omega)\Delta(\mathbf{K}) \quad (\text{I}.2.37)$$

and

$$S^1(\mathbf{K},\omega)=\sum_{\mathbf{q}jj'}\Delta(\mathbf{K}+\mathbf{q})F(\mathbf{K},\mathbf{q}j)F(-\mathbf{K},-\mathbf{q}j')\mathscr{S}(\mathbf{q}jj',\omega), \quad (\text{I}.2.38)$$

where $\mathscr{S}(\mathbf{q}jj',\omega)$ is the displacement–displacement correlation function

$$\mathscr{S}(\mathbf{q}jj',\omega)=\frac{1}{2\pi}\int_{-\infty}^{\infty}\langle Q(\mathbf{q}j,\tau)Q(-\mathbf{q}j',0)\rangle\exp(i\omega\tau)d\tau. \quad (\text{I}.2.39)$$

This displacement–displacement correlation function can be related to the imaginary part of the dynamic susceptibility introduced in the previous section as shown for example by Cowley (1963). The result is that

$$(n(\omega)+1)\chi^{\text{I}}(\mathbf{q}jj',\omega)=\frac{\pi}{\hbar}\mathscr{S}(\mathbf{q}jj',\omega), \quad (\text{I}.2.40)$$

where $n(\omega)$ is the Bose–Einstein factor. Furthermore the static displacement–displacement correlation function

$$\mathscr{S}(\mathbf{q}jj')=\int_{-\infty}^{\infty}\mathscr{S}(\mathbf{q}jj',\omega)\,d\omega \quad (\text{I}.2.41)$$

is within the classical approximation, $\hbar\omega\ll k_{\text{B}}T$, related to the static susceptibility, by use of eqn. (I.2.27), to give

$$\beta\mathscr{S}(\mathbf{q}jj')=\chi(\mathbf{q}jj'). \quad (\text{I}.2.42)$$

Since all X-ray and many neutron scattering experiments measure the integrated scattering, this relationship is of considerable importance.

These results for the susceptibility and scattering show the one-phonon parts are determined by the normal mode susceptibility multiplied by the appropriate coefficients. Provided, then, that these coefficients can be treated as wave-vector-, frequency- and temperature-independent constants, the behaviour of the susceptibility or scattering cross-section close to a phase transition will be very similar, irrespective of the probe used to perform the measurements.

A similar analysis can be carried through for an order–disorder system. For example in an AB alloy the coherent scattering lengths may be b_A and b_B, which are not usually the same. The structure factor, eqn. (I.2.33), then becomes

$$F(\mathbf{K},\tau)=\sum_l\sum_\kappa(\bar{b}_\kappa+S(l\kappa,\tau)b'_\kappa)\exp(i\mathbf{K}\,.\,\mathbf{r}(l\kappa,\tau)), \tag{I.2.43}$$

where $\bar{b}_\kappa=\frac{1}{2}(b_A+b_B)$ and $b'_\kappa=b_A-b_B$. The expansion of $F(\mathbf{K},\tau)$ is now given by eqn. (I.2.34) with $\bar{b}_\kappa$ replacing b_κ. The leading term in the spin variables is given by

$$\left(\frac{1}{N}\right)^{1/2}\sum_{\mathbf{q}j}F_s(\mathbf{K},\mathbf{q}j)\Delta(\mathbf{K}+\mathbf{q})S(\mathbf{q}j,\tau), \tag{I.2.44}$$

where

$$F_s(\mathbf{K},\mathbf{q}j)=\sum_\kappa b'_\kappa\exp(-W_\kappa)E(\kappa,\mathbf{q}j)\exp(i(\mathbf{K}+\mathbf{q})\,.\,\mathbf{R}(\kappa)). \tag{I.2.45}$$

Substitution of eqn. (I.2.44) into eqn. (I.2.31) and repeating the development given above then shows that the cross-section mirrors the behaviour of the spin correlation function given by eqn. (I.2.26).

Finally to determine the Raman scattering of laser light requires an expansion of the polarizability of the material in powers of the displacements. Formally the expansion can be written

$$P_{\alpha\beta}(\mathbf{K})=P^0_{\alpha\beta}(\mathbf{K})\Delta(\mathbf{K})+\left(\frac{1}{N}\right)^{1/2}\sum_{\mathbf{q}j}[P_{\alpha\beta}(\mathbf{q}j)Q(\mathbf{q}j)\Delta(\mathbf{K}+\mathbf{q})$$
$$+P^{(s)}_{\alpha\beta}(\mathbf{q}j)S(\mathbf{q}j)\Delta(\mathbf{K}+\mathbf{q})]+\ldots. \tag{I.2.46}$$

Substituting this expansion into eqn. (I.2.31) yields an expression for the Raman scattering tensor $I_{\alpha\beta\gamma\delta}$, which determines the scattering into a polarization $(\beta\delta)$ from an incident beam polarized in the $(\alpha\gamma)$ direction. The cross-section is thus dependent upon the displacement–displacement or spin–spin correlation functions, as in the case of X-ray and neutron scattering. Unlike the neutron or X-ray scattering case, however, in Raman scattering the wave-vector transfer $\mathbf{K}$, is restricted to be much smaller than a reciprocal lattice vector.

§I.3. Order parameters and fluctuations

I.3.1. *Primary order parameters*

At a structural phase transition the symmetry of the crystal changes: some of the symmetry operations of the high-symmetry phase, which is usually but not always the high-temperature phase (see §I.4.5), are not symmetry operations of the low-symmetry phase. Above the ferroelectric phase transition of $BaTiO_3$, fig. I.1, each Ba and Ti ion is at a site having the full cubic symmetry but in the distorted phase the symmetry at each site is only tetragonal. This breaking, in the low-symmetry phase, of some of the symmetry operations of the high-symmetry phase is the essential characteristic of most phase transitions. (However, cf. §I.4.3.2.)

The displacements of the atoms which are associated with the breaking of the symmetry are characteristic of the distorted phase, and, furthermore, at a continuous phase transition decrease continuously to zero as the high-symmetry phase is approached. In a single domain of the distorted phase, the displacements in any one unit cell have a definite relationship to those occurring in any other unit cell. In $BaTiO_3$ the displacements are equal in every unit cell, while in $SrTiO_3$, fig. I.2, they are of equal magnitude but are systematically in opposite directions in neighbouring unit cells. This long-range order is characteristic of distorted phases, and suggests that the displacements can be described in terms of the normal modes, or spin variables, introduced in § I.2. This is indeed the case at the majority of structural phase transitions—one exception being the lock-in phase transitions discussed in § I.4.5.

The simplest example of a structural phase transition occurs when the distortions may be described by a single normal mode; the amplitude of the distortion is then given by the order parameter

$$Q_0 = N^{-1/2} \langle Q(\mathbf{q}_s j) \rangle, \tag{I.3.1}$$

where $(\mathbf{q}_s j)$ is the particular normal mode which describes the atomic displacements in the distorted phase, and the factor of $N^{1/2}$ is introduced so that Q_0 is independent of the crystal size.

If the phase transition is continuous the order parameter will approach zero as $T \to T_c$ and is assumed to be given by a power law;

$$Q_0 \sim (T_c - T)^{\beta}, \tag{I.3.2}$$

where β is known as the critical exponent for the order parameter. In classical theories of phase transitions, such as those discussed in §§ I.4, and I.5 the exponent $\beta = 0{\cdot}5$, while in practice β varies between 0·25 and 0·5 as described in part II.

Frequently the possible atomic displacements of the distorted phase cannot be described solely in terms of a single normal mode, but only in terms of linear combinations of n normal mode coordinates; n is then known as the number of components of the order parameter (or the order parameter dimensionality) and in the high-symmetry phase all these n normal modes have the same frequency. We shall write the amplitudes of these normal modes

$$Q_i = N^{-1/2} \langle Q(\mathbf{q}_{si} j_i) \rangle$$

and the order parameter in the distorted phase as

$$Q_i = Q_0 \zeta_i, \tag{I.3.3}$$

where

$$\sum_i \zeta_i^2 = 1, \tag{I.3.4}$$

and the parameters ζ_i determine the structure of the distorted phase, while Q_0 varies with temperature as described by eqn. (I.3.2).

Both $BaTiO_3$ and $SrTiO_3$ are examples of systems where $n = 3$. In the case of $BaTiO_3$ the ordered phase may be characterized by displacements along the x, y or z axes, each of which corresponds to a different normal mode, while in $SrTiO_3$ the oxygen octahedra can rotate about any of three cube axes. In both of these cases the wave-vectors $\mathbf{q}_{si}$ are the same for all three values of i; $\mathbf{q}_s = 0$ for $BaTiO_3$, $\mathbf{q}_s = (\frac{1}{2}, \frac{1}{2}, \frac{1}{2}) 2\pi/a$ for $SrTiO_3$. For this type of phase transition n cannot be larger than 3. Recently, however, there has been considerable interest in incommensurate phase transitions, where (in cubic materials) the value of n may, in principle, be as large as 48, as described in § I.4.4.

Although we have introduced the concept of an order parameter through the displacive phase transitions shown in figs. I.1 and I.2, the same approach may be used to specify an order parameter in the order–disorder case when

$$Q_0 = N^{-1/2}\langle S(\mathbf{q}_s j)\rangle. \tag{I.3.5 a}$$

Thus, for example in the case of $NaNO_2$, the order parameter gives a measure of the difference between the numbers of NO_2 groups displaced along the positive and negative b axis.

A more formal description of the order parameter at a phase transition may be given in the language of group theory (Landau and Lifshitz 1959). The density distribution of a crystal $\rho(\mathbf{r})$ can be expressed as a linear combination of normalized functions, ϕ, which transform into one another under the symmetry operations of the group, G, of the high-symmetry phase. It is convenient to choose the functions ϕ so that they divide into a sum of sets of functions which transform only within the set under the symmetry operations. These functions, ϕ_i^p, then form the bases of the irreducible representations of G, where p labels the irreducible representation and i the number of the functions in that representation. Then amongst the functions ϕ_i^p there is one which is invariant under G, and this describes the structure $\rho_0(\mathbf{r})$ of the high-symmetry phase. Any distortion $\delta\rho(\mathbf{r})$ can then be written as

$$\delta\rho = \sum_p{}' \sum_i C_i^p \phi_i^p, \tag{I.3.5 b}$$

where Σ' omits the term describing the structure of the high-symmetry phase. Now at a continuous phase transition the distortion in structure can be described in terms of only one irreducible representation, $p=s$, say; the coefficients C_i^s then describe the structure of the distorted phase.

The displacements of the atoms about their equilibrium positions are a set of functions which transform under the symmetry operations of the group, G. In setting up the normal modes in § I.2 the symmetry of the crystal ensures that each normal mode transforms as a particular irreducible representation of the space group. The functions $Q(\mathbf{q}j)$ are then proportional to the functions ϕ_i^p introduced above and the coefficients C_i^s are proportional to the order parameters Q_i. The number of components of the order parameter, n, is then the dimensionality of the sth irreducible representation of the full space group.

I.3.2. *Secondary order parameters*

The term secondary order parameters is used, alas, in two different ways. In the first instance, it is used to describe distortions (in the low-symmetry phase), which arise at the phase transition, but which do not break as many of the symmetry elements of the high-symmetry phase as are broken by the primary ordering. This concept may, perhaps, be clarified by considering $BaTiO_3$ and $SrTiO_3$, for both of which there is a macroscopic strain in the distorted phase in addition to the primary distortion. In the case of $BaTiO_3$ the point group of the high-temperature phase has the full cubic symmetry, m3m, and the distorted phase has the tetragonal point group 4 mm. If, however, the distortion to a tetragonal structure arose purely from the tetragonal strain, $\eta_{zz} - (\eta_{xx} + \eta_{yy}/2)$, then the distorted phase would have the symmetry 4/mmm. The point group 4/mmm can be obtained from 4 mm by adding a mirror plane perpendicular to the tetragonal axis. The symmetry of this mirror plane is broken by the distortions described by the primary order parameter but is not broken by the secondary order parameter, the tetragonal strain.

In the case of $SrTiO_3$ the symmetry of the low-temperature phase is 4/mmm but contains two formula units in each unit cell, whereas if the only deformation was the strain the point group symmetry would be 4/mmm but there would be only one formula unit in each unit cell. Consequently the primary order parameter breaks the translational symmetry between the two formula units in each unit cell of the distorted phase whereas the secondary order parameter (the tetragonal strain) does not.

Since the distortions produced by a secondary order parameter break the symmetry of the high-symmetry phase they must decrease to zero as $T \to T_c$. We therefore write the temperature dependence of a secondary order parameter, $\bar{Q}$, whether it be the strain or another normal mode, as

$$\bar{Q} \sim (T_c - T)^{\tilde{\beta}}. \tag{I.3.6}$$

Since the secondary order parameter does not break all the symmetry elements that are broken at T_c, it must, when close to T_c have a vanishingly small amplitude compared with the primary order parameter; thus one expects that $\tilde{\beta} > \beta$.

The other use of the term secondary order parameter arises when there is more than one normal mode ($\mathbf{q}j$) which has the symmetry of the primary order parameter. For example, in $BaTiO_3$ there are three different frequencies of the transverse optic modes of long wavelength. Although we can choose the mode eigenvectors so that the displacements close to T_c are given by only one of these modes, as the distortions become larger the pattern of the atomic displacements may alter and this can be represented as a coupling to the other normal modes.

I.3.3. *Experimental determination of the order parameter*

I.3.3.1. *Macroscopic measurements*

If the order parameter has wavevector $\mathbf{q}_s = 0$, a ferrodistortive transition, then its temperature dependence may be determined by macroscopic measurements. For example, in ferroelectric materials the spontaneous polarization, $\mathbf{P}_s$, is proportional to the amplitude of the distortion of a long wavelength infra-red active mode. In the notation developed above

$$\mathbf{P}_s = \frac{1}{vN^{1/2}} \sum_j \mathbf{M}(0j) \langle Q(0j) \rangle, \tag{I.3.7}$$

where v is the volume of the unit cell and the dipole moment operator of the mode is given by eqn. (I.2.20).

If there is only one mode j, eqn. (I.3.1) shows that

$$\mathbf{P}_s = \frac{1}{v} \mathbf{M}(0j) Q_0, \tag{I.3.8}$$

showing that a measurement of $\mathbf{P}_s$ determines the order parameter Q_0. In fig. I.5 we show the measured spontaneous polarization for $BaTiO_3$ (Merz 1953); note that the phase transition in $BaTiO_3$ is of first order.

The spontaneous polarization gives a measure of the primary order parameter only if the phase transition is to a ferroelectric phase. If the order parameter

Fig. I.5

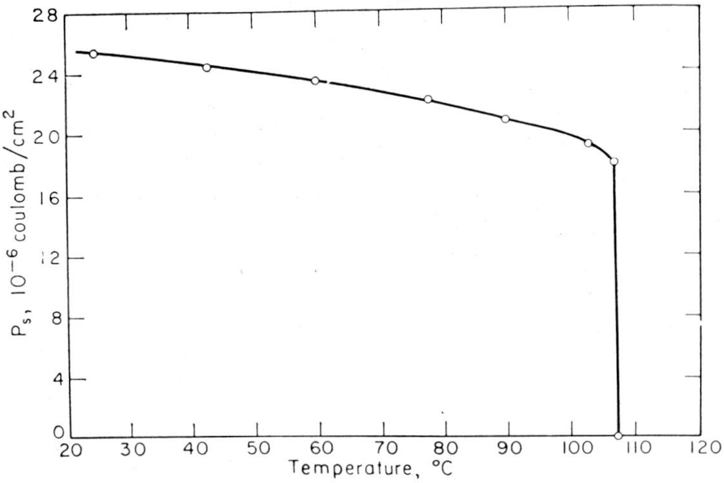

The spontaneous polarization of $BaTiO_3$ as a function of temperature (Merz (1953) after Jona and Shirane (1962)).

corresponds to a $\mathbf{q}_s=0$ mode of even symmetry, then there will be a change in the refractive index or optical polarizability of the crystal, proportional to the order parameter Q_0. For example, the optical polarizability is given by eqn. (I.2.46) as an expansion in powers of the normal mode coordinates. If the leading term is non-zero the change in polarizability is given by

$$\delta P_{\alpha\beta}=P_{\alpha\beta}(0j)Q_0, \tag{I.3.9}$$

if there is only one mode j. Harley and MacFarlane (1975) have measured the order parameter in $DyVO_4$ and $TbVO_4$ using the birefringence of the crystal as illustrated in fig. II.15 of part II.

Finally if the primary order parameter is the macroscopic strain, it can be measured macroscopically, or possibly more reliably by using neutron or X-ray diffraction.

Macroscopic techniques have the disadvantage that they can only measure the primary order parameter at ferrodistortive, $\mathbf{q}_s=0$, phase transitions, and furthermore require single domain samples to be available. Unfortunately single domain samples can usually be produced only by applying a field to the specimen which may then influence the results. Also close to T_c, the susceptibility becomes large, see § I.3.4, and the measurement of a small spontaneous polarization, in the presence of a large dielectric constant is far from being a technically easy experiment.

Secondary order parameters can be measured using macroscopic techniques, but the results for the spontaneous polarization or refractive index may be difficult to interpret. This is because the two-phonon term in the expansion of the dipole moment or polarizability, eqn. (I.2.46), is of the form,

$$P^{(2)}_{\alpha\beta}(\mathbf{K})=\frac{1}{N}\sum_{\mathbf{q}_1\mathbf{q}_2j_1j_2} P_{\alpha\beta}\begin{pmatrix}\mathbf{q}_1 & \mathbf{q}_2\\ j_1 & j_2\end{pmatrix} Q\begin{pmatrix}\mathbf{q}_1\\ j_1\end{pmatrix} Q\begin{pmatrix}\mathbf{q}_2\\ j_2\end{pmatrix}\Delta(\mathbf{K}+\mathbf{q}_1+\mathbf{q}_2).$$

If we approximate $\langle Q(\mathbf{q}_1j_1)Q(\mathbf{q}_2j_2)\rangle$ by $\langle Q(\mathbf{q}_sj)\rangle\langle Q(-\mathbf{q}_sj)\rangle$, then this term will contribute a term to the change in polarizability

$$\delta P_{\alpha\beta} \approx P_{\alpha\beta}\begin{pmatrix} \mathbf{q}_s & -\mathbf{q}_s \\ j & j \end{pmatrix} Q_0^2.$$

This term may well be of the same order of magnitude as the term arising from the one-phonon contribution of the secondary order parameter $\bar{Q}$, which is given, eqn. (I.2.46), by

$$\delta P_{\alpha\beta} \approx P_{\alpha\beta}\begin{pmatrix} 0 \\ j \end{pmatrix} \bar{Q}.$$

Without a detailed knowledge of the polarizability parameters it is impossible to distinguish between these two processes experimentally.

I.3.3.2. *Scattering measurements*

The theory of the scattering of X-rays and neutrons from crystals was briefly reviewed in §I.2.3. The structure of the crystal is determined by the Bragg reflections, which are dependent upon the infinitely long time part of the correlation function, eqn. (I.2.31). In the long time limit

$$\langle F(\mathbf{K},\tau)F(\mathbf{K},0)\rangle = |\langle F(\mathbf{K},0)\rangle|^2, \tag{I.3.10}$$

so that the Van Hove scattering function has the form

$$S^0(\mathbf{K},\omega) = |\langle F(\mathbf{K},0)\rangle|^2 \delta(\omega). \tag{I.3.11}$$

In the high-symmetry phase these results have already been given by eqns. (I.2.35) and (I.2.37) and enable the structure of the phase to be determined experimentally.

In the distorted phase the scattering can be obtained if the positions $\mathbf{R}(\kappa)$ are replaced by the positions in the distorted structure. It is, however, more instructive to expand the structure factor $\langle F(\mathbf{K},0)\rangle$ in powers of the displacement as in eqn. (I.2.34). When the expectation value of eqn. (2.34) is taken one obtains

$$\langle F(\mathbf{K},0)\rangle = F^0(\mathbf{K})\Delta(\mathbf{K}) + \sum_{i=1}^{n} F(\mathbf{K},\mathbf{q}_{si}j_i)\langle Q_i\rangle\Delta(\mathbf{K}+\mathbf{q}_{si}) \tag{I.3.12}$$

where we have neglected, for the present, higher powers in the primary order parameter and all secondary order parameters.

Now if $\mathbf{q}_s$ is non-zero, eqn. (I.3.12) shows that there will be *new* Bragg reflections present in the distorted phase with an intensity given by $N|F(\mathbf{K},\mathbf{q}_{si}j_i)|^2\langle Q_i\rangle^2 \Delta(\mathbf{K}+\mathbf{q}_{si})$. The intensity of these Bragg reflections enables the relative sizes of the different components of the order parameters, the $\langle Q_i\rangle$, to be found while their temperature dependence gives a measure of the exponent β describing the temperature dependence of the primary order parameter. The temperature dependence of the new Bragg reflection in the antiferrodistortive phase of $SrTiO_3$ is shown in fig. I.6 as determined with neutron scattering techniques by Riste *et al.* (1971).

The behaviour is less straightforward for ferrodistortive phase transitions, $\mathbf{q}_s = 0$. The intensity is given by

$$S^0(\mathbf{K},\omega) = N\delta(\omega)|F^0(\mathbf{K}) + \sum_i F(\mathbf{K},0j_i)\langle Q_i\rangle|^2\Delta(\mathbf{K}). \tag{I.3.13}$$

The change in the intensity of these reflections close to T_c is proportional to $\langle Q_i\rangle^2$ because the cross terms cancel, provided that $F^0(\mathbf{K})$ is not zero or strongly

Fig. I.6

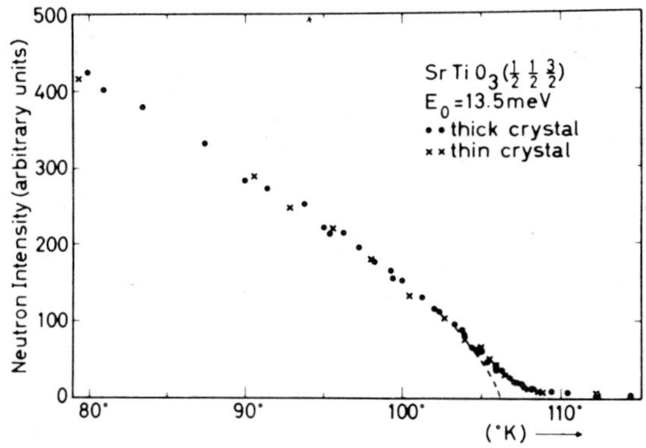

The temperature dependence of the superlattice reflection $\mathbf{K}=\frac{2\pi}{a}(\frac{1}{2},\frac{1}{2},\frac{3}{2})$ in $SrTiO_3$ after Riste *et al.* (1971). The dotted line indicates that T_c for the specimen was $106{\cdot}1 \pm 0{\cdot}3$ K.

temperature dependent. (We discuss the temperature dependence of the Debye–Waller factors in §§ I.3.5 and I.4.)

The use of scattering techniques to determine secondary order parameters is subject to the same difficulties as those discussed for macroscopic properties. In the expansion of the structure factor $\langle F(\mathbf{K},0)\rangle$, eqn. (I.2.38), there are not only linear terms in $Q(\mathbf{q}j)$ but second-, third-, and higher-order terms. The second-order terms will give rise to additional contributions to the scattering with the wave-vector conservation conditions given by $\Delta(\mathbf{K})$ and $\Delta(\mathbf{K}+2\mathbf{q}_s)$. The intensity arising from these diffraction effects caused by the primary order parameters, must be included when the amplitude of the secondary order parameters are being deduced from the intensities measured at these wave-vector transfers. Note that, whereas the relative sizes of the one- and two-phonon terms in the expansion of the *polarizability* are largely unknown, in the case of displacive phase transitions these relative amplitudes for scattering experiments are given exactly in terms of the supposedly known eigenvectors and scattering lengths. It is therefore possible to separate these two effects at least in principle. This advantage of diffraction is frequently rather illusory in practice, because the effects of the two-phonon terms in a diffraction experiment are usually relatively larger than in the polarizability experiment.

At strictly order–disorder phase transitions, such as the AB alloy, there are only linear terms in $S(l\kappa)$ in the expansion of $F(\mathbf{K},\tau)$, eqn. (I.2.43). Consequently the interpretation of the scattering is simpler in these systems than in the displacive materials. In the molecular systems, such as $NaNO_2$, however, the situation is more complex because of a coupling between the spin and the displacement variables, so that it again becomes difficult to extract secondary order parameters reliably.

There are two experimental problems in the application of diffraction techniques to the determination of order parameters. Firstly Bragg reflections from real crystals frequently suffer from extinction effects. At antiferrodistortive phase transitions close to T_c, the new Bragg reflections of the distorted phase are weak and probably not greatly influenced by extinction, as shown (fig. I.6) by the agreement between results of experiments on different sizes of crystal; however, as the system is cooled extinction will become more and more important. Extinction provides a particularly difficult problem at ferrodistortive transitions because the crystal will usually break up into domains changing the extinction effects on the normal Bragg reflections. Changes in these Bragg intensities will not then reflect the order parameter in the way suggested after eqn. (I.3.13).

A second problem (to be discussed in detail in the next section) is that, close to T_c, the diffuse scattering is large for small frequency transfers and at those wave-vector transfers at which the Bragg reflections of the distorted phase appear. This is the cause (at least in part) of the high temperature 'tail' to the Bragg reflection intensity shown in fig. I.6 (to which we shall devote much attention in III).

The exponent β, can be accurately obtained only to the extent that the diffuse scattering can be accurately subtracted out. Since the Bragg scattering is a delta function in both energy and wave-vector transfer whereas the diffuse scattering is broad the subtraction is made easier if the experiment is performed with very good resolution in either $\mathbf{K}$ or ω. In the case of conventional X-ray scattering it is impossible to energy discriminate against the diffuse scattering, but in principle it is possible to use very good $\mathbf{K}$ space resolution. It is possible to energy discriminate using Mössbauer γ-rays as a source for the scattering and then to perform energy analysis (Darlington and O'Connor 1976) using a similar Mössbauer absorber. It is surprising that this technique has not been applied to measure β.

By using neutron scattering techniques, the energy analysis is more easily performed. Conventional inelastic scattering techniques give an energy resolution of about 0·1 meV which is not normally good enough to permit the neglect of the diffuse scattering. The development of new techniques with a resolution of 1 μeV, such as the backscattering spectrometer (Alefeld *et al.* 1969) and the spin–echo spectrometer (Mezei 1972) holds out the possibility of eliminating the diffuse scattering more completely. The wave-vector resolution is, however, inferior to that which can be obtained with the best X-ray spectrometers.

I.3.3.3. *Resonance methods*

The order parameter at a structural phase transition may be determined by using electron paramagnetic resonance, E.P.R., or nuclear quadrupole resonance, N.Q.R. (Müller 1971, Blinc 1971). The E.P.R. method depends upon the presence in the crystal of a magnetic ion. The most suitable ions are those with a half-filled closed shell, Mn^{2+}, Fe^{3+}, Gd^{3+} or Tb^{4+} because; (i) they are spherically symmetric and so do not introduce local distortions in symmetry and (ii) they have very long E.P.R. relaxation times even at room temperature. One of these ions must then be present in the crystal at a low concentration, say, 10^{18} cm^{-3}; Fe^{3+} is usually present in 'pure' $SrTiO_3$ at about this concentration.

The crystal field hamiltonian for a $3d$ transition metal ion can be expressed as a spin hamiltonian in terms of the normalized magnetic spin operators, O_l^m (Hutchings 1964), which transform in the same way as the corresponding cartesian polynomials;

$$H_{\rm cf} = \sum_{m=-2}^{2} B_2^m O_2^m + \sum_{m=-4}^{4} B_4^m O_4^m.$$

The coefficients in this expansion B_2^m, B_4^m may then be obtained at least approximately in terms of the ionic charges, bonding and nearest neighbour distances (Hutchings 1964). Typically the B-coefficients are about 10^{-3} meV.

In the high-symmetry phase the site symmetry is such that some of the B_l^m coefficients are zero; for example in $SrTiO_3$ with an Fe^{3+} ion replacing the Ti^{4+} ion only the coefficients B_4^0 and $B_4^{\pm 4}$ are non-zero. In the distorted phase, one (or more) of the B-coefficients becomes non-zero and increases in a manner dependent upon the primary order parameter, Q_0, and the secondary order parameter $\bar{Q}$

$$B = A_1 Q_0 + A_2 Q_0^2 + \ldots + \bar{A}_1 \bar{Q} + \bar{A}_2 \bar{Q}^2 + \ldots \quad \text{(I.3.14)}$$

In the case of Fe^{3+} in $SrTiO_3$, B_2^0 is allowed in the distorted phase. However, since the Fe^{3+} ion is situated at a centre of symmetry even in the distorted phase, the coefficient A_1 in eqn. (I.3.14) is zero. The secondary order parameter for $SrTiO_3$ is the strain and so $\bar{A}_1$ is non-zero. Consequently B_2^0 for $SrTiO_3$ has two contributions proportional to Q_0^2 and to $\bar{Q}$, and as with the macroscopic polarizability it is difficult to distinguish the effects of these two different processes.

In contrast Gd^{3+} ions in $SrTiO_3$ are situated at the Sr^{2+} site, which is not a centre of symmetry in the distorted phase, and so A_1 of eqn. (I.3.14) is non-zero (Unoki and Sakudo 1967).

Müller and Berlinger (1971) were, however, able to directly measure the primary order parameter in $SrTiO_3$ using the Fe^{3+} centre. In a single domain specimen of distorted $SrTiO_3$, fig. I.2, the two Ti^{4+} sites have their environments twisted in opposite senses. Hence, by determining the orientation of the principal axes of the spin hamiltonian for these two types of sites, the rotation angle of the oxygen octahedra could be determined. Using this approach with the Fe^{3+}-vacancy centre in $SrTiO_3$ Müller and Berlinger were able to determine the rotation angle in $SrTiO_3$ to $\pm 0{\cdot}01°$, fig. II.17, but more typically the accuracy of such experiments is $0{\cdot}1°$.

In an E.P.R. experiment the line width of the resonance is typically about 10^{-4} T, and so there will be a splitting once the local distortion of the structure persists for times longer than 10^{-6} s. Unfortunately, the E.P.R. method cannot be used when the material is magnetic and it depends on being able to produce material with the necessary impurities situated at known sites in the host material. It also suffers from the disadvantage that there is always a niggling uncertainty as to whether the behaviour near the E.P.R. impurity site is the same as that of the bulk material.

Nuclear quadrupole resonance arises from the transitions between the different nuclear energy levels caused by the interaction between the nuclear quadrupole moment and the electric field gradient. The electric field gradient may be expanded in a similar manner to the electrostatic potential which determines the E.P.R. spectrum, only the selection rules for the coefficients are different. In particular, if the ion is situated at a centre of symmetry in the high-symmetry phase, the linear term, A_1, in eqn. (I.3.14) is zero even in the distorted phase and the B coefficient is proportional to Q_0^2. Examples of this behaviour are provided by the work of Borsa (1971), using ^{139}La and ^{27}Al resonances in $LaAlO_3$, which undergoes a similar phase transition to $SrTiO_3$.

In materials of lower symmetry the linear term in eqn. (I.3.14) is non-zero and the B coefficients are proportional to the order parameter. This is illustrated by the ^{133}Cs resonance in CsH_2AsO_4 as shown in fig. I.7 (Blinc and Mali 1969).

N.M.R. has the advantage over E.S.R. that it does not necessitate the introduction of an impurity, but the selection rules are such that the B parameters are more commonly proportional to Q_0^2 than to Q_0 and so cannot be used to determine the exponent β. The frequency resolution of the spectrometers is such that the local distortion must persist for 10^{-3} s to give a sharp line characteristic of the distorted phase.

Finally the Mössbauer effect may be used to determine the order parameter at a structural phase transition (Shenoy 1973). At a phase transition the Mössbauer line may split because of a change in the nuclear energy levels caused by the interaction of the nuclear quadrupole moment with the electric field gradient. The selection rules and dependence of the splitting on the order parameter are then the same as those described for N.M.R. experiments. The accuracy of Mössbauer experiments is usually less than that of N.M.R. experiments and furthermore the experiments frequently require that the Mössbauer nucleus be introduced into the material as an impurity.

Fig. I.7

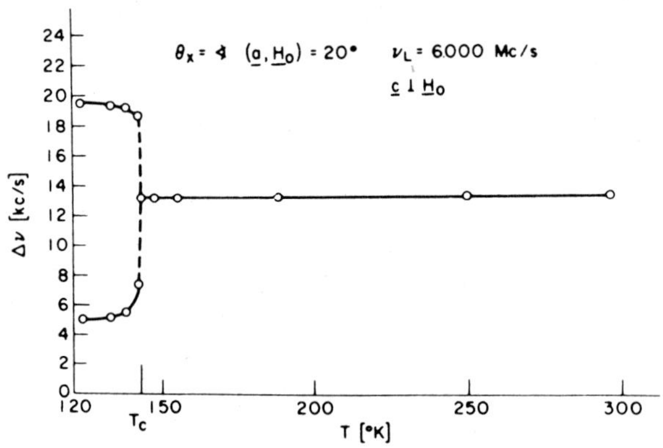

Temperature dependence of the nuclear quadrupole ^{133}Cs magnetic resonance in CsH_2AsO_4 after Blinc and Mali (1969).

I.3.4. *Static susceptibility*

In § I.2, the susceptibility of a one-phonon process was shown to be dependent upon the coupling between the probe and the phonon (described by the operators such as $B(\mathbf{K}, \mathbf{q}j)$ or $F(\mathbf{K}, \mathbf{q}j)$) and the one-phonon response functions, $\chi(\mathbf{q}jj')$, eqn. (I.2.22). In the distorted phase $\langle Q(\mathbf{q}_s j)\rangle$ is non-zero, so that, at a continuous phase transition, it is reasonable to expect that $\langle Q(\mathbf{q}_s j)Q(-\mathbf{q}_s j)\rangle$, a measure of the fluctuations of the normal mode coordinate $Q(\mathbf{q}_s j)$, will diverge at T_c giving rise, eqn. (I.2.22), to a divergence of the susceptibility $\chi(\mathbf{q}_s jj)$, as illustrated for $BaTiO_3$ in fig. I.8. We shall therefore study the behaviour of the susceptibility $\chi(\mathbf{q}jj)$ as the temperature $T \to T_c$ and wave-vector $\mathbf{q} \to \mathbf{q}_s$, for the particular modes whose expectation values are non-zero in the distorted phase. It is convenient to introduce a notation to describe the fluctuations for the normal modes with wave-vectors close to $\mathbf{q}_s$. We define

$$\left.\begin{aligned} \hat{Q}_i(\mathbf{k}) &= Q(\mathbf{k}+\mathbf{q}_{si}\, j_i) - \langle Q(\mathbf{k}+\mathbf{q}_{si}\, j_i)\rangle \\ \text{and} \qquad & \\ Q_i(\mathbf{k}) &= Q(\mathbf{k}+\mathbf{q}_{si} j_i). \end{aligned}\right\} \qquad (\text{I}.3.15\,a)$$

Note that in the fluctuating parts of the order parameter we have not introduced the factor of $N^{-1/2}$ included in eqn. (I.3.1); the $Q_i(\mathbf{k})$ are proportional to the phonon amplitude.

We introduce a corresponding notation for the susceptibility

$$\begin{aligned} \chi_i(\mathbf{k}) &= \chi(\mathbf{k}+\mathbf{q}_{si} j_i j_i), \\ &= \beta\langle \hat{Q}_i(\mathbf{k})\hat{Q}_i(-\mathbf{k})\rangle. \end{aligned} \qquad (\text{I}.3.15\,b)$$

Fig. I.8

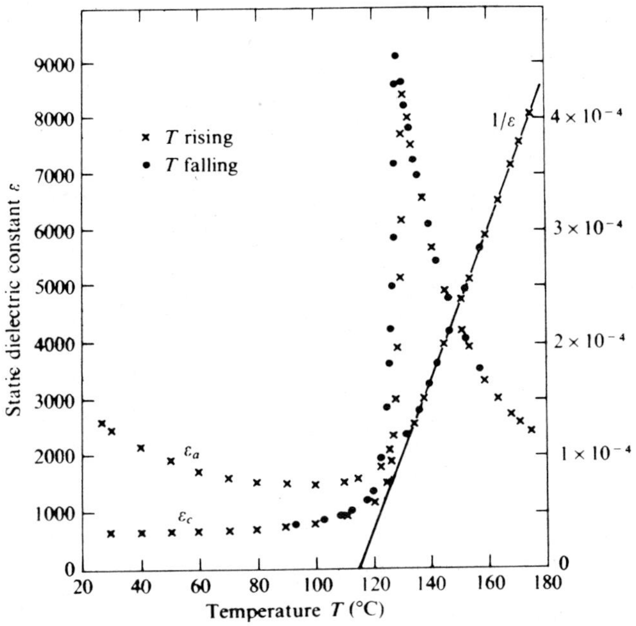

Static dielectric constant in $BaTiO_3$, both above and below T_c. Below T_c there are two dielectric constants, parallel to (ε_c) and perpendicular to (ε_a) the ferroelectric c axis. (After Johnson 1965.)

In the high-symmetry phase $\chi_i(\mathbf{k})$ is independent of i and its divergence is given by scaling theory (cf. Stanley 1971, §II.4.3.5) for $T \to T_c$, $\mathbf{k} \to 0$, as

$$\chi_i(\mathbf{k}) \approx C_+ t^{-\gamma} D(k_\alpha^2 \xi_\alpha^2), \tag{I.3.16}$$

where

$$t = (T - T_c)/T_c, \tag{I.3.17}$$

and γ is the susceptibility exponent which is unity for classical theories, but can be much larger in real systems. The function $D(x^2)$ is a scaling function, (normalized such that $D(0) = 1$) describing the wave-vector dependence of the susceptibility. ξ_α is the correlation length in the α direction. The correlation length diverges at a continuous phase transition, and the divergence is described by the exponent ν

$$\xi_\alpha = f_+(\alpha) t^{-\nu}, \tag{I.3.18}$$

where classical theories yield $\nu = 0{\cdot}5$, but more exact theories again give rather larger values.

There are two special cases of eqn. (I.3.16) of particular importance.

$$\chi_i(0) = C_+ t^{-\gamma}, \quad |\mathbf{k}| = 0 \tag{I.3.19}$$

and

$$\chi_i(k_\alpha) = D_0 / k_\alpha^{2-\eta}, \quad T = T_c, \tag{I.3.20}$$

where the exponent, $\eta = 2 - (\gamma/\nu)$, is zero according to classical theory, and remains small in more refined theory. This exponent is particularly hard to measure and we shall not discuss it further.

In the distorted phase, the symmetry is broken so that the different susceptibility functions are neither independent nor equal. It is, however, possible to choose basis functions from the $Q_i(\mathbf{k})$ in terms of which the susceptibilities are independent. In $BaTiO_3$ with a ferroelectric c axis the dielectric susceptibility in the ferroelectric phase is different for the a and c axes as illustrated in fig. I.8.

In the distorted phase each of the different susceptibilities is divergent as $T \to T_c$, and the nature of that divergence may be described by eqns. (I.3.16)—(I.3.20) except that the amplitudes C_+ and $f_+(\alpha)$ must be replaced by C^i_- and $f^i_-(\alpha)$. In all but exceptional circumstances (cf. footnote to Table II.1 in part II) these divergences will be described by the *same* exponents (γ and ν) as characterize the divergences when $T \to T_c^+$. Note, however, that if a *continuous* symmetry is broken at the phase transition, a more careful treatment of the low-temperature phase susceptibilities is required (§§ I.4.2, I.4.4).

I.3.5. *Soft modes and the dynamic susceptibility*

The imaginary part of the dynamic susceptibility is related to the static susceptibility by the Kramers–Kronig relation, eqn. (I.2.27), according to which

$$\chi^R(0) = \frac{1}{\pi} \int_{-\infty}^{\infty} \chi^I(\omega')\omega'^{-1}\, d\omega'$$

The imaginary part of $\chi^I(\omega)$ gives the power dissipation of the system, and depends upon the matrix elements coupling the initial and final states of the system and on the density of initial and final states whose energies differ by $\hbar\omega$. The static susceptibility diverges at T_c for $\mathbf{k}=0$. Since it is unreasonable to assume that the matrix elements are very temperature dependent, this divergence must reflect a movement in the weight of $\chi^I(\omega)$ to smaller ω—the 'critical slowing down' discussed by Van Hove (1954).

Dynamic phenomena have played an especially important role in the development of our understanding of structural phase transitions. In particular the appealing concept of a soft mode, introduced independently by Anderson (1960) and Cochran (1960), has frequently proved helpful in describing structural phase transitions, and in illuminating their essential characteristics.

The idea was prompted by the form of the static susceptibility of a harmonic oscillator, given in eqn. (I.2.23) by $\chi_H(\mathbf{q}jj)$. Comparing this result with the behaviour of the static susceptibility in the classical theory (eqn. (I.3.19) with $\gamma=1$) suggests that

$$\omega(\mathbf{q}_s j)^2 = a(T - T_c), \quad T > T_c. \tag{I.3.21}$$

Thus, in the soft mode approach, the phase transition is viewed as an instability of the crystal against a particular normal mode, the frequency of which decreases to zero at T_c. In the low-symmetry phase the crystal is distorted so as to stabilize this mode, and the distortion is described by a 'frozen in' amplitude of this normal mode.

As an illustration of the soft-mode concept the result of inelastic neutron scattering measurements probing the soft mode in $SrTiO_3$ are shown in fig. I.9. There is clearly a peak in the scattering whose frequency decreases towards zero as $T \to T_c$. In fig. I.10 we show the square of these frequencies plotted against temperature and the agreement with eqn. (I.3.21) is clearly excellent. Below T_c the triplet soft mode splits into a singlet and a doublet and rises in frequency as described in § I.4.2. Similar results have now been obtained in many materials undergoing structural phase transitions (Scott 1974, Shirane 1974).

Fig. I.9

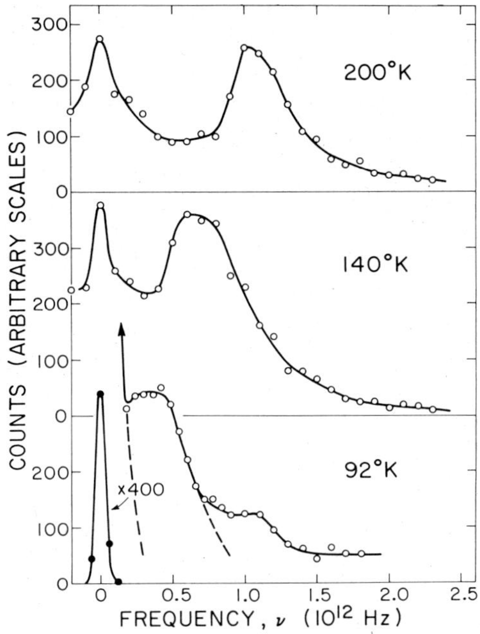

Neutron scattering observed at three temperatures from the soft mode in $SrTiO_3$. The Bragg reflection below 110 K is shown on a reduced scale. Below 110 K the triplet soft mode splits into a singlet and a doublet. (From Cowley *et al.* 1969.)

Fig. I.10

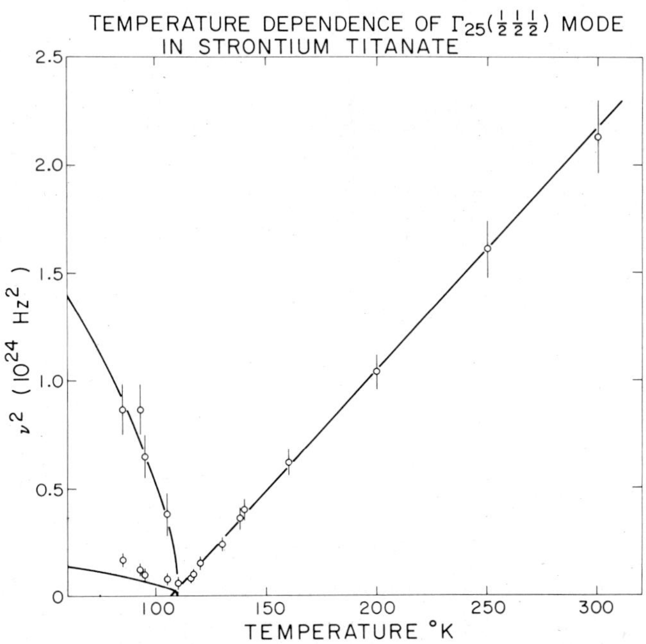

The temperature dependence of the frequencies of the soft mode in $SrTiO_3$ both above and below T_c. (From Cowley *et al.* 1969.)

Despite these successes, the soft-mode theory is in a strict sense inconsistent. The frequencies of the normal modes in a harmonic crystal are temperature independent. Furthermore soft modes are found to be damped and sometimes even overdamped showing also that the harmonic theory is inadequate. A soft-mode frequency must therefore be interpreted as some type of effective or renormalized frequency as we shall discuss further in § I.5.1.1.

Phenomenologically the damping has been incorporated by approximating the susceptibility by the susceptibility of a classically damped oscillator.

$$\chi^{-1}(\mathbf{q}jj,\omega)=\omega(\mathbf{q}j)^2-\omega^2-2i\omega\gamma_0, \tag{I.3.22}$$

where γ_0 is the damping constant.

The poles of the susceptibility in the complex ω plane are given by

$$\begin{aligned}\omega^{\pm} &= \pm(\omega(\mathbf{q}j)^2-\gamma_0^2)^{1/2}+i\gamma_0, \quad \omega(\mathbf{q}j)>\gamma_0,\\ &= i(\gamma_0\pm(\gamma_0^2-\omega(\mathbf{q}j)^2)^{1/2}), \quad \omega(\mathbf{q}j)<\gamma_0,\end{aligned} \tag{I.3.23}$$

showing that the soft mode is underdamped when $\omega(\mathbf{q}j)>\gamma_0$ but overdamped when $\gamma_0>\omega(\mathbf{q}j)$. If γ_0 is temperature independent the mode always becomes overdamped as $T\to T_c$, and the pole of the complex frequency susceptibility approaches the origin of the frequency plane along the imaginary ω axis, as shown in fig. I.11.

Fig. I.11

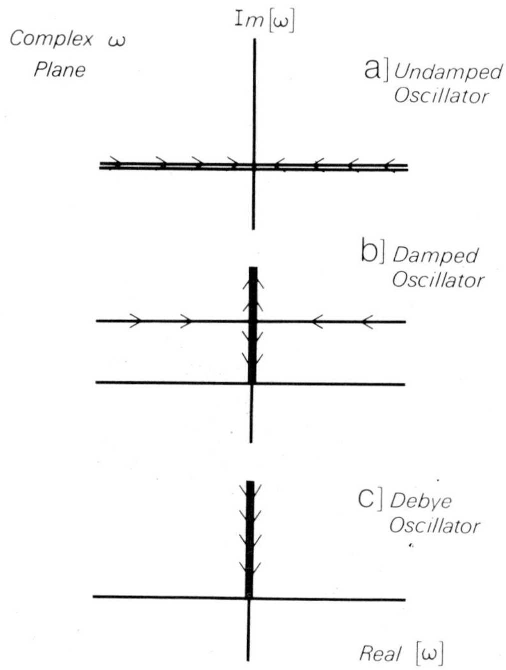

The complex frequency plane ω, showing the motion of the poles as $T\to T_c$ for a classical oscillator, both (*a*) undamped and (*b*) damped, and (*c*) for a Debye relaxation response.

When the damping becomes very large the mode is overdamped and ω^2 may be neglected compared with $2\omega\gamma_0$ in eqn. (I.3.22): the response is then given by

$$\chi^{-1}(\mathbf{q}jj,\omega)=\omega(\mathbf{q}j)^2(1-i\omega\tau_D), \tag{I.3.24}$$

where τ_D is known as the Debye relaxation time and is given by $\tau_D = 2\gamma_0/\omega(\mathbf{q}j)^2$, and so diverges as $T \to T_c$. In this case the pole of the susceptibility in the complex ω plane is on the imaginary frequency axis, fig. I.11, and the measured response is concentrated at very low frequencies.

A continuous phase transition involves a divergence of the static susceptibility at T_c. This is always associated with the approach of a pole of the susceptibility to the origin of the complex frequency plane, as T approaches T_c. This pole can be called a soft mode even when it is always heavily overdamped, so that a continuous phase transition always has a soft mode. Alternatively, the use of the term soft mode can be restricted to those cases where the mode shows at least some vestiges of resonant behaviour, the frequency of which decreases as $T \to T_c$. This restriction would confine the use of the term to the displacive class of materials such as $SrTiO_3$ as illustrated by figs. I.9 and I.10. In the order–disorder materials, the modes describing the small oscillations of the atoms or molecules are not strongly temperature dependent as illustrated in fig. I.12 for the optically active modes in $NaNO_2$. The ferroelectric

Fig. I.12

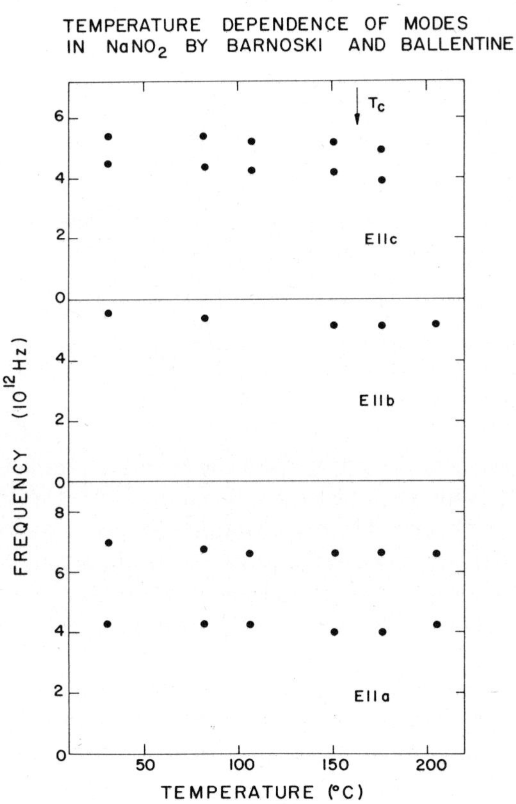

The frequencies of the infra-red active modes in $NaNO_2$ as a function of temperature (Barnoski and Ballentyne 1968).

fluctuations in $NaNO_2$ are associated with the relatively slow reorientation of the NO_2 groups and, as shown in fig. I.13 by the dielectric susceptibility measurements, the anomalous behaviour of the susceptibility is entirely absent for probe frequencies in excess of 10^9 Hz. The soft mode in this case is not one of the normal modes of

Fig. I.13

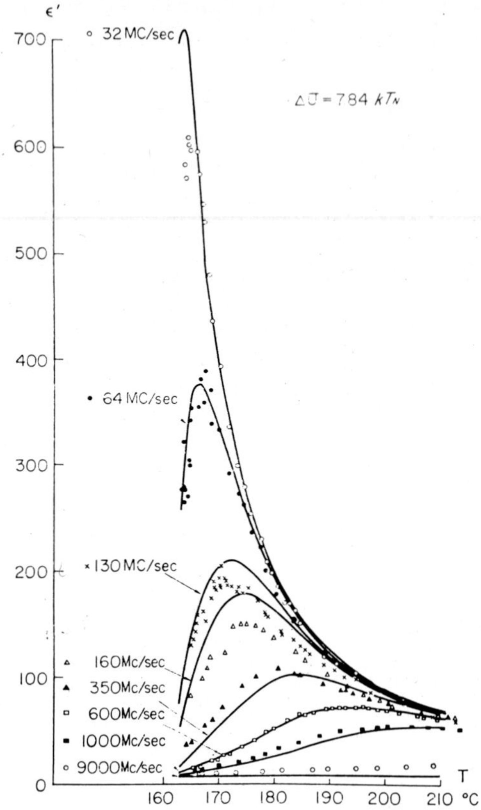

The dielectric susceptibility of $NaNO_2$ as a function of frequency as measured by Hatta (1968) after Yamada *et al.* (1968).

vibration describing the small oscillations of the atoms and molecules, but one of the modes described by the spin variables introduced in § I.2.1. Since this hopping is not treated in the harmonic theory of lattice dynamics we prefer to restrict the use of the term 'soft mode' to displacive materials. Alas, since there is not a clear distinction between displacive and order–disorder materials, this inevitably leaves a number of materials, such as KDP, where it is uncertain if a soft-mode description is appropriate.

In this section we have discussed the dynamical susceptibility in terms of phenomenological models which have one or perhaps two simple poles in the complex ω plane. In § I.5.1, and in part III, we shall discuss a more general approach to this problem. Nevertheless, a characteristic feature of the phase transition is that the dynamical response becomes slower as $T \to T_c$. It is often convenient to describe the fluctuations of the order parameter by a characteristic frequency ω_Q where

$$\frac{1}{\pi}\int_{-\omega Q}^{\omega Q}\frac{1}{\omega}\operatorname{Im}\left[\chi(\mathbf{q}jj,\omega)\right]d\omega=\frac{1}{2}\chi(\mathbf{q}jj).$$

On comparing this definition with eqn. (I.2.27), it is clear that $\omega_Q \to 0$ as $T \to T_c$ no matter how complicated the spectral response.

I.3.6. *Measurement of susceptibilities*

I.3.6.1. *Macroscopic measurements*

Macroscopic measurements probe the susceptibilities for $\mathbf{K}=0$, and consequently measure the susceptibility for the primary order parameters only if $\mathbf{q}_s=0$. The dielectric susceptibility can be measured by dielectric measurement or by infra-red absorption techniques. The susceptibility can be obtained from eqn. (I.2.21) if both operators A and B are taken to be the dipole moment operator $\mathbf{M}$. If only one mode is temperature dependent then

$$\chi_{\alpha\beta}(\omega)=\chi_{\alpha\beta}^{\infty}+\frac{1}{v}M_{\alpha}(0j)M_{\beta}(0j)\chi(0jj,\omega), \qquad \text{(I.3.25)}$$

where $\chi_{\alpha\beta}^{\infty}$ is the dielectric susceptibility of the other modes and the electronic excitations. Static dielectric measurements then give a measure of the soft-mode static susceptibility as illustrated in fig. I.8 for $BaTiO_3$. Using electrical techniques it is possible to measure the dynamical susceptibility for frequencies up to 5×10^{10} Hz as illustrated in fig. I.13. In principle, a measurement of the infra-red reflectivity gives the dielectric response at higher frequencies and in fig. I.14, we show the dielectric response of $BaTiO_3$, as measured using these techniques. In practice, however, these techniques are difficult to apply because the range of frequency of most interest in displacive systems is frequently between 10^{11} and 10^{12} Hz, and this is a very difficult range experimentally. Secondly, the measurements yield the reflectivity, and when the susceptibility is large, the reflectivity is close to unity and large differences in the susceptibility give only very small changes in the reflectivity.

Fig. I.14

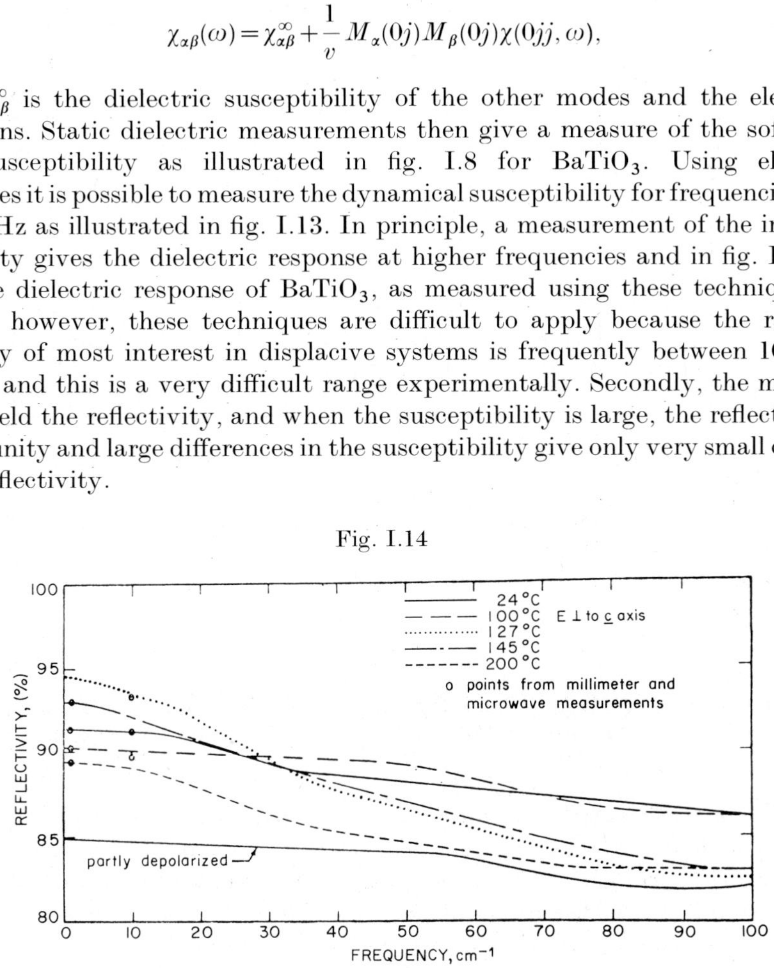

The reflectivity of $BaTiO_3$ as a function of temperature. (After Ballentyne 1964.)

Ultrasonic measurements at structural phase transitions have been reviewed by Rehwald (1973). When the strain is a primary order parameter, the ultrasonic velocity plays the role of $\omega(\mathbf{q}_s j)^2$, and hence the velocity decreases to zero as $T\rightarrow T_c$ as shown for KDP (KH_2PO_4) in fig. I.15. More frequently the strain is a secondary order parameter and the ultrasonic velocity measures the inverse of their correlation function whose behaviour (to be discussed in more detail in II.5.1.2) is typified by the results shown in fig. I.16.

Fig. I.15

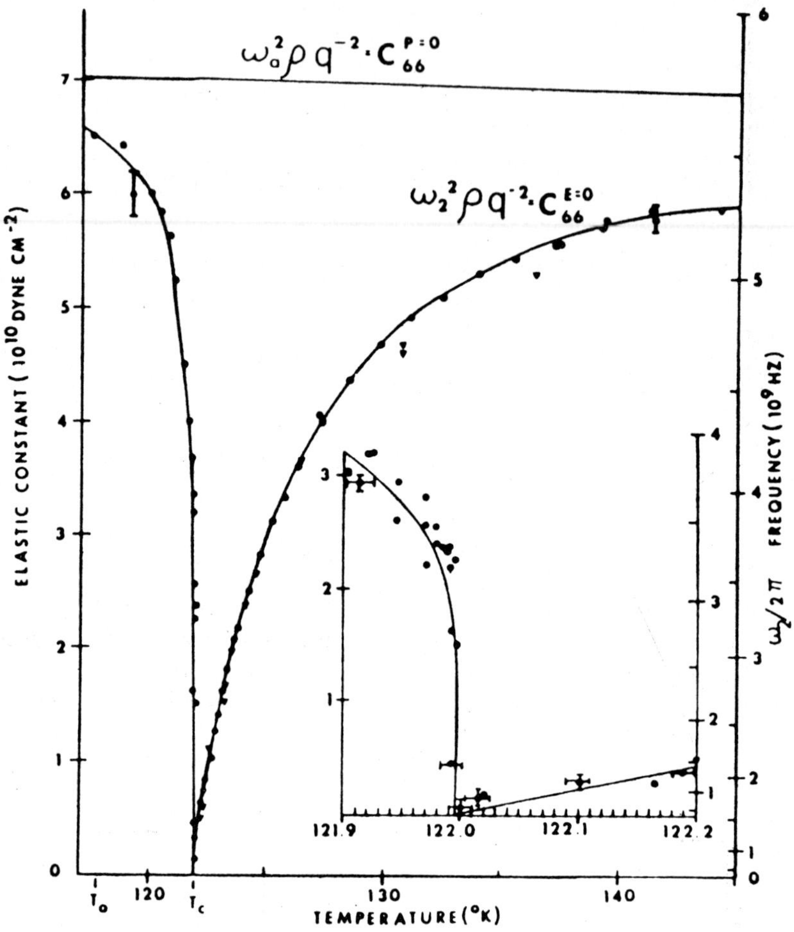

The elastic constants of KDP, C_{66}^{E}, and $C_{66}^{P}=C_{66}$ as a function of temperature as measured by Brody and Cummins (1968) using Brillouin scattering techniques.

Ultrasonic experiments also measure the attenuation of the elastic waves; the critical behaviour of the attenuation is discussed briefly in III.3.2.2.

The specific heat may also exhibit singular behaviour near a structural phase transition. In practice the change in the specific heat is frequently very small and difficult to measure accurately as shown in fig. I.17 for $SrTiO_3$.

I.3.6.2. *Scattering measurements*

By far the most detailed information about susceptibilities can be obtained using the scattering of X-rays, neutrons or light. In the case of X-ray scattering, it is not possible to obtain sufficient energy resolution to measure the dynamic susceptibility; consequently the cross-section for a wave-vector transfer $\mathbf{K}$ is given in the one-phonon approximation by eqns. (I.2.38) and (I.2.42) as

$$\frac{d\sigma(\mathbf{K})}{d\Omega}=k_{\mathrm{B}}T\sum_{\mathbf{q}jj'}F(\mathbf{K},\mathbf{q}j)F(-\mathbf{K},-\mathbf{q}j')\chi(\mathbf{q}jj')\Delta(\mathbf{K}+\mathbf{q}). \tag{I.3.26}$$

Fig. I.16

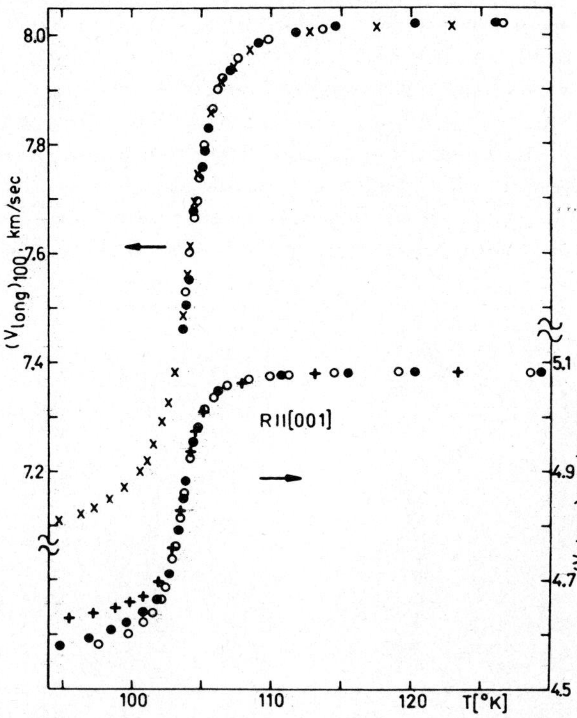

Velocities of longitudinal and shear waves propagating in a [100] direction in $SrTiO_3$ from Lüthi and Moran (1970). Key: ○ shows results at 50 MHz; ● at 30 MHz; × at 16 MHz; and + at 10 MHz.

Fig. I.17

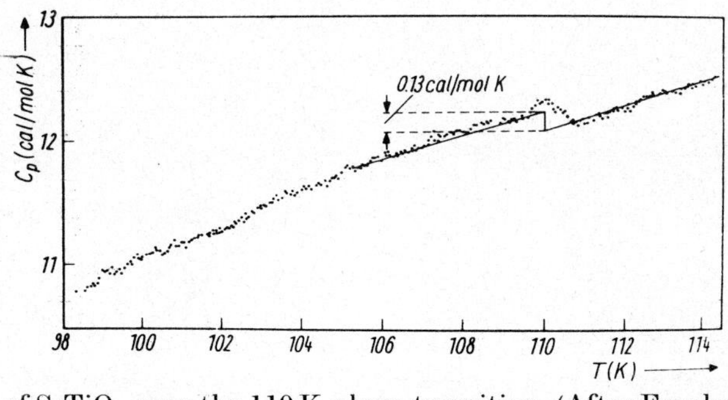

The specific heat of $SrTiO_3$ near the 110 K phase transition. (After Franke and Hegenbarth 1974.)

If we separate out the term involving the order parameter susceptibilities $\chi_i(\mathbf{k})$

$$\frac{d\sigma(\mathbf{K})}{d\Omega}=\sigma_B(\mathbf{K})+k_B T\sum_{i\mathbf{k}}|F(\mathbf{K},\mathbf{q}_{si}+\mathbf{k}j_i)|^2\chi_i(\mathbf{k})\Delta(\mathbf{K}+\mathbf{q}_{si}+\mathbf{k}), \qquad (I.3.27)$$

where $\sigma_B(\mathbf{K})$ represents the other terms which produce a background scattering. The experimental results are frequently interpreted by neglecting the **K**-dependence of the structure factors $F(\mathbf{K}, \mathbf{q}j)$, when the observed intensity is, after subtraction of the background, proportional to the susceptibility. This approximation is probably reasonably accurate for small **k** and can be tested by measuring $\chi_i(\mathbf{k})$ for several different wave-vector transfers **K** giving equivalent **k**.

There are two different techniques which have been used to determine the X-ray scattering cross-section. In both cases it is necessary to use monochromatic radiation to reduce the background. The technique, developed at Orsay by Comès and his collaborators (Comès *et al.* 1970) employs photographic detection and hence is invaluable for a rapid qualitative survey of the nature of the scattering. In fig. I.18

Fig. I.18

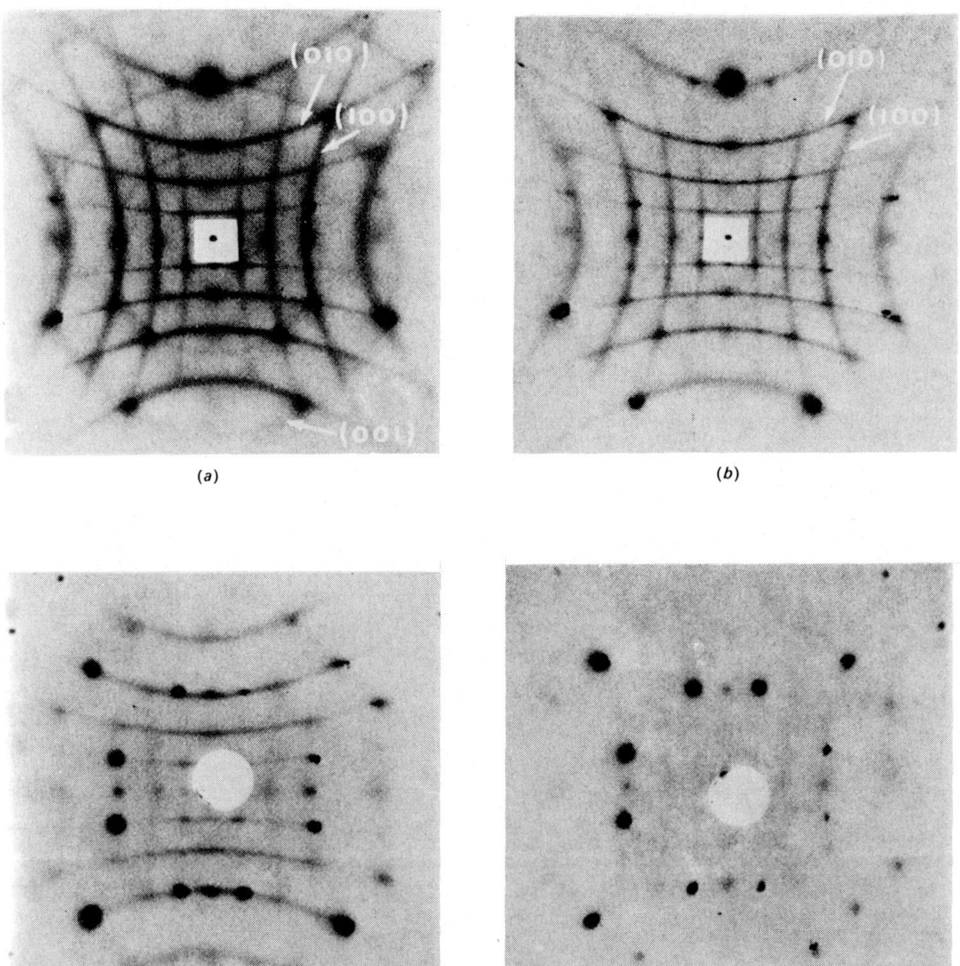

The diffuse X-ray scattering in $KNbO_3$: (*a*) cubic phase at 770 K; (*b*) tetragonal phase at 520 K; (*c*) orthorhombic phase at 220 K; and (*d*) rhombohedral phase at 220 K. Note the disappearance of the streaks of diffuse scattering as the material changes phase. (After Comès *et al.* 1970.)

we show a typical result obtained for $KNbO_3$ which undergoes very similar transitions to $BaTiO_3$. Notice the diffuse streaks joining the Bragg reflections. These show that $\chi_i(\mathbf{k})$ is anisotropic, its value being considerably larger for the wave-vectors, $\mathbf{k}$, along the [100] directions than for $\mathbf{k}$ along other directions. The other technique involves the use of counters and so in principle provides more quantitative information but is slower. In fig. I.19 we show the diffuse scattering obtained using counter techniques in $NaNO_2$ by Yamada and Yamada (1966). This technique is capable of very high resolution in wave-vectors $\sim 10^{-4}\,\text{Å}^{-1}$, and so can be used to give very accurate measurements of the exponents, γ and ν. It is rather surprising that there has not been more effort made to exploit this technique in the field of structural phase transitions. (There is, however, some difficulty in constructing accurately controlled temperature environments for specimens for X-ray scattering.)

Fig. I.19

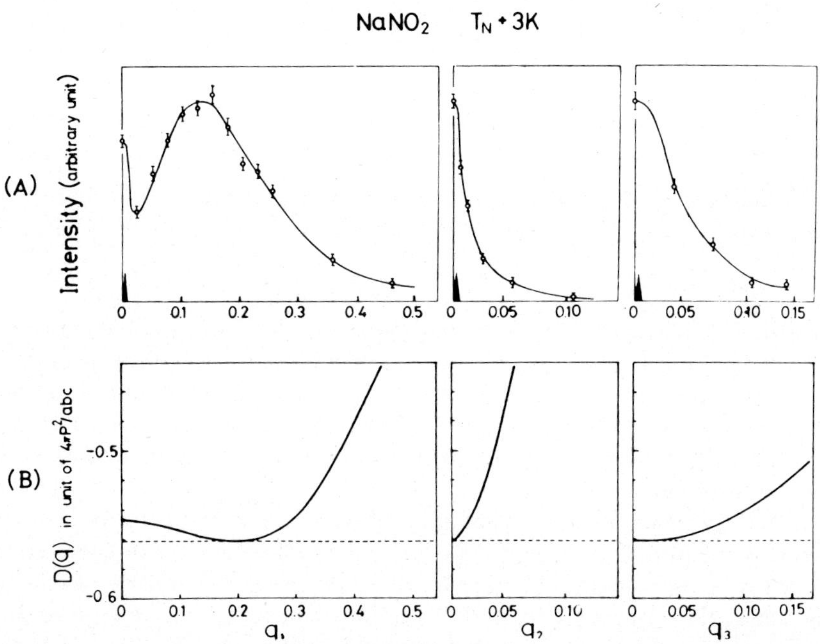

The diffuse X-ray scattering observed in $NaNO_2$ at $T_c+3\,\text{K}$ for scattering for $\mathbf{K}=\tau+\zeta_1 a^*$, $\mathbf{K}=\tau+1/8a^*+\zeta_2 b^*$ and $\mathbf{K}=\tau+1/8a^*+\zeta_3 c^*$. The lower part of the figure shows $J(k)=-D(q)$ calculated using the electrostatic approximation. (From Yamada and Yamada 1966.)

It is frequently possible to measure the differential cross-section using neutron scattering, since neutron energies are comparable with the energies of the phonons unlike X-ray photon energies. However, this feature leads to difficulties in the determination of the *total* scattering cross-section using neutron scattering. In a two-axis spectrometer the neutrons scattered in a particular direction are recorded irrespective of their energy. As shown in fig. I.20 neutrons with different energy transfers have different wave-vector transfers. The integrated intensity is not then an integral over all frequencies for a particular wave-vector transfer, but an integral along a particular path in $(\mathbf{K}, \omega)$ space. This will correspond to the true integrated intensity only if the variation in wave-vector transfer $\mathbf{K}$, is negligible for all the

Fig. I.20

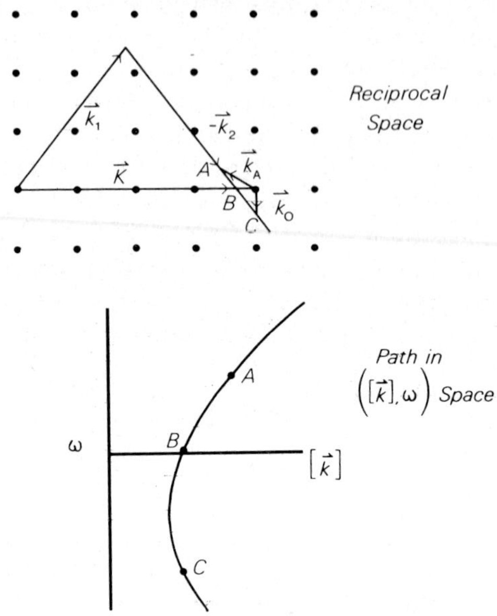

Sketch in reciprocal space of a two-axis neutron diffraction experiment. Note that the length of $\mathbf{k}_2$ is not fixed. In the lower part of the figure the path of integration in $|\mathbf{k}|, \omega$ space is sketched.

frequency transfers occurring in the experiment, $\bar{\omega}$. This can be achieved when the energy of the incident neutron beam is very much larger than $\hbar\bar{\omega}$. In principle, the accuracy of the integration can be tested by performing a differential scattering experiment in which the spectrometer is controlled so as to obtain a frequency distribution for constant wave-vector transfer, $\mathbf{K}$, but this requires more complex experimental equipment.

One of the cases where there is no difficulty with the wave-vector dependence in the integration is in order–disorder systems where the energy transfers from the diffuse scattering are negligible. In fig. I.21 we show the diffuse scattering of neutrons, due to the ordering of Cu and Zn atoms in brass as measured by Als-Nielsen and Dietrich (1967). The scattering becomes more intense as $T \to T_c$ and also more concentrated in $\mathbf{k}$-space.

The differential scattering cross-section is given in the one-phonon approximation by eqn. (I.2.35). If the scattering from the fluctuations in the order parameter are well separated in frequency from the other modes of the same symmetry, then

$$S^1(\mathbf{K}, \omega) = \sum_{\tau} |F(\mathbf{K}, \mathbf{k} + \mathbf{q}_s j)|^2 \Delta(\mathbf{K} + \mathbf{q}_s + \mathbf{k}) \mathscr{S}(\mathbf{K}, \omega), \qquad \text{(I.3.28)}$$

where from eqn. (I.2.4)

$$\mathscr{S}(\mathbf{K}, \omega) = \frac{\hbar}{\pi}(n(\omega) + 1)\chi_1^1(\mathbf{k}, \omega). \qquad \text{(I.3.29)}$$

If $F(\mathbf{K}, \mathbf{k} + \mathbf{q}_s j)$ is independent of $\mathbf{k}$, then the scattering will have the same form at different wave-vectors $\mathbf{K}$ corresponding to the same $\mathbf{k}$, and will give a direct measure

Fig. I.21

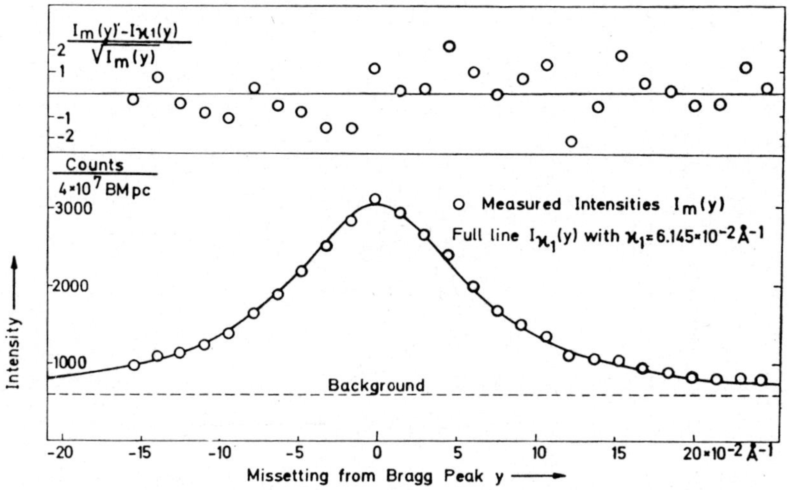

The neutrons scattered from β-brass 8·9 K above T_c, and a fit to a Lorentzian cross-section as a function of k, or misset from the Bragg peak. (From Als-Nielsen and Dietrich 1967.)

of $\chi_i^{\mathrm{I}}(\mathbf{k}, \omega)$ as shown in fig. I.9. If, however, the frequency of the fluctuations in the order parameter is not well separated from the frequencies of the other modes of the same symmetry, the interference terms in eqn. (I.2.38) become important and the shape of the observed spectrum for different but equivalent $\mathbf{K}$ become different. We shall examine this point in detail in § I.4.1.

Neutron scattering then provides a very powerful way of studying the critical fluctuations at structural phase transitions. Normal techniques enable an energy resolution of about 5×10^{10} Hz and a wave-vector resolution of about 10^{-2} Å^{-1}. A novel technique employed by Collins and Teh (1973) to measure the differential cross-section for order–disorder alloys was to measure the decay of the fluctuations in real time. This technique is applicable for phase transitions where the characteristic frequency is about 10^{-2} Hz.

Raman scattering is restricted to very small wave-vectors $\mathbf{K}$ (effectively $\mathbf{K}=0$ in most cases) but enables detailed measurements to be made of the frequency dependence of $\chi(\mathbf{K}, \omega)$. There are two distinct cases which must be discussed. Firstly there are cases, such as KDP, for which there is a linear coupling between the light and the order parameter fluctuations (the first term in eqn. (I.2.46)). If this is the only mode of importance in the sum over j then the Raman scattering tensor is given by

$$I_{\alpha\beta\gamma\delta} = P_{\alpha\beta}(\mathbf{K}j)P_{\gamma\delta}(-\mathbf{K}j)\mathscr{S}(\mathbf{K}, \omega). \qquad \text{(I.3.30)}$$

The results of the Raman scattering in KDP are shown in fig. I.22, and show that the spectral width becomes progressively narrower, as $T \to T_c$, showing the phenomenon of critical slowing down.

This type of result can only be obtained in materials for which the soft mode is Raman active in the high-symmetry phase.

If $\mathbf{q}_s = 0$ but the mode is inactive it is possible to break the Raman scattering selection rules by applying an external field to the material and measuring the field

Fig. I.22

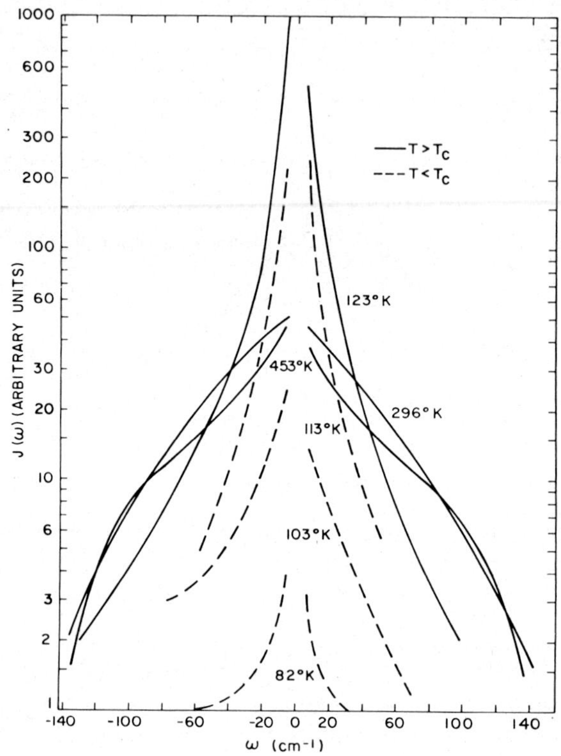

Raman scattering from the ferroelectric fluctuations of KDP. (After Kaminow and Damen 1968.)

induced scattering. This technique has been used to make the infra-red-active (but Raman-inactive) modes visible in the Raman scattering from $SrTiO_3$ and $KTaO_3$ (Fleury and Worlock 1968).

The two-phonon Raman scattering has a contribution arising from the dependence of the polarizability on

$$P_{\alpha\beta}\begin{pmatrix}\mathbf{q} & -\mathbf{q}\\ j & j'\end{pmatrix} Q\begin{pmatrix}\mathbf{q}\\ j\end{pmatrix} Q\begin{pmatrix}-\mathbf{q}\\ j'\end{pmatrix}.$$

Close to the phase transition we consider the particular part of this term for which the normal mode coordinates are those of the order parameters to obtain

$$I^{(2)}_{\alpha\beta\gamma\delta} = \sum_{\mathbf{k}_1\mathbf{k}_2 ii'} P_{\alpha\beta}\begin{pmatrix}\mathbf{q}_{si}+\mathbf{k}_1 & -\mathbf{q}_{si}-\mathbf{k}_1\\ j_{i'} & j_i\end{pmatrix}$$
$$\times P_{\gamma\delta}\begin{pmatrix}\mathbf{q}_{si'}+\mathbf{k}_2 & -\mathbf{q}_{si'}-\mathbf{k}_2\\ j_{i'} & j_{i'}\end{pmatrix} \frac{1}{2\pi}\int \langle Q_i(\mathbf{k}_1,\tau)Q_i(-\mathbf{k}_1,\tau)$$
$$\times Q_{i'}(\mathbf{k}_2,0)Q_{i'}(-\mathbf{k}_2,0)\rangle_c \exp(i\omega\tau)\,d\tau. \qquad \text{(I.3.31)}$$

If we neglect the wave-vector dependence of the coefficients, this cross-section gives a measure of the frequency-dependent four-point correlation function. In the interpretation of Raman scattering results this correlation function has almost invariably been approximated by writing

$$\langle Q_i(\mathbf{k}_1,t)Q_i(-\mathbf{k}_1,t)Q_{i'}(\mathbf{k}_2,0)Q_{i'}(-\mathbf{k}_2,0)\rangle$$
$$\approx 4\langle Q_i(0)\rangle\langle Q_{i'}(0)\rangle\,\langle Q_i(0,\tau)Q_{i'}(0,0)\rangle. \qquad \text{(I.3.32)}$$

The approximation (I.3.32) should be reasonable when Q_i is large, but will fail when it is comparable in size to the fluctuations $\hat{Q}_i(\mathbf{k})$, i.e. close to T_c. With this approximation the Raman scattering becomes $I^{(2)}_{\alpha\beta\gamma\delta}$

$$I^{(2)}_{\alpha\beta\gamma\delta} \approx N\sum_i P_{\alpha\beta}\begin{pmatrix}\mathbf{q}_{si} & -\mathbf{q}_{si}\\ j_i & j_i\end{pmatrix}P_{\gamma\delta}\begin{pmatrix}\mathbf{q}_{si} & -\mathbf{q}_{si}\\ j_i & j_i\end{pmatrix}\mathscr{S}_i(0,\omega)Q_i^2. \qquad \text{(I.3.33)}$$

In fig. I.23 we illustrate Raman scattering results from $SrTiO_3$ for $T \to T_c$; there is clearly a soft mode whose frequency agrees with that obtained from neutron scattering measurements below T_c (fig. I.10) showing that Raman scattering provides a valuable way of measuring $\mathscr{S}_i(0,\omega)$ in the distorted phase. The approximation of eqn. (I.3.32) should be reasonable when $\langle Q_i\rangle$ is large, but will fail when the fluctuations in Q_i become comparable with $\langle Q_i\rangle$—i.e. close to T_c (see §III.3.2.2). The analysis of Raman scattering results close to T_c, in terms of eqn. (I.3.33) is therefore inappropriate. We suspect that some of the difficulties experienced in the analysis of results close to T_c derive from this erroneous use of eqn. (I.3.33).

Fig. I.23

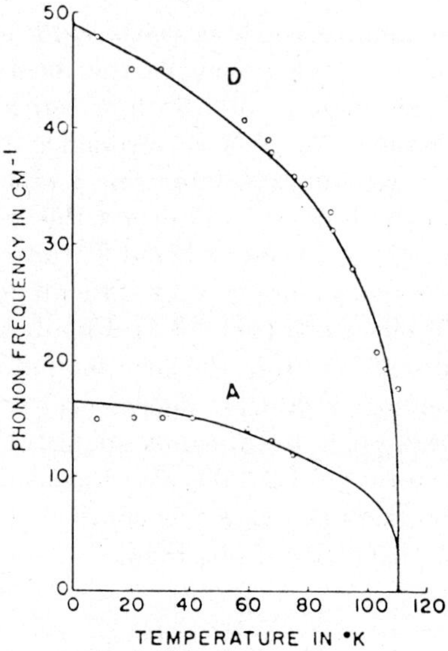

Frequencies of the two soft modes in $SrTiO_3$ below T_c as observed with Raman scattering techniques by Fleury *et al.* (1968). A is the transverse mode and D the longitudinal mode.

I.3.6.3. *Local susceptibilities*

In §I.3.3.3 we discussed the measurement of order parameters using electron spin resonance and nuclear magnetic resonance techniques. In both cases the method involved determining the resonance frequency of either a particular electron or nucleus in the electric field produced by its environment. This environment is not static but fluctuating, and hence there will be a line width of the resonance which is determined by the rate at which the spin (electronic or nuclear) can lose its energy. This rate can be calculated and within second-order perturbation theory will depend upon the density of states of the same frequency as the resonance frequency, and the coupling between the spin and these states.

If we extend the expansion eqn. (I.3.14) of the coefficients of the electrostatic potential B to include the fluctuations in the order parameter then, remembering that B is a local quantity,

$$B=\left(\frac{1}{N}\right)^{1/2}\sum_{\mathbf{k}}A_1(\mathbf{k})Q(\mathbf{k})+\frac{1}{N}\sum_{\mathbf{k}_1\mathbf{k}_2}A_2(\mathbf{k}_1\mathbf{k}_2)Q(\mathbf{k}_1)Q(\mathbf{k}_2)+\ldots \tag{I.3.34}$$

where $A_1(0)$ is the A_1 of eqn. (I.3.14) and we have assumed that there is only one component of the order parameter. The line width for a resonance frequency ω is proportional to (Blinc 1971)

$$\Gamma(\omega)=\frac{1}{2\pi}\int_{-\infty}^{\infty}\langle B(\tau)B(0)\rangle\exp(i\omega\tau)\,d\tau,$$

where ω is the resonance frequency and $B(\tau)$ is the operator B at time τ. Substituting the first term of eqn. (I.3.34) for $B(\tau)$ yields

$$\Gamma(\omega)=\frac{1}{N}\sum_{\mathbf{k}}|A_1(\mathbf{k})|^2\mathscr{S}(\mathbf{k},\omega), \tag{I.3.35}$$

where $\mathscr{S}(\mathbf{k},\omega)$ is the one-phonon Van Hove scattering function for the fluctuations. This expression has a sum over $\mathbf{k}$ owing to the local nature of the resonance technique. Nevertheless, an increase in the line width is found close to T_c as illustrated in fig. I.24 for the ^{133}Cs N.M.R. resonance in CsD_2AsO_4 (Blinc 1971). Unfortunately, since these experiments determine a *sum* over $\mathbf{k}$ (for one particular frequency, ω) it is frequently difficult to interpret the results without relying on a particular model for $\mathscr{S}(\mathbf{k},\omega)$. Furthermore, below T_c the resonance line is split by the distortion and it may be experimentally very difficult to distinguish between the splitting and a line width effect (cf. §III.4.3.2). Finally if $A_1(0)$ of eqn. (I.3.34) is zero, the line width results partly from the two-phonon terms in eqn. (I.3.34) and partly from the terms involving $A_1(\mathbf{k})\approx\mathbf{k}^2$. The line width increase at T_c is then small and in practice almost impossible to interpret quantitatively.

The Debye–Waller factor (eqn. I.2.35) is also a measure of single site properties. To the extent that the motions of the atoms can be described by non-interacting normal modes (i.e. in the displacive limit) then

$$W_\kappa=\sum_{\alpha\beta}\frac{K_\alpha K_\beta}{2}\langle u_\alpha(\kappa)u_\beta(\kappa)\rangle,$$

where the displacement–displacement correlation function is

$$\langle u_\alpha(\kappa)u_\beta(\kappa)\rangle=\frac{1}{M_\kappa}\sum_{\mathbf{q}jj'}e_\alpha(\kappa,\mathbf{q}j)e_\beta(\kappa,-\mathbf{q}j')\langle Q(\mathbf{q}j)Q(-\mathbf{q}j')\rangle. \tag{I.3.36}$$

The normal-mode correlation function is given at high temperatures in terms of $\chi(\mathbf{q}jj')$ by eqn. (I.2.22). Since the susceptibility diverges as $T\to T_c$ for $\mathbf{q}\to\mathbf{q}_s$, W_κ will increase as $T\to T_c$. This increase will, however, be less marked than that of the susceptibility $\chi(\mathbf{q}_s jj)$ due to the sum over $\mathbf{q}j$ and j' in eqn. (I.3.36).

Fig. I.24

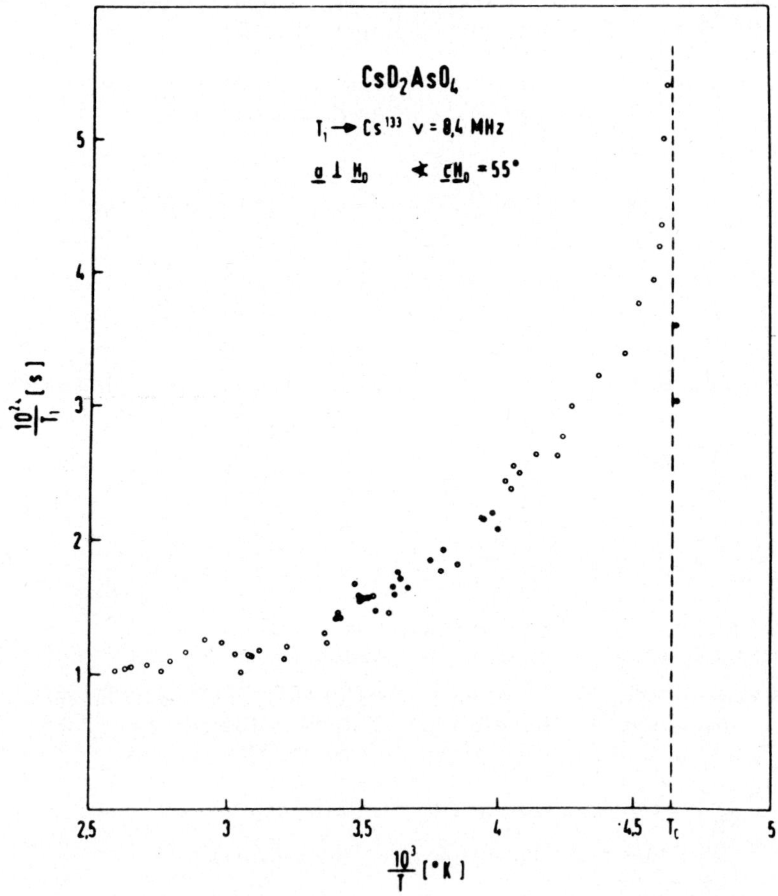

Temperature dependence of the ^{133}Cs spin–lattice relaxation rate in CsD_2AsO_4. (After Blinc 1971.)

The behaviour of the Debye–Waller factor in an order–disorder system is different. Consider as an example the N ion in NO_2 which has two positions $\pm\boldsymbol{\delta}$ for the equilibrium position. In the high-temperature phase, both sites are equally populated so that the Debye–Waller factor is given by $\cos \mathbf{K}.\boldsymbol{\delta}$. In the low-temperature phase the amplitude of the scattering is proportional to $\cos \mathbf{K}.\boldsymbol{\delta}+2i\langle S\rangle \sin \mathbf{K}.\boldsymbol{\delta}$, so that the Debye–Waller factor due to the hopping is given by

$$\exp(-W_{\kappa})=(\cos^2(\mathbf{K}.\boldsymbol{\delta})+4\langle S\rangle^2 \sin^2(\mathbf{K}.\boldsymbol{\delta}))^{1/2}. \qquad (I.3.37)$$

This shows that the Debye–Waller factor is constant above T_c, and increases below T_c reaching unity when $\langle S\rangle=\frac{1}{2}$. Since most real materials are neither strictly order–disorder nor displacive in character, the interpretation of the temperature dependence of Debye–Waller factors is far from unambiguous.

The Debye–Waller factor can be measured, for a particular $\mathbf{K}$, from the amplitude of the Mössbauer γ-rays emitted from a suitable source. In fig. I.25 we show the Debye–Waller factor of ^{57}Fe in $PbTiO_3$ as measured using Mössbauer techniques by Bhide and Hegde (1972). The Debye–Waller factor can also be measured using the incoherent scattering of thermal neutrons or from the temperature variation of the intensity of Bragg reflections.

Fig. I.25

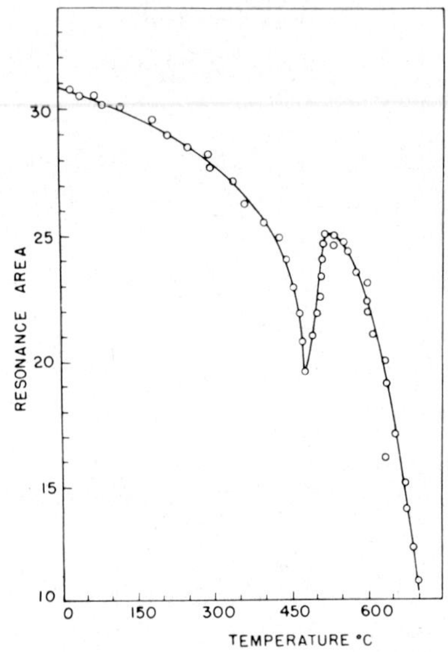

Temperature variation of the Debye–Waller factor in $PbTiO_3$ as measured from the resonance area of the ^{57}Fe Mössbauer line by Bhide and Hegde (1972).

§ I.4. Phenomenological Landau theory

I.4.1. *Single-component order parameters*

I.4.1.1. *The Landau theory*

In 1937, Landau (1937) developed a theory of phase transitions which he applied to structural phase transitions. In this section this theory is applied to a wide variety of different systems, because the first step in the understanding of any new material is the development of the appropriate Landau theory. We begin with the simplest familiar (Landau and Lifshitz 1959) example when the order parameter is a singlet so that there is only one component of $Q_i = Q$ of § I.3 which can then always be chosen to be real. Landau *assumes* that the free energy of one unit cell of the crystal may be expanded in a power series in this order parameter;

$$G = G_0 + \frac{1}{2} r Q^2 + dQ^3 + uQ^4, \tag{I.4.1}$$

where the coefficients may be functions of temperature.

A phase is stable only if

$$\left.\frac{\partial G}{\partial Q}\right|_{Q_0} = 0, \quad \left.\frac{\partial^2 G}{\partial Q^2}\right|_{Q_0} > 0. \tag{I.4.2}$$

The high-symmetry phase $\langle Q \rangle = Q_0 = 0$ is then stable only if the linear term in Q in eqn. (I.4.1) is absent and if r is positive. The system is distorted if r is negative.

Furthermore at a continuous phase transition G must increase with $|Q|$ at T_c, $r=0$, so that $d=0$ and $u>0$. Landau's theory is then completed by the further *assumption* that since r changes sign at T_c, it can be written as an analytic function of T near T_c

$$r=a(T-T_c),$$

while the coefficient u is assumed to be independent of temperature for temperatures sufficiently close to T_c. The free energy of the simplest model then follows as

$$G=G_0+\frac{1}{2}a(T-T_c)Q^2+uQ^4. \tag{I.4.3}$$

The stability conditions for the low-temperature phase then give for $T<T_c$ that the order parameter will have the value

$$Q_0^2=\frac{a}{4u}(T_c-T), \tag{I.4.4}$$

while the free energy for the equilibrium displacement is given by

$$G_s=\langle G\rangle=G_0-\frac{a^2}{16u}(T_c-T)^2, \quad T<T_c \tag{I.4.5 a}$$

and consequently the change in specific heat is given by

$$\begin{aligned}\tilde{C} &= -T\frac{\partial^2 G}{\partial T^2}=\frac{a^2 T}{8u}, & T<T_c; \\ &=0 & T>T_c.\end{aligned} \tag{I.4.5 b}$$

The susceptibility $\chi(\mathbf{k}=0)=\left(\dfrac{\partial^2 G}{\partial Q^2}\right)^{-1}$, is then

$$\begin{aligned}\chi(0) &=\frac{1}{a(T-T_c)}, & T>T_c; \\ &=\frac{1}{2a(T_c-T)}, & T<T_c.\end{aligned} \tag{I.4.6}$$

These results give, by assumption, that the exponents describing the critical behaviour are $\beta=\frac{1}{2}$ and $\gamma=1$. In fig. I.26 we show the free energy for this model both above and below T_c, and in fig. I.27 the various properties as a function of temperature.

In eqn. (I.4.3) we assumed that $d=0$. This is always the case when there is only one component of the order parameter, but is not necessarily the case for multi-component systems, for example at some elastic instabilities (Cowley 1976 a).

When d is non-zero the transition is of first order and so the expansion of eqn. (I.4.1) is suspect. Nevertheless, in many systems the transition is nearly continuous and the phase transition can be analysed using the Landau expansion. The transition occurs when $T=T_I$ where

$$T_I=T_c+\frac{d^2}{2au}, \tag{I.4.7}$$

Fig. I.26

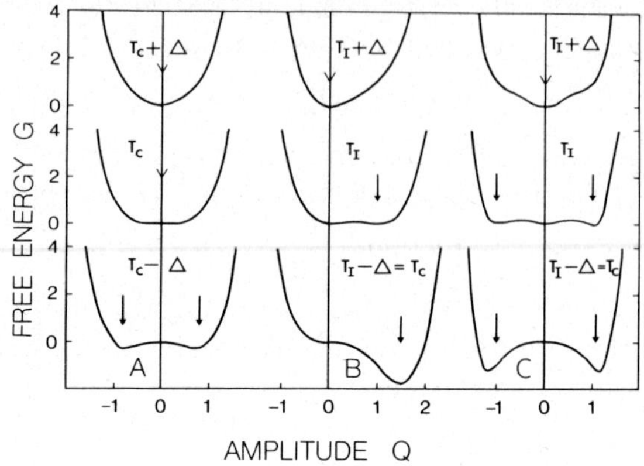

The free energy $G-G_0$ as a function of the coordinate Q for three models: Model A is the simplest given by eqn. (I.4.3) with $a=2$, $u=4$ and Δ is one in these units. Model B incorporates the cubic term d which was chosen to be 6 when the first-order transition is at $T_c+\Delta$. Model C also has a first-order transition but in eqn. (I.4.9) $a=2$, $u=-8$ and $h=6$ when $T_I=T_c+\Delta$.

Fig. I.27

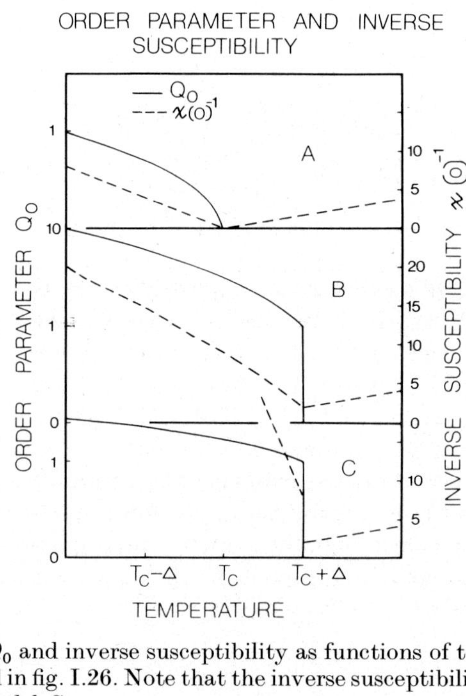

The order parameter Q_0 and inverse susceptibility as functions of temperature for the three models described in fig. I.26. Note that the inverse susceptibility increases very rapidly for $T<T_I$ for model C.

while the order parameter takes on a value at T_I given by

$$Q_0 = -\frac{d}{2u},$$

and below T_I is given by

$$Q_0 = \frac{-3d-(9d^2-16ua(T-T_c))^{1/2}}{8u}. \tag{I.4.8}$$

In figs. I.26 and I.27 we illustrate the properties of this model in more detail. In this case the first-order transition occurs whenever the symmetry allows a non-zero cubic term, d. It is therefore a symmetry-driven first-order transition.

A first-order transition also occurs in the simplest model if $u<0$, but in this case it is a consequence not of the symmetry but of the particular sign of the u coefficient in the Landau expansion. If u is negative, G is unbounded and the expansion must be extended beyond quartic terms. We therefore add an extra term to the free energy, when it is given by

$$G = G_0 + \frac{1}{2}a(T-T_c)Q^2 + uQ^4 + hQ^6 + \ldots, \tag{I.4.9}$$

where u is now negative and h is *assumed* to be positive. The transition is of first order and occurs at T_I where

$$T_I = T_c + \frac{u^2}{2ah}, \quad u<0. \tag{I.4.10}$$

The step in the order parameter at T_I is

$$Q_0 = \pm\left(\frac{a(T_I-T_c)}{|u|}\right)^{1/2}$$

while below T_I the order parameter is given by

$$Q_0^2 = \frac{-4u+(16u^2-24ah(T-T_c))^{1/2}}{12h}. \tag{I.4.11}$$

The properties of this type of first-order transition, of which KDP(KH_2PO_4) is an example, are illustrated in figs. I.26 and I.27.

I.4.1.2. *Coupling to strains*

At all structural phase transitions the order parameter is coupled to the elastic strains or possibly other secondary order parameters by terms in the free energy which are linear in the strain and quadratic in the order parameter. In the simplest case we extend the model of the previous section by including the coupling to a single strain parameter, η. The free energy, given by eqn. (I.4.3), is then extended to include the strains as

$$G = G_0 + \frac{1}{2}a(T-T_c)Q^2 + uQ^4 + eQ^2\eta + \frac{1}{2}C\eta^2 + \ldots, \tag{I.4.12}$$

where e is the coupling and C is the appropriate elastic constant. Close to T_c there is no reason to expect that C will be particularly temperature dependent, except by an unlikely accident. The condition for the material to be stress-free is that $\partial G/\partial\eta=0$, which gives

$$\eta=-\frac{eQ^2}{C}, \tag{I.4.13}$$

which can be used to eliminate η from eqn. (I.4.12) with the result

$$G=G_0+\frac{1}{2}a(T-T_c)Q^2+\tilde{u}Q^4, \tag{I.4.14}$$

where

$$\tilde{u}=u-\frac{e^2}{2C}. \tag{I.4.15}$$

Since eqn. (I.4.14) for G has exactly the same form as eqn. (I.4.3) except that u is replaced by $\tilde{u}$, the development goes through unchanged except for the replacement of u by $\tilde{u}$. Equation (I.4.15) shows that $\tilde{u}$ is less than u: consequently transitions under conditions of constant stress are more likely to be of first order than ones at constant strain.

The elastic strain is proportional to the square of the order parameter, eqn. (I.4.13), and so using eqn. (I.4.4), is expected to be proportional to $T-T_c$. The susceptibility can be measured in two different ways: at low frequencies the strain will be able to follow the order parameter, Q and the susceptibility is written χ^0, but at high frequencies the strain will be unable to follow and the susceptibility is written χ^∞. For $T>T_c$ both susceptibilities are equal to $(a(T-T_c))^{-1}$, but for $T<T_c$

$$\chi^0(0)=[2a(T_c-T)]^{-1}, \tag{I.4.16}$$

while $\chi^\infty(0)$ is given by $\partial^2G/\partial Q^2$ with the derivatives evaluated with η held constant when the result is

$$\chi^\infty(0)=\frac{\tilde{u}}{2ua(T_c-T)}. \tag{I.4.17}$$

Since $\tilde{u}$ is less than u, $\chi^\infty<\chi^0$, although they both diverge at T_c.

It is also necessary to distinguish between the high- and low-frequency response for the elastic constants below T_c. The elastic constant under conditions of constant order parameter, Q, is $\partial^2G/\partial\eta^2=C$, whereas under conditions where Q can follow the changes in η it is given by $\mathrm{d}^2G/\mathrm{d}\eta^2=C-e^2/4u$. Hence, at low frequencies when the order parameter can follow the elastic fluctuations, the elastic constant is predicted to decrease by $e^2/4u$ at the transition.

In the low-temperature phase there is also a non-zero susceptibility describing the response of the strain to a field which couples to the order parameter, Q. This is given by $[\partial^2G/\partial\eta\partial Q]^{-1}=1/2eQ_0$, and hence diverges at T_c.

I.4.1.3. *Fluctuations in systems with short-range forces*

The expansion of the free energy in powers of the order parameter Q is not sufficient to give a description of the wave-vector dependence of the susceptibility. We now extend the discussion to incorporate the effects of the fluctuations $Q(\mathbf{k})$.

Formally eqn. (I.4.3) can be extended to give

$$G = G_0 + \frac{1}{2N}\sum_{\mathbf{k}} R(\mathbf{k})Q(\mathbf{k})Q(-\mathbf{k}) + \frac{1}{N^2}\sum_{\mathbf{k}_1\mathbf{k}_2\mathbf{k}_3} u(\mathbf{k}_1,\mathbf{k}_2,\mathbf{k}_3)Q(\mathbf{k}_1)Q(\mathbf{k}_2)Q(\mathbf{k}_3)Q(-\mathbf{k}_1-\mathbf{k}_2-\mathbf{k}_3)+\ldots,$$

where the coefficients in the expansion are wave-vector dependent and use has been made of wave-vector conservation. This expression for the free energy (effective hamiltonian) includes the interactions between the excitations with different wave-vectors and is difficult to solve as discussed in detail in part II. Here the effect of the fluctuations is neglected except for the leading terms. This is then the Landau–Ginzburg theory and is equivalent to neglecting the interactions, the u term, except when at least two of the Qs are replaced by the static order parameter, Q_0.

For the simplest model of a transition $u(0,0,0)$ is non-zero at T_c, and so we neglect the $\mathbf{k}$ dependence of the anharmonic coefficient u. $R(0)$ is, however, small at T_c and so we include in $R(\mathbf{k})$ the leading terms in an expansion in $\mathbf{k}$. Since if the order parameter has a single component $\mathbf{q}_s$ is a 'special point' within the Brillouin zone, there are no linear terms in $\mathbf{k}$. Consequently for an isotropic system with short-range interactions the free energy becomes

$$G = G_s + \frac{1}{2N}\sum_{\mathbf{k}} [a(T-T_c) + fk^2 + 12Q_0^2u]\hat{Q}(\mathbf{k})\hat{Q}(-\mathbf{k}),$$

where G_s is the static part of the free energy, eqn. (I.4.5 a). The static susceptibility for wave-vector, $\mathbf{k}$, is given by;

$$\chi(\mathbf{k})^{-1} = [a(T-T_c) + fk^2] \quad \text{for} \quad T > T_c,$$

and by

$$\chi(\mathbf{k})^{-1} = [2a(T_c-T) + fk^2] \quad \text{for} \quad T < T_c. \tag{I.4.18}$$

It is sometimes useful to define the inverse of these static susceptibilities as the squares of frequencies,

$$\chi(\mathbf{k})^{-1} = \omega(\mathbf{k})^2,$$

by analogy with the result for a harmonic crystal, eqn. (I.2.23). These frequencies may or may not correspond to the frequencies of resonant normal modes or 'soft modes' as discussed in § I.3.5. In terms of these 'frequencies' the free energy may be rewritten as

$$G = G_s + \frac{1}{2N}\sum_{\mathbf{k}} \omega(\mathbf{k})^2 Q(\mathbf{k})Q(-\mathbf{k}), \tag{I.4.19}$$

where

$$\omega(\mathbf{k})^2 = a(T-T_c) + fk^2 + 12Q_0^2u.$$

The expansion coefficient f must be positive at T_c for the wave-vector of the distorted phase $\mathbf{q}_s$ to be stable, so that to a first approximation it is independent of temperature. These assumptions then give the critical exponents for the correlation function, eqn. (I.3.18), $\nu = 0{\cdot}5$, and $\eta = 0$, and the line shape of the scattering as a Lorentzian in wave-vector, $\mathbf{k}$, at fixed temperature. This latter result is consistent with experimental results on, for example, β-brass as shown in fig. I.21.

The temperature dependence of the Debye–Waller factor can be obtained from eqns. (I.3.36) and (I.4.19) for a displacive system as, (Borsa and Rigamonti 1972);

$$\sum_{\mathbf{k}} \chi(\mathbf{k}) = K_1 - K_2|T - T_c|^{1/2}, \tag{I.4.20}$$

where K_1 and K_2 are constants which depend upon a and f. This result shows that the Debye–Waller factor does not diverge at T_c but exhibits a cusp-like behaviour (see, however, § II.5.4.1). A similar behaviour is predicted at other structural phase transitions except when the dispersion relation at small $\mathbf{k}$ cannot be expanded in powers of $\mathbf{k}$ as we discuss below.

Only in cubic materials is the coefficient of $\mathbf{k}$ in eqn. (I.4.19) independent of the direction of $\mathbf{k}$. More generally eqn. (I.4.19) becomes

$$\omega(\mathbf{k})^2 = a(T - T_c) + 12Q_0^2 u + \sum_{\alpha\beta} f_{\alpha\beta} k_\alpha k_\beta.$$

The critical scattering is then Lorentzian in wave-vector, but the width of the Lorentzian depends on the direction of the wave-vector as observed close to the α–β transition in quartz by Dorner *et al.* (1974). The exponent is, however, independent of the direction of $\mathbf{k}$, as tacitly assumed in eqn. (I.3.18).

A Lifshitz point (Hornreich *et al.* 1975) is the point in the phase diagram such that for one or more directions of $\mathbf{k}$, the term in k^2 in eqn. (I.4.18) has a coefficient $f = 0$. This point then separates phase transitions to a phase described by $\mathbf{q}_s$ being a 'special wave-vector' from phases where $\mathbf{q}_s$ is incommensurate with the underlying lattice. These special points are discussed in more detail in § II.5.3.2.

I.4.1.4. *Uniaxial ferroelectrics*

The formal expansion of the $R(\mathbf{k})$ term in powers of k^2 in § I.4.1.3 is valid only if there are no long-range forces in the material. The commonest example of a long-range force is the long-range electrostatic force in ferroelectric materials. In these materials the macroscopic electric field associated with the ferroelectric fluctuations makes the response dependent on the direction of the wave-vector; there is a splitting between the longitudinally and transversely polarized modes. Taking acount of this electrostatic interaction, and taking the z direction as the unique ferroelectric axis, the free energy is given by eqn. (I.4.19) but with the frequencies given by

$$\omega(\mathbf{k})^2 = a(T - T_c) + \sum_{\alpha\beta} f_{\alpha\beta} k_\alpha k_\beta + g\frac{k_z^2}{k^2} + 12uQ_0^2, \tag{I.4.21}$$

where the coefficient g is given in terms of the dipole moment operator as

$$g = \frac{1}{v\varepsilon_0}|M_z(\mathbf{k})|^2,$$

and is independent of $\mathbf{k}$ if the wave-vector dependence of $M_z(\mathbf{k})$ is neglected.

The susceptibility or scattering cross-section, for wave-vectors $\mathbf{k}$ in the x and y direction, is similar to that of the isotropic systems discussed above. For other directions of $\mathbf{k}$ the limit as $\mathbf{k} \to 0$ is not divergent even at T_c. This gives rise to a characteristic doughnut shape to the critical scattering, a slice through which is shown in fig. I.28 as measured in DKDP (KD_2PO_4) by Skalyo *et al.* (1970). An analysis of this data enables the parameters f and g to be obtained as shown by Paul *et al.* (1970).

Fig. I.28

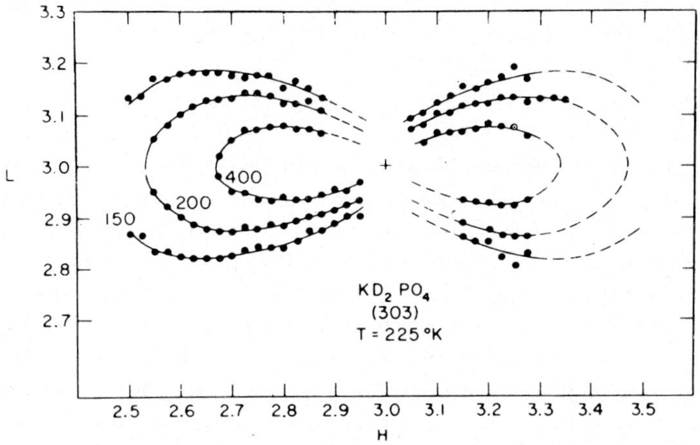

The critical scattering intensity contours in the (h0l) zone of KD_2PO_4 from the ferroelectric fluctuations polarized along the z-axis. T_c for the crystal was 221 K. (After Skalyo *et al.* 1970.)

The singularity in $\chi(\mathbf{k})$ as $\mathbf{k} \rightarrow 0$ cannot occur in real systems: it is prevented by a more detailed treatment of the electrostatic interactions including retardation or polariton effects, and by the effect of the crystal surfaces (Born and Huang 1954).

The Debye–Waller factor of a uniaxial ferroelectric has an even weaker singularity at T_c than systems with only short-range forces, eqn. (I.4.20). The result was obtained by Meyer and Cowley (1976) as

$$\sum_{\mathbf{k}} \chi(\mathbf{k}) = K_1 + K_2|T - T_c| \log (K_3|T - T_c|), \qquad \text{(I.4.22)}$$

where the Ks are constants.

I.4.1.5. *Acoustic mode instabilities*

The stability of a crystal against macroscopic strains is most conveniently discussed in the Voigt notation for the strains:

$$\eta_{xx} = \eta, \eta_{yy} = \eta_2, \eta_{zz} = \eta_3, \eta_{xy} + \eta_{yx} = \eta_6, \eta_{xz} + \eta_{zx} = \eta_5, \eta_{yz} + \eta_{zy} = \eta_4.$$

The elastic matrix, eqn. (I.2.10), then has the form

$$H_D = \frac{v}{2} \sum_{\rho\sigma} C_{\rho\sigma} \eta_\rho \eta_\sigma.$$

The crystal is stable (Born and Huang 1954) only if all the eigenvalues of $C_{\rho\sigma}$ are positive, leading to stability conditions which are dependent upon the particular crystal class (Cowley 1976 a).

The Landau theory of these transitions then follows §I.4.1.1 but with the appropriate eigenvalue of the elastic constant playing the role of the parameter r.

Since the point groups of the crystal classes have at most triply degenerate irreducible representations, the order parameter may have one, two or three components.

An example of a one component order parameter occurs for an η_6, strain in the point groups $\bar{4}2$ m, 422 or 4/mmm. The eigenvalue of the elastic constant matrix is C_{66} and the symmetry of the distorted phase is 2 mm, 222, or mmm respectively. The high-symmetry phase will be stable if $C_{66}>0$ and a phase transition to the distorted phase will occur if $C_{66}<0$. We therefore *assume* that $C_{66}/\rho=a(T-T_c)$.

The fluctuations associated with the strains are the acoustic modes of vibration. The velocity of any particular acoustic mode is given by a linear combination of elastic constants, but that combination is dependent upon the polarization and the direction of the wave-vector. In the case of our example, the largest fluctuations close to T_c will arise from the modes with velocities determined solely by $(C_{66}/\rho)^{1/2}$. Since C_{66} is the smallest eigenvalue of the matrix $C_{\rho\sigma}$ every other acoustic mode must have a larger velocity. Now in our example the only acoustic modes with velocities given only by C_{66} are those transverse modes with $\mathbf{k}$ along the [100] or [010] directions and with polarization vectors along [010] or [100] respectively.

The frequencies of the acoustic modes with $\mathbf{k}$ close to the [100] direction are given by

$$\omega(\mathbf{k})^2=a(T-T_c)k_x^2+g_1k_y^2+g_2k_z^2+f_1k_x^2(k_y^2+k_z^2)+f_2k_x^4+\ldots, \qquad (\text{I}.4.23)$$

and for $\mathbf{k}$ nearly parallel to [010] by a similar expression with $\mathbf{k}_y$ and $\mathbf{k}_x$ interchanged. This expression shows that the critical fluctuations are confined to particular lines in $\mathbf{k}$ space, whereas they are restricted to planes for uniaxial ferroelectrics.

At some phase transitions there are no acoustic modes with velocities given by the eigenvalue of the elastic constant matrix. In these cases, for example a uniform compression, there are no fluctuations, but they appear in practice to always be of the first order.

I.4.1.6. *Interaction between acoustic modes and ferroelectric fluctuations*

The macroscopic effects of the coupling between the strains and the electric polarization in ferroelectric materials are described, for example, by Jona and Shirane (1962). In this section we shall emphasize the effect of the coupling on the fluctuations. In materials, like KH_2PO_4, which are piezoelectric in the high-symmetry phase there is a linear coupling between the strain and the electric polarization. Consequently the order parameter is a linear combination of the strain and polarization. It is convenient, however, to extend the analysis of § I.4.1.1 by explicitly including in the theory both strain coordinates and the polarization. For the case of KH_2PO_4 we label the polarization along the z axis as Q_1 and this is coupled to the strains η_{xy}. If for simplicity we restrict $\mathbf{k}$ to the x direction then the dielectric fluctuations are coupled only to the transvers acoustic modes polarized along [010] and we label them Q_2. The quadratic term in the expansion of the free energy is for these modes given by

$$G_2=\frac{1}{2N}\sum_{ij=1,2} R_{ij}(k_x)Q_i(k_x)Q_j(k_x),$$

where

$$R_{11}(k)=a(T-T_c^x)+fk^2, \quad R_{22}(k)=C_{66}k^2/\rho,$$

$$R_{12}(k)=R_{21}(k)=kD.$$

Here D is given in terms of the piezoelectric constant h_{36} as $D=h_{36}\,M_z(0)\,(\rho v)^{-1/2}$, where $M_z(0)$ is the dipole moment operator of the ferroelectric mode, eqn. (I.2.20).

The effect of the coupling is most readily seen by calculating the dynamical susceptibility within the harmonic approximation. The response to a field which couples only to mode 1 is given by

$$\chi_1(k,\omega)=\frac{R_{22}(k)-\omega^2}{(R_{11}(k)-\omega^2)(R_{22}(k)-\omega^2)-R_{12}(k)^2}. \qquad \text{(I.4.24)}$$

Now, for small ω, this expression has the form

$$\chi_1(k,0)=[a(T-T_c)+fk^2]^{-1},$$

where the free Curie temperature is given by

$$T_c=T_c^x+\frac{\rho D^2}{aC_{66}}. \qquad \text{(I.4.25)}$$

The static $k=0$ dielectric susceptibility diverges at the Curie temperature of the free crystal which is a higher temperature than the Curie temperature of the clamped crystal, T_c^x. At high frequencies, $\omega^2 \gg R_{22}(k)$, eqn. (I.4.24) reduces to the susceptibility of the clamped crystal—that obtained by neglecting the coupling to the acoustic modes.

Similar results are obtained for the acoustic mode response. At low frequencies the response is given by

$$\chi_2(k,\omega)=\frac{\rho}{C_{66}^E-\rho\omega^2},$$

where

$$C_{66}^E=C_{66}-\frac{\rho D^2}{a(T-T_c)}, \qquad \text{(I.4.26)}$$

and the change in the velocity arises because the dielectric fluctuations follow the acoustic mode fluctuations and the crystal responds so as to keep the electric field, $\mathbf{E}$, constant. At high frequencies where the ferroelectric fluctuations cannot follow the acoustic response, the latter reduces to C_{66} and the electric polarization is constant in the crystal. This behaviour is illustrated in KD_2PO_4 where ultrasonic measurements yield the velocity, C_{66}^E, which decreases to zero at T_c, fig. I.15, while neutron scattering measurements at frequencies above the dielectric response give the relatively temperature independent, C_{66} (Paul *et al.* 1970).

Throughout this section we have assumed that the dielectric fluctuations are intrinsically temperature dependent. The same analysis can be made if we assume in that C_{66} is temperature dependent, while $R_{11}(k)$ is relatively temperature independent. This behaviour then corresponds to an elastically driven ferroelectric transition of which lithium ammonium tartrate is an example (Sawada *et al.* 1977).

In a neutron or X-ray scattering experiment there may be a coupling to both the ferroelectric and acoustic modes, so that the cross-section is given by

$$\begin{aligned} S(\mathbf{K},\omega)=Nk_BT\,\mathrm{Im}\,[\{&|F_1(\mathbf{K})|^2(R_{22}(\mathbf{k})-\omega^2)\\ &-(F_1(\mathbf{K})F_2(-\mathbf{K})+F_2(\mathbf{K})F_1(-\mathbf{K}))R_{12}(\mathbf{k})\\ &+|F_2(\mathbf{K})|^2(R_{11}(\mathbf{k})-\omega^2)\}/\{(R_{11}(\mathbf{k})-\omega^2)(R_{22}(\mathbf{k})-\omega^2)\\ &-R_{12}(\mathbf{k})^2\}], \end{aligned} \qquad \text{(I.4.27)}$$

where the wave-vector transfer $\mathbf{K} = \mathbf{k} + \boldsymbol{\tau}$, with $\boldsymbol{\tau}$ a reciprocal lattice vector and $F_1(\mathbf{K})$ and $F_2(\mathbf{K})$ are the structure factors of the modes. Now the unusual feature about this result is that since $R_{12}(\mathbf{k})$ is proportional to k, the interference term between the scattering by modes 1 and 2 will change sign with k, provided that $F_1(\mathbf{K})$ and $F_2(\mathbf{K})$ do not also change sign. The scattering is then markedly asymmetric in k. This effect has been observed in KD_2PO_4 by Skalyo *et al.* (1970). In KD_2PO_4 the ferroelectric fluctuations are overdamped so that inelastic neutron scattering measurements recording the quasi-elastic response gives the scattering by the ferroelectric

Fig. I.29

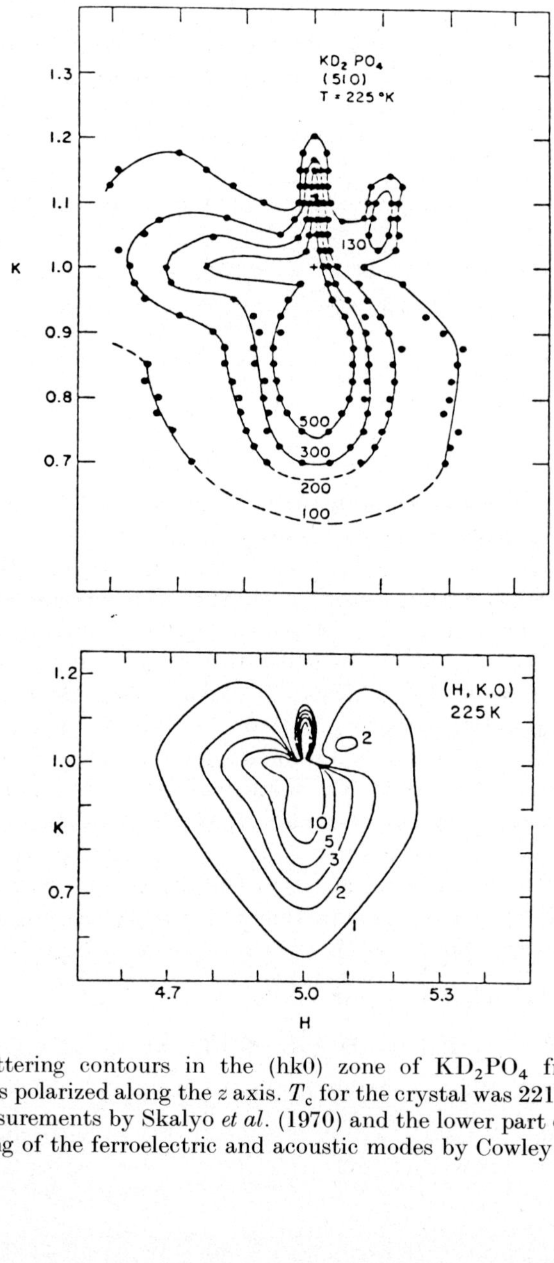

The critical scattering contours in the (hk0) zone of KD_2PO_4 from the ferroelectric fluctuations polarized along the z axis. T_c for the crystal was 221 K and the upper part shows measurements by Skalyo *et al.* (1970) and the lower part calculations including the coupling of the ferroelectric and acoustic modes by Cowley (1976 b).

fluctuations. In fig. I.29 the highly asymmetric critical scattering observed around the $(5,1,0)$ lattice point is shown, together with parameter-free calculations of the pattern, performed using the theory illustrated above (Cowley 1976 b).

In non-piezoelectric materials the leading coupling between acoustic modes and the ferroelectric fluctuations is proportional to k^2—there is then no difference above T_c between the clamped and free dielectric behaviour as shown in § I.4.1.2, and the scattering is almost symmetric in k. There are, however, interesting interference effects between the acoustic and optic modes in the differential scattering cross-section as a function of frequency ω. These effects are particularly marked if the ferroelectric mode 1 is heavily damped so that its spectral response overlaps the frequency of the acoustic mode 2. In this case the term in eqn. (I.4.27) proportional to $|F_2(\mathbf{K})|^2$ gives a Lorentzian, centred approximately on $\omega^2 = R_{22}(\mathbf{k})$; the interference terms, proportional to $F_1(\mathbf{K})F_2(-\mathbf{K})$, give two contributions—a Lorentzian centred at the same position and a term, asymmetric about that frequency, due to the imaginary part of $R_{11}(\mathbf{k})$. This asymmetric term gives rise to a characteristic line shape and a shift in the apparent frequency of the peaks in the scattering.

This type of behaviour, which is common to all anharmonic materials, is illustrated particularly strikingly in the uniaxial ferroelectric lead germanate, as shown in fig. I.30. Near the lattice point (006) the ferroelectric fluctuations have low intensity and only a well-defined transverse acoustic mode is observed. Near the lattice points (005) and (004) both modes scatter strongly and the interference terms give rise to a characteristic asymmetric behaviour at frequencies close to the acoustic

Fig. I.30

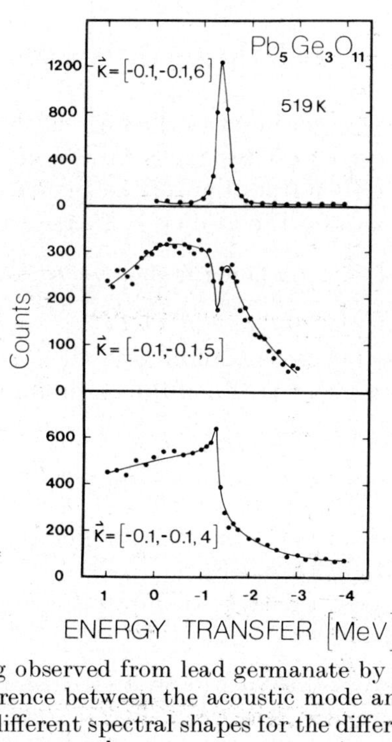

The neutron scattering observed from lead germanate by Cowley *et al.* (1976). The results show the interference between the acoustic mode and a heavily damped optic mode resulting in the different spectral shapes for the different measurements corresponding to the same wave-vector, **k**.

mode. Near (005) the interference enhances the high-frequency part of the scattering, and near (004) it has the opposite sign and enhances the low-frequency part. Clearly as with the scattering shown in fig. I.29 the detailed shape of the scattering is dependent upon the ratio $F_1(\mathbf{K})/F_2(\mathbf{K})$ and so is qualitatively different around different lattice points, $\boldsymbol{\tau}$.

I.4.2. *Multiple-component order parameters*

I.4.2.1. *The Landau theory*

The order parameter at a structural phase transition may have more than one component (§ I.3.1): for example, in $BaTiO_3$ and $SrTiO_3$ there are $n=3$ components. Throughout this section we shall consider the case where the Q_i are real. A necessary condition for this is that the soft-mode wave-vector $\mathbf{q}_s$, must satisfy $\mathbf{q}_s=\boldsymbol{\tau}/2$ or 0. A sufficient condition can be obtained with the aid of group theory: it is that the antisymmetric square of the order parameters be zero. This condition appears to be the same as the requirement that

$$\nabla_{\mathbf{k}}\omega(\mathbf{q}_s+\mathbf{k}j)\Big|_{\mathbf{k}=0}=0$$

which is satisfied by modes with $\mathbf{q}_s=0$ and for $\mathbf{q}_s=\boldsymbol{\tau}/2$ for the symmorphic space groups, but for the non-symmorphic groups it is not necessarily true. For example, in the diamond structure the longitudinal modes at the [100] zone boundary do not have zero gradient. We shall discuss systems in which the order parameter is complex in §§ I.4.4 and I.4.5.

We begin by discussing a system with $n=2$ components for which the free energy, analogous to eqn. (I.4.3) is

$$G=G_0+\frac{1}{2}a(T-T_c)(Q_1^2+Q_2^2)+u(Q_1^2+Q_2^2)^2+v(Q_1^4+Q_2^4)+\ldots \quad \text{(I.4.28)}$$

The symmetry of the order parameter imposes the form of the quadratic term and the third-order term in the Q-coordinates must be absent if the transition is continuous. Equation (I.4.28) differs from that of the singlet model eqn. (I.4.3) in that there are two coefficients u and v. Equation (I.4.28) reduces to the form of eqn. (I.4.3) if we write $Q_i=\zeta_i Q$ (eqn. (I.3.3)), where the eigenvector is normalized $\sum_i \zeta_i^2=1$, when $\tilde{u}=u+v(\zeta_1^4+\zeta_2^4)$, and $\tilde{u}$ replaces u in eqn. (I.4.3).

The free energy of the ordered phase is a minimum if the constant $\tilde{u}$ is a minimum, eqn. (I.4.5 *a*), and this determines the nature of the ordered phase. There are three cases of continuous phase transitions:

I. $v<0$ but $v+u>0$. In this case $\tilde{u}=u+v>0$ and the ordered phase has one of four domains ($\zeta_1=\pm 1$, $\zeta_2=0$, or $\zeta_2=\pm 1$, $\zeta_1=0$).

II. $v=0$ and $u>0$. In this case $\tilde{u}=u$ and the ordered phase may take up any orientation in the space of Q_1 and Q_2.

III. $v>0$ and $\tilde{u}=u+\frac{1}{2}v>0$. In this case there are four ordered domains with $\zeta_1=\pm 1/\sqrt{2}$, $\zeta_2=\pm 1/\sqrt{2}$.

These features, a multiplicity of domains and a structure which is dependent upon the particular values of certain of the interactions, are characteristic of all $n>1$ systems.

The analysis of the model follows exactly as in the $n=1$ system for the order parameter, Q_0, specific heat, free energy and susceptibility above T_c. Some additional analysis is, however, necessary to obtain the behaviour of the susceptibility below T_c. The static susceptibilities are given by the inverse of the second derivatives of the free energy with respect to Q_1 and Q_2, evaluated at the equilibrium state.

For a distorted phase of type I, with $\zeta_1=1$ and $\zeta_2=0$ the second derivatives of the free energy yield the susceptibilities,

$$\chi_1(0)=\left[\frac{\partial^2 G}{\partial Q_1^2}\right]^{-1} \quad \text{and} \quad \chi_2(0)=\left[\frac{\partial^2 G}{\partial Q_2^2}\right]^{-1}.$$

Below T_c the 'longitudinal' susceptibility is given by

$$\chi_1(0)=[2a(T_c-T)]^{-1}, \qquad \text{(I.4.29)}$$

the same expression as for the $n=1$ system eqn. (I.4.6), while the 'transverse' susceptibility is given by

$$\chi_2(0)=-\frac{u+v}{av(T_c-T)}, \qquad \text{(I.4.30)}$$

which since this phase is only stable if $v<0$ is always positive. The terms 'longitudinal' and 'transverse', are used by analogy with the Heisenberg model of magnetism: χ_1 describes the fluctuations parallel to the order parameter, and χ_2 the fluctuations perpendicular to that direction. For case II, $v=0$ and $\chi_2(0)$ is infinite for all $T\leqslant T_c$. Physically this is because the hamiltonian has continuous rotational symmetry in Q_1 and Q_2 so that it costs no energy to rotate the order parameter. The existence of the infinite susceptibility is then an example of Goldstone's theorem (see. e.g., Wagner 1966).

In phase III, with $\zeta_1=\zeta_2=1/\sqrt{2}$, the derivatives of the free energy are as follows:

$$\frac{\partial^2 G}{\partial Q_1^2}=\frac{\partial^2 G}{\partial Q_2^2}=2a(T_c-T)\frac{u+v}{2u+v}, \qquad \text{(I.4.31)}$$

and

$$\frac{\partial^2 G}{\partial Q_1 \partial Q_2}=2a(T_c-T)\frac{u}{2u+v}.$$

The susceptibility can be brought into a diagonal form as discussed in § I.3.3, by considering the susceptibility for fields along the ordering direction, $\chi_L(0)$, and transverse to that direction, $\chi_T(0)$. Then $\chi_L(0)$ is given by eqn. (I.4.6) or (I.4.29) while

$$\chi_T(0)=\frac{2u+v}{2av(T_c-T)}, \qquad \text{(I.4.32)}$$

which is necessarily positive when phase III is stable.

This analysis can be repeated for an $n=3$ component system (Slonczewski and Thomas 1970) when the free energy is

$$G=G_0+\frac{1}{2}a(T-T_c)(Q_1^2+Q_2^2+Q_3^2)+u(Q_1^2+Q_2^2+Q_3^2)^2 + v(Q_1^4+Q_2^4+Q_3^4). \qquad \text{(I.4.33)}$$

There are again three possible phases:

I. $v<0$; the ordered phase is tetragonal (for example $\zeta_1=\pm 1, \zeta_2=\zeta_3=0$) and $\tilde{u}=u+v$. An ordered phase of this type occurs in $BaTiO_3$ and $SrTiO_3$.

II. $v=0$; the orientation of the distortion is undefined.

III. $v>0$; the ordered phase has trigonal symmetry with $\zeta_1=\pm\zeta_2=\pm\zeta_3=\pm 1/\sqrt{3}$ and $\tilde{u}=u+1/3v$. This phase occurs in $LaAlO_3$ for which the transition is otherwise similar to that occurring in $SrTiO_3$ (Müller *et al.* 1968).

In all of these phases there is a longitudinal susceptibility given by

$$\chi_L(0)=[2a(T_c-T)]^{-1}, \tag{I.4.34}$$

and a doubly degenerate transverse susceptibility given, for phase I, by

$$\chi_T(0)=-\frac{u+v}{av(T_c-T)}, \tag{I.4.35 a}$$

and for phase III by

$$\chi_T(0)=\frac{3u+v}{2av(T_c-T)}. \tag{I.4.35 b}$$

Finally, we briefly consider the more complicated case of an $n=4$ system. The most general form of the free energy is then

$$\begin{aligned} G=G_0+\frac{1}{2}a(T-T_c)(Q_1^2+Q_2^2+Q_3^2+Q_4^2) \\ +u(Q_1^2+Q_2^2+Q_3^2+Q_4^2)^2+v(Q_1^4+Q_2^4+Q_3^4+Q_4^4) \\ +w_1(Q_1^2Q_2^2+Q_3^2Q_4^2)+w_2(Q_1^2Q_3^2+Q_2^2Q_4^2) \\ +w_3Q_1Q_2Q_3Q_4. \end{aligned} \tag{I.4.36}$$

In real systems not all the fourth-order invariants may be allowed by the symmetry of the problem as we shall discuss later for $BaMnF_4$ and NbO_2.

There are five different possible phases of the most general $n=4$ model and we list in table I.1 a typical $\boldsymbol{\zeta}$ vector, for each phase. The most stable phase is that with the smallest value of

$$\tilde{u}=u+v\sum_i \zeta_i^4 v+w_1(\zeta_1^2\zeta_2^2+\zeta_3^2\zeta_4^2)+w_2(\zeta_1^2\zeta_3^2+\zeta_2^2\zeta_4^2)+w_3\zeta_1\zeta_2\zeta_3\zeta_4.$$

Table I.1. The ordered phases of the $n=4$ model.

	$\boldsymbol{\zeta}$	$\tilde{u}$
I	(1000)	$u+v$ —
II	$(1111)\frac{1}{2}$	$u+\frac{1}{4}v+\frac{1}{8}(w_1+w_2)+\frac{1}{4}w_3$
III	$(1100)\frac{1}{\sqrt{2}}$	$u+\frac{1}{2}v+\frac{1}{4}w_1$
IV	$(1010)\frac{1}{\sqrt{2}}$	$u+\frac{1}{2}v+\frac{1}{4}w_2$
V	$(1001)\frac{1}{\sqrt{2}}$	$u+\frac{1}{2}v$

The susceptibilities in the ordered phase may be calculated as discussed, above. In each case one finds a longitudinal susceptibility given by eqn. (I.4.34), and three transverse susceptibilities. These transverse susceptibilities are dependent on the anisotropic coefficients, but unlike the $n=3$ systems, they may not be all equivalent.

I.4.2.2. *Coupling to strains*

In multi-component systems the coupling to secondary order parameters, such as the strain is more complicated than for $n=1$ systems described in § I.4.1.2. It is necessary to include the different strain components so that for $BaTiO_3$ and $SrTiO_3$, for example, there are six components of the strain. The free energy of both of these materials is then given by eqn. (I.4.33) with the addition of an elastic contribution G_e, which is given by (Devonshire 1954, Slonczewski and Thomas 1970)

$$\begin{aligned} G_e = e_1(\eta_1+\eta_2+\eta_3)(Q_1^2+Q_2^2+Q_3^2)+e_2[\eta_1(2Q_1^2-Q_2^2-Q_3^2) \\ +\eta_2(2Q_2^2-Q_1^2-Q_3^2)+\eta_3(2Q_3^2-Q_1^2-Q_2^2)] \\ +e_3(Q_1Q_2\eta_6+Q_1Q_3\eta_5+Q_2Q_3\eta_4)+\frac{1}{2}\sum_{\alpha\beta} C_{\alpha\beta}\eta_\alpha\eta_\beta, \end{aligned} \tag{I.4.37}$$

where we used the Voigt notation for the strains. Minimizing $G+G_e$ with respect to the strains gives

$$\eta_1 = -\frac{e_1}{C_{11}+2C_{12}}(Q_1^2+Q_2^2+Q_3^2)-\frac{e_2}{C_{11}-C_{12}}(2Q_1^2-Q_2^2-Q_3^2),$$

and

$$\eta_4 = -\frac{e_3}{C_{44}}Q_2Q_3.$$

When these results are substituted back into eqns. (I.4.33) and (I.4.37), the free energy has the same form as eqn. (I.4.33) for the strain-free case except that u and v are replaced by

$$\tilde{u} = u-\frac{3}{2}\left[\frac{e_1^2}{C_{11}+2C_{12}}-\frac{e_2^2}{C_{11}-C_{12}}+\frac{e_3^2}{C_{44}}\right],$$

$$\tilde{v} = v-\frac{9}{2}\frac{e_2^2}{C_{11}+C_{12}}+\frac{e_3^2}{C_{44}}.$$

This result shows that the coupling to the strain modifies the parameters of the Landau theory, as found in § I.4.1.2, and so may drive the transition to be of first order, or change the structure of the distorted phase. As discussed for the $n=1$ system, it is necessary to distinguish between the high- and low-frequency susceptibilities below T_c (Slonczewski and Thomas 1970). Experimental confirmation of the necessity to distinguish between the high- and low-frequency susceptibilities is provided by Raman scattering and neutron scattering measurements in $SrTiO_3$. The ratios of the squares of the 'longitudinal' frequencies below and above T_c is $2{\cdot}4 \pm 0{\cdot}2$ (Worlock and Olsen 1971) in excess of the factor 2 which would be expected in the absence of the coupling to strain.

As described in § I.4.1.2, the theory can also be used to predict a jump in the elastic constant at T_c. This is qualitatively the correct behaviour as shown by the measurements on $SrTiO_3$ illustrated in fig. I.16. In detail, however, the behaviour is more complex as described in §II.5.2.2.

In this section the temperature dependence of the strain and elastic constants is found to be independent of its symmetry. This is correct only within Landau theory. In a more complete theory discussed in §§II.5.1.2 and II.5.2.2. it is essential to distinguish between strains which are secondary order parameters and break the symmetry of the high-symmetry phase, and strains which have the complete symmetry of that phase. In $SrTiO_3$ an isotropic strain $\eta_1=\eta_2=\eta_3$ does not break any symmetry operation of the high-symmetry phase, whereas a tetragonal strain, $\eta_1=\eta_2=\eta_1/2$ does. The latter strains must necessarily be zero above T_c. The former strains need not be zero above T_c and in a better theory than Landau theory the behaviour of these two types of strains is different.

I.4.2.3. *Improper ferroelectrics*

Improper ferroelectrics are examples of systems for which the secondary order parameter is a ferroelectric distortion. Their properties are reviewed by Dvorak (1974). Since in a ferroelectric material there are different domains corresponding to the different directions of the polarization, there cannot be terms in the free energy of the form $\bar{Q}Q^2$, where Q is the primary order parameter of an $n=1$ system. Improper ferroelectrics can occur only when the primary order parameter has at least two components. The simplest model (Cochran 1971) of an improper ferroelectic is that of an $n=2$ system with a coupling to the ferroelectric polarization of the form $Q_1Q_2\bar{Q}$; when the free energy is

$$G=G_0+\frac{1}{2}a(T-T_c)(Q_1^2+Q_2^2)+u(Q_1^2+Q_2^2)^2$$

$$+v(Q_1^4+Q_2^4)+eQ_1Q_2\bar{Q}+\frac{1}{2}b\bar{Q}^2, \qquad (I.4.38\,a)$$

where b is the inverse dielectric susceptibility which is assumed to be only weakly temperature dependent. Minimization of the free energy with respect to $\bar{Q}$ yields

$$\bar{Q}=-\frac{e}{b}Q_1Q_2, \qquad (I.4.38\,b)$$

which substituted back into eqn. (I.4.38) yields an equation identical to eqn. (I.4.38) but with u and v replaced by

$$\tilde{u}=u-\frac{e^2}{4b} \quad \text{and} \quad \tilde{v}=v-\frac{e^2}{4b}.$$

If $\tilde{v}<0$, the ordered structure is case I of §I.4.2.1, which then with eqn. (I.4.38) gives $\langle\bar{Q}\rangle=0$ and the phase is not ferroelectric. On the other hand if $\tilde{v}>0$ the ordered structure is case III and $\bar{Q}=\pm(e/2b)Q_0^2$ with the plus sign if $\zeta_1=-\zeta_2$ and the minus sign if $\zeta_1=\zeta_2$.

In practice the phase transitions of improper ferroelectrics are of first order (Dvorak 1974), possibly because the interaction $\tilde{u}$ is always less than u, but possibly also because of the effects discussed in §II.5.2.2. There is, however, evidence that $\bar{Q}$ is proportional to Q_0^2 as shown in fig. I.31 for $Tb_2(MoO_4)_3$ (Dorner *et al.* 1972).

Fig. I.31

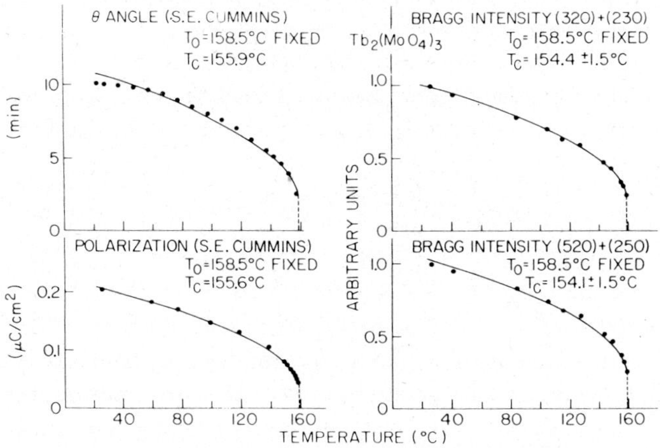

The temperature dependence of the strain distortion θ, the spontaneous polarization and of the square of the primary order parameter as measured by the Bragg reflections in the improper ferroelectric $Tb_2(MoO_4)_3$. (After Dorner *et al.* 1972.)

Fig. I.32

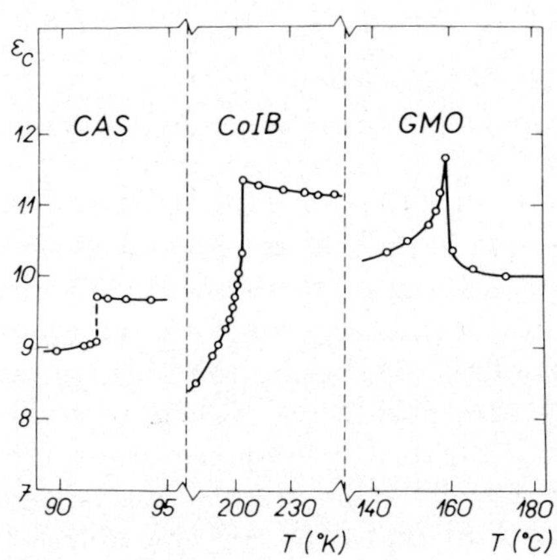

The temperature dependence of the dielectric constant in three improper ferroelectrics. (After Dvorak 1974.)

The development described in § I.4.1.2 can be repeated to show that the dielectric susceptibility in an improper ferroelectric should, within Landau theory, have a step at T_c. In fig. I.32, we show measurements of the susceptibility in three improper ferroelectrics. This simple theory describes the results in $(NH_4)_2Cd_2(SO_4)_3$, but is less satisfactory for CoI boracite or $Gd_2(MoO_4)_3$. In part this undoubtedly reflects the fact that the theory given above is very much too simple to describe real improper ferroelectrics. In particular, there is a coupling between the strain and the primary order parameter and between the strain and the spontaneous polarization, as illustrated in fig. I.31. It is then necessary to distinguish between the clamped and free dielectric constant as described in detail by Dvorak (1974).

I.4.2.4. *Fluctuations and short-range interactions*

The wave-vector dependence of the fluctuations in systems with more than one component for the order parameter can be obtained in much the same manner as was described in detail in § I.4.1.3 for $n=1$ component systems.

At a transition for which the order parameter is neither an electrical polarization nor an elastic strain, the quadratic term in the expansion of the free energy is

$$G_2=\frac{1}{2N}\sum_{ij\mathbf{k}} R_{ij}(\mathbf{k})Q_i(\mathbf{k})Q_j(-\mathbf{k}),$$

where

$$R_{ij}(\mathbf{k})=a(T-T_c)\delta_{ij}+f_{\alpha\beta}(ij)k_\alpha k_\beta. \qquad \text{(I.4.39)}$$

The number of coefficients $f_{\alpha\beta}(ij)$ may be reduced by symmetry so that, for example, in $SrTiO_3$ there are three components of the order parameter corresponding to rotations of the oxygen octahedra about the x, y, z directions, respectively, and three independent $f_{\alpha\beta}(ij)$ coefficient. In this case eqn. (I.4.39) can conveniently be rewritten as

$$R_{ij}(\mathbf{k})=[a(T-T_c)+\lambda(k^2+fk_i^2)]\delta_{ij}+\lambda hk_ik_j(1-\delta_{ij}) \qquad \text{(I.4.40)}$$

and the coefficients λ, f and h have been measured for $SrTiO_3$ by Stirling (1972) from the anisotropy in the phonon dispersion relations about the R point, with the results

$$\lambda=216\pm20(\text{THz Å})^2, \quad f=-0{\cdot}97\pm0{\cdot}01, \quad h=0{\cdot}19\pm0{\cdot}04$$

showing a very large degree of anisotropy.

The anisotropy and coupling between the modes in these systems means that the eigenvectors of the matrix in eqn. (I.4.39), are very dependent upon the direction of $\mathbf{k}$. A detailed analysis of the form of the susceptibility is then difficult. An unexpected result of this behaviour occurs in the scattering cross-section around the incipient ordering wave-vector, $\mathbf{q}_s$, which may not exhibit the full symmetry of the crystal. This is illustrated by the form of the scattering for wave-vector transfers $\mathbf{K}$, close to $\mathbf{K}_0=2\pi/a\ (3/2, 1/2, 1/2)$ in $SrTiO_3$. At this position the structure factors of the three modes are $F(\mathbf{K}, 1)=0$, $F(\mathbf{K}, 2)=F(\mathbf{K}, 3)=F$. Because f is close to -1 the susceptibility and scattering is large for the ith component for wave-vectors along the ith direction. Consequently there will be a large scattering observed for wave-vectors $\mathbf{K}=\mathbf{K}_0+(0, k, 0)$ and $\mathbf{K}=\mathbf{K}_0+(0, 0, k)$ but no corresponding scattering along $\mathbf{K}=\mathbf{K}_0+(k, 0, 0)$ because $F(\mathbf{K}, 1)=0$.

I.4.2.5. *Ferroelectrics*

The long-range electrostatic forces modify the form of the fluctuations for small $\mathbf{k}$ in ionic materials as explained for uniaxial ferroelectrics in § I.4.1.4. In the case of cubic ferroelectrics, such as $BaTiO_3$, we choose the ith component of the order parameter to give the electrical polarization along the ith direction. The expansion of the second term in the free energy is then given by eqn. (I.4.40) but with the additional electrostatic term;

$$R_{ij}(\mathbf{k})=[a(T-T_c)+\lambda(k^2+fk_i^2)]\delta_{ij}+\lambda hk_ik_j(1-\delta_{ij})+g\frac{k_ik_j}{k^2}. \qquad \text{(I.4.41)}$$

The matrix $R_{ij}(\mathbf{k})$ can then be diagonalized for $\mathbf{k}\rightarrow0$ to give the 'frequencies' of a longitudinal fluctuation above T_c as $\omega_L^2 = a(T - T_c) + g$ and two transverse fluctuation 'frequencies' as $\omega_T^2 = a(T - T_c)$. If the structure factors of the three modes are proportional to K_i, eqn. (I.2.36), then the scattering intensity for large g is given (Cochran 1969) by $|K^2| \sin^2 \theta$ where θ is the angle between **K** and **k** as illustrated in fig. I.33.

Fig. I.33

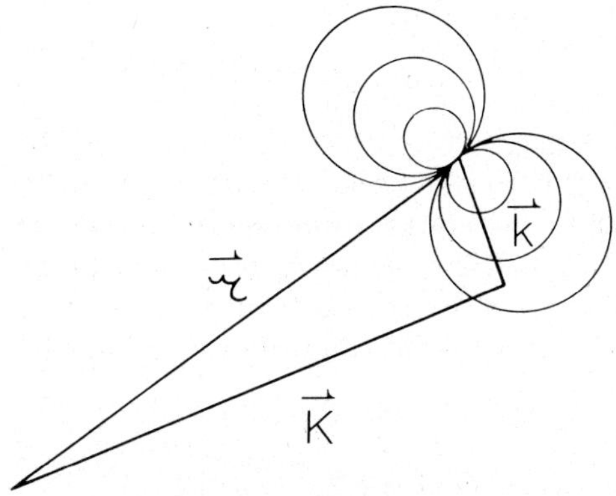

Sketch of constant intensity contour lines as function of wave-vector transfer **K** for a cubic ferroelectric or for **K** within the plane of ferroelectric fluctuations for a [2*d*] ferroelectric. (After Cochran 1969.)

In the case of ferroelectrics with two components polarized in say the x and y directions, there are two transverse fluctuations for **k** along the z direction, but a longitudinal and a transverse fluctuation for **k** in the (x, y) plane. For **K** in the (x, y) plane the scattering has the same form as shown in fig. I.33. Although the tetragonal phase of $BaTiO_3$ is a two-component ferroelectric system, we are unaware of any detailed measurements of the critical scattering.

I.4.2.6. *Acoustic instabilities*

Nearly all the acoustic instabilities with $n \geqslant 2$ have cubic terms in the Landau expansion of the free energy, and so are predicted to be of first order. The only example (Cowley 1976a) with $n = 2$ and no cubic terms is an instability against η_4 and η_5 in the tetragonal system. The velocities of the acoustic waves are zero at T_c for modes with wave-vector **k** perpendicular to the [001] direction and polarized along (001) or with **k** along [001] and polarized transversely. The former set give rise to a similar scattering pattern to that of uniaxial ferroelectrics, while that of the latter set is similar to that of the acoustic instabilities discussed in § I.4.1.6.

I.4.3. *Phase diagrams*

I.4.3.1. *Tricritical points*

In § I.4.1.1 the free energy, eqn. (I.4.9), was shown to give rise to a continuous phase transition if $u > 0$, and a first-order transition if $u < 0$. The point $u = 0$ terminates the line (in the space of u and T) of continuous phase transitions and is known as a tricritical point. The Landau expansion of the free energy at this point is

$$G=G_0+\frac{1}{2}a(T-T_c)Q^2+hQ^6, \tag{I.4.42}$$

so that the order parameter is given by

$$Q_0^4=\frac{a}{6h}(T_c-T) \tag{I.4.43 a}$$

and the susceptibility below T_c by

$$\chi(0)=[4a(T_c-T)]^{-1}. \tag{I.4.43 b}$$

The results show that within Landau theory the principal tricritical point exponents have values $\beta=1/4$, $\gamma=1$ (to be compared with the Landau results for an ordinary critical point, $\beta=1/2$, $\gamma=1$). It also follows from (I.4.43 *b*) that the ratio of the susceptibility above T_c to that below T_c for a constant $|T-T_c|$ is 4 within the classical tricritical theory—twice the Landau result for an ordinary critical point. A fuller description of tricritical points and a comparison with experimental results is given in § II.5.3.1.

I.4.3.2. *Phase transitions without symmetry breaking*

In § I.3.1 we emphasized the importance of the concept of symmetry breaking at structural phase transitions. There are, however, phase transitions where (as at the liquid–gas critical point) the symmetry is unchanged. Consider, to be specific, the Landau free energy for an $n=1$ system in an applied field:

$$G=G_0+\frac{1}{2}a(T-T_0)Q^2+uQ^4+hQ^6-EQ$$

For $u>0$ there is no phase transition when $E\neq 0$. For $u<0$ there is a first-order phase transition for small E, and no phase transition for large E. There exists a critical value of the field E_c at which the transition is continuous and beyond which no transition occurs. The critical field, and the corresponding transition temperature T_c are located by the conditions

$$\left(\frac{\partial G}{\partial Q}\right)_E=\left(\frac{\partial^2 G}{\partial Q^2}\right)_E=\left(\frac{\partial^3 G}{\partial Q^3}\right)_E=0,$$

as

$$E_c=16h\left(\frac{|u|}{5h}\right)^{5/2}$$

and

$$T_c=T_0+\frac{6u^2}{5ha}$$

There is no symmetry breaking at the critical point, since (by virtue of the applied field) the coordinate Q already has the non-zero value

$$Q_0^c=\left(\frac{|u|}{5h}\right)^{1/2}$$

The order parameter for the phase transition must thus be defined as the *difference* between the two possible values of Q_0 characterizing the two phases that coexist on the phase boundary which terminates at the critical point.

This type of behaviour has been observed in KH_2PO_4 by Eberhard and Horn (1975). At atmospheric pressure the phase transition is first order and shows a temperature hysteresis of 0·025 K. On increasing the applied field the hysteresis decreases and extrapolation of their data suggests that $E_c = 6{\cdot}5 \times 10^5$ V/m.

I.4.3.3. *Competing interactions*

One of the characteristic features of an $n > 1$ component system is the possible occurrence of different ordered phases: the $n = 2$ system has an ordered phase, I, with $\zeta_1 = 1$, $\zeta_2 = 0$ and another, II, with $\zeta_1 = \zeta_2 = 1/\sqrt{2}$ whose relative stability is determined by the sign of v, § I.4.2.1. More than one phase transition may occur if there are competing interactions one of which favours one of these phases, while another favours the other phase. This type of competition is responsible for the successive phase transitions in $BaTiO_3$ (Devonshire 1954) and we illustrate the behaviour by extending the free-energy equation, (I.4.33), to include sixth-order terms in the order parameters;

$$G = G_0 + \frac{1}{2}a(T - T_c)(Q_1^2 + Q_2^2 + Q_3^2) + u(Q_1^2 + Q_2^2 + Q_3^2)^2$$
$$+ v(Q_1^4 + Q_2^4 + Q_3^4) + h_1(Q_1^2 + Q_2^2 + Q_3^2)^3$$
$$+ h_2(Q_1^6 + Q_2^6 + Q_3^6) + h_3[Q_1^4(Q_2^2 + Q_3^2)$$
$$+ Q_2^4(Q_1^2 + Q_3^2) + Q_3^4(Q_1^2 + Q_2^2)]. \qquad \text{(I.4.44)}$$

The quadratic term and the terms in u and h_1 are isotropic and so do not determine the nature of the ordered phase. We shall determine the structure of the ordered phase by using perturbation theory to calculate the energy of the anisotropic terms. Consider three possible structures for which (Q_1, Q_2, Q_3) are given by: I$(1, 0, 0)Q_0$, II$(1/\sqrt{2}, 1/\sqrt{2}, 0)Q_0$ and III $(1/\sqrt{3}, 1/\sqrt{3}, 1/\sqrt{3})Q_0$. The contribution of the anisotropic terms to the free energy is

$$G_{AN}(\text{I}) = (v + h_2 Q_0^2)Q_0^4,$$
$$G_{AN}(\text{II}) = \left(\frac{1}{2}v + \frac{1}{4}(h_2 + h_3)Q_0^2\right)Q_0^4,$$
$$G_{AN}(\text{III}) = \left(\frac{1}{3}v + \frac{1}{9}(h_2 + 2h_3)Q_0^2\right)Q_0^4. \qquad \text{(I.4.45)}$$

If $v < 0$ then phase I will tend to be more stable than phase II or III. If, however, h_2 and h_3 are negative then they will tend to stabilize phase III rather than phases I or II. Since, however, the h-coefficients are multiplied by Q_0^2, there may be a temperature at which it is energetically favourable for a transition to occur from the structure favoured by v to that favoured by the h-coefficients. If we neglect h_3 for simplicity then three transitions are predicted by eqns. (I.4.45) if $v < 0$ and $h_2 > 0$: disordered → phase I when $T = T_c$; phase I → phase II when $Q_0^2 = -2v/3h_2$; phase II → phase III when $Q_0^2 = -6v/5h_2$. Whereas the first transition may be continuous, the second and third transitions are necessarily of first order. It is interesting to note that phase II can never be stable under the action of the parameters v or h_2 alone, but can be stabilized by a competition between these terms. Devonshire (1954) applied this theory in detail to $BaTiO_3$ and was able to show that it explained the occurrence of the three transitions. However, not surprisingly, a satisfactory quantitative description could be obtained only when the parameters were allowed to be temperature dependent.

If the parameter v is positive, phase III is stable below T_c and, if h_2 is negative there is a first-order transition to phase I, when $Q_0^2 = -3v/4h$. In this case phase II is never stable. In fig. I.34 the phase diagrams of these $n=3$ systems are illustrated.

Fig. I.34

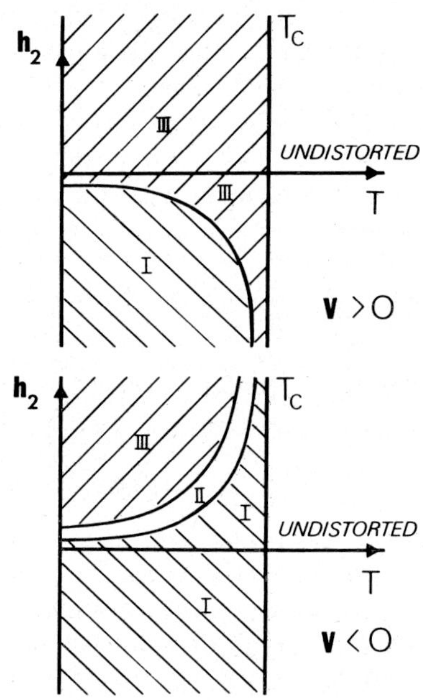

Phase diagrams for an $n=3$ system with Q^4 and Q^6 anisotropies.

Another type of competition between different phases arises if a stress or strain is applied to the crystal favouring one of the possible lower temperature phases, rather than the other (Bruce and Aharony 1975).

The effect of a [100] stress on $SrTiO_3$ is to split the degeneracy of the three modes Q_1, Q_2 and Q_3. The free-energy equation, (I.4.33), may then be rewritten as

$$G = G_0 + \frac{1}{2}[a(T-T_c)(Q_1^2+Q_2^2+Q_3^2) + 2p_1Q_1^2 + p_2(Q_2^2+Q_3^2)] + u(Q_1^2+Q_2^2+Q_3^2)^2 + v(Q_1^4+Q_2^4+Q_3^4), \qquad \text{(I.4.46)}$$

where the new terms p_1 and p_2 are proportional to the stress applied along the [100] direction. In the interests of simplicity we shall assume that $p_1 = -p_2 = p_0$, neglect the strain dependence of u and v and assume v is negative (e.g. $SrTiO_3$). In this case both the applied stress and v favour ordering along the cube axis, Q_1, if $p_0 < 0$, and Q_2

or Q_3 if $p_0>0$. The phase transitions occur when

$$T=T_c-\frac{2p_0}{a} \quad \text{for} \quad p_0<0$$

$$T=T_c+\frac{p_0}{a} \quad \text{for} \quad p_0>0,$$

and there is a first-order transition separating the two ordered phases, as shown in fig. I.35 (*a*).

Fig. I.35

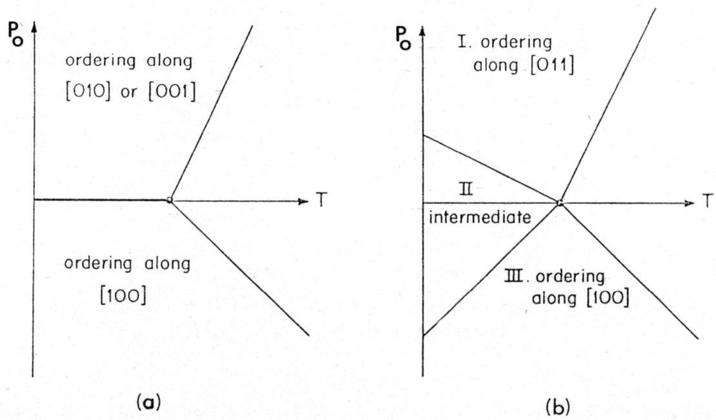

Phase diagrams for a [100] stressed $n=3$ system. (After Bruce and Aharony 1975.) (*a*) The case $v<0$ (e.g. $SrTiO_3$). (*b*) The case $v>0$ (e.g. $LaAlO_3$).

The behaviour is more complicated if $v>0$, (as in the case of $LaAlO_3$), because then the applied field favours ordering along the cube axes while v favours the [1, 1, 1] type of ordering. The essential form of the phase diagram may be determined by assuming that the anisotropic terms are small and using perturbation theory to calculate their effects: a more complete treatment is given by Bruce and Aharony (1975). If the order parameter of the ordered phase is written $(Q_1, Q_2, Q_3) = (\cos\theta, \zeta_2 \sin\theta, \zeta_3 \sin\theta)Q_0$ where $\zeta_2^2+\zeta_3^2=1$, then the anisotropic contribution to the free energy is

$$G_{AN}=\frac{3p_0}{2}Q_0^2\cos^2\theta+v(\cos^4\theta+\sin^4\theta(\zeta_2^4+\zeta_3^4))Q_0^4.$$

Now since $v>0$ and we wish to minimize G_{AN}, $\zeta_2^4+\zeta_3^4=\frac{1}{2}$, and hence we can minimize G_{AN} with respect to the angle θ to obtain three solutions:

$$\text{I.} \quad \cos\theta=0,$$

$$\text{II.} \quad \cos^2\theta=\frac{1}{3}-\frac{p_0}{2vQ_0^2}$$

$$\text{III.} \quad \sin\theta=0$$

Phase I is stable when $p_0>2vQ_0^2/3$ and phase III stable when $p_0<-4vQ_0^2/3$ and an intermediate phase in which the ordering vector is more general is stable for intermediate values of p_0. The phase diagram is sketched in fig. I.35 (*b*).

A similar analysis may be applied to the case of [111]-stressed materials, though the treatment is complicated by the non-orthogonality of the different [111] directions. This time, as one might expect, an 'intermediate' phase occurs when v is *negative*, as shown schematically in Fig. I.36. A comparison of these results with experimental data is given in § II.5.3.3.

Fig. I.36

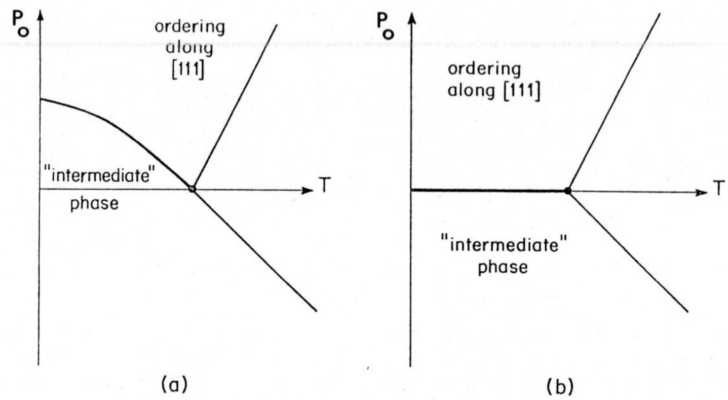

Phase diagrams for a [111] stressed $n=3$ system. (After Bruce and Aharony 1975.) (*a*) The case $v<0$. (*b*) The case $v>0$.

I.4.4. *Incommensurate phase transitions*

I.4.4.1. *The simplest incommensurate phase*

As yet we have described only those phase transitions where the order parameters, Q_i, are real, but in general they may be complex. Consider a tetragonal material in which the order parameter, Q_i, has a wave-vector $\mathbf{q}_s=(0,0,\zeta)$. Then unless the wave-vector is the zone centre or zone boundary, the order parameter Q_i will be complex. Initially it seems surprising that $\mathbf{q}_s$ can have a value between 0 and the zone boundary because it is only at these points that symmetry may impose that $\nabla_{\mathbf{q}}\omega(\mathbf{q}j)=0$. There is nothing, however, to prevent $\nabla_{\mathbf{q}}\omega(\mathbf{q}j)=0$ for an arbitrary point in the zone for some particular combination of interatomic forces. However, in these cases the wave-vector $\mathbf{q}_s$ of the minimum in the dispersion relation is not fixed by symmetry and so is usually temperature dependent. Both of these features are illustrated in figs. I.37 and I.38 for K_2SeO_4 (Iizumi *et al.* 1977), where we show the measured dispersion relation and the temperature dependence of $\mathbf{q}_s$.

In the theory of incommensurate phase transitions the order parameter Q_1 is complex and there is necessarily another coupled order parameter, Q_2, with wave-vector $-\mathbf{q}_s$. Since, however, the atomic displacements, $\mathbf{u}(l\kappa)$ are real, eqn. (I.2.1) shows that Q_1 and Q_2 must be complex conjugates of one another, and it is frequently useful to define new variables to explicitly take account of this condition. The most direct way of doing this is to write

$$Q_1=\frac{1}{\sqrt{2}}A\exp(i\phi),\quad Q_2=\frac{1}{\sqrt{2}}A\exp(-i\phi), \tag{I.4.47}$$

where the coordinate A is the amplitude of the distorted wave and ϕ is its phase and

Fig. I.37

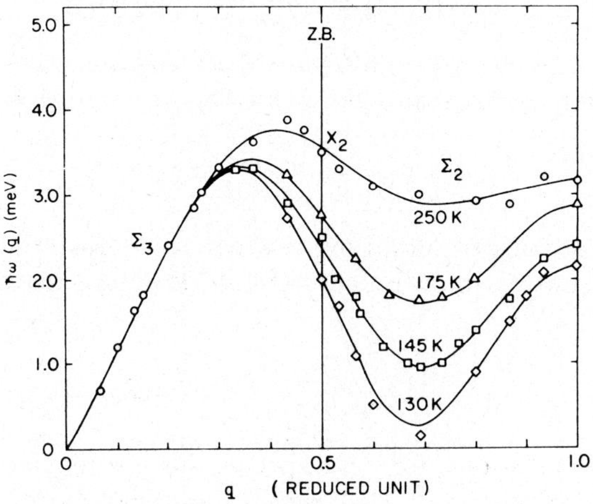

The dispersion relation of the soft mode in K_2SeO_4 at various temperatures. (After Iizumi *et al.* 1977.)

Fig. I.38

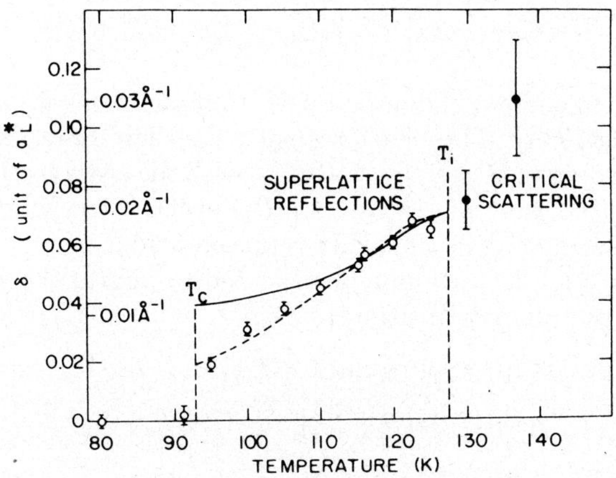

The temperature dependence of $q_s = \frac{1}{3}(1-\delta)\frac{2\pi}{a}$ in K_2SeO_4 compared with calculation by Iizumi *et al.* (1977).

both are real. An alternative approach is to introduce new coordinates which are written as

$$P_1 = \frac{1}{\sqrt{2}}(Q_1 + Q_2),$$

$$P_2 = \frac{i}{\sqrt{2}}(Q_2 - Q_1), \qquad \text{(I.4.48}\,a\text{)}$$

where P_1 and P_2 are again real.

The relationship between these two coordinate systems is

$$P_1 = A\cos\phi \quad \text{while} \quad P_2 = -A\sin\phi. \qquad \text{(I.4.48}\,b\text{)}$$

The free energy of this system is

$$G = G_0 + a(T - T_c)Q_1Q_2 + 4uQ_1^2Q_2^2, \tag{I.4.49}$$

where the terms in Q_1^2, Q_2^2, Q_1^4, and Q_2^4 are absent because they fail to satisfy wave-vector conservation. Substituting eqn. (I.4.47) into eqn. (I.4.48) we obtain;

$$G = G_0 + \frac{1}{2}a(T - T_c)A^2 + uA^4; \tag{I.4.50}$$

an expression involving the amplitude A which is identical to the free energy of the singlet system discussed initially, eqn. (I.4.3). We immediately obtain, therefore, that

$$A_0^2 = \frac{a}{4u}(T_c - T), \quad T < T_c; \tag{I.4.51}$$

and analogous expressions for the specific heat and free energy. The other noteworthy point about eqn. (I.4.50) is that it does not explicitly involve the phase angle ϕ. The free energy is therefore independent of the choice of ϕ.

In the alternative representation with the order parameters, P_1 and P_2, the free-energy equation, (I.4.49), becomes;

$$G = G_0 + \frac{1}{2}a(T - T_c)(P_1^2 + P_2^2) + u(P_1^2 + P_2^2)^2. \tag{I.4.52}$$

This form of the free energy is identical with that of the $n = 2$ system, discussed in § I.4.2.1, except that $v = 0$. The phase transition is therefore to a phase of type II, § I.4.2.1, corresponding to $(P_1, P_2) = A_0(\zeta_1, \zeta_2)$ with the eigenvector $\boldsymbol{\zeta}$ arbitrary. This continuously variable direction in the (P_1, P_2) space corresponds to the arbitrary choice of the phase angle ϕ in the other representation.

The analysis of § I.4.2.1 can now be repeated to obtain the susceptibility both above and below the phase transition to give

$$\chi_L(0) = \chi_T(0) = [a(T - T_c)]^{-1}, \quad T > T_c; \tag{I.4.53 a}$$

$$\chi_L(0) = [2a(T_c - T)]^{-1}, \quad T < T_c; \tag{I.4.53 b}$$

$$\chi_T(0) = \infty, \quad T < T_c, \tag{I.4.53 c}$$

where χ_L is determined by the fluctuations parallel to the ordering vector, $\boldsymbol{\zeta}$, which are frequently called amplitude fluctuations, and χ_T corresponds to fluctuations perpendicular to that direction which have an infinite susceptibility. These fluctuations were called phasons by Overhauser (1971) because they correspond to changes in the phase of the order parameter.

I.4.4.2. *More complex incommensurate phases*

Not all incommensurate phases have only two components of the order parameter. In order to illustrate the type of behaviour which occurs we examine the properties of two $n = 4$ systems. The first corresponds to an incommensurate phase in an orthorhombic crystal with wave-vector $\mathbf{q}_s = (\pm\zeta_1, \pm\zeta_2, 0)$ or a tetragonal crystal with $\mathbf{q}_s = (\pm\zeta, 0, 0)$ and $(0, \pm\zeta_1 0)$. It is then convenient to introduce two amplitudes and phases by the requirement that

$$Q_1 = \frac{1}{\sqrt{2}} A_1 \exp(i\phi_1), \quad Q_2 = Q_1^*,$$
$$Q_3 = \frac{1}{\sqrt{2}} A_2 \exp(i\phi_2), \quad Q_4 = Q_3^*. \tag{I.4.54}$$

The free energy is then of the form

$$G = G_0 + a(T - T_c)(Q_1Q_2 + Q_3Q_4) + 4(u+v)(Q_1^2Q_2^2 + Q_3^2Q_4^2) + 8uQ_1Q_2Q_3Q_4. \tag{I.4.55}$$

With the aid of the transformation, eqn. (I.4.54), the free energy becomes

$$G = G_0 + \frac{1}{2}a(T - T_c)(A_1^2 + A_2^2) + u(A_1^2 + A_2^2)^2 + v(A_1^4 + A_2^4), \tag{I.4.56}$$

which is independent of the phase angles ϕ_1 and ϕ_2. Now, since eqn. (I.4.56) is identical with the $n=2$ system described in § I.4.2.1, there are three solutions

I. $v<0$ A_1 or $A_2 = A_0$, and the other is equal to zero.

II. $v=0$ $A_1^2 + A_2^2 = A_0^2$.

III. $v>0$ $A_1 = A_2 = A_0/\sqrt{2}$.

The susceptibilities can be obtained in the usual way. For structure I, with say $A_1 = A_0$, there is a longitudinal susceptibility corresponding to fluctuations in A_1 given by

$$\chi_1(0) = [2a(T_c - T)]^{-1}, \quad T < T_c, \tag{I.4.57}$$

while the transverse susceptibility arising from the fluctuations in the phase ϕ_1 is infinite. The susceptibilities of the Q_3 and Q_4 modes are degenerate and given by

$$\chi_3(0) = \chi_4(0) = -\frac{u+v}{av(T_c - T)}, \quad T < T_c.$$

If the ordered phase is of type III then there is a longitudinal susceptibility given by eqn. (I.4.57), and another amplitude susceptibility corresponding to anti-phase fluctuations of the two amplitudes with a susceptibility given by

$$\chi_{A^-}(0) = \frac{2u+v}{2va(T_c - T)}, \quad T < T_c.$$

and two infinite susceptibilities, corresponding to fluctuations in the phase angles ϕ_1 and ϕ_2. In this system the number of phasons with infinite susceptibility depends on the number of pairs of $\mathbf{q}_s$ and $-\mathbf{q}_s$ present in the ordered structure. If there is only one pair, phase I, then there is only one infinite phase susceptibility, but if both modes are present in the ordered structure, two susceptibilities are infinite.

Finally let us consider another case of an $n=4$ system; $BaMnF_4$ which has an 'a' face centred orthorhombic structure and an incommensurate phase described by the four order parameters;

$Q_1(\mathbf{q}_s = (0{\cdot}39, \frac{1}{2}, \frac{1}{2}))$, $Q_2(\mathbf{q}_s = (-0{\cdot}39, \frac{1}{2}, \frac{1}{2}))$,

$Q_3(\mathbf{q}_s = (0{\cdot}39, -\frac{1}{2}, \frac{1}{2}))$ and $Q_4(\mathbf{q}_s = (-0{\cdot}39, \frac{1}{2}, \frac{1}{2}))$, (Shapiro *et al.*, 1976).

The free energy is now of the same form as eqn. (I.4.55) but with an additional term given by $4w(Q_1^2Q_3^2+Q_4^2Q_2^2)$. When this is transformed to the form of the free energy given in eqn. (I.4.56), we obtain an additional term $wA_1^2A_2^2\cos(2\phi_1+2\phi_2)$. This term has the effect of fixing the relative phases between the two pairs of incommensurate wave-vectors. Consequently for the structure of type I, discussed above, this term splits the degeneracy between $\chi_3(0)$ and $\chi_4(0)$, while for the structure of type III, there is only one infinite susceptibility which arises from fluctuations which preserve $\phi_1+\phi_2$, while the other phase susceptibility is finite and corresponds to fluctuations in $\phi_1+\phi_2$. The effect of the Umklapp interaction in $BaMnF_4$ is therefore to reduce the degeneracy of the susceptibilities, and in particular to ensure that there is only one phase mode with an infinite susceptibility. The number of infinite phase susceptibilities is then the number of pairs of wave-vectors in the ordered structure, $n/2$, less the number of phase locking terms. There are always phase locking terms which restrict the number of infinite phase susceptibilities to be a maximum of three.

I.4.4.3. *Fluctuations of incommensurate phases*

We now extend the analysis, particularly of the simplest incommensurate phase, to include the wave-vector dependence of the fluctuations. The free-energy equation, (I.4.49), is extended to include the wave-vector dependence of the quadratic term in Q_1Q_2 when for small $\mathbf{k}$ expanding about the temperature dependent, $\mathbf{q}_s$, gives

$$G=G_0+\frac{1}{N}\sum_{\mathbf{k}}(a(T-T_c)+f_1(k_x^2+k_y^2)+f_2(k_z^2))Q_1(\mathbf{k})Q_2(-\mathbf{k})$$

$$+\frac{4u}{N^2}\sum_{\mathbf{k}_1}\sum_{\mathbf{k}_2}\sum_{\mathbf{k}_3}Q_1(\mathbf{k}_1)Q_2(\mathbf{k}_3)Q_1(\mathbf{k}_2)Q_2(-\mathbf{k}_1-\mathbf{k}_2-\mathbf{k}_3). \qquad (I.4.58)$$

The fluctuations can be found by transforming to the $P_1(\mathbf{k})$, $P_2(\mathbf{k})$ coordinates when for $T>T_c$ the quadratic terms in the $P(\mathbf{k})$ become

$$\frac{1}{2N}\sum_{\mathbf{k}}\sum_{i=1}R_{ii}(\mathbf{k})P_i(\mathbf{k})P_i(-\mathbf{k})$$

with

$$R_{ii}(\mathbf{k})=a(T-T_c)+f_1(k_x^2+k_y^2)+f_2k_z^2, \qquad (I.4.59)$$

so that the 'frequencies' of the fluctuations described by $P_1(\mathbf{k})$ and $P_2(\mathbf{k})$ are equal and given by

$$\omega_1(\mathbf{k})^2=\omega_2(\mathbf{k})^2=R_{11}(\mathbf{k}). \qquad (I.4.60)$$

The corresponding frequencies below T_c can be found by assuming a particular phase for the ordered structure so that $\phi=0$ in eqn. (I.4.48 *b*), when the longitudinal or amplitude frequency is given by

$$\omega_L(\mathbf{k})^2=2a(T_c-T)+f_1(k_x^2+k_y^2)+f_2k_z^2 \qquad (I.4.61)$$

and the transverse or phase frequency by

$$\omega_T(\mathbf{k})^2=f_1(k_x^2+k_y^2)+f_2(k_z^2) \qquad (I.4.62)$$

In fig. I.39 we illustrate these frequencies close to T_c.

The phase frequencies are linear with k below T_c and their slope is independent of temperature in this model.

Fig. I.39

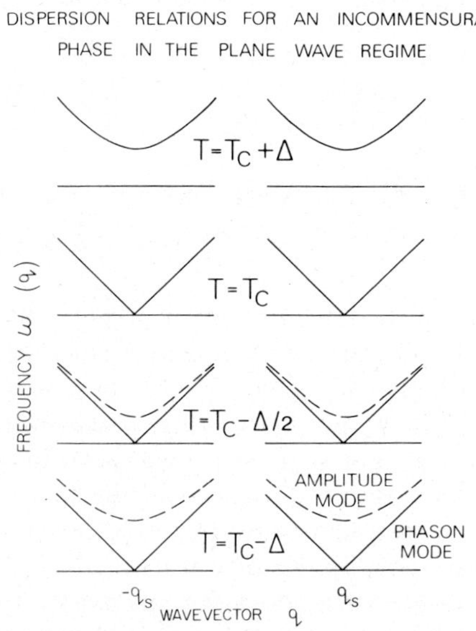

The dispersion relations for the excitation frequencies of an incommensurate system with a plane-wave ground state. (After Bruce and Cowley 1978.)

Before discussing experimental observations of phasons, it is worthwhile discussing their nature in more detail. An alternative approach for describing the fluctuations of an incommensurate phase is to consider fluctuations in the amplitude and phase. By analogy with the treatment of the fluctuations in the order parameter Q as wave-vector dependent $Q(\mathbf{k})$, we introduce a wave-vector-dependent fluctuation in the amplitude, $A(\mathbf{k})$, and phase $\phi(\mathbf{k})$ so that from eqn. (I.4.47) we obtain for small fluctuations

$$Q_1(\mathbf{k})=\frac{1}{\sqrt{2}}(A(\mathbf{k})\exp(i\phi_0)+iA_0\exp(i\phi_0)\phi(\mathbf{k})) \qquad \text{(I.4.63)}$$

where ϕ_0 is the phase of the ordered structure. If the ordered structure is chosen as above to have $\phi_0=0$, then to leading order in the fluctuations

$$P_1(\mathbf{k})=A(\mathbf{k}), \quad P_2(\mathbf{k})=A_0\phi(\mathbf{k}). \qquad \text{(I.4.64)}$$

$P_1(\mathbf{k})$ and $P_2(\mathbf{k})$ can now be identified with the amplitude and phase fluctuations, but the identification is accurate only for leading-order terms in the expansion of the complex exponential, $\exp(i\phi(\mathbf{k}))$; once $\phi(\mathbf{k})$ becomes sufficiently large to invalidate this expansion, then the two developments are no longer identical. In particular, the second-order terms in the expansion of $\exp(i\phi(\mathbf{k}))$ give a real contribution to the right-hand side of eqn. (I.4.63) which must be incorporated with the $P_1(\mathbf{k})$. This leads to a large correction to the longitudinal susceptibility: in fact we have shown (Bruce and Cowley 1978) that it varies as $1/k$ for small k below T_c, cf. eqn. (I.4.53*b*). This is an exactly analogous problem to the well-known divergence of the longitudinal susceptibility of a Heisenberg ferromagnet below T_c on the coexistence curve (Vaks *et al.* 1967). In our notation it arises because fluctuations in the direction of the order parameter in (P_1, P_2) space are relatively easy and to second order give rise to fluctuations in the length of any component of that parameter.

The scattering cross-section for X-ray or neutron scattering is readily obtained using the techniques developed in § I.2.3 and the definition of the P coordinates in eqn. (I.4.48). The result is

$$S^1(\mathbf{K},\omega)=\frac{\hbar N}{2\pi}(n(\omega)+1)\sum_{\mathbf{k}}[\Delta(\mathbf{K}+\mathbf{q}_s+\mathbf{k})\times|F(\mathbf{K},\mathbf{q}_s+\mathbf{k})|^2+\Delta(\mathbf{K}-\mathbf{q}_s+\mathbf{k})|F(\mathbf{K},-\mathbf{q}_s+\mathbf{k})|^2]\times\mathrm{Im}\,[\chi_L(\mathbf{k},\omega)+\chi_T(\mathbf{k},\omega)], \qquad (I.4.65)$$

where $\chi_L(\mathbf{k},\omega)$ and $\chi_T(\mathbf{k},\omega)$ are the frequency dependent susceptibilities of the longitudinal and transverse fluctuations, and the Fs are the mode structure factors defined by eqn. (I.2.36). Above T_c, eqn. (I.4.65) reduces to the usual one-phonon cross-section from the soft mode, but below T_c,χ_L and χ_T differ so that there are two response functions for each $\mathbf{K}$. The scattering cross-section in principle enables both the phase and amplitude modes to be observed as illustrated in fig. I.39. Alas we are unaware of any neutron scattering measurements which convincingly vindicate the existence of phasons in real incommensurate systems. In K_2SeO_4 (Iizumi *et al.* 1977) and $BaMnF_4$ (Shapiro *et al.* 1976) considerable scattering is observed for the appropriate wave-vector transfers, but it has not, as yet, been possible to distinguish the scattering from the phase modes from that of the normal elastic waves seen emanating from the Bragg reflection of the incommensurate phase. In fig. I.40 we show experimental results obtained for the one-dimensional conductor, KCP, (Carneiro *et al.* 1976) and this too definitely shows scattering at low energies close to the incommensurate wave-vector, $2k_F$, but it is not clearly of phason character. In $TaSe_2$ (Moncton *et al.* 1975) and $NaNO_2$ (Sakurai *et al.* 1970) the transitions are more closely of the order–disorder type and so it is not possible with conventional neutron scattering techniques to determine the dispersion relations.

The amplitude and phase modes can also be observed using light scattering techniques. The polarizability tensor is expanded in powers of the coordinates Q_1 and Q_2 as

$$P_{\alpha\beta}(\mathbf{K})=\sum_{\mathbf{k}}P_{\alpha\beta}(\mathbf{q}_s+\mathbf{k},\,-\mathbf{q}_s-\mathbf{k}+\mathbf{K})Q_1(\mathbf{k})Q_2(-\mathbf{k}+\mathbf{K}).$$

Now since the modes can be determined most directly by one-phonon scattering we are interested in the terms in which either $Q_1(\mathbf{k})$ or $Q_2(\mathbf{k})$ takes on its average value. It is useful to expand the polarizability for small $\mathbf{K}$ so that

$$P_{\alpha\beta}(\mathbf{q}_s,\,-\mathbf{q}_s+\mathbf{K})=P^0_{\alpha\beta}+\sum_{\gamma}P^1_{\alpha\beta,\gamma}K_\gamma+\ldots$$

and then since $P_{\alpha\beta}(\mathbf{k}_1,\mathbf{k}_2)$ is symmetric in $\mathbf{k}_1$ and $\mathbf{k}_2$

$$P_{\alpha\beta}(\mathbf{q}_s+\mathbf{K},\,-\mathbf{q}_s)=P^0_{\alpha\beta}-\sum_{\gamma}P^1_{\alpha\beta,\gamma}K_\gamma+\ldots.$$

When the polarizability is expressed in terms of the P coordinates and the ordered structure is assumed to have $\phi=0$ (eqn. I.4.48 b) then

$$P_{\alpha\beta}(\mathbf{K})=A_0P^0_{\alpha\beta}P_1(\mathbf{K})+iA_0\sum_{\gamma}P^1_{\alpha\beta,\gamma}K_\gamma P_2(\mathbf{K}).$$

Fig. I.40

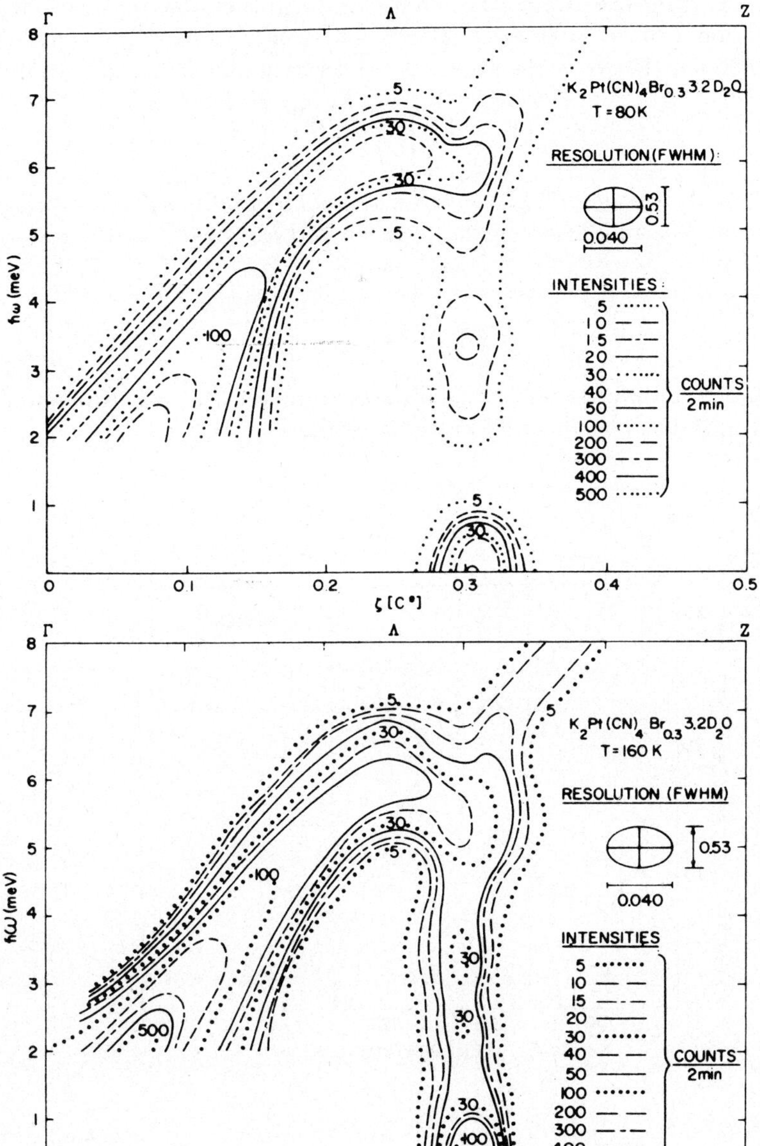

Neutron scattering intensity counters for scattering vectors $K = \zeta C^* + \tau$ in KCP. (From Carneiro *et al.* 1976.)

This equation shows that the amplitude fluctuations are observable with Raman scattering techniques, while the phase fluctuations can be observed by Brillouin scattering techniques. The frequencies and line widths of the amplitude mode in K_2SeO_4, measured by Raman scattering techniques (Wada *et al.* 1977) are shown in fig. I.41. Since the phase modes can be observed in Brillouin scattering they must have the same symmetry as acoustic modes, and will therefore interact with them. This interaction can be evaluated (Bruce and Cowley 1978) by expanding the coupling coefficient between the strain $\eta_{\alpha\beta}$ and the modes ($\mathbf{q}j$) and ($-\mathbf{q}j$),

$$V_{\alpha\beta}\begin{bmatrix} \mathbf{q} & -\mathbf{q} \\ j & j \end{bmatrix}$$

in eqn. (I.2.10), as was described above for the polarizability. The result is similar; there is a linear coupling between the phasons and the strain parameters which is determined by

$$iA_0 V^1_{\alpha\beta,\gamma} = \frac{\partial}{\partial k_\gamma} V_{\alpha\beta}\begin{bmatrix} \mathbf{q}_s & \mathbf{q}_s+\mathbf{k} \\ j & j \end{bmatrix}.$$

This coupling is proportional to the order parameter A_0, and will modify the velocities of both the phasons and acoustic modes.

Fig. I.41

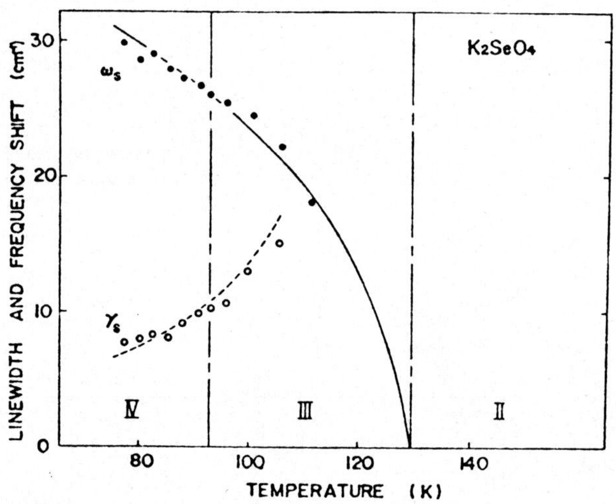

Raman scattering measurements of the longitudinal or amplitude mode in K_2SeO_4. Phase (2) is paraelectric, (3) is incommensurate and (4) is a locked-in phase. (After Wada *et al.* 1977.)

Alas we do not, as yet, know of any direct observation of phasons by either Brillouin or ultrasonic techniques. Bechtle and Scott (1977) have suggested that some anomalies in the intensity of the Brillouin scattering of the acoustic modes in $BaMnF_4$ arise because of coupling to the phasons but this is not yet proven.

Finally, we discuss the effect of the phasons on the Debye–Waller factor of the incommensurate phase. The Debye–Waller factor is determined by the displacement–displacement correlation function for the particular site in the crystal. When wave-vector is conserved this can be expressed in terms of the normal mode correlation function $\sum_{\mathbf{q}jj'} \langle Q(\mathbf{q}j)Q(-\mathbf{q}j')\rangle$ as shown by eqn. (I.3.36). In an incommensurate phase, this correlation function is given by

$$\langle u_\alpha(l\kappa)u_\beta(l\kappa)\rangle = \frac{1}{M_\kappa N} \sum_{\mathbf{q}_1\mathbf{q}_2 j_1 j_2} e_\alpha(\kappa, \mathbf{q}_1 j_1)$$
$$\times e_\beta(\kappa, \mathbf{q}_2 j_2)\langle Q(\mathbf{q}_1 j_1)Q(\mathbf{q}_2 j_2)\rangle \exp i(\mathbf{q}_1 - \mathbf{q}_2) \,.\, \mathbf{R}(\kappa). \qquad \text{(I.4.66)}$$

We shall now consider the part of the Debye–Waller factor dependent upon the correlation function of the amplitude and phase modes described above. When the transformation is made to the P_1 and P_2 coordinates the relevant contribution to (I.4.66) is proportional to

$$\frac{1}{2N}\sum_{\mathbf{k}} [(1 - \cos(2\mathbf{q}_s \,.\, \mathbf{R}(l\kappa)))/\omega_{\mathrm{T}}(\mathbf{k})^2 + (1 + \cos(2\mathbf{q}_s \,.\, \mathbf{R}(l\kappa)))/\omega_{\mathrm{L}}(\mathbf{k})^2] \qquad \text{(I.4.67)}$$

where the frequencies are given by eqns. (I.4.60)–(I.4.62). This result is unusual below T_c in that it varies with position, $\mathbf{R}(l\kappa)$. In an incommensurate phase each atom is displaced a different amount and since the fluctuations correspond to fluctuations in that displacement it is to be expected that the Debye–Waller factor varies with position. The intensity of the Bragg reflections corresponding to the underlying lattice $\mathbf{K} = \boldsymbol{\tau}$ and to the incommensurate phase, $\mathbf{K} = \boldsymbol{\tau} + \mathbf{q}_s$, have the same Debye–Waller factor determined by

$$\bar{W} = \frac{1}{2N}\sum_{\mathbf{k}}\left(\frac{1}{\omega_{\mathrm{T}}(\mathbf{k})^2} + \frac{1}{\omega_{\mathrm{L}}(\mathbf{k})^2}\right). \qquad \text{(I.4.68)}$$

An alternative approach to the scattering properties and the Debye–Waller factor was developed by Overhauser (1971) using the phase variable $\phi(\mathbf{k})$ directly. The component of the Bragg scattering from the $(l\kappa)$th atom is determined by $\langle \exp i\mathbf{K} \,.\, \mathbf{u}(l\kappa)\rangle \exp i\mathbf{K} \,.\, \mathbf{R}(l\kappa)$. If we consider only the displacements due to the phase fluctuations then

$$\mathbf{u}(l\kappa) = \left(\frac{2}{M_\kappa}\right)^{1/2} A_0 \mathbf{e}(\kappa, \mathbf{q}_s j) \cos(\mathbf{q}_s \,.\, \mathbf{R}(l\kappa) + \phi_0 + \phi(l\kappa)), \qquad \text{(I.4.69)}$$

where $\phi(l\kappa)$ is the fluctuating part of the phase

$$\phi(l\kappa) = \frac{1}{\sqrt{N}}\sum_{\mathbf{k}} \phi(\mathbf{k}) \exp(i\mathbf{k} \,.\, \mathbf{R}(l\kappa)). \qquad \text{(I.4.70)}$$

The Bessel function expansion for a complex exponential is

$$\exp(iz\cos\psi) = \sum_{n=-\infty}^{\infty} i^n J_n(Z) \exp(in\psi),$$

and gives

$$\langle \exp(i\mathbf{K} \,.\, \mathbf{u}(l\kappa))\rangle = \sum_{n=-\infty}^{\infty} i^n J_n(y) \exp(in(\mathbf{q}_s \,.\, \mathbf{R}(l\kappa) + \phi_0)))$$
$$\times \langle \exp(in\phi(l\kappa))\rangle, \qquad \text{(I.4.71)}$$

where

$$y = \sqrt{2} A_0 \mathbf{K} \,.\, \mathbf{e}(\kappa, \mathbf{q}_s j)/M_\kappa^{1/2}.$$

The effective Debye–Waller factor for the different terms $d_n = \langle \exp(in\phi(l\kappa)) \rangle$ is then given, using eqns. (I.4.70) and (I.4.64), as

$$\exp(-\bar{w}) = d_n = \exp\left(-\frac{n^2}{2A_0^2 N}\sum_{\mathbf{k}} \omega_T(\mathbf{k})^{-2}\right). \tag{I.4.72}$$

This result shows that the Bragg reflections of the underlying lattice, $\mathbf{K} = \boldsymbol{\tau}$, are given by the terms in eqn. (I.4.71) with $n = 0$. Since $d_0 = 1$, the phase fluctuations do not contribute to their Debye–Waller factor, while the primary Bragg reflections of the incommensurate phase $\mathbf{K} = \boldsymbol{\tau} - \mathbf{q}_s$ have a Debye–Waller factor given by d_1. This is an unusual Debye–Waller factor because it is independent of $\mathbf{K}$ and decreases to zero rapidly as $T \to T_c$ due to the dependence on A_0^2. The higher terms in the expansion (I.4.71) correspond to the two-phonon, three-phonon terms, etc., in the expansion of the scattering cross-section as discussed in § I.3.1.

These results are clearly very different from those derived using the expansion in terms of P_1 and P_2, eqn. (I.4.68), and so we are left with two questions. Firstly why are they so different, and secondly which, if either, is correct? The reason for the difference is that the expansion in P_2 and the phase fluctuations, ϕ, are only exactly equivalent to first order. Since the Debye–Waller factor is a summation over infinitely many-phonon processes the two expansions give different results because they include terms beyond the first order in a different way. It is clearly physically appealing to use the phase fluctuation model of Overhauser (1971) because the energy is unchanged by a change in the phase, ϕ_0. Consequently it would seem that eqn. (I.4.72) is to be preferred. Close to T_c, however, the fluctuating displacement, $\phi(l\kappa)$, can become large, because the amplitude A_0 becomes very small. Our development, however, has treated the fluctuations in $\phi(l\kappa)$ as a harmonic oscillator and this is clearly inappropriate when $\phi(l\kappa) \sim \pi$. Consequently we expect the phase mode contribution to the Debye–Waller factor to be given accurately by eqn. (I.4.72) for $T \ll T_c$, but close to T_c there will be deviations from this behaviour and it will possibly revert more closely to the form obtained in eqn. (I.4.68), which yields the normal Debye–Waller factor for $T > T_c$.

I.4.4.4. *Secondary order parameters*

At incommensurate phase transitions there is always a coupling between the strain and pairs of order parameters. In the simplest model the coupling is given by $2eQ_1Q_2\eta$, which when transformed into the phase and amplitude coordinates becomes $eA^2\eta$. Consequently the coupling alters the magnitude of the amplitude and the character of the amplitude fluctuations but has no direct effect on the phase fluctuations. These effects can be treated using the methods outlined in § I.4.1.2, and the results are essentially the same as those obtained there.

In principle it is possible to have an improper ferroelectric in which the primary order parameter has an incommensurate wave-vector, if the irreducible representation of the order parameter is doubly degenerate. In this case the ferroelectric ordering will be accompanied by a particular choice of the relative phases of the two systems (Cochran 1971). Since, however, we are unaware of any real systems of this type we shall not discuss them in more detail.

Of more interest in incommensurate phases are secondary order parameters associated with wave-vectors of the form

$$\mathbf{q}_m = m\mathbf{q}_s + \boldsymbol{\tau}_m, \tag{I.4.73}$$

where m is any integer. A simple mean field argument shows that if there is a term in the hamiltonian of the form

$$V\begin{pmatrix} \mathbf{q}_s & \mathbf{q}_s & -2\mathbf{q}_s \\ j & j & j' \end{pmatrix},$$

then in an incommensurate phase with a distortion of wave-vector $\mathbf{q}_s$, there will also be a distortion with wave-vector $2\mathbf{q}_s$. Extending this argument shows that distortions will occur for all the wave-vectors given by eqn. (I.4.73). We may expect, however, that increasing j will involve increasing orders of the anharmonic interactions so that the amplitudes of these distortions decrease rapidly as j increases. Despite this it is necessary to consider if these secondary distortions can influence the qualitative features discussed above.

The free energy of the system within a completely general Landau theory may be written (Kwok and Miller 1966)

$$G=G_0+\sum_n \sum_{\substack{\mathbf{q}_1\ldots\mathbf{q}_n \\ j_1\ldots j_n}} V_n\begin{pmatrix} \mathbf{q}_1 & \cdots & \mathbf{q}_n \\ j_1 & \cdots & j_n \end{pmatrix}\langle Q(\mathbf{q}_1 j_1)\rangle\ldots\langle Q(\mathbf{q}_n j_n)\rangle\Delta(\mathbf{q}_1+\ldots+\mathbf{q}_n)$$

where the $\mathbf{q}_m$ belong to the set given in eqn. (I.4.73) and the V_n coefficients may be temperature dependent. Now consider a change in the phase of the distorted waves to

$$\langle Q(\mathbf{q}_m j_m)\rangle_\phi = \exp(im\phi)\langle Q(\mathbf{q}_m j_m)\rangle, \tag{I.4.74}$$

then the free energy of the new phase is

$$G_\phi=G_0+\sum_n \sum_{\substack{\mathbf{q}_1\ldots\mathbf{q}_n \\ j_1\ldots j_n}} V_n\begin{pmatrix} \mathbf{q}_1 & \cdots & \mathbf{q}_n \\ j_1 & \cdots & j_n \end{pmatrix}\exp(i(m_1+\ldots+m_n)\phi)$$
$$\times\langle Q(\mathbf{q}_1 j_1)\rangle\ldots\langle Q(\mathbf{q}_n j_n)\rangle\Delta(\mathbf{q}_1+\ldots+\mathbf{q}_n).$$

But if $\mathbf{q}_s$ is truly incommensurate, wave-vector conservation implies that $m_1+m_2+\ldots+m_n=0$. Consequently $G_\phi=G$ and the free energy of the incommensurate state is unchanged by the transformation (I.4.74). This then suggests that the phason behaviour, namely $\chi_T(\mathbf{k})\rightarrow\infty$ as $k+0$, will occur even when the effects of these secondary order parameters are included. An explicit example has been detailed by us elsewhere (Bruce and Cowley 1978).

These secondary order parameters break the same symmetry operations of the high-temperature phase as the primary wave-vector, $\mathbf{q}_s$. They are therefore secondary order parameters of the second type discussed in § I.3.2. From a somewhat pedantic point of view, eqn. (I.4.73) defines a set of $\sim N^{1/3}$ possible components of the order parameter with different values of $\mathbf{q}_m$, in the same direction, corresponding to different components of the same set of order parameters. A different type of combination of this set of order parameters is employed in the next section to discuss lock-in phase transitions.

I.4.5. *Lock-in phase transitions*

I.4.5.1. *Phase-locked phases*

In §§ I.4.1 to I.4.3 we discussed the properties of phase transitions for which symmetry imposed $\nabla_\mathbf{q}\omega(\mathbf{q}_s j)=0$. We shall now discuss the properties of commensurate phases for which $\nabla_\mathbf{q}\omega(\mathbf{q}j)$ is not necessarily zero. For example, a tetragonal material for which $\mathbf{q}_s=(0,0,\zeta)$ and $\zeta=2\pi/pc$ has a commensurate wave-vector but if $p>2$ the gradient is in general non-zero. The additional terms in the free-energy equation, (I.4.49), are given by an Umklapp term involving pth order anharmonicity so that

$$G=G_0+a(T-T_c)Q_1Q_2+4uQ_1^2Q_2^2+2^{(\frac{p}{2}+1)}V(Q_1^p+Q_2^p). \qquad \text{(I.4.75)}$$

When this expression is transformed to the amplitude and phase coordinates

$$G=G_0+\frac{1}{2}a(T-T_c)A^2+uA^4+VA^p\cos p\phi. \qquad \text{(I.4.76)}$$

This expression depends upon ϕ and so the phase of the distorted wave is one of the p different choices of ϕ which will minimize eqn. (I.4.76). Strictly speaking within Landau theory, and for $T\simeq T_c$, only the values of $p=3$ and 4 are of interest. When $p=3$ there are cubic terms in the free energy and so the transition is necessarily of first order, § I.4.1.1, while when $p=4$ the Umklapp term has the form $A^4V\cos 4\phi$. Consequently the amplitude of the distorted wave is given by

$$A_0^2=\frac{a(T_c-T)}{4(u-|V|)}$$

and the phases of the different possible domains of the distorted wave are $\phi_0=0, \pi/2, \pi, 3\pi/2$ if V is negative and $\pm\pi/4$, $\pm 3\pi/4$ if V is positive. The longitudinal susceptibility can be obtained below T_c as

$$\chi_L(0)=(2a(T_c-T))^{-1},$$

while the transverse or phase susceptibility is given by

$$\chi_T(0)=\frac{u-|V|}{4|V|a(T_c-T)}, \qquad \text{(I.4.77)}$$

which can be written in terms of the amplitude, A_0, to give the transverse frequency as

$$\omega_T(0)^2=(p^2|V|A_0^{p-2}). \qquad \text{(I.4.78)}$$

These results show that the interaction locks the phase of the distortion so that the susceptibility of the phase mode is finite and of magnitude inversely proportional to that interaction.

Since there is no reason to expect that the phonon dispersion relation will have a minimum exactly at $\boldsymbol{\tau}/p$, even if it is close to that wave-vector, a continuous phase transition will normally give an incommensurate phase with $\mathbf{q}_s$ close to $\boldsymbol{\tau}/p$. This will be followed at a lower temperature by a lock-in phase transition to the commensurate phase with wave-vector, $\boldsymbol{\tau}/p$, as discussed below.

In NbO_2, however, the transition is apparently directly from a disordered phase to a commensurate phase described by a wave-vector $\mathbf{q}_s=(\frac{1}{4},\frac{1}{4},\frac{1}{2})$ in reciprocal lattice units (Pynn and Axe 1976). Finally, when $\mathbf{q}_s$ is commensurate, there are only a finite set of the secondary order parameters with different wave-vectors, and these are then secondary order parameters of the first type because they do not break all the symmetry elements. In NbO_2, for example, the secondary order parameters have $\mathbf{q}_j=(\frac{1}{2},0,0)$, or $(0,\frac{1}{2},0)$ or $(0,0,0)$, none of which break all the translational symmetry elements which are broken by the distortion with $\mathbf{q}_s=(\frac{1}{4},\frac{1}{4},\frac{1}{2})$.

I.4.5.2. *Soft solitons*

In the preceding section we have shown that Umklapp terms in the free energy lock the phase of the distorted wave for commensurate wave-vectors $\mathbf{q}_s$ and hence lower the free energy. We now consider the tetragonal system discussed in §§ I.4.4.1 and I.4.5.1 but with a phase transition described by an incommensurate wave-vector $\mathbf{q}_s$, which is close to the commensurate wave-vector $\mathbf{q}_0 = \tau/p$. The quadratic term in the expansion of the free energy is

$$G_2 = \frac{1}{N}\sum_{\mathbf{k}}[a(T-T_c)+fk^2]Q_1(\mathbf{k})Q_2(-\mathbf{k}), \tag{I.4.79}$$

where

$$\mathbf{k} = \mathbf{q} - \mathbf{q}_s \quad \text{and} \quad \boldsymbol{\delta}_0 = \mathbf{q}_0 - \mathbf{q}_s.$$

The system undergoes a phase transition at T_c to an incommensurate phase, whose free energy is given (neglecting the higher Fourier components of the distortion) by eqn. (I.4.5*a*), as

$$G_{\text{inc}} = G_0 - \frac{a^2}{16u}(T_c - T)^2. \tag{I.4.80}$$

The free energy of the commensurate phase with $p=4$ is

$$G_{\text{com}} = G_0 - \frac{a^2}{16(u-|V|)}(T_{\text{co}} - T)^2 \tag{I.4.81}$$

where T_{co} is the phase transition temperature that the commensurate phase would have had if the incommensurate phase did not occur first, $T_{\text{co}} = T_c - f\delta_0^2/a$ and V is the lock-in energy, eqn. (I.4.76). For temperatures greater than T_I the free energy (I.4.80) is smaller than (I.4.81) and the incommensurate phase is stable, but at T_L these equations predict a first-order transition to the commensurate phase where

$$T_L = T_c - \frac{2f\delta_0^2 u}{|V|a}.$$

This analysis demonstrates the essential qualitative features of a lock-in phase transition. There is a competition between the quadratic term in the free energy (the frequency dispersion relation energy) which favours ordering at $\mathbf{q}_s$, and the Umklapp interaction energy which favours a commensurate ordering, $\mathbf{q}_0$. At high temperatures the second term dominates, but at lower temperatues, A_0 becomes large, and the interaction term dominates.

Although the above analysis contains the essential physics, it neglects some interesting and novel features of these phase transitions. These arise because the interaction energy responsible for the lock-in also makes the secondary order parameters very large, and neglecting them as we have done so far in this section is invalid (Moncton *et al.* 1975).

The effect of the secondary order parameters is most easily calculated by rewriting the free energy in a continuum form as an integral over position, $\mathbf{r}$. In principle we should introduce a position dependent amplitude $A(\mathbf{r})$ and phase $\phi(\mathbf{r})$; in practice, however, the analysis is more tractable if we assume $A(\mathbf{r}) = A_0$. This phase-modulation-only approximation is justified because the phase or transverse susceptibility is much larger than the amplitude or longitudinal susceptibility, eqn. (I.4.53), and consequently it is much easier to create local distortions of the phase than of the amplitude. The free energy, eqns. (I.4.76) and (I.4.79), in a real-space representation then becomes (Bruce *et al.* 1978).

$$G=G_0-\frac{a^2}{16u}(T_c-T)^2+\frac{fA_0^2}{Nv}\int G(\mathbf{r})\,d\mathbf{r}, \qquad \text{(I.4.82}\,a\text{)}$$

where v is the volume of the unit cell and the position dependent free energy is

$$G(\mathbf{r})=\frac{1}{2}(\nabla\phi(\mathbf{r})-\boldsymbol{\delta}_0)^2-\bar{V}\cos p\phi(\mathbf{r}), \qquad \text{(I.4.82}\,b\text{)}$$

where

$$\bar{V}=VA_0^{p-2}/f$$

If the vector $\boldsymbol{\delta}_0$ is along the z axis, eqn. (I.4.82) is minimized when $\phi(\mathbf{r})$ is a function only of z and the phase dependent part of the free energy becomes (for a crystal of length L)

$$\frac{G_\phi}{fA_0^2}=\frac{1}{L}\int_0^L G(z)\,dz-\delta_0\bar{k}+\frac{1}{2}\delta_0^2-\bar{V}, \qquad \text{(I.4.83}\,a\text{)}$$

where now

$$G(z)=\frac{1}{2}\left(\frac{d\phi}{dz}\right)^2-\bar{V}[\cos p\phi(z)-1], \qquad \text{(I.4.83}\,b\text{)}$$

and

$$\bar{k}=\frac{1}{L}(\phi(L)-\phi(0)) \qquad \text{(I.4.84)}$$

is the effective wave-vector for the incommensurate phase. The problem of finding the nature of the incommensurate phase is then reduced to finding the forms of $\phi(z)$ which minimize eqns. (I.4.83). McMillan (1976) solved this problem by writing $\phi(z)$ as a Fourier series

$$\phi(z)=\delta z+\sum_n H_n \sin(pn\delta z),$$

and then solving numerically for the coefficients δ and H_n. More elegantly, Bak and Emery (1976) observed that the functions for which the integral in eqn. (I.4.83 *a*) is an extremum satisfy the sine–Gordon equation

$$\frac{d^2\phi(z)}{dz^2}=p\bar{V}\sin p\phi(z),$$

one solution of which is the soliton or domain wall (Scott *et al.* 1973)

$$\phi(z)=\theta(z)\equiv\frac{4}{p}\tan^{-1}(\exp(p\bar{V}^{1/2}z)), \qquad \text{(I.4.85)}$$

which separates a region with $\phi(z)=0$, say, from one in which $\phi(z)=2\pi/p$. Substituting this solution back into eqn. (I.4.83) gives the energy of a single domain wall as $1/L\int G(z)\,dz=8\bar{V}^{1/2}/p$. Consequently the free energy of a regular array of M

solitons with spacing b can be found, taking account of their long distance repulsion, as (Bak and Emery 1976)

$$\frac{G_\phi}{fA_0^2}=\frac{M}{L}\frac{8\bar{V}^{1/2}}{p}[1+4\exp(-p\bar{V}^{1/2}b)]-\bar{k}\delta_0-\bar{V}+\frac{\delta_0^2}{2}.$$

Now using the fact that $\bar{k}=2M\pi/pL=2\pi/pb$ the free energy becomes

$$\frac{G_\phi}{fA_0^2}=\frac{M}{L}\left(\frac{8\bar{V}^{1/2}}{p}-\frac{2\pi\delta_0}{p}\right)+\frac{4M}{L}\frac{8\bar{V}^{1/2}}{p}\exp\left(-\frac{p\bar{V}^{1/2}L}{M}\right)-\bar{V}+\frac{\delta_0^2}{2}. \qquad \text{(I.4.86)}$$

The first term is the free energy of M non-interacting solitons and if $4\bar{V}^{1/2}>\pi\delta_0$ this is positive and the commensurate phase with $M=\bar{k}=0$ is favoured. If, however, $4\bar{V}^{1/2}<\pi\delta_0$ a state with non-zero $\bar{k}$ and M exists which is the incommensurate phase. Thus, there will be a continuous transition from the commensurate to incommensurate state when $4\bar{V}^{1/2}=\pi\delta_0$, which using the relationship between $\bar{V}$ and the bare interaction V, occurs when

$$A_0=(\pi^2\delta_0^2f/16V)^{\frac{1}{p-2}}$$

For larger temperatures, smaller A_0, M steadily increases but is limited by the interaction term in eqn. (I.4.86). This interaction has an exponential dependence on $1/M$ so that the temperature dependence of the order parameter, M, is given by

$$\frac{2\pi M}{pL}=\bar{k}=\frac{2\pi\bar{V}^{1/2}}{\log\left|\frac{\pi\delta_0}{16\bar{V}^{1/2}}-\frac{1}{4}\right|}. \qquad \text{(I.4.87)}$$

This form is very difficult to distinguish experimentally from the discontinuity of a first-order phase transition as shown in fig. I.42.

Fig. I.42

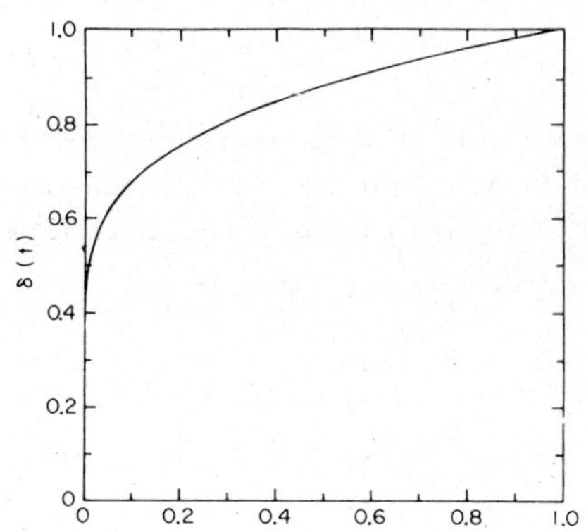

Sketch of the order parameter $\bar{k}$, units δ_0, against reduced temperature for a lock-in phase transition. (After McMillan 1976.)

The nature of the structure of the phase function $\phi(z)$ in the incommensurate phase is given by

$$\phi(z)=\phi_0+\frac{2\pi m}{p}+\theta(z-mb), \qquad \text{(I.4.88)}$$

where m is the closest integer to z/b. This structure was calculated by McMillan (1976) for the case of $p=3$ and is illustrated in fig. I.43.

This type of phase transition is quite different from those discussed earlier, because the order parameter is not the amplitude of a single Fourier component of the atomic displacements but is the number of domain walls, each of which is well localized in space. This can occur for an incommensurate phase because of the large number of possible order parameters for an incommensurate phase, as discussed in § I.4.4.4. Whenever there is a very large number of order parameters, there is always the possibility of forming linear combinations, such as domain walls, with unexpected or non-perturbative characteristics.

Fig. I.43

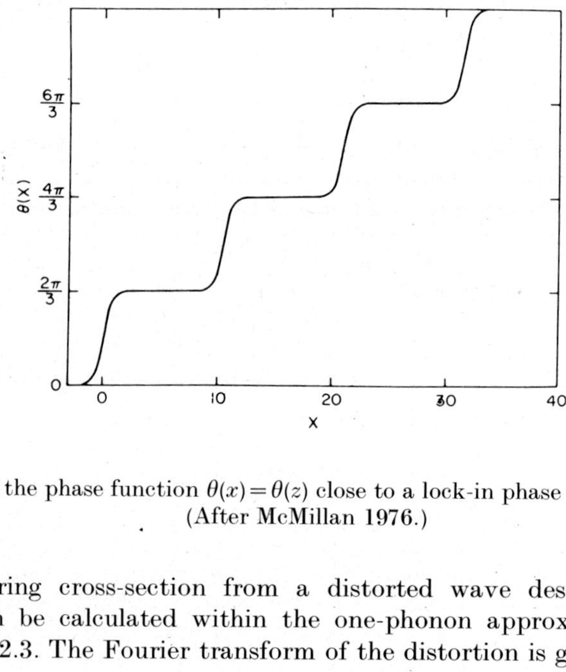

The structure of the phase function $\theta(x)=\theta(z)$ close to a lock-in phase transition with $p=3$. (After McMillan 1976.)

The scattering cross-section from a distorted wave described by $2A_0 \cos (q_0 z+\phi(z))$ can be calculated within the one-phonon approximation using the methods of § I.2.3. The Fourier transform of the distortion is given by

$$F_s(K)=F_+(K)+F_-(K), \tag{I.4.89}$$

where K is the z component of $\mathbf{K}$, and

$$F_-(\mathbf{K})=\frac{A_0}{2a}\int \exp\,(i(K-q_0)z-\phi(z))\,dz, \tag{I.4.90}$$

which on substituting for $\phi(z)$ from eqn. (I.4.88) becomes

$$F_-(\mathbf{K})=\frac{bA_0}{2a}f(K-q_0)\exp\,(-i\phi_0)\sum_{m=1}^{M}\exp\left(i(K-q_0)mb-i\frac{2\pi}{p}m\right),$$

where $f(K)$ is the form factor of the soliton

$$f(K)=\frac{1}{b}\int_{-b/2}^{b/2}\exp\,(i(Kz-\theta(z)))\,dz. \tag{I.4.91}$$

The summation over m yields the total number of domain walls M provided that the wave-vector K is given by

$$K=T_I=q_0+\frac{2\pi}{b}\left(\frac{1}{p}+I\right), \tag{I.4.92}$$

where I is any positive or negative integer. The location of these Bragg reflections is illustrated schematically in fig. I.44.

Fig. I.44

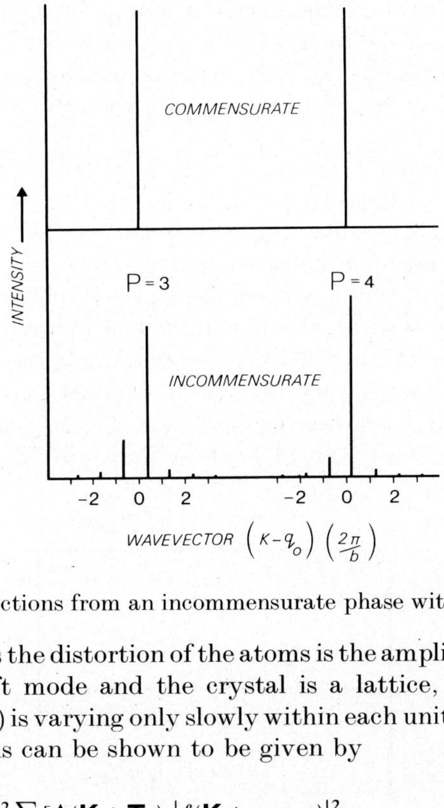

Sketch of the Bragg reflections from an incommensurate phase with a soliton ground state.

Now in real systems the distortion of the atoms is the amplitude multiplied by the eigenvector of the soft mode and the crystal is a lattice, not a continuum. If, however, the phase $\phi(z)$ is varying only slowly within each unit cell then the intensity of the Bragg reflections can be shown to be given by

$$I(\mathbf{K})=\frac{NA_0^2}{4}|F(\mathbf{K},\mathbf{q}_0j)|^2\sum_I[\Delta(\mathbf{K}+\mathbf{T}_I)\,|f(\mathbf{K}+\mathbf{q}_0-\boldsymbol{\tau})|^2$$

$$+\Delta(\mathbf{K}-\mathbf{T}_I)\,|f(\mathbf{K}-\mathbf{q}_0-\boldsymbol{\tau})|^2], \tag{I.4.93}$$

where $\boldsymbol{\tau}$ is a reciprocal lattice vector of the underlying lattice and $\Delta(\mathbf{K})$ is given by eqn. (I.2.19).

In principle the form factor of the soliton, eqn. (I.4.91), depends on the detailed structure of the soliton, but if the spacing between them, b, is much larger than the thickness, $1/p\bar{V}^{1/2}$, then they may be approximated as phase step functions, when

$$f(K-q_0)=f(T_I)=\left[\pi\left(\frac{1}{p}+I\right)\right]^{-1}\sin(\pi/p)\exp(-i\pi/p). \tag{I.4.94}$$

The relative intensities of the Bragg reflections calculated using this form factor are shown in fig. I.44.

Using the form factor eqn. (I.4.94), the sum over I in eqn. (I.4.93) can be performed to give the integrated intensity of these Bragg reflections. This is found to be identical with the intensity of the Bragg reflection of the commensurate phase. Consequently on approaching the lock-in phase transition, the many Bragg reflections of the incommensurate phase become closer and closer together in $\mathbf{K}$ as the number of solitons decreases, but their integrated intensity remains constant.

The phase excitation spectrum of a system of solitons has been calculated by us elsewhere (Bruce and Cowley 1978). In view of the fact that there are no direct measurements with which to compare these results we only briefly comment on the properties. There are two types of phase excitations. The first of these corresponds to oscillations of the solitons about their mean positions. Since the solitons are a regular array in one dimension, z, and interact only with nearest neighbours, the dispersion relation for these excitations is given by the usual expression for the excitations of a one-dimensional chain. A detailed calculation gives these frequencies as

$$\omega_p(\mathbf{k})^2 = 8fp^2\bar{V}\exp(-p\bar{V}^{1/2}b)(1-\cos(k_z b)) + f(k_x^2+k_y^2). \qquad (I.4.95)$$

This dispersion relation is illustrated in fig. I.45, and for small k_z, $\omega_p(\mathbf{K})$ is linear in k due to the translational invariance of the soliton array. The velocity of the phasons decreases as their separation b increases, and as $T \to T_L$ varies as $(T-T_L)^{1/2}$. The scattering cross-section of these excitations can be evaluated and also decreases to zero as $T \to T_L$, because the number of solitons is decreasing.

The other type of excitation consists of fluctuations in the phase of the essentially commensurate regions of the crystal. These excitations have nearly the same dispersion relation and scattering cross-section as the phase excitations of the commensurate phase discussed in § I.4.5.1, as shown in fig. I.45.

Fig. I.45

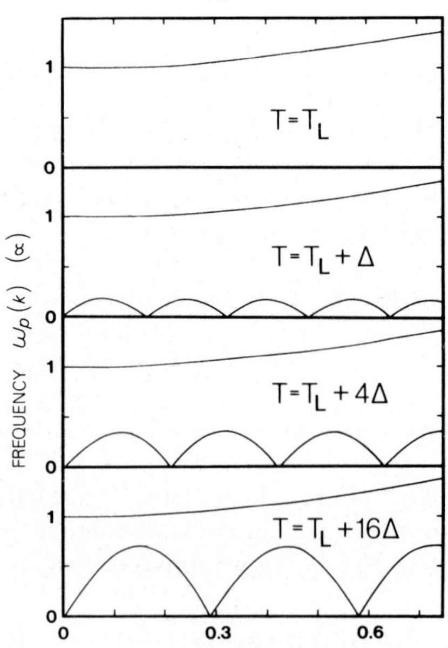

The phase excitation spectrum close to a lock-in phase transition. (From Bruce and Cowley 1978.)

This type of lock-in phase transition occurs in $TaSe_2$ (Moncton *et al.* 1975), TaS_2 (Nakanishi *et al.* 1977) and in TTF–TCNQ (Comès *et al.* 1976). The theory accounts qualitatively for the strong satellites observed in $TaSe_2$ and TaS_2, but in all cases the transitions are not continuous but of first order, in contradiction with the prediction of the soliton theory.

There are two main defects of the theory (Bruce *et al.* 1978). Firstly, there is a coupling to the macroscopic strains. Inclusion of this coupling shows that it leads to different effects dependent upon whether or not the strain can follow the phase soliton profile. This is dependent upon whether there are acoustic modes of the correct symmetry to describe the changes in the strains with distance. In our example, $\partial\eta_{zz}/\partial z$ is given by the longitudinal acoustic modes propagating along the z axis and so η_{zz} strains can follow the phase solitons and give rise to a change in the effective interaction, a modified $\bar{V}$ (Bak and Timonen 1978). In contrast there cannot be $\partial\eta_{yy}/\partial z$ components because there are no acoustic modes propagating along z corresponding to η_{yy} strains. Consequently there can only be a macroscopic η_{yy} strain, which leads to a term in the free energy which is proportional to M^2 and negative; exactly analogous to the negative contribution to u discussed in § I.4.3.1 for conventional phase transitions. In the soliton case, however, eqn. (I.4.86) shows that there is not an intrinsic positive term proportional to M^2, so this coupling to the strain necessarily drives this transition to be of first order. We are of the opinion that this latter type of coupling may always occur in real systems.

Secondly, if the soliton thickness is only a few lattice constants, it will be energetically favourable for the solitons to be centred on a particular position within the unit cell. The spacing between the solitons will then be an integral number of lattice spacings so that $b = Ja$, and the wave-vectors of the Bragg reflections will be given by

$$T_I = q_0\left(1 + \frac{1}{J}(1 + pI)\right).$$

Under these conditions the system will undergo a sequence of first-order transitions between these different commensurate states until it eventually reaches the primary commensurate phase with $T_I = 2\pi/ap$. This sequence of first-order transitions is sometimes known as a Devil's staircase (Aubry 1977).

The lattice may also produce a very large lock-in energy even though it does not favour any particular J. This may then have a large effect on the phase excitations of the system, as the phasons or soliton modes, fig. I.45, will then have a gap which can be very large compared with the dispersion in the energy.

Throughout this section we have assumed that there are only two components of the order parameter for the commensurate phase $\pm\mathbf{q}_0$. In practice the lock-in phase transition in $TaSe_2$ has three pairs of wave-vectors and so there is the possibility of three different types of solitons with interactions between them. As yet comparatively little work has been done on these more complicated systems. Nakanishi and Shiba (1977) suggest that the phase transitions are of first order. Bak (1978) argues that the interaction energy between the different sets of solitons gives rise to a term in the free energy which is proportional to M^2. If this interaction energy is negative, attractive, the transition is always of first order. If it is repulsive then there will probably be a transition to a phase with only one set of these solitons frozen into the crystal followed by another first-order transition to a phase with all three sets of solitons.

Long period structures also occur in alloy systems; for example, Oriani and Murphy (1958) showed that in CuAu there was a structure with long periodicity between 650 and 690 K. At 650 K this undergoes a lock-in phase transition to the well-known ordered CuAu structure. The structure with the long periodicity consists of ordered regions of ordered CuAu, of about 5 unit cells in length, separated by domain walls from the next phase. This transition was discussed in terms of an instability of the crystal against domain walls by Inglesfeld (1972). A similar structure has been postulated by Pynn (1978) for the structure of the ω-phase in Zr and Ti alloys. In his model the ω-phase in $Zr_{0\cdot8}Nb_{0\cdot2}$ consists of the bcc structure distorted by a commensurate wave-vector,

$$q_0=\frac{2\pi}{a}\left(\frac{2}{3},\frac{2}{3},\frac{2}{3}\right).$$

The three different domains corresponding to this deformation are then separated by domain walls which are pinned by the local fluctuations in the concentrations of Nb and Zr atoms and so are varying distances apart.

I.4.5.3. *Lock-in without soft solitons*

There are lock-in phase transitions to those wave-vectors for which the dispersion relations $\omega(\mathbf{q})$ are necessarily symmetric about $\mathbf{q}_0$. In these cases the linear term in $\nabla\phi(\mathbf{r})$ in eqns. (I.4.82 *a*, *b*) is absent and so the theory developed in § I.4.5.2 must be reworked. In the neighbourhood of one of these more symmetric wave-vectors the dispersion relation can be written in the form

$$\omega(\mathbf{q})^2=\omega(\mathbf{q}_0)^2+f_1\left[-\delta_0^2k_z^2+\frac{1}{2}k_z^4\right]+f_2(k_x^2+k_y^2). \tag{I.4.96}$$

which exhibits a minimum for $\mathbf{k}$ along the z axis at the point $k_z=\pm\delta_0$. Bruce *et al.* (1978) have discussed this type of phase transition using the type of approach developed in the previous section. The equation, analogous to the sine–Gordon equation is a fourth-order differential equation, which does not, to our knowledge, have a simple analytic solution. The solutions were, however, studied numerically by looking, in particular, for a solution of a soliton-like character. Such solutions were found but the energy of these solutions is such that the commensurate phase is unstable against the formation of an incommensurate phase with a nearly plane-wave structure before it is unstable against the formation of these soliton-like structures. Accordingly a first order phase transition is predicted.

There are several examples of this type of lock-in phase transition. $NaNO_2$ (Yamada *et al.* 1968) has an incommensurate phase with a wave-vector $\mathbf{q}_s\sim2\pi/8$ and a lock-in transition to a ferroelectric phase. The transition is of first order and no harmonics of the incommensurate wave-vector have been observed. In thiourea (McKenzie 1975) a similar phase transition is again first order and occurs at 180 K, while weak first harmonics (about 2% of the primary distortion) were found.

I.4.5.4. *Improper lock-in transitions*

Some lock-in phase transitions are associated with the onset of ferroelectricity, as found in K_2SeO_4 (Iizumi *et al.* 1977) and $(ND_4)_2BeF_4$ (Iizumi and Gesi 1977). In these cases the lock-in terms in the free energy are of the form

$$2^{(\frac{p}{2}+1)} W(Q_1^p + Q_2^p)\bar{Q}, \tag{I.4.97}$$

where $\bar{Q}$ is the ferroelectric secondary order parameter.

In an incommensurate phase with wave-vector $\mathbf{q}_s$ terms of the form of eqn. (I.4.97) will necessarily give rise to distortions of the secondary order parameter with a wave-vector $\pm\mathbf{q} = \pm(p\mathbf{q}_s + \tau)$. This gives rise to a lock-in energy only if $\pm\mathbf{q}$ are the same wave-vector when there will be a constructive interference between these two distortions resulting in a lower energy for the resulting commensurate phase. Of course, there will also be the possibility of lock-in due to terms of the form

$$2^{(p+1)} V'(Q_1^{2p} + Q_2^{2p}),$$

for these wave-vectors. This term is usually non-zero but has frequently been neglected in the theory of these materials.

When a plane-wave ground state for the incommensurate phase is assumed the theory can be evaluated as given in detail by Iizumi *et al.* (1977) for K_2SeO_4, and the analysis is very similar to the treatment of secondary order parameters in § I.4.3 and of lock-in phase transitions at the beginning of § I.4.5.2. Since, however, the dispersion relation is not symmetric about $\mathbf{q}_s$, this analysis requires further development to examine if soliton-like structures can modify the behaviour. This approach has been developed by Bruce *et al.* (1978) and soliton-like structures may occur in the phase of the distorted wave. Since, however, the coupling is through a secondary order parameter the free energy includes a negative term in M^2, where M is the number of solitons. Consequently, as discussed in § I.4.5.2, the transition is always of first order because there is no positive term in M^2 in the soliton interactions.

Shiba and Ishibashi (1978) have also analysed this problem using a Fourier series expansion of the distortions Q_1, Q_2 and $\bar{Q}$. They find a continuous transition with properties very similar to that of soliton theory given in § I.4.5.2. We believe this is because their numerical work is not sufficiently accurate.

The lock-in transitions in both K_2SeO_4 (Iizumi *et al.* 1977) and $(ND_4)_2BeF_4$ (Iizumi and Gesi 1977) are of first order. In the former case, $p=3$, relatively strong first harmonics of the primary distortion were observed, while in the latter case, $p=2$, no harmonics have as yet been reported.

§ I.5. Microscopic theories

In this section we review the different microscopic models which may be used to calculate the phenomenological parameters of Landau theory. The statistical mechanics of the models is discussed in only the most elementary way, but we focus on the extent to which the models provide a more microscopic picture than Landau theory. Alas the very success of Landau theory in correlating experimental data means that any microscopic model with a few parameters may well be able to fit a large amount of data, but the model may or may not be a correct microscopic description. We therefore examine the different models critically and find that very few can be said to be a justifiable microscopic and quantitative description.

I.5.1. *The weakly anharmonic crystal models*

I.5.1.1. *The soft mode*

In crystals with large and temperature independent band gaps, such as $SrTiO_3$, structural phase transitions of the displacive type are almost certainly correctly described by the anharmonic crystal model. The theory of the weakly anharmonic crystal was developed in the early 1960s by Maradudin and Fein (1962) and by Maradudin and Flynn (1963) as reviewed by Cowley (1963). It was applied to ferroelectrics by Silverman and Joseph (1963) and by Cowley (1964).

The basic result is that the one-phonon susceptibility or scattering cross-section can be obtained from the expressions for a harmonic crystal if the harmonic frequencies $\omega(\mathbf{q}j)$ are replaced by

$$\tilde{\omega}(\mathbf{q}j,\omega)^2 = \omega(\mathbf{q}j)^2 + \sum(\mathbf{q}j,\omega), \qquad (I.5.1)$$

where $\Sigma(\mathbf{q}j,\omega)$ is the anharmonic part of the self-energy and depends on the frequency, ω, used to study the mode in an experiment.

There are three contributions to the self-energy to lowest order in the anharmonicity as shown in fig. I.46. The first contribution arises from the thermal strain, $\eta^T_{\alpha\beta}$, and is obtained from eqn. (I.2.10) as

$$\Sigma^T(\mathbf{q}j) = 2\sum_{\alpha\beta} V_{\alpha\beta}\begin{pmatrix}\mathbf{q} & -\mathbf{q}\\ j & j\end{pmatrix}\eta^T_{\alpha\beta}. \qquad (I.5.2)$$

This contribution can be obtained from measurements of the strain dependence of the frequencies of the mode and a knowledge of the thermal strain. Recently Samara *et al.* (1975) have shown that $\Sigma^T(\mathbf{q}j)$ decreases with increasing temperature for all the ferroelectric materials they studied, while it increases for all other types of transition. We are unaware of any reason to expect this to be a universally correct observation.

Fig. I.46

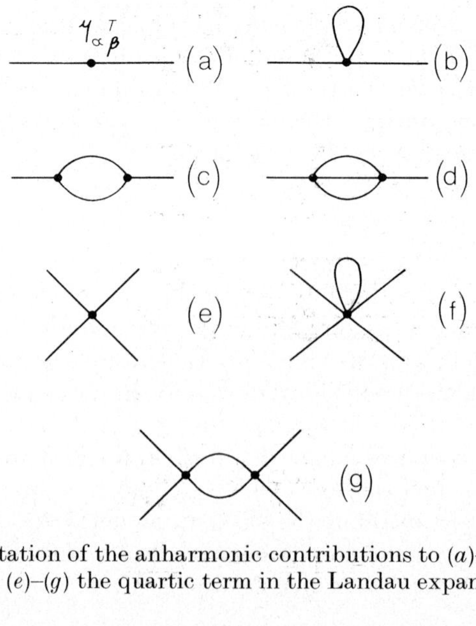

Diagrammatic representation of the anharmonic contributions to (*a*)–(*d*) the self-energy of a normal mode and (*e*)–(*g*) the quartic term in the Landau expansion of the free energy.

The second contribution to the self-energy in fig. I.46 arises from the quartic terms in the anharmonic expansion and contributes, using eqns. (I.2.5) and (I.2.7),

$$\Sigma^4(\mathbf{q}j)=6\hbar\sum_{\mathbf{q}_1 j_1} V\begin{pmatrix}\mathbf{q} & -\mathbf{q} & \mathbf{q}_1 & -\mathbf{q}_1\\ j & j & j_1 & j_1\end{pmatrix}\frac{(2n(\mathbf{q}_1 j_1)+1)}{\omega(\mathbf{q}_1 j_1)}, \tag{I.5.3}$$

which at high temperatures $\hbar\omega(\mathbf{q}_1 j_1)\ll k_B T$ is proportional to T. This term usually gives a contribution to the self-energy which increases with increasing temperature.

The third contribution shown in fig. I.46 arises from the cubic anharmonicity and yields

$$\Sigma^3(\mathbf{q}j,\omega)=-4{\cdot}5\hbar\sum_{\mathbf{q}_1\mathbf{q}_2 j_1 j_2}\left|V\begin{pmatrix}\mathbf{q} & \mathbf{q}_1 & \mathbf{q}_2\\ j & j_1 & j_2\end{pmatrix}\right|^2$$
$$\times(\omega(\mathbf{q}_1 j_1)\omega(\mathbf{q}_2 j_2))^{-1}[R(\omega)+i\pi S(\omega)], \tag{I.5.4}$$

where

$$R(\omega)=(n_1+n_2+1)((\omega_1+\omega_2+\omega)_p^{-1}-(\omega-\omega_1-\omega_2)_p^{-1})$$
$$+(n_1-n_2)((\omega-\omega_1+\omega_2)_p^{-1}-(\omega+\omega_1-\omega_2)_p^{-1})$$

and

$$S(\omega)=(n_1+n_2+1)(\delta(\omega-\omega_1-\omega_2)-\delta(\omega+\omega_1+\omega_2))$$
$$+(n_1-n_2)(\delta(\omega+\omega_1-\omega_2)-\delta(\omega-\omega_1+\omega_2)),$$

where $\omega_1=\omega(\mathbf{q}_1 j_1)$, etc. This term has both real and imaginary parts which depend on the frequency ω. The static real part $\Sigma^3(\mathbf{q}j,0)$ is negative and increases in magnitude as the temperature increases. The imaginary part is odd in ω and so at low frequencies can be written as proportional to ω.

The change in the free energy of the crystal/unit cell due to a static distortion, $Q(\mathbf{q}j)$, is given by (Cowley 1964)

$$G=G_0+\frac{1}{2}\tilde{\omega}(\mathbf{q}j,0)^2 Q(\mathbf{q}j)Q(-\mathbf{q}j). \tag{I.5.5}$$

This result shows that the free energy of the distorted crystal is determined by the same frequencies as the static susceptibility. They may therefore be equated with the frequencies of eqn. (I.4.16) as

$$\omega(\mathbf{k})=\tilde{\omega}(\mathbf{k}+\mathbf{q}_s j,0)$$

The calculation of the static frequencies, $\tilde{\omega}(\mathbf{q}j,0)$, is then a calculation of the parameters of the quadratic term in the Landau expansion, eqn. (I.4.1). The Landau theory is recovered if $\tilde{\omega}(\mathbf{q}_s j,0)^2=0$ at T_c, and is positive for higher temperatures; the change in $\tilde{\omega}(\mathbf{q}_s j,0)$ with temperature gives the parameter a.

These results imply firstly that the harmonic frequencies are imaginary, so that the crystal is only stabilized by the anharmonic terms. Secondly the Landau theory is obtained only if the anharmonic self-energy is analytically well behaved at T_c. The former is the case if the terms in the summations over $(\mathbf{q}_1 j_1)$ in eqns. (I.5.3) and (I.5.4) for which $(\mathbf{q}_1 j_1)$ are the soft modes make a negligible contribution to these summations. Strictly, this approximation is invalid close to T_c as discussed in detail in paper II, but Landau theory is valid if these terms are relatively small which is probably the case for temperatures not too close to T_c.

At low frequencies we can make use of the low frequency expansion of $\Sigma(\mathbf{q}_s j, \omega)$ to give

$$\tilde{\omega}(\mathbf{q}_s j, \omega)^2 = \omega(\mathbf{q}_s j)^2 + \Sigma(\mathbf{q}_s j, 0) - 2i\omega\gamma_0,$$

where γ_0 is a constant. This gives the one-phonon response function as a classical oscillator, eqn. (I.3.22), but is only valid for the range of ω for which it is possible to replace $\Sigma(\mathbf{q}_s j, \omega)$ by its low-frequency expansion. We shall discuss this further in paper III, but briefly eqn. (I.5.4) suggests that it is valid for frequencies $\omega <$ a typical acoustic mode zone boundary frequency provided that in eqn. (I.5.4) there are not many terms for which $\omega(\mathbf{q}_1 j_1) = \omega(\mathbf{q}_2 j_2)$.

The higher order terms in the Landau expansion can be obtained in a similar way as shown schematically in fig. I.45 for the quartic term. Since the leading terms in these expressions are independent of temperature, the Landau theory approximation, that u is independent of temperature, is clearly a reasonable approximation close to T_c. Weakly anharmonic perturbation theory provides explicit expressions for the parameters of Landau theory in terms of the anharmonic interactions and normal modes. There is, however, still a quantitative problem because in most simple materials, the anharmonic corrections to the frequencies are a few per cent for temperatures up to the Debye temperature. We need therefore to understand why the corrections are larger in materials undergoing structural phase transitions and to do so with reasonable interatomic interactions. In the next section we describe attempts to do this for $SrTiO_3$.

I.5.1.2. *Models of* $SrTiO_3$

There have been many attempts to develop microscopic models to describe the behaviour of $SrTiO_3$. Since, however, we are concerned in this section to examine if the behaviour of $SrTiO_3$ can be obtained using reasonable values of the interatomic forces we shall discuss only the most detailed model used, which happens (!) to have been developed largely by Bruce and Cowley (1973). Stirling (1972) measured the phonon dispersion relations for $SrTiO_3$ and fitted these to obtain the parameters of a shell model for the interatomic forces in $SrTiO_3$. This model which had short-range Ti–O and Sr–O interatomic forces and much weaker forces between neighbouring oxygen atoms, provided an accurate description of the frequencies and eigenvectors of the normal modes in $SrTiO_3$. The model was then extended to include the third and fourth derivatives of the interatomic forces between the Ti–O and Sr–O atoms. The third and fourth derivatives of the Sr–O potential and the third derivative of the Ti–O potential were fitted to the anharmonic terms in the Landau theory of the 110 K transition in $SrTiO_3$ as developed in §§ I.4.2 and I.4.3 and initially by Slonczewski and Thomas (1970). With these three parameters it was possible to obtain reasonable values of the Landau parameters and to calculate the different contributions to the self-energy of the soft mode. In fig. I.47 we show the frequency dependence of $\Sigma^3(\mathbf{q}_s j, \omega)$ at 300 K. It shows considerable structure but for $\omega < 3$ THz, the real part is well approximated by a constant and the imaginary part is almost linear in frequency as required for the classically damped oscillator to be an accurate description of the soft mode. The different contributions to the self-energy of the soft-mode are shown as a function of temperature in fig. I.48. As expected the thermal strain contribution is small, and Σ^3 and Σ^4 are both large and to a large extent cancel. It is clearly inappropriate to calculate one of these contributions without also including the others. The agreement between experiment and

Fig. I.47

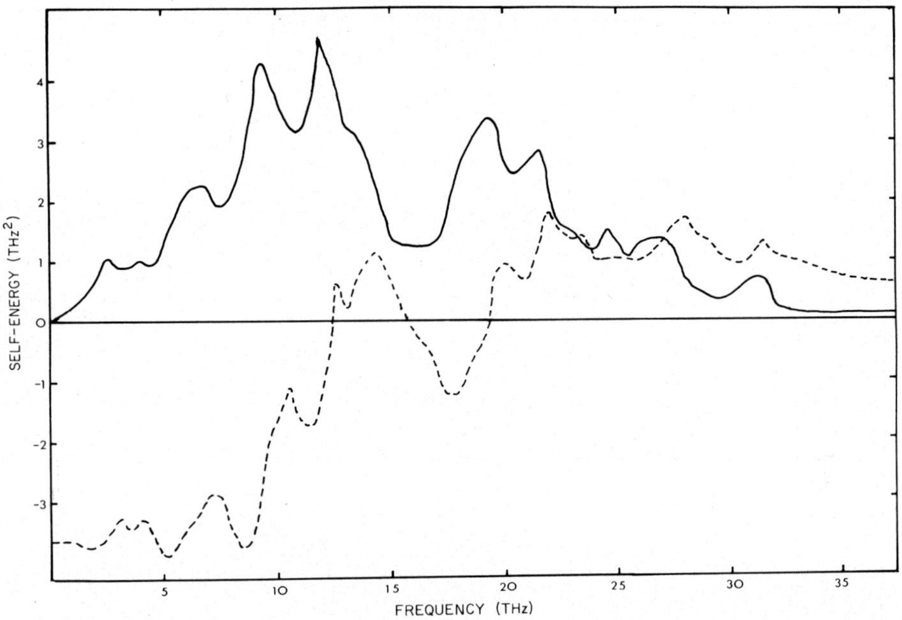

$\Sigma^3(\mathbf{q}j, \omega)$ of the R_{25} mode as a function of ω for $SrTiO_3$ at 300 K (after Bruce and Cowley 1975). The real part is a dotted line and the imaginary part a solid line.

calculation shown by fig. I.48 is very gratifying, as is the agreement of several other properties shown in table I.2. The theory also gives a good description of the widths of the infra-red active $\mathbf{q}=0$ modes as shown by the calculated reflectivity in fig. I.49. Since the numerical values of the anharmonic force constants were not larger than would be expected from the usual $1/r^{12}$ behaviour of short-range forces, the results of these calculations were very satisfactory and showed that at least for $SrTiO_3$, the weakly anharmonic model provided a satisfactory description of the zone boundary soft mode.

The temperature dependence of the $\mathbf{q}=0$ modes also involves the fourth derivative of the Ti–O force. When this was obtained from the observed temperature dependence its value turned out to be large and furthermore the other properties which depend upon it, such as the non-linear dielectric constant, were not given accurately by the model. Recently Migoni *et al.* (1976) have suggested that the strong two-phonon Raman scattering arises in a shell model from a very anharmonic shell-core coupling and further that this coupling contributes to the temperature dependence of the ferroelectric mode in perovskites. This may explain why our calculations of those effects which depend upon the fourth derivative of the Ti–O forces were unsatisfactory, but as yet Migoni *et al.* have not included the other anharmonic effects in their theory.

Although further calculations will no doubt be performed for $SrTiO_3$, we believe that these results have shown that anharmonicity is the origin of the temperature dependence of the soft modes in these materials. Furthermore the anharmonic parameters for the interatomic forces are very much the size to be expected from work on alkali halides. We are left then with the basic question as to why the effect of the anharmonicity is so much larger in $SrTiO_3$ than in, say, KBr. In $SrTiO_3$ the mean

Fig. I.48

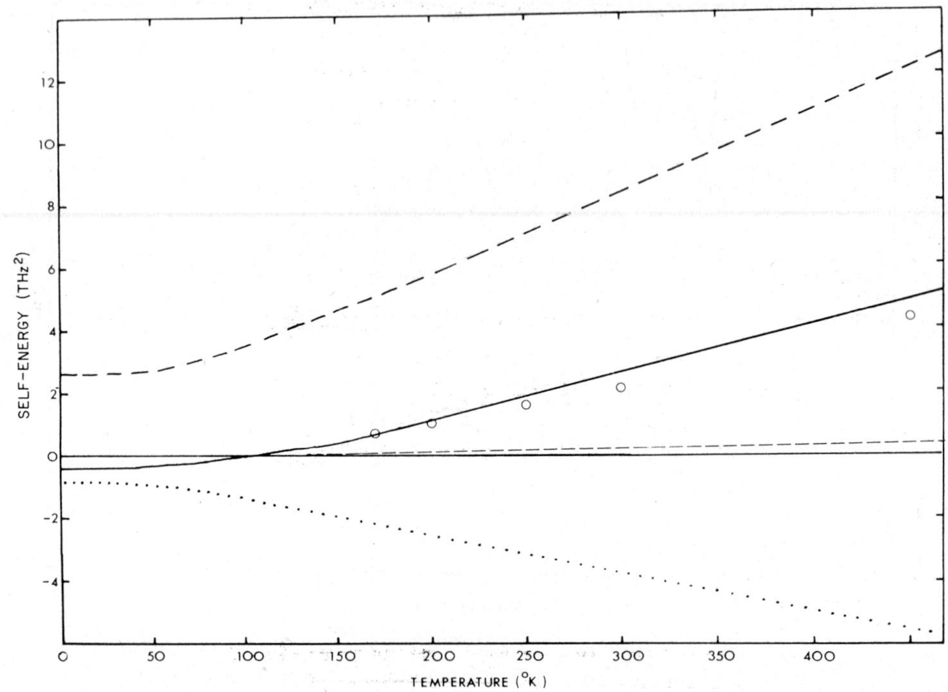

The different contributions to the self-energy of the R_{25} soft mode in $SrTiO_3$. The dashed line gives Σ^4, dotted line Σ^3, fine dashed line Σ^T, and the solid line $\tilde{\omega}^2$. The circles are experimental results. (After Bruce and Cowley 1975.)

Table I.2. Properties of $SrTiO_3$. (After Bruce and Cowley 1973.)

Quantity	Units	Experimental	Calculated
Order parameter, $Q_0^2/T_c - T$	10^{-5} Å^2 K^{-1}	0·165	0·167
Raman frequency ratio, ω_L^2/ω_T^2		9·4	9·25
Ratio of Raman frequency to frequency above T_c, $\omega_L^2/\omega(qj)^2$		2·4	2·33
$\dfrac{dT_c}{dp}$	10^{-2} K bar^{-1}	0·17	0·178
$\dfrac{dT_c}{dp_{(111)}}$	10^{-2} K bar^{-1}	0·33	0·294
Thermal expansion at 300 K	10^{-6} K^{-1}	10·0	8·76
Elastic distortion, $\dfrac{\eta}{T_c - T}$	10^{-5} K^{-1}	0·83	0·77
Gruneisen γ	10^3	5·67	5·17
Change in soft-mode self energy between 110 and 450 K	THz^2	4·3	4·86
Soft-mode damping constant γ_0 at 110 K	THz	0·1	0·06
Line width of $q=0$ optic modes at 85 K and 47 K	THz	0·09	0·08
		<0·06	0·01
		0·15	0·17

Fig. I.49

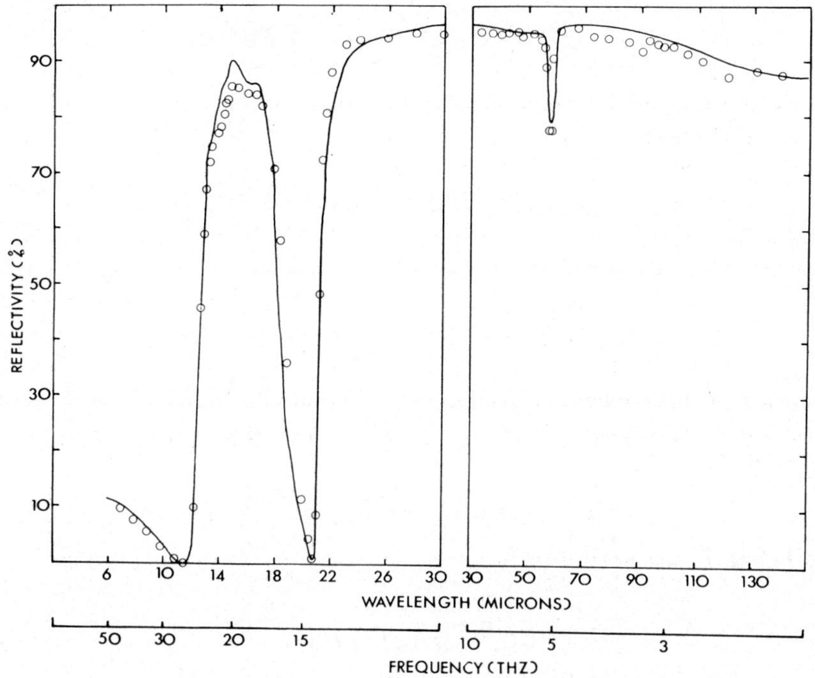

The reflectivity of $SrTiO_3$ compared with measurement of Spitzer *et al.* (1962). (After Bruce and Cowley 1975.)

frequency of the normal modes is about 8 THz, so that the self-energy effect for the soft mode at 450 K, fig. I.47, is only about 8% of the mean frequency squared. This is comparable to the anharmonic self-energy in alkali halides at these temperatures. Consequently in $SrTiO_3$ the anharmonic effects are *not* large, but the harmonic frequencies of the self mode are very small. Possibly this arises because perovskites have three types of ion and only one free parameter, the lattice constant. It is then very likely that one of the atoms is only a loose fit in its site and so can easily be displaced from its high-symmetry position. This type of argument would seem to be nearly as far as we can go in explaining why many perovskites exhibit structural phase transitions.

I.5.2. *Order–disorder systems*

I.5.2.1. *The time-dependent Ising model*

A spin representation for describing order–disorder systems was introduced in § I.2. We shall now discuss the calculation of the parameters of the Landau theory for this model and the time-dependent properties using the theory of Glauber (1963) and as developed by Suzuki and Kubo (1968). In their model each spin interacts with a heat bath, which induces spontaneous spin flips. It is then assumed that each spin relaxes back towards its local equilibrium configuration with a time constant, τ_0, which in the simplest form of the theory is independent of the configuration of the neighbouring spins.

In the interest of simplicity we consider a system with one spin in each unit cell, $S(l) = \pm\frac{1}{2}$, when the hamiltonian of the system in an external field $E(l)$ can be obtained from eqn. (I.2.11) as

$$H=2\sum_{ll'}\phi_2(ll')S(l)S(l')-\sum_{l}E(l)S(l),$$

where we have assumed that $\phi_1(l)$ is included in the external field. The local molecular field on the site l is

$$E_m(l)=E(l)-4\sum_{l'}\phi_2(ll')\langle S(l')\rangle$$

and hence within the simplest time-dependent model the equation of motion is

$$\tau_0\frac{d}{dt}\langle S(l)\rangle=-\langle S(l)\rangle+\frac{1}{2}\tanh(\beta E_m(l)/2). \tag{I.5.6}$$

The solution of this equation is well known. The susceptibility above T_c is obtained by expanding the tanh function to leading order when, after a Fourier transformation,

$$\chi(\mathbf{q},\omega)=\chi(\mathbf{q})(1+i\omega\tau(\mathbf{q}))^{-1},\quad T>T_c, \tag{I.5.7}$$

where the static susceptibility is

$$\chi(\mathbf{q})=\frac{1}{4(k_BT-J(\mathbf{q}))},$$

and the interaction $J(\mathbf{q})$ is given by

$$J(\mathbf{q})=-\sum_{l'}\phi_2(ll')\exp(i\mathbf{q}\,.\,(\mathbf{R}(l)-\mathbf{R}(l'))).$$

The relaxation time is given by

$$\tau(\mathbf{q})=\tau_0(1-\beta J(\mathbf{q}))^{-1}. \tag{I.5.8}$$

The static susceptibility, eqn. (I.5.7), shows that the phase transition will occur for the wave-vector $\mathbf{q}_s$ for which $J(\mathbf{q}_s)$ is a maximum and at the temperature T_c such that $k_BT_c=J(\mathbf{q}_s)$. The Debye relaxation form has a relaxation time, $\tau(\mathbf{q})$, which for $\mathbf{q}=\mathbf{q}_s$ diverges as $(T-T_c)^{-1}$.

We can obtain the parameters of the Landau theory for these systems by comparing these results for the susceptibility with those obtained in § I.4, and also by comparing the expressions for the order parameter, $Q_0=N^{-1/2}\langle S(\mathbf{q}_s)\rangle$ for $T<T_c$. The latter is obtained from eqn. (I.5.6) by expanding the tanh to third order terms to give

$$Q_0^2=-\frac{3(k_BT-J(\mathbf{q}_s))k_B^2T^2}{4J(\mathbf{q}_s)^3},$$

when we can identify the parameters of the Landau theory as

$$R(\mathbf{k})=4(k_BT-J(\mathbf{q}_s+\mathbf{k})), \tag{I.5.9}$$

and close to T_c

$$u=\frac{4}{3}k_BT_c. \tag{I.5.10}$$

Consequently the parameters of Landau theory can be obtained with this model once $J(\mathbf{q})$ is known, and the relaxation time τ_0. A feature of this model which is simpler than the weakly anharmonic crystal model is the way in which the same interaction $J(\mathbf{q}_s)$ determines both T_c and u, whereas in the other model there is no direct relationship between them.

I.5.2.2. *The tunnelling model*

In the introduction we described an order–disorder system by using a spin variable $S(l)$ which, for example, described the orientation of the NO_2 molecular groups. In some systems quantum-mechanical tunnelling between the two wells lifts the degeneracy of the two orientations and the wavefunctions are either the symmetric or antisymmetric linear combinations. This type of two level system can always be described by an $S=\frac{1}{2}$ spin system but the spin is now a vector. In this case the S^z component describes the location of the ion and the S^x component the tunnelling between the sites. A careful discussion of the derivation of the effective spin hamiltonian for the KH_2PO_4 system has been given by Tokunaga and Matsubara (1966) and is reviewed by Blinc and Zeks (1974). In the simplest case the hamiltonian of the tunnelling model with one spin/unit cell is

$$H=2\sum_{ll'}\phi_2(ll')S^z(l)S^z(l')+\sum_{l}(2\Omega S^x(l)-E(l)S^z(l)) \tag{I.5.11}$$

where Ω is the tunnelling integral, and we have assumed that the external field couples only to the z component of the spin.

This hamiltonian can be solved within the molecular field approximation as described in the last section and by Blinc and Zeks (1974). The molecular field is a vector

$$\mathbf{E}_m(l)=(-2\Omega,0,-4\sum_{l'}\phi_2(ll')\langle S^z(l')\rangle+E(l))$$

and the static susceptibility becomes

$$\chi(\mathbf{q})=\frac{1}{4(\Omega\coth\beta\Omega-J(\mathbf{q}))} \tag{I.5.12}$$

This shows that a phase transition occurs when $\coth\beta_c\Omega=J(\mathbf{q}_s)/\Omega$, and that the ordered phase is described by the wave-vector $\mathbf{q}_s$ for which $J(\mathbf{q})$ is a maximum. This result reduces to that of § I.5.2.1, namely $k_BT_c=J(\mathbf{q}_s)$ where Ω is small, and as the tunnelling integral becomes larger there is a decrease in T_c. This explanation for the transition temperature being much lower in KH_2PO_4 than in KD_2PO_4 was the original motivation for the development of the tunnelling model (Blinc 1960).

The parameters of the Landau theory are given by

$$\omega(\mathbf{k})^2=R(\mathbf{k})=4(\Omega\coth\beta\Omega-J(\mathbf{q}_s+\mathbf{k})), \tag{I.5.13}$$

which close to T_c becomes

$$R(\mathbf{k})=4\left[\left(\frac{\beta_c\Omega}{\sinh(\beta_c\Omega)}\right)^2k_B(T-T_c)-J(\mathbf{q}_s+\mathbf{k})+J(\mathbf{q}_s)\right], \tag{I.5.14}$$

showing that the parameter a depends on the tunnelling integral.

The quartic parameter, u, in eqn. (I.4.1) can be found by calculating the temperature dependence of the order parameter using the molecular field theory.

The result, after some algebra, is that

$$u = \frac{2J^3(\mathbf{q}_s)}{\Omega^2}\left[1 - \frac{\beta_c J(\mathbf{q}_s)}{\cosh^2(\beta_c\Omega)}\right] \tag{I.5.15}$$

which reduces to the result (I.5.10) in the $\Omega \to 0$ limit.

These results show that expressions for the Landau parameters may be obtained from the tunnelling model. The effect of the tunnelling is to decrease T_c and a, and to increase u.

In addition to these static results, the tunnelling model has also been used to calculate dynamical properties. Within the random phase approximation the Heisenberg equations of motion for the spins lead to a tunnelling frequency for $T > T_c$ (Brout *et al.* 1966, Blinc and Zeks 1974) given by

$$\hbar^2\omega_T(\mathbf{k})^2 = 4\Omega(\Omega - J(\mathbf{q}_s + \mathbf{k})\tanh\beta\Omega). \tag{I.5.16}$$

For $\mathbf{k} = 0$ this frequency decreases to zero as $T - T_c$, but is given by a different expression to the 'frequencies' of the Landau theory, eqn. (I.5.13). These tunnelling frequencies are quantum mechanical in origin and not the same as the Landau frequencies. This is because in the tunnelling model the spin is a vector and the tunnelling corresponds to the application of a field in the x direction. This causes a part of the spin to align along x, with a consequential reduction in the fraction which aligns along z, and hence in the susceptibility which depends directly only on the z component.

Before proceeding it is interesting to compare the tunnelling model with the Glauber model discussed in the previous section. In both models a mechanism is put forward by which the particle at site l changes its spin. In the Glauber model, this is by contact with a heat bath when a thermal fluctuation permits the particle to surmount the potential barrier between the two spin values, while in the tunnelling model the particle tunnels through the barrier quantum mechanically. Clearly the former process will tend to be the dominant process at high temperatures and with heavy molecular groups, while the latter will be the more likely at low temperatures and if only light atoms are moving at the transition. This is the case at a number of orientational transitions.

I.5.2.3. *Spin–phonon interaction*

In order–disorder materials there is often a coupling between the spin and the displacements of the atoms. For example, in $NaNO_2$, the force on a displaced Na ion is clearly dependent upon the orientation of the neighbouring NO_2 groups (fig. I.3). The interaction can be expanded as a double power series in the spins and the displacements, but we *assume* (although it is certainly not always valid) that only the leading term is important,

$$H_I = \sum_{\alpha l l'} W_\alpha(ll')u_\alpha(l)S(l'), \tag{I.5.17}$$

where for simplicity we are not including tunnelling effects in this section, and $W_\alpha(ll')$ gives the change in the force on atom l when the l' orientation is altered.

If the motion of the displacements is very much more rapid than that of the spins the displacements will take up average positions which are dependent on the particular spin coordinates. These average positions are most readily found by rewriting the interaction in terms of Fourier coordinates

$$H_I = \sum_{\mathbf{q}j} W(\mathbf{q}j) Q(\mathbf{q}j) S(-\mathbf{q}), \qquad (I.5.18)$$

when the instantaneous equilibrium positions of the displacements are given by

$$\tilde{Q}(\mathbf{q}j) = -\frac{W(\mathbf{q}j)}{\omega(\mathbf{q}j)^2} S(\mathbf{q}), \qquad (I.5.19)$$

and the normal modes of vibration are now oscillations about these displaced positions, but in this approximation their frequencies are unchanged.

Using eqn. (I.5.19) to eliminate the $Q(\mathbf{q}j)$ in the hamiltonian for H_I and the phonon hamiltonian, leads to an effective interaction between the spins which in Fourier space is given by

$$\tilde{J}(\mathbf{q}) = J(\mathbf{q}) - \frac{1}{8}\sum_j \frac{|W(\mathbf{q}j)|^2}{\omega(\mathbf{q}j)^2}. \qquad (I.5.20)$$

Unfortunately, this result is not completely correct because $J(\mathbf{q})$ is derived from the interactions between different sites l and l'. An interaction between a spin and itself cannot contribute to the ordering of the spins and consequently that part of the phonon modulated interaction must be subtracted from the effective interaction. The correct result (Kanamori 1960) is

$$\tilde{J}(\mathbf{q}) = J(\mathbf{q}) - \frac{1}{8}\sum_j \frac{|W(\mathbf{q}j)|^2}{\omega(\mathbf{q}j)^2} + \frac{1}{8N}\sum_{\mathbf{q}'j} \frac{|W(\mathbf{q}'j)|^2}{\omega(\mathbf{q}'j)^2}. \qquad (I.5.21)$$

The effect of the spin–phonon coupling is then to lead to a modification of the interaction between the spins. This is of particular importance in that the acoustic mode part of eqn. (I.5.21) may lead to a long-range interaction between the spins which falls off only as $1/R^3$ (Sugihara 1959).

Another consequence of the coupling is that the susceptibility or scattering is now a coupled phonon–spin problem. For example, the scattering cross-section from the spin variables, eqn. (I.2.45), is now modified to give an effective structure factor

$$\tilde{F}_s(\mathbf{K}, \mathbf{q}j) = F_s(\mathbf{K}, \mathbf{q}j) - \frac{W(\mathbf{q}j)}{\omega^2(\mathbf{q}j)} F(\mathbf{K}, \mathbf{q}j),$$

showing that there will be an interference between the scattering from the spin and the phonon modes.

Much of the theory discussed in this section was developed initially for coupled spin–phonon problems in magnetism. It was applied to order–disorder systems by Sakurai *et al.* (1970) and by Yamada *et al.* (1972) and to the tunnelling model by Kobayashi (1968).

Finally, in this section we have assumed that the normal modes of vibration could follow the spin coordinates. This is not possible for the long wavelength acoustic modes and in the tunnelling model may not be possible for other modes. The dynamics can then only be discussed once the dynamical response of both spin and normal mode systems is known and their coupling will give rise to effects similar to those described for coupled acoustic and optic modes in § I.4.1.5.

I.5.2.4. *Applications to real systems*

Throughout this article $NaNO_2$ has been used as an example of an order–disorder system. The first application of the order–disorder theory to $NaNO_2$ was by Yamada and Yamada (1966) to explain their diffuse X-ray scattering measurements (fig. I.19). The NO_2 groups have a dipole moment which may be aligned along the positive or negative b axis (fig. I.3). They calculated the dipolar interaction between the rigid dipoles and found that this part of $J(\mathbf{q}) = -D(\mathbf{q})$ gave a contribution with a maximum very close to the wave-vector of the incommensurate phase, and a transition temperature only 50% higher than the observed one (fig. I.19). When they introduced short-range forces between neighbouring dipoles they were able to produce a very good agreement with their experimental results with only three parameters specifying the three nearest neighbour forces.

Unfortunately further developments have questioned the validity of this agreement. In the ferroelectric phase of $NaNO_2$ the Na are displaced from the mid-points between the NO_2 ions (Kay 1972). This means that in addition to the direct dipole—dipole interaction there is also a phonon-modulated interaction as described in the previous section. When Sakurai *et al.* (1970) used models describing the normal modes of $NaNO_2$ to include this extra interaction between the NO_2 groups, they were no longer able to obtain a stable incommensurate phase. The situation is therefore unsatisfactory.

The dynamics of the ordering in $NaNO_2$ was studied by Hatta (1968) using dielectric measurements (fig. I.13). These results are well described by the model described in § I.5.2.1 with a relaxation time, τ_0, given by Eyring's reaction rate theory as

$$\tau_0 = \frac{2\pi\hbar}{k_B T} \exp(U_B/k_B T),$$

where U_B is the potential barrier between the two positions of the NO_2 groups. Assuming that U_B is weakly dependent upon temperature but approximately $7{\cdot}4 k_B T_N$ and taking $J(\mathbf{q})$ from the work of Yamada and Yamada (1966) described above, Yamada *et al.* (1968) obtained a good description of the experimental results as shown in fig. I.13.

The ammonium halides crystallize in the CsCl structure, and the tetrahedral NH_4 molecule has two possible orientations depending upon which four of the neighbouring halide ions are closest to the H atoms. These two orientations can be described by a spin variable as introduced in § I.2. The distorted phase of NH_4Br and NH_4I is the γ-phase which has an antiferrodistortive arrangement of the spins described by a wave-vector $\mathbf{q}_s = (\frac{1}{2}, \frac{1}{2}, 0)2\pi/a$. On the other hand NH_4Cl and NH_4Br at low temperatures form the δ-phase which is a ferrodistortive phase, $\mathbf{q}_s = 0$.

The direct interaction between the NH_4 ions is an octupole–octupole force and the corresponding $J(\mathbf{q})$ was calculated by Nagamiya (1943). The result is shown in fig. I.50 and favours the ferrodistortive phase. The X-ray structure determination of NH_4Br by Terauchi *et al.* (1972) showed that the Br ions become displaced at the transition. Consequently Yamada *et al.* (1972) developed the theory using the spin–phonon approach described in § I.5.2.3. Although the resulting theory inevitably involved a large number of parameters, it did show that the spin–phonon coupling favoured an antiferrodistortive ordering as illustrated in fig. I.50. The total effective exchange interaction $\tilde{J}(\mathbf{q})$ might then favour either ferrodistortive or antiferrodistortive distortions dependent upon the size of these two terms.

Fig. I.50

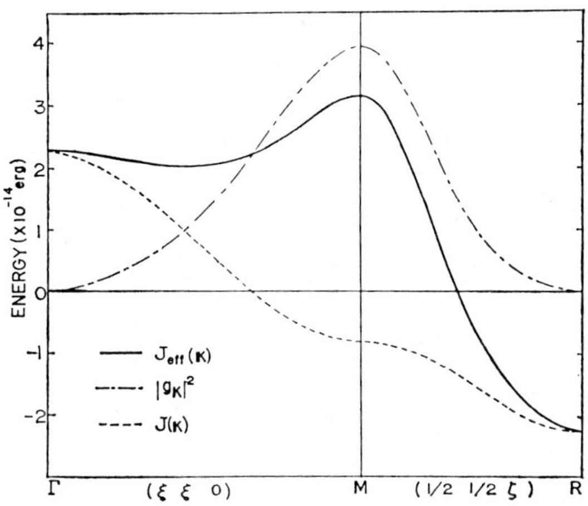

The exchange interaction between spins in the ammonium halide NH_4Br after Yamada *et al.* (1972). The direct interaction is given by - - - -, the spin-phonon by . . . ., and the total effective coupling by ————.

Another mechanism favouring the antidistortive structure was proposed by Hüller (1972), who suggested that the octupole on the NH_4 ion produced an electronic dipole on the neighbouring halide which then produced an electric field on the other NH_4 ions. This is clearly similar to the phonon mechanism except that the phonon mechanism produces an ionic dipole instead of an electronic dipole. At present the relative magnitude of these different effects is still uncertain, as reviewed by Vaks and Schneider (1976).

Possibly even more successful has been the theory of the phase transitions in methane. There are several phase transitions at which the different methane molecules within the unit cell order in position. James and Keenan (1959) suggested that the force was an octupole–octupole force and this gives the correct structure (Press 1972), and the anisotropy of the critical scattering (Hüller and Press 1972). Although the agreement with experiment is not perfect, it is sufficiently satisfactory to suggest that the nature of the force between the methane molecules is largely understood.

In other molecular materials such as N_2 the form of the intermolecular interaction and the effect of the electrostatic forces in polarizing the ions is less well understood. For some recent work on the problem of obtaining a potential for N_2, we refer to the work of Kjems and Dolling (1975) and of Raich and Gillis (1977).

Finally we comment on the validity of the tunnelling model for describing the KH_2PO_4 (KDP) class of materials. The original motivation behind the development of the tunnelling model was the ease with which it explains the large isotope effect (T_c for KH_2PO_4 is 123 K, and is 220 K for KD_2PO_4) since the change in mass will clearly make a large difference to the tunnelling integral. This was supported also by the results of dynamical Raman scattering experiments which showed for KH_2PO_4 an overdamped mode but with a considerable spectral width (fig. I.22), while for KD_2PO_4 the spectral width was at least an order of magnitude smaller (White *et al.* 1970). This is in agreement with the predictions of the tunnelling model if Ω is much larger for KH_2PO_4 than for KD_2PO_4, although the simple model fails to explain why the mode is overdamped. These successes were also supported by the result of the measurements of the decrease in the transition temperature with increasing pressure by Samara (1971) and of an underdamped soft mode at high pressure by Peercy (1973). These results were assumed to arise because the H–H distance decreased with increasing pressure and hence the overlap integral rapidly increases.

The difficulties for the theory are that *all* the atoms move in the ferroelectric motion (Skalyo *et al.* 1970), consequently there is a strong coupling between the hydrogen tunnelling and the other normal modes. Although this can in principle be included using the theory of Kobayashi (1968), this of necessity requires the introduction of many parameters, which makes agreement almost inevitable. Paul *et al.* (1970), in fact, conclude that it is not possible to obtain a good agreement between experiment and theory by assuming that the only effect of deuteration is to change the tunnelling integral. Furthermore Nelmes *et al.* (1978) have performed a structure determination under pressure which suggests that the difference between the minima in the H potential is not pressure dependent. If this result is confirmed it invalidates the simple tunnelling-model explanation of the pressure dependent results. In conclusion Nelmes *et al.* (1972) show that the structures of KH_2PO_4 and KD_2PO_4 differ so it is hardly to be expected that the model will be appropriate if only the tunnelling integral is assumed to depend on deuteration.

This leads us to the question of whether the tunnelling model is a good description of the KDP materials. The motion of the H atoms is certainly correlated with the motion of all the other atoms in the structure. This coupling alters the effective exchange coupling, § I.5.2.3 (Kobayaski 1968), but its effect on the tunnelling integral Ω is never discussed. Since the neighbours of an H atom move, the potential is locally not symmetric about the centre making the tunnelling to the other side more difficult. If, however, a cooperative tunnelling model is adopted the effective mass of all the atoms is clearly sufficiently large that tunnelling will be relatively unimportant compared with the hopping described by the Glauber model, § I.5.2.1. Certainly the results on KDP could be described by a soft mode with a large and very temperature dependent damping, and a frequency which was altered greatly on deuteration or by the application of pressure. Clearly further work is needed before we have a microscopic theory of these systems which lie between the displacive and order–disorder extremes.

I.5.3. *Electron–phonon models*

I.5.3.1. *The formalism*

It is frequently convenient to divide the interatomic force between two ions into two parts; a direct ion–ion interaction and an interaction mediated by the electrons, although the distinction between the ions and the other electrons is dependent upon the particular problem and assumptions. The calculation of the frequencies of the normal modes, or of the energies in the spin representation, is then a very complex subject as shown by the reviews in Horton and Maradudin (1974). In this section we attempt only to outline the essential points of the development with particular reference to the effects which may lead to structural phase transitions.

The potential seen by an electron in the crystal at the point, $\mathbf{r}$, may be expanded as a power series in the normal-mode coordinates

$$W(\mathbf{r}) = W^0(\mathbf{r}) + \sum_{\mathbf{q}j} W(\mathbf{q}j, \mathbf{r}) Q(\mathbf{q}j) + \dots \qquad \text{(I.5.22)}$$

The change in the potential at $\mathbf{r}$ due to a normal mode $(\mathbf{q}j)$, then causes a change in the electron distribution at the point $\mathbf{r}'$ which results in a change in the energy of the normal mode. If we assume that the symmetry is such that mode $\mathbf{q}j$ couples only to itself, then the electron–phonon contribution to its self-energy is

$$\Sigma^E(\mathbf{q}j) = -\iint W(\mathbf{q}j, \mathbf{r}) W(-\mathbf{q}j, \mathbf{r}') \chi^E(\mathbf{r}, \mathbf{r}')\, d\mathbf{r} d\mathbf{r}', \tag{I.5.23}$$

where $\chi^E(\mathbf{r}, \mathbf{r}')$ is the electron density response function. The frequency of the normal mode is given by

$$\omega(\mathbf{q}j)^2 = \omega^B(\mathbf{q}j)^2 + \Sigma^E(\mathbf{q}j), \tag{I.5.24}$$

where $\omega^B(\mathbf{q}j)$ is the contribution to the frequency arising from the direct ion–ion interaction, and also from the second-order terms in the expansion of $W(\mathbf{r})$ in powers of $Q(\mathbf{q}j)$.

It is often convenient to rewrite eqn. (I.5.23) in terms of Fourier components by making use of translational invariance when

$$\Sigma^E(\mathbf{q}j) = -\sum_{\boldsymbol{\tau}, \boldsymbol{\tau}'} W(\mathbf{q}+\boldsymbol{\tau}, j) W(-\mathbf{q}-\boldsymbol{\tau}', j) \chi^E(-\mathbf{q}-\boldsymbol{\tau}, \mathbf{q}+\boldsymbol{\tau}'), \tag{I.5.25}$$

where

$$W(\mathbf{q}+\boldsymbol{\tau}, j) = \int W(\mathbf{q}j, \mathbf{r}) \exp(-i(\mathbf{q}+\boldsymbol{\tau}) \cdot \mathbf{r})\, d\mathbf{r}. \tag{I.5.26}$$

It is of interest to compare this result with the developments of §§ I.4.1.6 and I.5.2.3. In the latter case $\omega(\mathbf{q}j)^2$ is the inverse susceptibility of the phonon normal mode with which the spins are interacting; eqn. (I.5.20) for $\tilde{J}$ is then of the same structure as eqn. (I.5.25). The extra complication of the sum over $\boldsymbol{\tau}$ and $\boldsymbol{\tau}'$ arises because the electron distribution is defined for all $\mathbf{r}$, whereas the phonon displacements or spin variables are defined only at the lattice points.

Equation (I.5.24) shows that a structural phase transition will result if Σ^E becomes sufficiently negative, while eqn. (I.5.25) shows that this may occur if either the electron–phonon interaction, W, is sufficiently large or the electronic susceptibility is sufficiently large. In particular eqn. (I.5.25) shows that whenever there is an instability of the electron gas, $\chi^E \to \infty$, then unless the electron–phonon coupling happens to be zero, there will always be a structural phase transition which occurs at a temperature somewhat in excess of the temperature at which the electronic system alone is unstable (cf. the difference between the clamped and free transition temperatures in § I.4.1.6). We are therefore concerned with the possible instabilities of the electronic system, which may take on a variety of different forms; for example, Jahn–Teller instabilities, charge density wave instabilities, spin density waves, Wigner crystallization and the Mott metal–insulator transition.

Before discussing some of these transitions in more detail, it is, however, useful to collect together some expressions (Sham 1974) for the electron–phonon interaction and the electronic susceptibility for systems with electron energy bands. If the electron–ion pseudo-potential is $v_\kappa(\mathbf{r} - \mathbf{r}(l\kappa))$ then the electron–phonon interaction is given by

$$W(\mathbf{q}j, \mathbf{r}) = \sum_{l\kappa} \left(\frac{1}{NM_\kappa}\right)^{1/2} \mathbf{e}(\kappa, \mathbf{q}j) \cdot \nabla_{\mathbf{R}(l\kappa)} v_\kappa(\mathbf{r} - \mathbf{R}(l\kappa)) \exp(i\mathbf{q} \cdot \mathbf{R}(l\kappa)). \tag{I.5.27}$$

If the electrons do not interact with one another and are described by Bloch waves with a wave-vector $\mathbf{k}$ and band index λ and have energy $\nu(\mathbf{k}\lambda)$, then the electronic susceptibility becomes

$$\chi^0(\mathbf{q}+\tau,\,-\mathbf{q}-\tau) = -\sum_{\lambda\lambda'}\sum_{\mathbf{k}} \frac{n(\mathbf{k}\lambda)-n(\mathbf{k}+\mathbf{q}\lambda')}{\nu(\mathbf{k}\lambda)-\nu(\mathbf{k}+\mathbf{q}\lambda')}$$

$$\langle\lambda\mathbf{k}|\exp(i(\mathbf{q}+\tau).\mathbf{r})|\mathbf{k}+\mathbf{q}\lambda'\rangle\langle\mathbf{k}+\mathbf{q}\lambda'|\exp(-i(\mathbf{q}+\tau).\mathbf{r})|\mathbf{k}\lambda\rangle, \quad \text{(I.5.28)}$$

where $n(\mathbf{k}\lambda)$ is the Fermi occupation number.

In practice electrons interact with one another through a potential whose Fourier transform we write as $v^E(\mathbf{q})$. Within the random-phase approximation the electronic susceptibility is then given by a matrix equation in $\boldsymbol{\tau}$ and $\boldsymbol{\tau}'$ which we can write schematically as

$$\chi^E = \chi^0 - \chi^0 v^E \chi^E. \quad \text{(I.5.29)}$$

The form of $v^E(\mathbf{q})$ is in general unknown: within the Hartree approximation $v^E(\mathbf{q})$ is given by the Fourier transform of the Coulomb interaction, but usually modifications are made to this form of $v^E(\mathbf{q})$ to incorporate the effects of exchange and correlation at least approximately.

I.5.3.2. *Jahn–Teller systems*

The simplest and most elegant examples of the cooperative Jahn–Teller effect occur in the rare earth zircons as reviewed in detail by Gehring and Gehring (1975). In these systems, $DyVO_4$, $TmVO_4$ and $TbVO_4$ for example, the $4f$ electrons on the rare earth ions are well localized and have a strong spin–orbit splitting. The lowest J multiplet is then split by the crystal field which, in the high-symmetry phase of these materials, is a tetrahedral field. In the simplest case, $TmVO_4$, the ground state of the Tm ion is a degenerate state whose degeneracy can be lifted by a distortion of the tetrahedral crystal field.

Since the $4f$ electrons are well localized this problem is particularly simple because we can calculate the electronic response, χ^E, for each rare earth ion independently when, using eqn. (I.5.23), we obtain

$$\Sigma^E(\mathbf{q}j) = -|W(\mathbf{q}j)|^2\chi^E.$$

In § I.5.2.1 we calculated the susceptibility of a two-level, spin $S=\frac{1}{2}$ system. We can use the results for $TmVO_4$ if we neglect the interactions between the different spins, when $\chi^E = \frac{1}{4}\beta$, where we have labelled the two electronic levels by the spin variable. The phase transition then occurs at the temperature T_c given by the maximum value of $|W(\mathbf{q}j)|^2/(4k_B\omega^B(\mathbf{q}j)^2)$, which also determines the wave-vector and branch index of the soft mode.

In the rare earth zircons the degeneracy of the electronic ground state is split by a shear of the surrounding tetrahedron, and the structure is such that the most energetically favourable mode is the acoustic shear mode determined by the elastic constant $C_{11}-C_{12}$. In other instances the soft mode is more complicated as for the Jahn–Teller distortion of the Cu ions in $K_2PbCu(NO_2)_6$ for which the unstable mode has an incommensurate wave-vector $\mathbf{q}_s = (0{\cdot}425, 0{\cdot}425, 0)\ 2\pi/a$ (Yamada 1977).

Jahn–Teller phase transitions may also occur even if the electronic ground state of the high-temperature phase is not degenerate. Consider a two-level system with a splitting Ω, while the distortion of crystal field causes transitions between these states. This is then an isomorphous problem with the tunnelling model discussed in § I.5.2.2, and the electronic susceptibility can be obtained from eqn. (I.5.12) with $J=0$. The high-temperature phase is stable only while

$$\Omega \coth \beta\Omega < \frac{W(\mathbf{q}j)^2}{4\omega^B(\mathbf{q}j)^2}.$$

which shows that there is a critical value of the splitting beyond which there is no ordering. This type of behaviour occurs in $DyVO_4$ and $TbVO_4$ although the crystal field structure is more complex than the doublet discussed above (Gehring and Gehring 1975). In magnetism this behaviour is known as an induced moment system (Grover 1965). This approach has been remarkably successful in explaining the properties of the rare earth zircons. Although it is not possible to calculate the electron–phonon interactions $W(\mathbf{q}j)$ from first principles, a very successful theory has been developed with relatively few adjustable parameters and this theory gives a good account of a very large quantity of experimental data, as reviewed by Gehring and Gehring (1975).

More complications arise in the theory of the rare earth pnictides because they are metals or semi-metals: the conduction electrons give rise to an interaction between the different rare earth ions. Qualitatively the theory can then be developed as above but using the full interacting susceptibilities of §§ I.5.2.1 and I.5.2.2 for the electronic susceptibility instead of the non-interacting susceptibilities used above, but it is then almost impossible to make calculations.

Finally this theory may be invalid for the $3d$ transition metal ions because in these systems there is a very strong coupling between the electronic levels and the surrounding ionic environment so that keeping only the leading terms in the expansion (I.5.22) is invalid. It is then possible to proceed by treating the interaction between an ion and its local environment using cluster methods, and coupling up the clusters using a mean field theory, as described by Thomas (1977). Of course difficulties then occur when the cluster around one $3d$ ion contains ions which are also part of the cluster surrounding another rare earth ion.

I.5.3.3. *Charge density waves*

In this section we discuss the structural phase transitions associated with charge density waves in metallic systems, as elegantly reviewed in more detail by Friedel (1977). The electronic susceptibility is dependent upon (eqn. (I.5.28))

$$I(\mathbf{q}) = -\sum_{\mathbf{k}\lambda\lambda'} \frac{n(\mathbf{k}\lambda) - n(\mathbf{k}+\mathbf{q}\lambda')}{\nu(\mathbf{k}\lambda) - \nu(\mathbf{k}+\mathbf{q}\lambda')}. \qquad \text{(I.5.30)}$$

This expression and hence χ^0 will become large if the electron energy denominators are small. In metallic systems this occurs particularly when the states $(\mathbf{k}+\mathbf{q}\lambda')$ and $(\mathbf{k}\lambda)$ are both at the Fermi level. Now, for any direction of $\mathbf{q}$, there will be a limiting value of $\mathbf{q}=\mathbf{K}_F$ such that for $\mathbf{q}>\mathbf{K}_F$ both $(\mathbf{k}+\mathbf{q}\lambda')$ and $(\mathbf{k}\lambda)$ may be at the Fermi surface but not for $\mathbf{q}<\mathbf{K}_F$ or vice versa. We therefore expect a special contribution to $I(\mathbf{q})$ when $\mathbf{q}$ is close to these wave-vectors. There are three cases of importance.

(*a*) If the two Fermi surfaces have a tangent plane but different curvatures, as found in a [$3d$] free-electron system, the $I(\mathbf{q})$ becomes

$$I(\mathbf{q}) = K_1 + K_2 p \log |p|, \qquad \text{(I.5.31)}$$

where K_1 and K_2 are constants while $\mathbf{q} \| \mathbf{K}_F$ and $p = |\mathbf{q} - \mathbf{K}_F|$. This shows that $I(\mathbf{q})$ has an infinite derivative (Kohn 1959) when $\mathbf{q} = \mathbf{K}_F$, and hence $I(\mathbf{q})$ cannot be a maximum for these wave-vectors. In this instance a charge density wave will not occur with $\mathbf{q}_s = \mathbf{K}_F$.

(*b*) If the two Fermi surfaces have a common tangent and one common curvature then

$$I(\mathbf{q}) = K_1 + K_2/(K_3 + p^{1/2}). \tag{I.5.32}$$

This is the form of the expression obtained for a $|2d|$ free-electron system and shows that there may be a maximum in $I(\mathbf{q})$ when $\mathbf{q} = \mathbf{K}_F$. This suggests that there may be a charge density wave instability with $\mathbf{q}_s = \mathbf{K}_F$.

(*c*) If the Fermi surfaces have a common tangent and two common curvatures, then

$$I(\mathbf{q}) = K_1 + K_2 \log |p|, \tag{I.5.33}$$

which is the divergent expression characteristic of a $[1d]$ electron system (Peierls 1955, Froehlich 1954). This suggests, neglecting electron–electron interactions, that a structural phase transition will always occur in this case, and the behaviour is known as an example of perfectly nesting Fermi surfaces.

These concepts originated with the work of Jones (1934) on the crystal structure of alloys. In our cases (*b*) and (*c*), the lattice distortion, if it occurs, opens up a gap in the electron energy spectra at the Fermi level. Consequently there are new Brillouin zone boundaries and a lowering of the total electron energy in the distorted phase.

Before discussing the application of these concepts in practice, it is necessary to mention the role of the electron–electron interactions. For simplicity we consider the case of nearly free electrons, when $\chi^E(\mathbf{q}+\boldsymbol{\tau}, -\mathbf{q}-\boldsymbol{\tau})$ becomes diagonal in the wave-vector, and furthermore we neglect the effect of Umklapp processes. We can then write eqn. (I.5.25) as

$$\Sigma^E(\mathbf{q}j) = -|W(\mathbf{q}j)|^2 \frac{I(\mathbf{q})}{1 + v^E(\mathbf{q})I(\mathbf{q})},$$

which then leads to a condition for the stability of a normal mode from eqn. (I.5.24) as (Chan and Heine 1973)

$$\frac{|W(\mathbf{q}j)|^2}{\omega^B(\mathbf{q}j)^2} - v^E(\mathbf{q}) < \frac{1}{I(\mathbf{q})}, \tag{I.5.34}$$

where $v^E(\mathbf{q})$ is the Fourier transform of the electron–electron interaction and in the Hartree approximation is positive and given by $e^2/\varepsilon_0 q^2$. Equation (I.5.34) has the unexpected property that in the absence of electron–phonon interaction, there would be no charge density instabilities of an electronic system because $I(\mathbf{q})$ is necessarily positive (Friedel 1977). It shows furthermore that charge density wave instabilities occur even in $[1d]$ systems only if $|W(\mathbf{q}j)|^2/\omega^B(\mathbf{q}j)^2$ is sufficiently large. A completely convincing theory of a structural phase transition can be carried out only if there is a careful calculation of the electron–phonon interaction $W(\mathbf{q}j)$, the susceptibility $I(\mathbf{q})$ and the electron–electron interactions $v^E(\mathbf{q})$.

In practice this programme has not been carried out to completion for any system. Many of the calculations have concentrated on calculating the core of the non-interacting electronic susceptibility, $I(\mathbf{q})$, by using simple models and more recently by using the results of detailed band structure calculations. In the case of the [1*d*] conductors such as KCP($K_2Pt(CN)_4Br_{0\cdot3}3H_2O$) (Comès *et al.* 1973 a) and TTF-TCNQ (Comès *et al.* 1973 b) the distorted structure has a wave-vector $\mathbf{K}_F$ along the chain direction which is exactly consistent with the expected value of $\mathbf{K}_F$ from the known electron density. Consequently there is no doubt that the distorted structure results from the peak in the electronic susceptibility for $\mathbf{q}=\mathbf{K}_F$ predicted by eqn. (I.5.33). Electron–electron interactions play a role however not only for the interactions along the chains, but also in giving rise to forces between the chains (Barisic 1971) without which there would not be a transition to a [3*d*] ordered structure. Clearly until the theory of the interacting [1*d*] electron gas is more perfectly understood we cannot hope to have a completely satisfactory theory of these materials.

The layered transition metal dichalcogenides have distorted phases described by incommensurate wave-vectors. These were initially assumed to arise from cylindrical parts of the Fermi surface associated with the layered structure, giving rise to a corresponding peak in $I(\mathbf{q})$. This model, however, fails to explain why different examples of the series have very similar periodicity for the charge density waves; $NbSe_2$ and $TaSe_2$ both have incommensurate phases with $\mathbf{q}_s=(1-\delta)a^*/3$ with $\delta\approx0{\cdot}02$. Furthermore calculations by Myron *et al.* (1977) of $I(\mathbf{q})$ using a self-consistent muffin tin potential for $TaSe_2$ show only a very weak peak for $\mathbf{q}\approx\mathbf{q}_s$, and similar calculations for TaS_2 give a peak at larger wave-vectors than $\mathbf{q}_s$ (fig. I.51). Clearly the nature of the electron–phonon interaction, and of the electron–electron interactions in these materials is of crucial importance.

In the A15 compounds, Nb_3Sn and V_3Si, the situation is also confused. The first model proposed to explain the transition was put forward by Labbe and Friedel (1966). Essentially their model describes the shear as arising from an inequivalence of the electron energy bands in the different directions of the distorted crystal. There is then a redistribution of the electrons and a consequent lowering of the energy. This effect is included in the calculation of $I(\mathbf{q})$ in eqn. (I.5.33), but is only large enough to give an instability if the density of states at the Fermi surface is large. Labbe and Friedel obtained a large enough density of states by postulating [1*d*]-like bands arising from the chains of metal atoms in the A15 structure. The symmetry of the A15 structure also causes a degeneracy of the electron energy bands at the [100] zone boundary. If the crystal distorts, this degeneracy is lifted and Gorkov (1973) suggested that if the Fermi surface was nearly [1*d*] and very close to the [100] zone boundary this splitting would account for the transition. Unfortunately Mattheiss's calculations (1965, 1975) of the band structure failed to substantiate these simplified models of the band structure and in more recent theories, such as those of Bhatt (1977), it is difficult to obtain a simple physical picture.

Calculations on other systems have been less ambitious.

I.5.3.4. *Other charge density waves*

Besides the charge density waves associated with the Fermi surface effects described above, there are a number of structural phase transitions associated with other types of electronic instability. In this section, we only briefly indicate some of these transitions.

The first example is the charge ordering which takes place in Fe_3O_4. In the high-temperature phase the iron ions have two sites; a tetrahedral site which is occupied

Fig. I.51

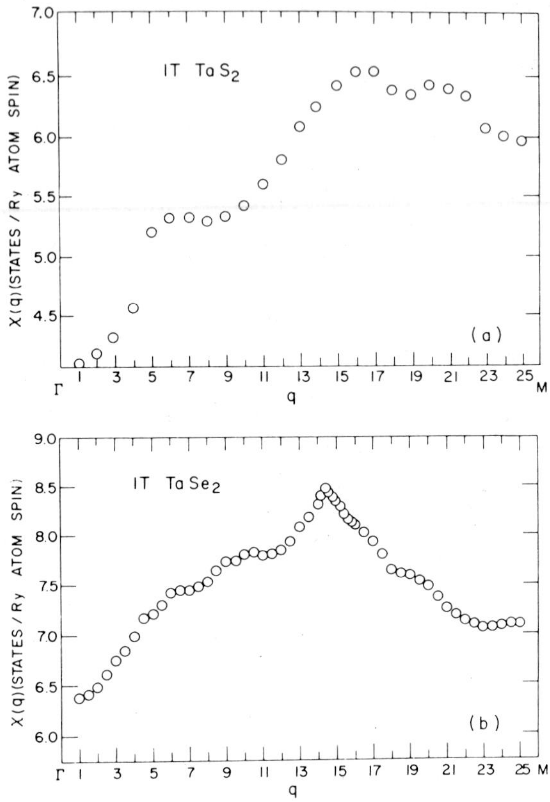

$I(\mathbf{q})=\chi(\mathbf{q})$ for $TaSe_2$ and TaS_2 for $\mathbf{q}$ along the ΓM direction. Whereas the curve for $TaSe_2$ shows a weak peak at the incommensurate wave-vector, that of TaS_2 does not. (After Myron *et al.* 1977.)

by an Fe^{3+} ion and two octahedral sites occupied by Fe^{2+} and Fe^{3+} ions. In the high-temperature phase the distribution of Fe^{2+} and Fe^{3+} ions is random but below the Verwey transition at 123 K the Fe^{2+} and Fe^{3+} ions take up an ordered arrangement (Verwey and Haayman 1941, Hamilton 1958). In both phases the electrons are localized but the hopping conductivity is drastically reduced at the phase transition. Since this phase transition is clearly associated with the charge distribution, there is a corresponding structural distortion. Both of these effects have been determined using neutron diffraction techniques as reviewed by Shirane (1977).

This transition is qualitatively not very different from a metal–insulator transition (Mott and Zinamon 1970). Basically in a system with one electron/atom, the correlation energy favours a situation in which each electron is situated on a different atom in an antiferromagnetic configuration. If the overlap with the neighbouring ions is sufficiently small there will be a gap between this insulating state and one in which the electrons can propagate through the crystal. This possible change in the electron wavefunctions causes a change in the electron density and hence may give rise to a structural phase transition.

Another example of where a change in the electronic properties gives rise to a structural phase transition is in the fluctuating valence systems, such as SmS,

Sm/YS or Ce/Th alloys. In these systems (Varma 1976) the atomic $4f$ levels are close in energy to the $5d$–$6s$ bands and as composition, temperature or pressure are varied, the electrons may take up either the localized $4f$ states or the band states. In SmS alloyed with YS, for example, at 19% of YS there is a first-order transition from a semiconducting phase with localized $4f$ electrons to a metallic phase of mixed $4f$ and $5d$–$6s$ states (Tao and Holtberg 1975). This transition is associated with a 12% decrease in volume. As discussed in § I.4.1.5, a bulk modulus instability has no critical fluctuations and is necessarily of first order due to cubic terms in the free energy. These conclusions are in excellent accord with the available experimental results for the acoustic modes in the Sm/YS system by Penney *et al.* (1976).

Finally we discuss magnetic transitions. If electron–phonon interactions are neglected then eqn. (I.5.34) suggests that electronic systems will only show phase transitions if $v^E(\mathbf{q})<0$. This is indeed the case for magnetic transitions because exchange interactions are attractive, and eqn. (I.5.34) with $W(\mathbf{q}j)=0$ is the well-known Stoner (1946) criterion for itinerant electron magnetism. The coupling to the phonons requires, however, more discussion. In the absence of spin–orbit interaction $W(\mathbf{q}j)=0$ and there is no linear coupling between the spin density and the phonon modes. The phonon modes play the role of secondary order parameters. Inclusion of the spin–orbit coupling allows a non-zero interaction, but this is frequently very much weaker than the corresponding coupling to charge density waves.

I.5.3.5. *Dynamic effects and temperature-dependent effects*

In § I.5.3.1 the expression for the electronic response was evaluated assuming that a static field was set up by the normal modes. In principle, it is comparatively easy to generalize this by expressing the results in terms of the dynamic electronic susceptibility, which is given by the same expression as eqn. (I.5.28) but with the denominator replaced by $\nu(\mathbf{k}\lambda)-\nu(\mathbf{k}+\mathbf{q}\lambda')+\omega$. This suggests that the dynamic response will only be significantly different from the static response if the frequencies ω of the normal modes are comparable with the difference in energy of the two electron states. In most instances nearly all the electronic energy denominators are comparable with the Fermi energy and consequently much larger than $\hbar\omega$ except for a very few electronic transitions between electron states close to the Fermi surface. In these cases there will be very little damping of the normal modes by the electronic excitations, and very little difference between the static and dynamic susceptibilities. There are two instances where this may not be the case. Firstly in [$1d$] conductors, KCP and TTF-TCNQ, the density of states at the Fermi surface is large and phonon damping may result as possibly illustrated in fig. I.40. Secondly in the insulating Jahn–Teller systems discussed in § I.5.3.2, the energies of the electronic excitations are relatively small and well defined (Gehring and Gehring 1975) so that coupled mode spectra are observed similar to the behaviour discussed in § I.4.1.6.

Finally we discuss the temperature dependence of the phonon frequencies due to the electronic susceptibility. In many theories such as those of the A15 materials and of the transition-metal dichalcogenides, attempts have been made to explain the temperature dependence of the normal modes as arising from the change in the Fermi factors in $\chi^E(\mathbf{q})$ (eqn. (I.5.28)). This would appear to be unreasonable except in [$1d$] conductors and insulating Jahn–Teller systems because the Fermi temperature is much higher than the transition temperature. It is surely more reasonable to explain the temperature dependence as arising from the anharmonicity described in § I.5.1 and to make use of the electronic terms only to explain the unstable zero temperature 'harmonic' frequencies of the normal modes.

This combined approach has now been shown to be the appropriate one in describing the temperature and dopant concentration dependence of the soft mode and phase transition in doped SnTe. Pure SnTe has the NaCl crystal structure at high temperature but below about 120 K has a phase transition to a ferroelectric rhombohedral phase (Kobayashi *et al.* 1976, Pawley *et al.* 1966). This phase-transition temperature is decreased as the carrier concentration is increased and reaches 0 K for a concentration of $8\times10^{26}\,\mathrm{m}^{-3}$. The soft mode has been studied below T_c for several samples by Sugai *et al.* (1977 b), fig. I.52. The first theory of the concentration dependence by Kristoffel and Konsin (1968) attempted to explain the results entirely in terms of the temperature dependence of the electronic response of the free carriers, but more recently Sugai *et al.* (1977 a) found it necessary to include anharmonicity as well. This is basically because the Raman spectra, fig. I.52, show that the frequency of the TO mode has much the same temperature dependence, irrespective of the concentration, and arises from the anharmonicity, while the frequency at zero temperature is carrier dependent. Finally we comment that phonon interactions were also found to be of crucial importance in explaining the observed temperature dependence in KCP (Sham and Patten 1976).

Fig. I.52

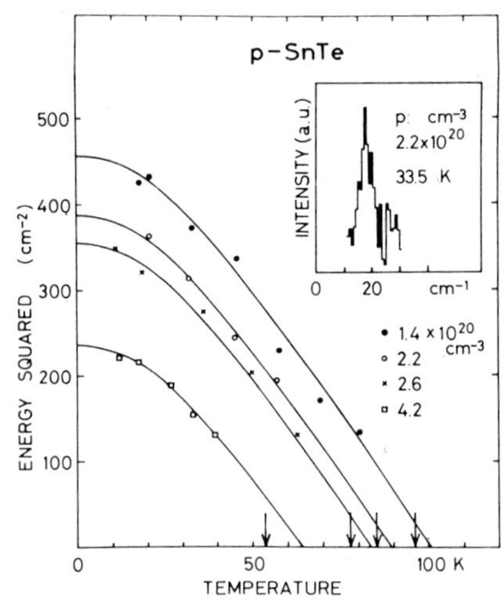

The temperature dependence of the TO frequency squared in doped SnTe. (After Sugai *et al.* 1977 a.)

I.5.4. *Conclusions*

The basic conclusions of this section have to be that the microscopic theory of structural phase transitions is in an unsatisfactory state. In part this reflects the very success of the Landau/soft mode theory which in terms of a few arbitrary parameters is able to correlate a very large amount of experimental data. It is then far too easy to invent a 'microscopic model' with a similar number of parameters which not surprisingly also correlates a large amount of experimental data. A microscopic model is only justified if either it predicts the parameters of a Landau theory T_c, a, $(\mathbf{q}_s j)$, etc., without introducing new parameters, or else also predicts the results of a large quantity of other microscopic data such as phonon-dispersion relations or optical reflectivity of a metal.

On these strict tests we have made very little progress. Except for the [$1d$] metals and, less convincingly, some order–disorder systems we can barely predict the wave-vector $\mathbf{q}_s$ with any confidence let alone the transition temperature, T_c. We hope that this review, by cataloguing our inadequacies, will stimulate further work to develop more satisfactory microscopic theories, in the hope not just of improving our understanding, but also that we may make better use of the important technological properties of materials associated with structural transitions.

References

Alefeld, B., Birr, M., and Heidemann, A., 1969, *Naturwissenschaften*, **8**, 410.

Als-Nielsen, J., and Dietrich, O. W., 1967, *Phys. Rev.*, **153**, 706, 711, 717.

Anderson, P. W., 1960, *Fizika Dielectrikov*, edited by G. I. Skanavi (Moscow: Acad. Nauk. SSR).

Aubry, S., 1977, *Proceedings of the Conference on Stochastic Behaviour in Classical and Quantum Systems*, Como, June 20–24.

Bak, P., 1978, *Proceedings of the Symposium on Soliton Structure and Dynamics in Condensed Matter*, Oxford, 22–29 June.

Bak, P., and Emery, V. J., 1976, *Phys. Rev. Lett.*, **36**, 978.

Bak, P., and Timonen, J., 1978, *J. Phys.* C, **11**, 4901.

Ballentyne, J. M., 1964, *Phys. Rev.* A, **429**, 136.

Barisic, S., 1971, *Solid St. Commun.*, **9**, 1507.

Barnoski, M. K., and Ballentyne, J. M., 1968, *Phys. Rev.*, **174**, 946.

Bechtle, D. W., and Scott, J. F., 1977, *J. Phys.* C, **10**, L209.

Bhatt, R. N., 1977, *Phys. Rev.* B, **16**, 1915.

Bhide, V. G., and Hegde, M. S., 1972, *Phys. Rev.* B, **5**, 3488.

Blinc, R., 1960, *J. Phys. Chem. Solids*, **13**, 204; 1971, *Structural Phase Transitions and Soft Modes*, edited by E. J. Samuelson, E. Anderson and J. Feder, (Oslo: Universitetsforlagel), pp. 97–132.

Blinc, R., and Mali, M., 1969, *Solid St. Commun.*, **7**, 1413.

Blinc, R., and Zeks, B., 1974, *Soft Modes in Ferroelectrics and Antiferroelectrics* (New York: Elsevier).

Born, M., and Huang, K., 1954, *Dynamical Theory of Crystal Lattices* (Oxford University Press).

Borsa, F., 1971, *Structural Phase Transitions and Soft Modes*, edited by E. J. Samuelson, E. Anderson and J. Feder, (Oslo: Universitetsforlaget), p. 371.

Borsa, F., and Rigamonti, A., 1972, *Physics Lett.* A, **40**, 399.

Brody, E. M., and Cummins, H. Z., 1968, *Phys. Rev. Lett.*, **21**, 1263.

Brout, R., Muller, K. A., and Thomas, H., 1966, *Solid St. Commun.*, **4**, 507.

Bruce, A. D., and Aharony, A., 1975, *Phys. Rev.* B, **11**, 478.

Bruce, A. D., and Cowley, R. A., 1973, *J. Phys.* C, **6**, 2422; 1978, *Ibid.*, **11**, 3609.

Bruce, A. D., Murray, A. F., and Cowley, R. A., 1978, *J. Phys.* C, **11**, 3591.

Carneiro, K., Shirane, G., Werner, S. A., and Kaiser, S., 1976, *Phys. Rev.*, **13**, 4258.

Chan, S. K., and Heine, V., 1973, *J. Phys.* F, **3**, 795.

Cochran, W., 1960, *Adv. Phys.*, **9**, 387; 1969, *Ibid.*, **18**, 157; 1971, *Structural Phase Transitions and Soft Modes*, edited by E. J. Samuelson, E. Anderson and J. Feder, (Oslo: Universitetforlaget), p. 1; 1973, *The Dynamics of Atoms in Crystals* (London: Edward Arnold).

Collins, M. F., and Teh, H. C., 1973, *Phys. Rev. Lett.*, **30**, 781.

Comès, R., Lambert, M., and Guinier, A., 1970, *Acta crystallogr.* A, **26**, 244.

Comès, R., Lambert, M., Launois, H., and Zeller, H. R., 1973 a, *Phys. Rev.* B, **8**, 571.

Comès, R., Lambert, M., and Zeller, H. R., 1973 b, *Phys. Stat. Sol.* (b), **58**, 587.

Comès, R., Shapiro, S. M., Shirane, G., Garito, A. F., and Heeger, A. J., 1976, *Phys. Rev.* B, **14**, 2376.

Cowley, R. A., 1963, *Adv. Phys.*, **12**, 421; 1964, *Phil. Mag.*, **11**, 673; 1968, *Rep. Prog. Phys.*, **31**, 123; 1976 a, *Phys. Rev.* B, **13**, 4877; 1976 b, *Phys. Rev. Lett.*, **36**, 744.

Cowley, R. A., Axe, J. D., and Iizumi, M., 1976, *Phys. Rev. Lett.*, **36**, 806.

COWLEY, R. A., and BRUCE, A. D., 1978, *J. Phys.* C, **11,** 3577.
COWLEY, R. A., BUYERS, W. J. L., and DOLLING, G., 1969, *Solid St. Commun.*, **7,** 181.
CUMMINS, S. E., 1970, *Ferroelectrics*, **1,** 11.
DARLINGTON, C. N. W., and O'CONNOR, D. A., 1976, *J. Phys.*, C, **9,** 3561.
DEVONSHIRE, A. F., 1949, *Phil. Mag.*, **40,** 1040; 1954, *Adv. Phys.*, **3,** 85.
DORNER, B., AXE, J. D., and SHIRANE, G., 1972, *Phys. Rev.* B, **6,** 1950.
DORNER, B., BAUER, K. H. W., and GRIMM, H., 1974, *Ferroelectrics*, **7,** 291.
DVORAK, V., 1974, *Ferroelectrics*, **7,** 1.
EBERHARD, J. W., and HORN, P. M., 1975, *Solid St. Commun.*, **16,** 1343.
FLEURY, P. A., SCOTT, J. F., and WORLOCK, J. M., 1968, *Phys. Rev. Lett.*, **21,** 16.
FLEURY, P. A., and WORLOCK, J. M., 1968, *Phys. Rev.*, **174,** 613.
FRANKE, V., and HEGENBARTH, E., 1974, *Phys. Stat. Sol.*, **25,** K17.
FRIEDEL, J., 1977, *Electron and Phonon Interactions and Phase Transitions*, edited by T. Riste (New York: Plenum).
FROEHLICH, H., 1954, *Proc. R. Soc.* A, **233,** 296.
GEHRING, G. A., and GEHRING, K. A., 1975, *Rep. Prog. Phys.*, **38,** 1.
GLAUBER, R. J., 1963, *J. Math. Phys.*, **4,** 294.
GORKOV, L. P., 1973, *Soviet Phys. JETP Lett.*, **17,** 379.
GROVER, B., 1965, *Phys. Rev.* A, **140,** 1944.
HAMILTON, W. C., 1958, *Phys. Rev.*, **110,** 1050.
HARADA, J., PEDERSEN, T., and BARNEA, Z., 1970, *Acta crystallogr.* A, **26,** 336.
HARLEY, R. T., and MACFARLANE, R. M., 1975, *J. Phys.* C, **8,** L451.
HATTA, I., 1968, *J. phys. Soc. Japan*, **24,** 1043.
HORNREICH, R. M., LUBAN, M., and SHTRIKMAN, S., 1975, *Phys. Rev. Lett.*, **35,** 1678.
HORTON, G. K., and MARADUDIN, A. A., 1974, *Dynamical Properties of Solids* (New York: Elsevier).
HOSHINO, S., and MOTEGI, H., 1967, *Jap. J. appl. Phys.*, **6,** 708.
HÜLLER, A., 1972, *Z. Phys.*, **254,** 456.
HÜLLER, A., and PRESS, W., 1972, *Phys. Rev. Lett.*, **29,** 266.
HUTCHINGS, M. T., 1964, *Solid State Physics 16* (New York: Academic Press), p. 227.
IIZUMI, M., AXE, J. D., SHIRANE, G., and SHIMAOKA, K., 1977, *Phys. Rev.* B, **15,** 4392.
IIZUMI, M., and GESI, K., 1977, *Solid St. Commun.*, **22,** 37.
INGLESFELD, J. E., 1972, *J. Phys.* F, **2,** 68.
JAMES, H. M., and KEENAN, T. A., 1959, *J. chem. Phys.*, **31,** 12.
JOHNSON, C. J., 1965, *Appl. Phys. Lett.*, **7,** 221.
JONA, F., and SHIRANE, G., 1962, *Ferroelectric Crystals* (New York: Pergamon).
JONES, H., 1934, *Proc. R. Soc.*, **144,** 225.
KAMINOW, I. P., and DAMEN, T. C., 1968, *Phys. Rev. Lett.*, **20,** 1105.
KANAMORI, J., 1960, *J. appl. Phys.*, **31,** 145.
KANZIG, W., 1951, *Helv. phys. Acta*, **24,** 175.
KAY, M. I., 1972, *Ferroelectrics*, **4,** 235.
KJEMS, J. K., and DOLLING, G., 1975, *Phys. Rev.* B, **11,** 1639.
KOBAYASHI, K. K., 1968, *J. phys. Soc. Japan*, **24,** 497.
KOBAYASHI, K. L. I., KATO, Y., KATAYAMA, Y., and KOMATSUBARA, K. F., 1976, *Phys. Rev. Lett.*, **37,** 772.
KOHN, W., 1959, *Phys. Rev. Lett.*, **2,** 393.
KRISTOFFEL, N. and KONSIN, P., 1968, *Phys. Stat. Sol.*, **28,** 731.
KWOK, P. C., and MILLER, P. B., 1966, *Phys. Rev.*, **151,** 387.
LABBÉ, J., and FRIEDEL, J., 1966, *J. Phys., Paris*, **27,** 153, 303, 708.
LANDAU, L. D., 1937, *Phys. Z. SowjUn*, **11,** 26.
LANDAU, L. D., and LIFSHITZ, E. M., 1959, *Statistical Physics* (London: Pergamon).
LINES, M. E., and GLASS, A. M., 1977, *Principles and Applications of Ferroelectrics and Related Materials* (Oxford University Press).
LÜTHI, B., and MORAN, T. J., 1970, *Phys. Rev.*, **2,** 1211.
MCKENZIE, D. R., 1975, *J. Phys.* C, **8,** 1607.
MCMILLAN, W. L., 1976, *Phys. Rev.* B, **14,** 1496.
MARADUDIN, A. A., and FEIN, A. E., 1962, *Phys. Rev.*, **128,** 2589.
MARADUDIN, A. D., and FLYNN, P. A., 1963 a, *Phys. Rev.*, **129,** 2529; 1963 b, *Ann. Phys.*, **22,** 223.

MATTHEISS, L. F., 1965, *Phys. Rev.* A, **138,** 112; 1975, *Phys. Rev.* B, **12,** 2161.
MEGAW, H. D., 1947, *Proc. R. Soc.*, **189,** 261.
MERZ, W. J., 1953, *Phys. Rev.*, **91,** 513.
MEYER, G. M., and COWLEY, R. A., 1976, *Ferroelectrics*, **11,** 479.
MEZEI, F., 1972, *Z. Phys.*, **255,** 146.
MIGONI, R. BILZ, H., and BAÜERLE, D., 1976, *Phys. Rev. Lett.*, **37,** 1155.
MONCTON, D. E., AXE, J. D., and DI SALVO, F. J., 1975, *Phys. Rev. Lett.*, **34,** 734.
MOTT, N. F., and ZINAMON, Z., 1970, *Rep. Prog. Phys.*, **33,** 881.
MÜLLER, K. A., 1971, *Structural Phase Transitions and Soft Modes*, edited by E. J. Samuelson, E. Anderson and J. Forbes, (Oslo: Universitetsforlaget), pp. 61–96.
MÜLLER, K. A., and BERLINGER, W., 1971, *Phys. Rev. Lett.*, **26,** 13.
MÜLLER, K. A., BERLINGER, W., and SLONCZEWSKI, J. C., 1970, *Phys. Rev. Lett.*, **25,** 734.
MÜLLER, K. A., BERLINGER, W., and WALDNER, F., 1968, *Phys. Rev. Lett.*, **21,** 814.
MYRON, H. W., RATH, J., and FREEMAN, A. J., 1977, *Phys. Rev.*, B, **15,** 885.
NAGAMIYA, T., 1943, *Proc. phys. math. Soc. Japan*, **24,** 137; 1944, *Ibid.*, **25,** 540.
NAKANISHI, K., and SHIBA, H., 1977, *J. phys. Soc. Japan*, **43,** 1839.
NAKANISHI, K., TAKATERA, H., YAMADA, Y., and SHIBA, H., 1977, *J. phys. Soc. Japan*, **43,** 1509.
NELMES, R. J., EIRIKSSON, V. R., and ROUSE, R. D., 1972, *Solid St. Commun.*, **11,** 261.
NELMES, R. J., KENNEDY, N. S. J., BAHARIE, E., and HEWAT, A. W., 1978, *Lattice Dynamics*, edited by M. Balkanski (Paris: Flammarion), p. 705.
ORIANI, R. A., and MURPHY, W. K., 1958, *J. Phys. Chem. Solids*, **6,** 277.
OVERHAUSER, A. W., 1971, *Phys. Rev.* B, **10,** 3173.
PAUL, G. L., COCHRAN, W., BUYERS, W. J. L., and COWLEY, R. A., 1970, *Phys. Rev.* B, **2,** 4603.
PAWLEY, G. S., COCHRAN, W., COWLEY, R. A., and DOLLING, G., 1966, *Phys. Rev. Lett.*, **17,** 753.
PEERCY, P. S., 1973, *Phys. Rev. Lett.*, **31,** 379.
PEIERLS, R. E., 1955, *Quantum Theory of Solids* (Oxford University Press).
PENNEY, T., MELCHER, R. L., HOLTZBERG, F. and GUNTHERODT, G., 1976, *A.I.P. Conference Proceedings*, **29,** 392.
PRESS, W., 1972, *J. chem. Phys.*, **56,** 2597.
PYNN, R., 1978, *J. Phys.* F, **8,** 1.
PYNN, R., and AXE, J. D., 1976, *J. Phys.* C, **9,** L199.
RAICH, J. C., and GILLIS, N. S., 1977, *J. chem. Phys.*, **66,** 846.
RAMAN, C. V., and NEDUNGADI, T. M. K., 1940, *Nature, Lond.*, **145,** 147.
REHWALD, W., 1973, *Adv. Phys.*, **22,** 721.
RISTE, T., SAMUELSON, E. J., and OTNES, K., 1971, *Structural Phase Transitions and Soft Modes*, edited by E. J. Samuelson, E. Anderson and J. Feder, (Oslo: Universitetsforlaget), pp. 395–408.
SAKSENA, B. D., 1940, *Proc. natn. Acad. Sci. India*, A, **12,** 93.
SAKURAI, J., COWLEY, R. A., and DOLLING, G., 1970, *J. phys. Soc. Japan*, **28,** 1426.
SAMARA, G. A., 1971, *Phys. Rev. Lett.*, **27,** 103.
SAMARA, G. A., SAKUDO, T., and YOSHIMITSU, K., 1975, *Phys. Rev. Lett.*, **35,** 16.
SAWADA, A., UDAGAWA, M., and NAKAMURA, T., 1977, *Phys. Rev. Lett.*, **39,** 829.
SCOTT, J. F., 1974, *Rev. mod. Phys.*, **46,** 83.
SCOTT, A. C., CHU, F. Y. E., and MCLAUCHLIN, D. W., 1973, *Proc. Inst. electron. Engrs*, **61,** 1443.
SHAM, L. J., 1974, *Dynamical Properties of Solids*, Vol. 1, edited by G. K. Horton and A. A. Maradudin (New York: Elsevier), p. 301.
SHAM, L. J., and PATTON, B. R., 1976, *Phys. Rev. Lett.*, **36,** 733; 1976, *Phys. Rev.* B, **13,** 315.
SHAPIRO, S. M., COWLEY, R. A., COX, D. E., and EIBSCHULTZ, M., 1976, Oak Ridge Laboratory Report Conf-760601-Pl, p. 399.
SHENOY, G. K., 1973, *Prospectives in Mössbauer Spectroscopy* (New York: Plenum Press), p. 141.
SHIBA, H., and ISHIBASHI, Y., 1978, *J. phys. Soc. Japan*, **44,** 1592.
SHIRANE, G., 1974, *Rev. Mod. Phys.*, **46,** 437; 1977, *Electron–Phonon Interaction and Phase Transitions*, edited by T. Riste (New York: Plenum).
SHIRANE, G., JONA, F., and PEPINSKY, R., 1955, *Proc. Inst. Radio Engrs*, **43,** 1738.

SHIRANE, G., and YAMADA, Y., 1969, *Phys. Rev.*, **177,** 858.

SILVERMAN, B. D., and JOSEPH, R. I., 1963, *Phys. Rev.*, **129,** 2062; 1964, *Ibid.*, **133,** A207.

SKALYO, J., FRAZER, B. C., and SHIRANE, G., 1970, *Phys. Rev.* B, **1,** 278.

SLONCZEWSKI, J. C., and THOMAS, H., 1970, *Phys. Rev.* B, **1,** 3599.

SPITZER, W. G., MILLER, R. C., KLEINMAN, D. A., and HOWARTH, L. E., 1962, *Phys. Rev.*, **126,** 1710.

STANLEY, H. E., 1971, *Introduction to Phase Transitions and Critical Phenomena* (Oxford University Press).

STIRLING, W. G., 1972, *J. Phys.* C, **5,** 2711.

STONER, E. C., 1946, *Rep. Prog. Phys.*, **11,** 43.

SUGAI, S., MURASE, K., KATAYAMA, S., TAKAOKA, S., NISHI, S. and KAWAMURA, H., 1977 a, *Solid St. Commun.*, **24,** 407.

SUGAI, S., MURASE, K., and KAWAMURA, H., 1977 b, *Solid St. Commun.*, **23,** 127.

SUGIHARA, K., 1959, *J. phys. Soc. Japan*, **14,** 1231.

SUZUKI, M., and KUBO, R., 1968, *J. phys. Soc. Japan*, **24,** 51.

TAO, L. J., and HOLTZBERG, F., 1975, *Phys. Rev.* B, **11,** 3842.

TERAUCHI, H., NODA, Y., and YAMADA, Y., 1972, *J. phys. Soc. Japan*, **32,** 1560.

THOMAS, H., 1977, *Electron–Phonon Interactions and Phase Transitions*, edited by T. Riste (New York: Plenum).

TOKUNAGA, M., and MATSUBARA, T., 1966, *Prog. theor. Phys., Kyoto*, **35,** 581.

UNOKI, H., and SAKUDO, T., 1967, *J. phys. Soc. Japan*, **23,** 546.

VAKS, V. G., LARKIN, A. I., and PIKIN, S. A., 1967, *Zh. éksp. teor. Fiz.*, **53,** 1089 (English translation: 1968 *Soviet Phys. JETP*, **26,** 647).

VAKS, U. G., and SCHNEIDER, J. E., 1976, *Phys. Stat. Sol.* (*a*), **35,** 61.

VAN HOVE, L., 1954, *Phys. Rev.*, **95,** 249.

VARMA, C. M., 1976, *Rev. mod. Phys.*, **48,** 219.

VERWEY, E. J. W., and HAAYMAN, P. W., 1941, *Physics*, **8,** 977.

WADA, M., SAWADA, A., ISHIBASHI, Y., and TAKAGI, Y., 1977, *J. phys. Soc. Japan*, **42,** 1229.

WAGNER, H., 166, *Z. Phys.*, **195,** 273.

WHITE, K. I., TAYLOR, W., KATIYAR, R. S., and KAY, S. M., 1970, *Physics Lett.* A, **33,** 175.

WORLOCK, J. M., and OLSON, D. H., 1971, *Proceedings of the International Conference on Light Scattering* (Paris: Flammarion), p. 410.

YAMADA, Y., 1977, *Electron–Phonon Interactions and Phase Transitions*, edited by T. Riste (New York: Plenum).

YAMADA, Y., FUJII, Y., and HATTA, I., 1968, *J. phys. Soc. Japan*, **24,** 1053.

YAMADA, Y., MORI, M., and NODA, Y., 1972, *J. phys. Soc. Japan*, **32,** 1565.

YAMADA, Y., and YAMADA, T., 1966, *J. phys. Soc. Japan*, **21,** 2167.

II

Static Critical Behaviour

By
A. D. BRUCE

Summary

The fundamental inconsistencies inherent in classical theories of structural phase transitions are discussed as a prelude to a review of the contemporary theory, emphasizing the concepts and techniques that constitute the renormalization group theory of static critical phenomena. The theory's predictions for critical behaviour at structural phase transitions are surveyed and compared with experiment.

§II.1. Introduction

Part I (hereafter referred to as I) surveyed the broad spectrum of systems exhibiting structural phase transitions, and showed how the gross features of their behaviour can be understood within the framework of classical Landau theory. In recent years more refined analyses of structural phase transitions have revealed a number of new features that have exposed the inadequacy of this theory on three fronts.

Firstly, electron paramagnetic resonance (E.P.R.) experiments by Müller and Berlinger (1971), on the structural phase transitions occurring in $SrTiO_3$ and $LaAlO_3$, provided the first unequivocal evidence of deviations from the classical predictions for static critical behaviour at structural phase transitions, and established a value for the order parameter exponent of $\beta = 0{\cdot}33 \pm 0{\cdot}02$, for these systems.

Secondly, inelastic neutron scattering experiments by Riste *et al.* (1971) and Shapiro *et al.* (1972), also on $SrTiO_3$, revealed a more complex behaviour in the soft-mode spectral function than had been suggested by previous experiments. In contrast to the simple soft-mode picture, the spectral function was found to exhibit, in addition to the anticipated soft-mode side band, a narrow peak centred on zero frequency-transfer. The intensity of this 'central component' was observed to diverge with the onset of criticality, while the frequency locating the soft-mode side band decreased to a *finite* value at T_c.

These observations were themselves partially responsible for the third development—renewed theoretical activity in the investigation of simple models for structural phase transitions, with particular emphasis on the strong non-linearity inherent in such systems (Schneider and Stoll 1973, 1975, Krumhansl and Schrieffer 1975, Aubry 1975, 1976, Varma 1976). These investigations confirmed that the behaviour of systems undergoing structural phase transitions can be considerably more subtle than is suggested by the soft-mode picture. They drew attention, in particular, to the role of the clusters of local (short-range) order that must inevitably form as the precursors (as $T \to T_c^+$) of the long-range order that sets in at the phase transition itself.

Though interrelated through the theme of non-linearity, these developments have opened up too extensive a territory to be surveyed in the one article. Thus we defer to Part III (hereafter referred to as III) a discussion of the predominantly dynamic issues associated with the central peak and cluster phenomena, and in this part focus upon the field of static critical behaviour.

We shall place particular emphasis on the insights into the critical region that have been acquired with the development of renormalization group (R.G.) methods (Wilson 1971, Wilson and Kogut 1974) whose impact on the field of structural phase transitions ranks with that of the developments cited above. For, within the R.G. framework, one finds a basis for the notion of universality, and with it, not only a unification of theories of critical phenomena at magnetic and structural phase changes, but also the power to make clear predictions for critical phenomena in real (complex!) systems—not just idealized (simple!) models.

More specifically the outline of this article is as follows. We begin, in § II.2, by introducing the simplest-conceivable microscopic model for a system exhibiting a structural phase transition. Its simplicity—belied by the richness of the problems it presents—is largely accountable for the frequency with which it has appeared in the literature of structural phase transitions: it has formed the basis of studies of the 'precursor cluster' problem and of the central peak phenomenon, and forms one of the basic ingredients of renormalization group theories of *all* structural phase transitions.

Having described those properties of the model that are established exactly, we shall use it, in § II.3, to illustrate the two classical decoupling schemes commonly employed to make the structural phase transition problem tractable, and to expose the origins of the failures that they share with the phenomenological theory of I—particularly, close to the critical point.

This will bring us naturally to a discussion (§ II.4.1) of the distinctive feature of the critical region—that of 'universality'—and of the way it is exploited in current theories of critical behaviour, specifically in the methods of the renormalization group. Our primary intention here (realized in §§ II.4.2–3) is to convey sufficient of the underlying principles and terminology so that the hitherto uninitiated reader will find the subsequent discussion intelligible. However, we shall also attempt, in the remaining part of § II.4, to indicate some of the practical problems encountered in implementing the theory—we outline one worked example—and to survey the bizarre assembly of techniques with which it has proved necessary to supplement the basic R.G. method.

The reader who is familiar with the language of the R.G., and whose primary interests lie in its applications, may reasonably skirt much of § II.4 and proceed directly to § II.5, where we shall survey the theory's predictions for static critical behaviour, and review the extent to which they have been borne out by experiments on structural phase transitions.

§ II.2. A microscopic model for a structural phase transition

The phenomena described briefly in the preceding section reveal a richness in the behaviour of systems undergoing structural phase transitions that is not illuminated by a purely phenomenological theory. In an effort to acquire the insights which the soft mode phenomenology fails to supply there have, in recent years, been numerous analyses of simple microscopic models exhibiting structural phase transitions. In this section we shall introduce the simplest and most widely-studied of these models. The insights which it affords make it worthy of study in its own right, and we shall have occasion to refer back to it in the discussion within III. Our primary purpose in introducing the model at this stage, however, lies in the fact that it provides a particularly simple framework in which to illustrate the two principal approximation schemes currently employed in microscopic theories, and to reveal the origin of their inadequacies close to a structural phase transition.

II.2.1. *Definition of the model*

The simple model whose properties we shall review is depicted schematically in fig. II.1, and is defined formally by the classical hamiltonian

$$H = T(\{p\}) + V(\{u\}), \qquad \text{(II.2.1 } a\text{)}$$

with

$$T = \frac{1}{2}\sum_l p^2(l) \qquad \text{(II.2.1 } b\text{)}$$

and

$$V = \sum_l V_s(u(l)) + \frac{C}{2}\sum_{l,l'}^{\text{n.n.}} (u(l) - u(l'))^2 - E\sum_l u(l). \qquad \text{(II.2.1 } c\text{)}$$

Here the coordinates $u(l)$ and $p(l)$ describe, respectively, the displacements and momenta of a set of atoms, of unit mass, whose high-temperature-phase equilibrium positions (with respect to which the displacements are defined) consist of the N sites of a d-dimensional hypercubic lattice, of period a. The atoms are supposed to interact through an harmonic nearest-neighbour coupling (the 2nd term in (II.2.1 c)) and to reside in a double-well single-particle potential V_s. This potential may be thought of

Fig. II.1

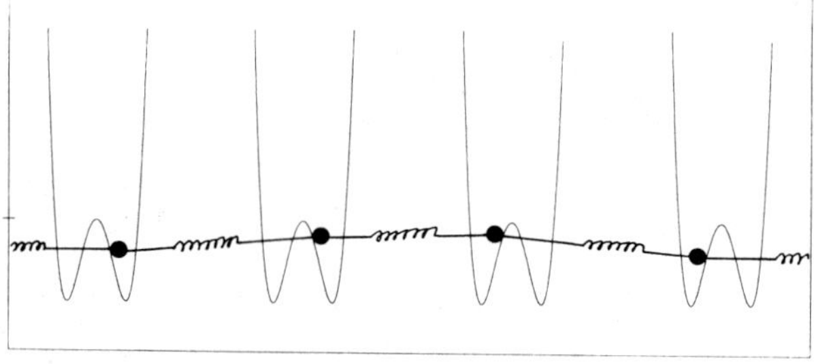

Schematic representation of the model defined by (II.2.1), in its one-dimensional form.

as arising from an underlying sublattice of atoms, which do not participate actively in the phase transition (cf. the Sr and Ti atoms in the case of the 105 K phase transition in $SrTiO_3$). The potential may be thought of as having a uniaxial anisotropy, which effectively restricts displacements to a single direction—hence the scalar character of the coordinates $u(l)$, $p(l)$. In keeping with the bulk of the literature we shall take this potential to have the simple form

$$V_s(u)=\frac{\bar{A}}{2}u^2+\frac{B}{4}u^4. \tag{II.2.1 d}$$

Finally we have, for later convenience, incorporated into the configurational energy (II.2.1 c) the effects of an ordering field E. Except where there is an explicit indication to the contrary, this field may be taken to be zero.

Equations (II.2.1) define a direct space representation of our model system. On occasion we will find it more convenient to consider a Fourier representation. Thus, following eqn. (I.2.1), and exploiting the simplifying features of our model, we define coordinates $Q(q)$ and $P(q)$ by

$$u(l)=\frac{1}{\sqrt{N}}\sum_{\mathbf{q}}Q(\mathbf{q})\exp[i\mathbf{q}\,.\,\mathbf{R}(l)] \tag{II.2.2 a}$$

$$p(l)=\frac{1}{\sqrt{N}}\sum_{\mathbf{q}}P(\mathbf{q})\exp[i\mathbf{q}\,.\,\mathbf{R}(l)]. \tag{II.2.2 b}$$

We may then rewrite the hamiltonian (II.2.1) in the form

$$H=T(\{P\})+V(\{Q\}), \tag{II.2.3 a}$$

with

$$T=\frac{1}{2}\sum_{\mathbf{q}}|P(\mathbf{q})|^2 \tag{II.2.3 b}$$

and

$$V=\frac{1}{2}\sum_{\mathbf{q}}[\bar{A}+4dCF(\mathbf{q})]\,|Q(\mathbf{q})|^2-E\sqrt{N}Q(0)$$
$$+\frac{B}{4N^2}\sum_{\mathbf{q}_1\mathbf{q}_2\mathbf{q}_3\mathbf{q}_4}Q(\mathbf{q}_1)Q(\mathbf{q}_2)Q(\mathbf{q}_3)Q(\mathbf{q}_4)\Delta(\mathbf{q}_1+\mathbf{q}_2+\mathbf{q}_3+\mathbf{q}_4). \tag{II.2.3 c}$$

Here the summations are over wavevectors in the first Brillouin zone, the Δ function reflects wavevector conservation, modulo a reciprocal lattice vector (cf. eqn. (I.2.19)

$$\Delta(\mathbf{q})=\begin{cases}N & \text{(for } \mathbf{q} \text{ a reciprocal lattice vector)}\\ 0 & \text{(otherwise)}\end{cases} \tag{II.2.3 d}$$

and $F(\mathbf{q})$ is defined by

$$F(\mathbf{q})=1-\frac{1}{d}\sum_{i=1}^{d}\cos(q_i a). \tag{II.2.3 e}$$

In the spirit of I, we shall ignore quantum-mechanical effects in treating the hamiltonian (II.2.1)—and, indeed, throughout this paper, except where there is an explicit indication to the contrary (cf. § II.5.4.3).

Model hamiltonians of the form (II.2.1) have been studied widely, for systems of dimensionality 1, 2 and 3, using both analytic and computer-aided techniques. We shall not attempt to provide an exhaustive bibliography of such analyses, but will refer to the work most relevant to our interests in the course of our discussion of the model—which we begin with a review of certain of its significant, general, features.

II.2.2. *Features of the model*

We consider first the question that is clearly of greatest importance to us—that of the existence of a phase transition in the system described by (II.2.1). That such a phase transition should, in fact, occur is immediately plausible: at high temperatures where inter-atom correlations may be neglected, $\langle u(l)\rangle$ must certainly vanish; at zero temperature minimization of the energy is achieved by a static displacement profile characterized by a non-zero order parameter Q_0 given by

$$Q_0^2 \equiv \frac{1}{N}\langle Q(\mathbf{q}_s=0)\rangle^2 = \langle u(l)\rangle^2 = -\frac{\bar{A}}{B}, \qquad \text{(II.2.4}\,a)$$

provided only that

$$\bar{A}<0. \qquad \text{(II.2.4}\,b)$$

This inequality—a *necessary* condition for the existence of a phase transition—is, of course, equivalent to the requirement that the single-particle-potential (II.2.1 *d*) be double-welled. In fact it has recently been shown (Sarbach 1977) that, for systems of dimensionality $d\geqslant 2$, (II.2.4 *b*) constitutes also a *sufficient* condition for the occurrence of a phase transition with critical temperature $T_c>0$.

For a system with $d=1$, of course, the familiar entropy/energy considerations (see Fisher 1972 for an illuminating discussion) rule out the possibility of long-range order at any $T>0$. Nevertheless, studies of the analytically-tractable $d=1$ model (cf. III) have proved particularly illuminating, suggesting in particular that some of the most interesting features of the behaviour of (II.2.1) are to be associated not so much with the existence of the phase transition itself, as with the double-well character of the potential V_s.

To forestall confusion, it is appropriate at this stage to note an alternative formulation of (II.2.1) favoured by some authors (see, e.g. Thomas 1971) who write the model potential energy (II.2.1 *c*) as

$$V=\sum_l \tilde{V}_s(u(l)) - C\sum_{l,l'}^{\text{n.n.}} u(l)u(l'), \qquad \text{(II.2.5)}$$

with the 'single particle potential' $\tilde{V}_s$ given by

$$\tilde{V}_s(u) = V_s(u) + 2dCu^2. \qquad \text{(II.2.6)}$$

This representation has certain advantages: firstly it emphasises the model's similarity to the Ising model of ferromagnetism; secondly, the potential $\tilde{V}_s(u)$ plays a central role in the 'mean field' theory of the hamiltonian (II.2.1) (cf. §II.3.1). Nevertheless, we feel that the former representation (in which V_s is described as 'the' single-particle potential) is more in keeping with the lattice-dynamical picture underlying our model.

It is helpful to notice that, associated with the hamiltonian (II.2.1) there are two characteristic energies. First of all the parameter

$$U_w \equiv \frac{\bar{A}^2}{4B} \tag{II.2.7 a}$$

describes the single-particle ('well') energy required to move an atom from the bottom to the top of one of the wells in the potential V_s, defined in (II.2.1 d). Secondly, the parameter

$$U_b = \frac{2dC|\bar{A}|}{B} \tag{II.2.7 b}$$

describes the incremental bond energy associated with this change, under the assumption of a complete and homogeneous ordering of all other atoms. The spectrum of systems described by (II.2.1) is thus spanned by the dimensionless quantity

$$s \equiv \frac{U_w}{U_b} = \frac{|\bar{A}|}{8dC}. \tag{II.2.8}$$

Two limiting values of s are worthy of particular mention since they characterize the two extremes ('order–disorder' and 'displacive': cf. § I.1) found among real systems exhibiting structural phase transitions.

Firstly, the *order–disorder* ($s \gg 1$) limit describes a system (typified by $NaNO_2$: cf. §I.1) in which a dominant well energy ensures that each particle resides close to the bottom of one or other of its potential wells. This is precisely the situation envisaged in the Ising model—and, indeed, taking $\bar{A} \to -\infty$, $B + \infty$ with $\bar{A}/B$ held constant, it is not hard to demonstrate formally the equivalence of the $s \to \infty$ limit of (I.2.1) with the familiar—and much studied—Ising model of ferromagnetism (thus motivating the pseudospin formalism described in § I.2.1).

Drawing upon these studies we note, in particular, that the transition temperature in this limit is given by (see, e.g. Fisher 1967)

$$k_B T_c(s \gg 1) = 4dk(d)C|\bar{A}|/B, \tag{II.2.9}$$

with $k(2) = 0{\cdot}5673\ \ldots$, and $k(3) \approx 0{\cdot}76$.

Although in doing so we anticipate the discussion of § I.4.1, it will also prove useful to note at this stage that, in the light of the universality hypothesis, this equivalence implies identical (Ising-model) values for the exponents, and certain other quantities, characterizing the critical behaviour of *all* the systems described by hamiltonians of the form (II.2.1). We now tabulate certain of these results (known either from the exact Onsager solution (1944) to the $d = 2$ Ising model, or from systematic expansions on the $d = 3$ model) since they will serve both as a touchstone for the approximation schemes discussed below, and as a reminder of the terminology of the critical region (cf. § I.3 and Fisher 1967).

Introducing the reduced temperature $t \equiv (T - T_c)/T_c$, the asymptotic behaviour ($t \to 0^-$) of the order parameter

$$Q_0 \equiv \frac{1}{\sqrt{N}} \langle Q(\mathbf{q}_s) \rangle \tag{II.2.10 a}$$

is given by

$$Q_0 \approx D|t|^{\beta}, \tag{II.2.10b}$$

with (see e.g. Fisher 1967)

$$\beta = \frac{1}{8}(d=2); \quad \beta \approx 0{\cdot}31(d=3). \tag{II.2.10c}$$

The specific heat may be written, asymptotically as $t \to 0^{\pm}$, in the form

$$\tilde{C} \approx \frac{A_{\pm}}{\alpha}(|t|^{-\alpha}-1), \tag{II.2.11a}$$

with (see e.g. Fisher 1967)

$$\left.\begin{array}{l}\alpha=0; \quad \text{logarithmic divergence } (d=2)\\ \alpha \approx 0{\cdot}125(d=3)\end{array}\right\} \tag{II.2.11b}$$

and (Brézin *et al.* 1974)

$$\frac{A_+}{A_-}=1(d=2); \quad \frac{A_+}{A_-} \approx 0{\cdot}55(d=3). \tag{II.2.11c}$$

The divergence of the order-parameter susceptibility is described by

$$\chi \approx C_{\pm}|t|^{-\gamma}, \tag{II.2.12a}$$

with (see e.g. Fisher 1967)

$$\gamma = 7/4(d=2); \quad \gamma \approx 1{\cdot}25(d=3) \tag{II.2.12b}$$

and (Tarko and Fisher 1975)

$$\frac{C_+}{C_-}=37{\cdot}693\ldots(d=2); \quad \frac{C_+}{C_-} \approx 5{\cdot}06(d=3). \tag{II.2.12c}$$

Finally, the correlation length (cf. § I.3.4) diverges according to

$$\xi \approx f_{\pm}|t|^{-\nu}, \tag{II.2.13a}$$

with (Tarko and Fisher 1975)

$$\nu = 1(d=2); \quad \nu \approx 0{\cdot}64(d=3) \tag{II.2.13b}$$

and

$$\frac{f_+}{f_-}=1{\cdot}414\ldots(d=2); \quad \frac{f_+}{f_-} \approx 1{\cdot}96(d=3). \tag{II.2.13c}$$

The applicability of these Ising ($s \gg 1$) results should, then, according to the universality hypothesis, extend even to the opposite end of the spectrum of systems (II.2.1), the so-called *displacive limit* ($s \ll 1$) which is peculiar to structural phase transitions. The contrasting physics of this limit (typified by the behaviour of $SrTiO_3$: cf. § I.1) is illuminated by the form of its transition temperature which, for the $d=3$ model, can be shown to be (see Oppermann and Thomas 1975 and further discussion in § II.3.2)

$$k_B T_c(s \ll 1) = \frac{4}{3q(3)} C|\bar{A}|/B, \tag{II.2.14}$$

with $q(3)=0{\cdot}5054$. Evidently the thermal energy at T_c in the displacive limit is large in comparison with the well energy U_w (eqn. (II.2.7 *a*)), while the latter is, of course, dominant for order–disorder systems. Indeed, it is this difference that is largely responsible for the relative richness in the behaviour—particularly the dynamic behaviour—of displacive systems, which will accordingly be the main focus of attention in discussions of time-dependent phenomena (cf. III).

Finally it should be mentioned that some authors (see e.g. Schneider and Stoll 1976 a; Thomas 1971) have tended to subdivide the spectrum of systems (II.2.1) into 'order–disorder' and 'displacive' regions according to whether the 'potential' $\tilde{V}_s$ (eqn. (II.2.6)) has two wells or a single well. It is clear from (II.2.1 *d*), (II.2.6) and (II.2.8) that this criterion is compatible with that adopted here. It is, furthermore, physically reasonable since, in the high-temperature limit, where intersite correlation may be ignored, the local dynamic behaviour is entirely governed by $\tilde{V}_s$, and exhibits a disordered (two-site) character, only if $\tilde{V}_s$ is double-welled (see e.g. Bruce and Schneider 1977). A subdivision according to this criterion also emerges naturally (as we shall see below) within the mean field theory of (II.2.1). However, it is probably less misleading to reserve the terms 'order–disorder' and 'displacive' for the two limiting regimes $s \gg 1$ and $s \ll 1$.

II.2.3. *Critique of the model*

We wish to emphasize from the outset that the chief merit of the model hamiltonian (II.2.1) does not lie in its immediate relevance to a structural phase transition in some particular physical system (cf., however, later discussions of critical phenomena). The attention devoted to this model system is motivated, rather, by the hope that its simplicity will enable one to identify—and to study—the distinctive properties common to systems undergoing structural phase transitions, all of which must, at some level of approximation, be characterized by a double- (or, more generally, a multiple-) welled single-particle potential.

Nevertheless it is to be noted that the cost of this simplicity is the omission of at least one potentially crucial feature of real systems—namely, acoustic modes, whose absence reflects the assumed incompressibility of the cage of 'passive' atoms that set up the single particle potential. In fact, we shall see in § II.5.1.2 that the coupling between elastic degrees of freedom, and those of the ordering variable, may play a particularly significant role in the critical region.

Naturally, this oversimplification—like the other, less drastic simplifying assumptions: the uniaxial anisotropy, the absence of third-order ('bond') anharmonicity, the short-range forces—can be relaxed (see e.g. Schneider and Stoll 1976 b). Most of the basic problems facing a theory of structural phase transitions are, however, already implicit in (II.2.1) as we shall see in succeeding sections.

Indeed, notwithstanding the simplicity of (II.2.1), these problems are already sufficiently acute that very few of the model's characteristics have been established exactly, and those that have are drawn entirely from the spectrum of equilibrium (time-independent) properties. In fact, with the exception of the existence of certain bounds on fluctuations, which may themselves be used to bound the transition temperature of the model (Fröhlich *et al.* 1976; Bunde and Lam 1977), we have already alluded above to all the exact results of which we are aware.

Bearing in mind these results, exemplified in eqns. (II.2.9) to (II.2.14), we now turn to examine the predictions of classical theories for the equilibrium properties of our model system.

§ II.3. Classical theories of structural phase transitions

In the course of this article we shall encounter several quite different approaches to the theoretical problems posed by hamiltonians of the form (II.2.1). We begin, here, with a review of the two simplest schemes, commonly referred to as 'mean-field theory' and 'self-consistent phonon theory'. In fact each of the theories employs a 'mean-field' approximation to reduce the many-body problem to a tractable single-body problem which is then solved 'self-consistently'. In the first of these schemes (discussed in the specific case of an order–disorder system in § I.5.2.1), the inter-*site* interaction is treated approximately. In the second of the schemes (in which the perturbation theory described in § I.5.1.1 is made self-consistent), it is the inter-*mode* interaction that is treated approximately. We shall, therefore, refer to the schemes as, respectively, the *independent-site* and *independent-mode* approximations.

II.3.1. *The independent-site approximation*

The essential character of the independent-site approximation (I.S.A.) is best revealed by considering the equation of motion of the direct space coordinates of the model, readily derived from (II.2.1) as

$$-\ddot{u}(l)=(\bar{A}+4dC)u(l)+Bu^3(l)-E-2C\sum_{l'}^{\text{n.n.}}u(l'). \tag{II.3.1}$$

The independent-site approximation replaces the last term in this equation by (Onodera 1970)

$$-2C\sum_{l'}^{\text{n.n.}}u(l')\approx -2C\sum_{l'}^{\text{n.n.}}\langle u(l')\rangle^{\text{I.S.}}\equiv -4dCQ_0^{\text{I.S.}}, \tag{II.3.2}$$

the implicit assumption being that the evolution of each local coordinate is influenced only by the average of its interactions with the other coordinates. The superscript on the average indicates that it is to be calculated self-consistently in the independent-site ensemble implied by the approximation (II.3.2). The set of equations (II.3.1) then decouple to give a set of N identical equations

$$-\ddot{u}=(\bar{A}+4dC)u+Bu^3-E-4dCQ_0^{\text{I.S.}} \tag{II.3.3}$$

describing a single particle moving in the potential $\tilde{V}_s$ (eqn. (II.2.6)) under the influence of a 'mean field'

$$E_m\equiv E+4dCQ_0^{\text{I.S.}}. \tag{II.3.4}$$

The thermal average of the displacement of such a particle may be written as

$$Q_0^{\text{I.S.}}\equiv\langle u\rangle^{\text{I.S.}}=\frac{\displaystyle\int_{-\infty}^{\infty}du\,u\exp\{-\beta[\tilde{V}_s(u)-E_mu]\}}{\displaystyle\int_{-\infty}^{\infty}du\exp\{-\beta[\tilde{V}_s(u)-E_mu]\}}\equiv\hat{U}(E_m). \tag{II.3.5}$$

For small E_m we may make use of the expansion

$$\hat{U}(E_m)=\hat{U}'(0)E_m+\frac{1}{6}\hat{U}'''(0)E_m^3+\dots, \tag{II.3.6}$$

where $\hat{U}'(0)$ may be identified as the linear response function χ_0 for the single-particle system,

$$\hat{U}'(0) \equiv \chi_0 \equiv \beta \langle u^2 \rangle^{(0)} \equiv \frac{\beta \int_{-\infty}^{\infty} du\, u^2 \exp[-\beta \tilde{V}_s(u)]}{\int_{-\infty}^{\infty} du \exp[-\beta \tilde{V}_s(u)]}. \tag{II.3.7}$$

The phase transition is located by the temperature at which, for zero applied field, it first becomes possible to satisfy eqns. (II.3.5)–(II.3.7) with non-zero $Q_0^{\text{I.S.}}$. With the aid of these equations one readily finds that the condition for criticality may be written variously as

$$4dC\chi_0 \Big|_{T_c^{\text{I.S.}}} = 1 \tag{II.3.8 a}$$

or

$$k_B T_c^{\text{I.S.}} = 4dC \langle u^2 \rangle^{(0)} \Big|_{T_c^{\text{I.S.}}}. \tag{II.3.8 b}$$

For general values of the model parameters, (II.3.7) and (II.3.8 *b*) have to be solved numerically: the results of such a calculation have been tabulated by Eisenriegler (1974). Analytic expressions for the transition temperature are, however, readily derived for the two limiting cases discussed in § II.2.2.

In the order–disorder ($s \gg 1$) limit, one has evidently

$$\langle u^2 \rangle^{(0)} \approx \frac{|A|}{B}, \tag{II.3.9}$$

leading to the familiar (Ising-model mean-field) result (cf. § I.5.2.1)

$$k_B T_c^{\text{I.S.}}\,(s \gg 1) = 4dC \frac{|\bar{A}|}{B}. \tag{II.3.10}$$

In the displacive limit we rewrite (II.3.7) in the form

$$\langle u^2 \rangle^{(0)} = \frac{1}{\beta(\bar{A} + 4dC)} \frac{\int_{-\infty}^{\infty} dv\, v^2 \exp\left(-\tfrac{1}{2}v^2 - \dfrac{\bar{B}}{4} v^4\right)}{\int_{-\infty}^{\infty} dv \exp\left(-\tfrac{1}{2}v^2 - \dfrac{\bar{B}}{4} v^4\right)}, \tag{II.3.11 a}$$

where

$$\bar{B} \equiv B/\beta(\bar{A} + 4dC)^2. \tag{II.3.11 b}$$

Anticipating that, in this limit, $\beta_c \to \infty$, the integrals in (II.3.11 *a*) can be evaluated by treating the $\bar{B}$ term as a perturbation—a natural consequence of the (relatively) weakly anharmonic character of the displacive limit. One readily finds

$$\langle u^2 \rangle^{(0)} = \frac{1}{\beta[\bar{A} + 4dC]} [1 - 3\bar{B} + 0(\bar{B}^2)], \tag{II.3.12}$$

so that, from (II.3.8 *b*), we have (Onodera 1970)

$$k_B T_c^{\text{I.S.}}(s \ll 1) = \frac{4dC}{3} \frac{|\bar{A}|}{B}. \tag{II.3.13}$$

Comparison of (II.3.10) and (II.3.13) with (II.2.9) and (II.2.14) reveals the mean-field theory's (customary) overestimate of the transition temperature. It is, however, interesting to note (S. Aubry, private communication) that the ratio $T_c(s)/T_c^{\text{I.S.}}(s)$ is relatively insensitive to s, varying (in the case $d=3$) from

$$T_c(s \gg 1)/T_c^{\text{I.S.}}(s \gg 1) \approx k(3) \approx 0{\cdot}76 \tag{II.3.14 a}$$

in the Ising limit, to

$$T_c(s \ll 1)/T_c^{\text{I.S.}}(s \ll 1) \approx 1/3q(3) \approx 0{\cdot}66 \tag{II.3.14 b}$$

in the displacive limit.

Developing, now eqns. (II.3.5) and (II.3.6), it is straightforward to show that the vanishing of the order-parameter as $T \to T_c^{\text{I.S.}}(s)$ from below obeys

$$Q_0^{\text{I.S.}} \approx D^{\text{I.S.}}(s)\,(1 - T/T_c^{\text{I.S.}}(s))^{\beta^{\text{I.S.}}} \tag{II.3.15 a}$$

with, for any dimensionality d (cf. eqn. (II.2.10 *c*))

$$\beta^{\text{I.S.}} = \frac{1}{2}, \tag{II.3.15 b}$$

while

$$\left.\begin{aligned} D^{\text{I.S.}}(s \gg 1) &\approx \sqrt{3}\left|\frac{\bar{A}}{B}\right|^{1/2} \\ D^{\text{I.S.}}(s \ll 1) &\approx \left|\frac{\bar{A}}{B}\right|^{1/2} \end{aligned}\right\}. \tag{II.3.15 c}$$

To determine the behaviour of the specific heat we combine the formal expression for the mean potential energy per atom,

$$U_v = \frac{1}{2}(\bar{A} + 4dC)\langle u(l)^2\rangle + \frac{B}{4}\langle u(l)^4\rangle - 2dC\langle u(l)u(l')\rangle, \tag{II.3.16 a}$$

(where l and l' are nearest neighbours) with the exact result (Schneider and Stoll 1976 a)

$$U_v = \frac{kT}{2} - \frac{B}{4}\langle u(l)^4\rangle, \tag{II.3.16 b}$$

to give

$$U_v = \frac{1}{4}(\bar{A} + 4dC)\langle u(l)^2\rangle - dC\langle u(l)u(l')\rangle + \frac{kT}{4}. \tag{II.3.16 c}$$

Within the independent-site approximation, then

$$U_v \approx \frac{1}{4}(\bar{A} + 4dC)\langle u^2\rangle^{\text{I.S.}} - dC(\langle u\rangle^{\text{I.S.}})^2 + \frac{kT}{4}, \tag{II.3.16 d}$$

where the averages are to be evaluated in the independent-site ensemble (as in eqn.

(II.3.5)). Exploiting the work of Onodera (1970) one can (with a little effort) recover from (II.3.16 *d*) the well-known mean-field behaviour of the specific heat, $\tilde{C}\equiv \mathrm{d}U/\mathrm{d}T$, in the Ising limit (see e.g. Stanley 1971):

$$\left.\begin{aligned} \tilde{C}^{\text{I.S.}} &\approx 0, & T &\gtrsim T_c^{\text{I.S.}} \\ \tilde{C}^{\text{I.S.}} &\approx \frac{3}{2}k_B, & T &\lesssim T_c^{\text{I.S.}} \end{aligned}\right\} (s \gg 1). \qquad \text{(II.3.17}\,a)$$

In the displacive limit, the techniques used above to establish the transition temperature may be employed to show that the potential-energy contribution to the specific heat behaves as

$$\tilde{C}^{\text{I.S.}} \approx \frac{k_B}{2}\,[1+O(s)], \quad T \approx T_c^{\text{I.S.}}(s \ll 1), \qquad \text{(II.3.17}\,b)$$

a result which, again, displays the quasi-harmonic character of this limit.

Thus, in conflict with the exact results (II.2.11), the independent-site approximation suggests that the specific heat will exhibit a discontinuity at $T_c^{\text{I.S.}}(s)$, whose magnitude tends to zero in the displacive limit.

The behaviour of the susceptibility also follows readily from (II.3.5) and (II.3.6):

$$\begin{aligned} \chi^{\text{I.S.}} = \frac{\partial Q_0^{\text{I.S.}}}{\partial E} &\approx \hat{U}'(0)\left[1 - 4dC\hat{U}'(0) - \frac{1}{2}\hat{U}'''(0)\,(4dC)^3 \times (Q_0^{\text{I.S.}})^2\right]^{-1} \\ &\approx C_\pm^{\text{I.S.}}(s)\left|1 - T/T_c^{\text{I.S.}}(s)\right|^{-\gamma^{\text{I.S.}}} \end{aligned} \qquad \text{(II.3.18}\,a)$$

Here the independent-site susceptibility exponent is, for every dimensionality d (cf. eqn. (II.2.12 *b*),

$$\gamma^{\text{I.S.}} = 1, \qquad \text{(II.3.18}\,b)$$

while the amplitudes $C_\pm^{\text{I.S.}}(s)$, characterizing the divergence of χ as $T \to T_c^{\text{I.S.}}$ from above (+) and below (−), satisfy

$$\left.\begin{aligned} C_+^{\text{I.S.}}(s \gg 1) &\approx 1/4dC \\ C_+^{\text{I.S.}}(s \ll 1) &\approx 1/|\bar{A}| \end{aligned}\right\}, \qquad \text{(II.3.18}\,c)$$

and (cf. eqn. (II.2.12 *c*))

$$C_+^{\text{I.S.}}(s)/C_-^{\text{I.S.}}(s) = 2. \qquad \text{(II.3.18}\,d)$$

It is worth noting that, while the amplitudes C_+ and C_- depend explicitly upon the microscopic parameters, their *ratio* is evidently parameter-independent—even within this approximate theory (although the explicit value of this ratio given in (II.3.18 *d*) is not in accord with the exact results (II.2.12 *c*)). These conclusions foreshadow the ideas of universality, touched on above, and developed in more detail in § II.4.1.

The correlation length, in the independent-site approximation, may be determined from the wavevector-dependent susceptibility (a natural generalisation of (II.3.18 *a*)), given in the high-temperature phase by the Landau–Ginzburg form

(Thomas 1971):

$$\chi^{\text{I.S.}}(\mathbf{q}) = \frac{\chi_0}{1 - 4dC(1 - F(\mathbf{q}))\chi_0} \approx \frac{\chi^{\text{I.S.}}}{1 + q^2(\xi^{\text{I.S.}})^2}, \tag{II.3.19a}$$

where

$$\xi^{\text{I.S.}} \equiv (2Ca^2\chi^{\text{I.S.}})^{1/2}, \tag{II.3.19b}$$

which has the form of (II.2.13 *a*), with exponent and amplitudes (cf. II.2.13 *b*)

$$\nu^{\text{I.S.}} = \frac{1}{2} \tag{II.3.19c}$$

$$\left.\begin{aligned} f_+(s \gg 1) &= a\,.\left(\frac{1}{2d}\right)^{1/2} \\ f_+(s \ll 1) &= a\,.\left(\frac{2C}{|\bar{A}|}\right)^{1/2} \end{aligned}\right\}. \tag{II.3.19d}$$

Evidently the displacive limit is characterized by a large value of the amplitude f_+ (cf. the definition (II.2.8))—a result which indicates the relatively greater coherence of displacive systems.

While offering certain qualitative insights which the phenomenological theory, by its very nature, cannot supply, the independent-site theory of our microscopic model evidently yields essentially the same (incorrect) predictions for the nature of critical singularities as does the Landau theory (cf. §§ I.4.1.1 and I.4.1.3). We shall return to examine the sources of the theory's deficiencies following a discussion of the second major decoupling scheme—the independent-mode approximation.

II.3.2. *The independent-mode approximation*

The assumptions underlying the independent mode (alias 'self-consistent phonon' or 'renormalized harmonic') approximation, and their relationship to those of the independent-site approximation, are presented most clearly in terms of the equation of motion of the *Fourier* coordinates, which follows from (II.2.3) as

$$-\ddot{Q}(\mathbf{q}) = (\bar{A} + 4dCF(\mathbf{q}))Q(\mathbf{q}) - \frac{E}{\sqrt{N}}\Delta(\mathbf{q}) + \frac{B}{N^3}\sum_{\mathbf{q}_1\mathbf{q}_2\mathbf{q}_3} Q(\mathbf{q}_1)Q(\mathbf{q}_2)Q(\mathbf{q}_3)\Delta(\mathbf{q}_1+\mathbf{q}_2+\mathbf{q}_3-\mathbf{q}). \tag{II.3.20}$$

The independent-mode approximation (I.M.A.) replaces the last term in this equation by

$$\begin{aligned} \frac{B}{N^2}\sum_{\mathbf{q}_1\mathbf{q}_2\mathbf{q}_3} Q(\mathbf{q}_1)Q(\mathbf{q}_2)Q(\mathbf{q}_3)\Delta(\mathbf{q}_1+\mathbf{q}_2+\mathbf{q}_3-\mathbf{q}) \\ \approx \frac{B}{N^2}\left\langle \sum_{\mathbf{q}_1\mathbf{q}_2\mathbf{q}_3} Q(\mathbf{q}_1)Q(\mathbf{q}_2)Q(\mathbf{q}_3)\Delta(\mathbf{q}_1+\mathbf{q}_2+\mathbf{q}_3-\mathbf{q})\right\rangle^{\text{I.M.}} \\ + \frac{3B}{N}[Q(\mathbf{q}) - \langle Q(\mathbf{q})\rangle^{\text{I.M.}}]\left\langle \sum_{\mathbf{q}_1} Q(\mathbf{q}_1)Q(-\mathbf{q}_1)\right\rangle^{\text{I.M.}} \end{aligned} \tag{II.3.21}$$

In effect the crucial approximation here also has a 'mean-field' character, the simplifying assumption being that the evolution of each *Fourier* coordinate is influenced only by the average of its interactions with the other coordinates. Once again the superscripts indicate that the averages are to be evaluated, self-consistently, in the 'independent-mode' ensemble implied by the approximation (II.3.21). In performing these averages one must allow explicitly for the non-vanishing expectation value of the coordinate $Q(0)$ in the ordered phase (or in the presence of a field). Thus we write

$$\begin{aligned} Q(\mathbf{q}) &\equiv \hat{Q}(\mathbf{q}) + \langle Q(\mathbf{q}) \rangle \\ &\equiv \hat{Q}(\mathbf{q}) + \frac{1}{\sqrt{N}} Q_0^{\mathrm{I.M.}} \Delta(\mathbf{q}), \end{aligned} \tag{II.3.22}$$

so that coordinates $\hat{Q}(\mathbf{q})$ denote the fluctuating parts of the coordinates $Q(\mathbf{q})$.

Suppressing the superscript on the order-parameter, the approximation to the equation of motion (II.3.20) then becomes

$$-\ddot{\hat{Q}}(\mathbf{q}) = (\bar{A} + 4dCF(\mathbf{q}) + 3B(I + Q_0^2))\hat{Q}(\mathbf{q}) + \frac{\Delta(\mathbf{q})}{\sqrt{N}}\{[\bar{A} + B(3I + Q_0^2)]Q_0 - E]\}. \tag{II.3.23}$$

Here we have introduced the quantity

$$I \equiv \frac{1}{N} \sum_{\mathbf{q}_1} \langle \hat{Q}(\mathbf{q}_1) \hat{Q}(-\mathbf{q}_1) \rangle^{\mathrm{I.M.}}, \tag{II.3.24}$$

and have exploited the fact that, in the independent-mode (effectively harmonic) ensemble, only *pairs* $\hat{Q}(\mathbf{q})\hat{Q}(-\mathbf{q})$ of the $\hat{Q}(\mathbf{q})$ coordinates can have non-vanishing expectation values.

Equation (II.3.23) describes a system of N uncoupled harmonic oscillators with frequencies given by

$$\omega^{\mathrm{I.M.}}(\mathbf{q})^2 \equiv \bar{A} + 4dCF(\mathbf{q}) + 3B(I + Q_0^2). \tag{II.3.25}$$

To appreciate the role of the additional terms present in (II.3.23) for the $\mathbf{q}=0$ coordinate, we equate the ensemble averages of each side of the equation, with the result

$$\langle \hat{Q}(\mathbf{q}=0) \rangle^{\mathrm{I.M.}} = \frac{\sqrt{N}}{\omega^{\mathrm{I.M.}}(0)^2} \{E - [\bar{A} + B(3I + Q_0^2)]Q_0\}. \tag{II.3.26}$$

However, according to our definition (II.3.22), the expectation value (II.3.26) is identically zero. Thus we have

$$E = [\bar{A} + B(3I + Q_0^2)]Q_0, \tag{II.3.27}$$

which defines the equation of state of the system, in the independent-mode approximation.

To develop the implications of eqns. (II.3.25) and (II.3.27)—the basic results of the independent-mode approximation—we note first that, in the independent-mode ensemble, the correlation function (II.3.24) may be written as

$$I = \frac{1}{N} \sum_{\mathbf{q}} \frac{k_B T}{\omega^{\mathrm{I.M.}}(\mathbf{q})^2}. \tag{II.3.28}$$

The self-consistent solution of the coupled eqns. (II.3.28) and (II.3.25) determines the frequency spectrum, and thence the order-parameter susceptibility, which follows from (II.3.27) as

$$[\chi^{\mathrm{I.M.}}]^{-1}=\left[\frac{\partial Q_0}{\partial E}\right]^{-1}=\bar{A}+\frac{3Bk_BT}{N}\sum_{\mathbf{q}}\frac{1}{[\bar{A}+4dCF(\mathbf{q})+3BI]}=\omega^{\mathrm{I.M.}}(0)^2 \tag{II.3.29}$$

in the disordered phase. We remark that this result's evident consistency with fluctuation-response theory (cf. eqn. (II.3.38) below) does not extend to the I.M.A. theory of the ordered phase.

It is clear from (II.3.29) that the stabilizing influence of the thermal fluctuations decreases as the temperature is lowered, and the system will reach its limit of stability (where the soft-mode frequency (II.3.29) vanishes) at a temperature $T_{\mathrm{c}}^{\mathrm{I.M.}}$ given by

$$0=\bar{A}+3BI\bigg|_{T_{\mathrm{c}}^{\mathrm{I.M.}}}=\bar{A}+3Bk_BT_{\mathrm{c}}^{\mathrm{I.M.}}\frac{1}{N}\sum_{\mathbf{q}}\frac{1}{4dCF(\mathbf{q})},$$

or

$$k_BT_{\mathrm{c}}^{\mathrm{I.M.}}=\frac{4}{3q(d)}\frac{C|\bar{A}|}{B}, \tag{II.3.30 a}$$

with

$$q(d)\equiv\frac{1}{N}\sum_{\mathbf{q}}\left(d-\sum_{i=1}^{d}\cos q_ia\right)^{-1} \tag{II.3.30 b}$$

However, it turns out that the temperature $T_{\mathrm{c}}^{\mathrm{I.M.}}$ does not locate the transition temperature of the model in the independent-mode approximation, as the softening of the zone-centre mode is preempted by the occurrence of a *first-order* phase transition.

To see this we turn to consider the equation of state (II.3.27), according to which an ordered-phase solution must satisfy

$$\bar{A}+B(3I+Q_0^2)=0, \tag{II.3.31}$$

so that, from (II.3.25),

$$\omega^{\mathrm{I.M.}}(\mathbf{q})^2=2BQ_0^2+4dCF(\mathbf{q}). \tag{II.3.32}$$

Substituting this result into (II.3.28), and evaluating the resulting q-space sum (as an integral) in the small Q_0 limit, the equation of state (II.3.31) assumes the form

$$Q_0^2-\frac{3k_BTB^{1/2}|Q_0|}{8\pi C^{3/2}}+\frac{\bar{A}}{B}(1-T/T_{\mathrm{c}}^{\mathrm{I.M.}})=0, \tag{II.3.33}$$

which has a real non-zero root for temperatures satisfying

$$\left(\frac{3k_BTB^{1/2}}{8\pi C^{3/2}}\right)^2>4\frac{|\bar{A}|}{B}(T/T_{\mathrm{c}}^{\mathrm{I.M.}}-1). \tag{II.3.34}$$

In the displacive limit this inequality is satisfied for temperatures $T<T^+$, where

$$T^+=T_{\mathrm{c}}^{\mathrm{I.M.}}\left(1+\frac{|\bar{A}|}{16\pi^2q^2(3)C}\right). \tag{II.3.35}$$

The fact that an ordered-phase solution of the equation of state exists, therefore, for temperatures higher than $T_c^{\mathrm{I.M.}}$ (specifically for $T_c^{\mathrm{I.M.}} < T < T^+$) indicates that, in this temperature range, the free energy must have the form depicted in fig. I.26*B*. Correspondingly the system will undergo a first-order phase transition at a temperature $T_0^{\mathrm{I.M.}}$, with

$$T_c^{\mathrm{I.M.}} < T_0^{\mathrm{I.M.}} < T^+. \tag{II.3.36}$$

To determine $T_0^{\mathrm{I.M.}}$ explicitly, one must obtain an expression for the free energy itself, and identify the temperature at which the ordered ($Q_0^{\mathrm{I.M.}} \neq 0$) phase solution is characterized by a free energy that is lower than that of the disordered ($Q_0^{\mathrm{I.M.}} = 0$) phase. Although this can, in fact, be readily accomplished by a careful integration of the equation of state (II.3.27), we shall not pursue the details here.

We conclude this section with a summary of the essential results that we have encountered, together with a brief guide to the key references where the reader will find these (and other) results of the independent-mode approximation discussed more fully.

Evidently, at least as far as our own interests are concerned, the most significant of the results discussed above is the prediction that the model (II.2.1) should exhibit a *first-order* phase transition. This conclusion stems originally from extensive numerical analysis of the independent-mode approximation (applied to a model analogous to (II.2.1)) performed by Gillis and Koehler (1971). (It may be noted parenthetically that the application of this approximation to models of the form (II.2.1) may be traced considerably further back, to the work of Boccara and Sarma (1965).) The interpretation of the predicted first-order character of the phase transition has been the subject of some controversy (Pytte 1972, Gillis and Koehler 1972), but there now seems little doubt that it is an artifact of the self-consistent phonon approximation. The justification of this assertion is not entirely trivial. It is now known (cf. the discussion in § II.4.3.8) that systems which, according to Landau theory (cf. § I.4.1.1) are permitted to exhibit a continuous phase transition (II.2.1 falls into this category, of course!) may, nevertheless, exhibit first-order phase transitions as a result of fluctuation effects. The model system (II.2.1), however, does not fall into this category, and it seems certain, therefore, that the prediction that the phase transition is first-order is quite spurious.

It is not quite so certain, however, that the first-order instability located by the independent mode argument is entirely devoid of physical content—a point to which we shall return in III. With that discussion in mind, we remark at this stage that the independent mode theory may be developed more systematically than we have chosen to do here, on the basis of variational arguments, which seek, formally, for the optimum *harmonic* representation of (II.2.1). This approach, expounded in considerable detail by Eisenriegler (1974), then establishes that this optimum representation (defined as the one which minimizes the corresponding 'trial' free energy) is characterized by the harmonic frequencies (II.3.25) and the equation of state (II.3.27).

Eisenriegler (1974) also applies these variational techniques to develop the independent-site approximation, and hence to compare the relative merits of the two mean-field theories. His analysis demonstrates that, not unnaturally, the independent-site approximation is to be preferred over (leads to a lower trial free energy than) the independent-mode approximation for models whose parameters lie in the order-disorder regime; for displacive systems, however, the independent mode approximation is the more satisfactory scheme.

Indeed, in the displacive limit ($s \to 0$, eqn. (II.2.8)) it is evident from (II.3.35) and (II.3.36) that (in $d=3$)

$$T_0^{\text{I.M.}}(s \ll 1) \simeq T_c^{\text{I.M.}} = T_c(s \ll 1), \qquad \text{(II.3.37)}$$

where the last step follows from (II.3.30 a) and (II.2.14). Thus, in $d=3$ and in the displacive limit, the independent-mode approximation to the transition temperature becomes exact (Oppermann and Thomas 1975) and the inequality, $T_c \geqslant T_c^{\text{I.M.}}$, satisfied by the transition temperature of the model (II.2.1) (Fröhlich *et al.* 1976, Sarbach 1977) holds as an equality.

Finally we remark that, in a formal way, one may define independent-mode critical exponents in terms of the theory's predictions for the disordered phase, continued through to the stability limit $T_c^{\text{I.M.}}$. As discussed, for example, by Benguigi (1975), these exponents coincide with those of the spherical model of Berlin and Kac (1952) and conflict with the exact results reviewed in § II.2.2.

II.3.3. *The failure of the classical approximation schemes*

The results presented in the preceding section expose, unequivocally, the inadequacies of the two classical approximation schemes. The independent-site approximation yields the same critical exponents as predicted by the phenomenological theory discussed in I—which are at variance with the exact results (for the model (II.2.1)) summarized in § II.2.2. The independent-mode approximation predicts, wrongly, that the phase transition exhibited by the model (II.2.1) is first-order.

In this section we turn to consider the sources of the failure of the approximation schemes, close to the critical point. We shall see that, in both cases, the failures can be traced to the growth (as $T \to T_c$) of fluctuations in the amplitudes of those coordinates $Q(\mathbf{q})$ with wavevectors near that of the soft mode. This superfluity of soft phonons implies a strong *correlation* of local (site) variables—a correlation which the independent-site approximation manifestly fails to accommodate satisfactorily. The large *fluctuations* in the soft Fourier coordinates enhance their effective interaction, thus invalidating the independent-mode approximation.

In the course of our discussion we shall see that the impact of these 'critical fluctuations' depends strongly upon the *dimensionality* of the system. Specifically, we shall show that there exists a 'borderline' or 'critical' dimensionality d_c above which their role is diminished to such an extent that the independent-site approximation becomes fully self-consistent (and, in fact, yields correct critical exponents), while the spuriously first-order character of the phase transition in the independent-mode approximation also disappears.

Our discussion of the independent-site approximation is in the spirit of many other arguments (Ginzburg 1960, Levanyuk 1959, Kadanoff *et al.* 1967) which *use* 'mean-field' theory to predict its own range of validity. Our approach will be to show that the level of correlations implied by the independent-site result for the wavevector-dependent *susceptibility* (eqn. (II.3.19 a)) is (for $d < d_c$) inconsistent with the independent-site result for the specific heat (eqn. (II.3.17)). Our arguments will focus exclusively upon the high-temperature phase, and (where it becomes necessary to make such a choice) we shall consider systems in the displacive, rather than the order–disorder, regime.

Combining the exact relationship (cf. eqn. (I.2.22))

$$\chi(\mathbf{q})=\beta\langle|Q(\mathbf{q})|^2\rangle \tag{II.3.38}$$

with the independent-site prediction for the wavevector-dependent susceptibility, eqn. (II.3.19 *a*), one finds

$$\langle|Q(\mathbf{q})|^2\rangle^{\text{I.S.}}=\frac{\beta^{-1}\chi_0}{[1-4dC(1-F(\mathbf{q}))\chi_0]}. \tag{II.3.39}$$

Exploiting the exact result

$$\langle u(l)u(l')\rangle=\frac{1}{N}\sum_{\mathbf{q}}\langle|Q(\mathbf{q})|^2\rangle\exp[i\mathbf{q}\,.\,(\mathbf{R}(l)-\mathbf{R}(l'))]. \tag{II.3.40}$$

we obtain from (II.3.39)

$$\begin{aligned}\langle u(l)u(l')\rangle^{\text{I.S.}}&=\frac{1}{N}\sum_{\mathbf{q}}\frac{\beta^{-1}\chi_0\exp[i\mathbf{q}\,.\,(\mathbf{R}(l)-\mathbf{R}(l'))]}{[1-4dC(1-F(\mathbf{q}))\chi_0]}\\&=\left(\frac{a}{2\pi}\right)^d\int_{\text{B.Z.}}\frac{\beta^{-1}\chi_0\exp[i\mathbf{q}\,.\,(\mathbf{R}(l)-\mathbf{R}(l'))]}{[1-4dC(1-F(\mathbf{q}))\chi_0]}d^dq.\end{aligned} \tag{II.3.41}$$

In a qualitative way this result already exposes the inconsistency of the independent-site approximation which, at its outset, ignores the very inter-site correlation implied by (II.3.41)! A more quantitative measure of this inconsistency is afforded by comparing the independent-site result for the local mean-square fluctuation (which is recovered from (II.3.41) if the interaction term in the denominator is ignored, cf. (II.3.7)),

$$\langle u^2(l)\rangle\approx\langle u^2(l)\rangle^{(0)}\equiv\langle u^2\rangle^{(0)} \tag{II.3.42 a}$$

with the modified result implied by (II.3.41)

$$\langle u^2(l)\rangle\approx\langle u^2(l)\rangle^{(1)}\equiv\left(\frac{a}{2\pi}\right)^d\int_{\text{B.Z.}}\frac{\langle u^2\rangle^{(0)}d^dq}{[1-4dC(1-F(\mathbf{q}))\beta\langle u^2\rangle^{(0)}]}. \tag{II.3.42 b}$$

Here the superscripts (0) and (1) indicate that (II.3.42 *a*) and (II.3.42 *b*) represent the first two in a hierarchy of increasingly refined approximations. The independent-site approximation can prove reliable only if the corrections to the leading result (II.3.42 *a*), implied by (II.3.42 *b*), may be ignored. In fact, the differences between (II.3.42 *a*) and (II.3.42 *b*) are illustrated more graphically through their implications for the specific heat. Making use of (II.3.16 *c*), one finds the correction to the specific heat

$$\Delta\tilde{C}\approx\frac{1}{4}\bar{A}\frac{d}{dT}(\langle u^2(l)\rangle^{(1)}-\langle u^2(l)\rangle^{(0)}), \tag{II.3.43}$$

the dominant part of which may be shown to be (in the displacive limit, and close to T_c)

$$\Delta\tilde{C}\approx\frac{k_B}{4}\int_{\mathbf{q}}^{\Lambda}\frac{d^dq}{(t+\alpha_0q^2)^2}. \tag{II.3.44 a}$$

Here we have written

$$\alpha_0\equiv 2Ca^2/|\bar{A}| \tag{II.3.44 b}$$

and have retained only the contributions to $\Delta\tilde{C}$ arising from those fluctuations with wavevectors near that of the soft mode (at $\mathbf{q}_s=0$), so that

$$\int_{\mathbf{q}}^{\Lambda} d^d q \equiv \left(\frac{a}{2\pi}\right)^d \int_{0<|\mathbf{q}|<\Lambda} d^d q \tag{II.3.44c}$$

with Λ some wavevector cutoff ($\lesssim \pi/a$).

Now, according to the independent-site approximation (cf. eqn. (II.3.17 b)), the specific heat exhibits a jump discontinuity (of order s in the displacive limit) at the phase transition: the approximation can prove reliable only if the correction term (II.3.44 a) is negligible on the scale of the predicted discontinuity. Thus a necessary condition for the self-consistency of the theory is that

$$\int_{\mathbf{q}}^{\Lambda} \frac{d^d q}{(t+\alpha_0 q^2)^2} \ll s, \tag{II.3.45a}$$

an inequality which will fail when (cf. the definition (II.2.8))

$$\frac{C}{|\bar{A}|}\int_{\mathbf{q}}^{\Lambda} \frac{d^d q}{(t+\alpha_0 q^2)^2} \sim 1. \tag{II.3.45b}$$

For the (physically unrealizable!) systems with dimensionality d larger than $d_c \equiv 4$ (the 'critical dimensionality' for the model (II.2.1)) the phase-space weighting factor in the integrand in (II.3.45) suppresses the contributions from the small q critical fluctuations, and the inequality in (II.3.45 a) is well-satisfied at all reduced temperatures. (Recall from (II.3.44 b) and (II.2.8) that $\alpha_0/a^2 \sim 1/s \gg 1$ in the displacive limit.) For systems with dimensionality d less than 4, however, the small q fluctuations promote a divergence of the quantity

$$\frac{C}{|\bar{A}|}\int_{\mathbf{q}}^{\Lambda} \frac{d^d q}{(t+\alpha_0 q^2)^2} \sim \frac{C}{|\bar{A}|} t^{-(2-d/2)} (\alpha_0/a^2)^{-d/2} \sim \left(\frac{C}{|\bar{A}|}\right)^{(1-d/2)} t^{-(2-d/2)}. \tag{II.3.46}$$

Evidently, for $d<d_c \equiv 4$, there will always exist a critical region where the inequality (II.3.45 a) cannot be satisfied, and the independent-site approximation cannot be self-consistent. Equations (II.3.45 b) and (II.3.46) suggest, moreover, that the reduced temperature $t^{\text{I.S.}}$ at which the inequality (II.3.45 a) fails,

$$t^{\text{I.S.}} \sim \left(\frac{|\bar{A}|}{C}\right)^{(d-2)/(4-d)} \tag{II.3.47}$$

becomes progressively smaller the more displacive is the system.

Let us now turn, briefly, to the independent-mode approximation. We have seen that this approximation results (in $d=3$) in a first-order phase transition, which (we argued) is entirely spurious. Thus it is reasonable to conclude that the approximation cannot be satisfactory over the range of temperatures between T^+ and $T_c^{\text{I.M.}}$ where it predicts the *coexistence* of ordered- and disordered-phase solutions. This range may be determined from the fact that the value of Q_0, characterizing the ordered phase at T^+, must itself satisfy

$$\frac{d}{dQ_0^2}(Q_0^2+3I)=0. \tag{II.3.48}$$

Making use of (II.3.32), and approximating the **q**-space sum in (II.3.28) by an integral over the dominant fluctuations, one finds that it will be possible to satisfy (II.3.48) only when

$$\frac{C}{|\bar{A}|}\int_{\mathbf{q}}^{\Lambda}\frac{d^d q}{\left(\frac{2B}{|\bar{A}|}Q_0^2+\alpha_0 q^2\right)^2}\sim 1, \tag{II.3.49}$$

a condition which closely resembles eqn. (II.3.45 *b*) characterizing the onset of the independent-site approximation's failure. Indeed, in dimensionalities $d>d_c=4$, (II.3.49) is never fulfilled: the temperatures T^+, $T_c^{\text{I.M.}}$ and $T_0^{\text{I.M.}}$ coincide, and the spurious first-order character of the phase transition is removed. For dimensionalities $d<d_c=4$, the small q fluctuations ensure that (II.3.48) and (II.3.49) can be satisfied at the reduced temperature

$$t^{\text{I.M.}}\equiv\frac{T^+}{T_c^{\text{I.M.}}}-1\sim\left(\frac{|\bar{A}|}{C}\right)^{(d-2)/(4-d)}, \tag{II.3.50}$$

a result which once again mirrors the conclusions (eqn. (II.3.47)) of our discussion of the independent-site approximation.

In summary, then, the arguments presented in this section not only establish the essential problem of the critical region, but also give a hint as to the way the problem may be resolved.

The problem itself, we have seen, may be traced to the 'critical fluctuations' associated with the group of modes with wavevectors adjoining that of the soft mode. These are the fluctuations that are responsible for the most serious deficiencies of the classical approximation schemes: the deficiencies persist even in more refined versions of these schemes (cf. the extension of the I.S.A. proposed by Lines (1974), and the 'second-order' I.M.A. calculations described by Sokolov and Vendik 1974), and ultimately undermine even such 'successes' as these schemes may appear to have (cf. Pietrass 1972) in the critical region.

One resolution of the problem will, we shall ultimately see, exploit the fact that there exists a critical ('borderline') dimensionality d_c, equal to 4 in the case we have discussed, above which these fluctuations do not carry enough phase-space weight to invalidate the classical schemes to such a serious extent.

§II.4. The contemporary theory of the critical region

With the inadequacies of the classical theories duly exposed, we can now turn to a discussion of the theoretical methods that have been developed to cope with the problems of the critical region. Not unreasonably, these methods prove to be more demanding than the classical approximation schemes—more demanding conceptually (to some extent) and more demanding technically (to a considerable extent). Our discussion will reflect this state of affairs in that we shall emphasize the essential concepts rather than the detailed techniques.

Specifically we begin (§II.4.1) with a discussion of the universality hypothesis (Kadanoff 1971) and of the way in which it is exploited to justify the modelling of the critical behaviour of very complex systems by very simple hamiltonians with (cf. §II.4.2) the same basic structure as that introduced in eqn. (II.2.1).

In § II.4.3 we describe the principles of the renormalization group method (Wilson 1971, Wilson and Kogut 1974) as it is applied to the problems of the critical region. In the course of this discussion we shall introduce enough of the theory's terminology to make this article self-contained. The reader interested in the problems encountered in putting the principles into practice will find a worked example in § II.4.4, where the remarkable idea of a perturbation expansion in $\varepsilon = d_c - d$ (Wilson and Fisher 1972) is introduced.

Finally, § II.4.5 surveys the spectrum of R.G. (and interrelated) techniques, currently employed in the theory of the critical region.

II.4.1. *The universality hypothesis*

The distinctive problem of the critical region—the existence of critical fluctuations—is offset to a significant extent by its distinctive simplifying feature, namely that it may be partially characterized by a set of quantities which are 'universal' (Kadanoff 1971).

The name 'universal' reflects the belief that these quantities (to be specified below) depend only upon certain essential, qualitative features (E.Q.F.s) of the system. Thus *all* the systems with the *same* E.Q.F.s (also specified below) are believed to exhibit critical behaviour characterized by the *same* set of universal quantities. Such systems (i.e. systems with the same E.Q.F.s) are said to belong to the same *universality class* (or 'critical equivalence class', in the language of Griffiths (1974)).

The class of universal quantities consists primarily, but not exclusively, of the set of critical exponents (most of which we encountered in § II.2.2) describing the critical singularities in thermodynamic and correlation functions. It includes, in addition, the *ratios* of certain amplitudes characterizing behaviour above and below the critical point (eqns. (II.2.11 *c*), (II.2.12 *c*) and (II.2.13 *c*) offer examples) and the *form* of certain *functions* (the equation of state provides one example; another will be encountered in § II.4.3.7) within the critical region.

It is appropriate to emphasize at this stage that there are many interesting quantities, characterizing critical behaviour, which are *not* universal. Perhaps the most prominent member of this group is the transition temperature, which reflects the details of the microscopic interactions and is thus 'system-specific' (compare the exact results (II.2.9) and (II.2.14)).

Not surprisingly it proves much harder to develop a theory of non-universal quantities, and (equally unsurprisingly!) it is in the determination of universal quantities that most theoretical effort has been expended.

When viewed against the backcloth of Landau theory and the independent-site approximation (which predict a *complete* universality of exponents, amplitude ratios...) the notion of universality is immediately plausible. However, for a satisfactory basis (both formally and intuitively) of the universality concept, one must look to the framework of the renormalization group, to be discussed in § II.4.3. It is, therefore, not appropriate at this stage to seek to justify either the idea itself or the identification (which we now proceed to make) of the essential features defining a universality class.

There are at least four 'essential qualitative features'.

The first E.Q.F. is the system *dimensionality* d. Indeed, the exact results cited in § II.2.2 (cf. eqns. (II.2.10 *c*), (II.2.11 *b*, *c*), (II.2.12 *b*, *c*)...) make it quite clear that universal quantities are dimensionality-dependent.

The second E.Q.F. is the *number of components of the order parameter*, n, or (in the language specific to structural phase transitions) the *degeneracy of the soft mode in the high-temperature (disordered) phase*. In the simple model introduced in § II.2 the (supposed) uniaxial anisotropy results in a single (non-degenerate) soft mode, and a single-component (scalar) order parameter, so that $n=1$ in this case. Physical realizations of systems with $n=2$ or 3 are easily found, and systems with still higher values of n also exist (cf. §§ I.4.2 and II.5.2).

The third E.Q.F. is the *range and angular dependence of the interatomic forces*. Since the character of these interactions is reflected in the phonon dispersion curves, this E.Q.F. may be re-expressed, in structural phase-transition language, as the *form of the dispersion of the soft mode*. Specifically, in systems undergoing transformations to ferroelectric phases, the long-range Coulomb forces induce the well-known Coulomb anomaly in the soft-mode dispersion. Consequently (cf. §§ II.5.1.3 and II.5.2.3) ferroelectric and non-ferroelectric transitions are characterized by different sets of universal quantities.

The fourth E.Q.F. is the *crystal symmetry*. Specifically, universal quantities (and, indeed, the very existence of a *continuous* phase transition—cf. § II.4.3.8) can depend upon the transformation properties of the order parameter under the space group of the high-temperature phase, i.e. the transformation properties that determine the number and nature of the invariants that appear in Landau–Devonshire expansions for the free energy (cf. § I.4.2.1).

Naturally, this list may, ultimately, prove to be incomplete, but there is little doubt that these four E.Q.F.s do constitute the principal characteristics reflected in universal critical behaviour. (It is not appropriate to dwell, here, upon those two-dimensional model systems in which the notion of universality fails altogether; see Baxter 1971, Kosterlitz 1974, José *et al*. 1977.) The importance of each of these features will be illustrated in the course of § II.5, where we shall see that it is the first and the third (the dimensionality and the soft-mode dispersion) that play the most crucial role in determining the magnitudes of the universal quantities, while the remaining features (the soft-mode degeneracy, and the overall symmetry) are, most crucially, instrumental in determining whether the phase transition is, in fact, continuous.

It is, of course, a task of the theory to establish the universality 'hypothesis' as a 'principle' and, to some extent, the renormalization group meets this challenge, as we shall see. In seeking to characterize the critical behaviour of real systems, however, the prevailing philosophy has been to *presuppose*, and *exploit*, the universality idea, rather than to demonstrate its validity.

The simplification of the problem that ensues with the presupposition of universality is considerable: for, provided one is concerned to calculate only the *universal* quantities characterizing the critical behaviour of some specific system, one may do so by studying the properties of the *simplest* system representative of the universality class to which the specific system of interest belongs.

The price of adopting this approach is that, in addition to forgoing the chance of 'demonstrating' universality (in its widest sense), one also forgoes any possibility of calculating (or even interrelating) the *non*-universal quantities associated with the system in interest. The merit of the approach lies in the fact that one may legitimately model the behaviour of very complex systems, with very simple hamiltonians, whose basic structure we now discuss.

II.4.2. *Models for universal critical behaviour*

The time-independent ('equilibrium') properties of a system undergoing a structural phase transition are defined, formally, by the canonical partition function

$$Z_c = \int_{(\{p\},\{u\})} \exp(-\beta[T(\{p\}) + V(\{u\})]), \tag{II.4.1 a}$$

where T and V are the kinetic- and potential-energy contributions to the crystal hamiltonian. Within classical mechanics (quantum effects are discussed briefly in § II.5.4.3) the integration over the momentum coordinates, implied in (II.4.1 a), is trivial, leading to an entirely regular 'background' contribution to the thermodynamic properties. One is left, therefore, with the problem of evaluating the configurational integral

$$Z = \int_{\{u\}} \exp[-\beta V(\{u\})], \tag{II.4.1 b}$$

which (usually) proves more tractable when rewritten as an integral over a set of Fourier coordinates:

$$Z = \int_{\{Q\}} \exp[-\beta V(\{Q\})]. \tag{II.4.1 c}$$

Here $\{Q\}$ denotes a set of coordinates $Q(\mathbf{q}j)$ giving a Fourier representation of the displacements of the atoms with respect to their high-temperature (disordered) phase equilibrium positions. The set may, formally, be chosen to diagonalize the strictly harmonic part of the crystal hamiltonian (cf. the Fourier representation (II.2.3) of the simple model introduced in § II.2). In this representation the coefficients of those terms in V that are quadratic in certain of the coordinates $Q(\mathbf{q}j)$ (specifically, those with $\mathbf{q} \approx \mathbf{q}_s$, the location of the soft mode, and j one of the n branches terminating in the n-fold degenerate soft mode) will be negative, expressing the inherent instability of the crystal in the harmonic approximation.

The configurational energy V will, in general, be a very complicated function of the coordinates $Q(\mathbf{q}j)$, and the integral (II.4.1 c) will be intractable for this reason—in addition to the inescapable problems arising from the critical fluctuations themselves. Invoking the universality hypothesis, however, the true configurational energy may be replaced by the simplest configurational energy possessing the same essential qualitative features as V itself. Thus we make the replacement

$$\beta V \rightarrow \mathscr{H}_0, \tag{II.4.2}$$

where $\mathscr{H}_0$ is a simplified configurational energy, or 'effective hamiltonian' in units of $k_B T$.

As an example we consider the construction of the simplest effective hamiltonian capable of yielding the universal critical behaviour of the model defined in § II.2. In fact, since this model itself represents a considerable simplification of any real physical situation, (II.2.3 c) is already (very nearly) the simplest configurational energy in its universality class. The construction of the effective hamiltonian thus requires only a few steps.

First of all we argue that, since the details of the Brillouin zone size and form do not constitute 'essential qualitative features', coordinates with wavevectors far from the soft mode (the 'harder' fluctuations) need not be retained explicitly in the effective hamiltonian. Thus we retain only the coordinates with wavevectors lying within some sphere of small (but otherwise arbitrary) radius Λ, centred on the soft-mode wavevector $\mathbf{q}_s = 0$. These are, of course, the 'softer', potentially critical, fluctuations discussed in §II.3.3.

The effects of the harder fluctuations cannot, however, be ignored entirely since they play an essential role in the stabilization of the high-temperature phase. Specifically they mediate interactions among the softer coordinates, thus renormalizing the coefficients of these interactions. In contrast to the effects arising from interactions among the critical fluctuations themselves, the effects of the non-critical modes (those outside the sphere) on the critical modes are *controlled*, since the mean-square amplitudes of the non-critical modes remain below a finite level (determined by the choice of Λ) even at the phase transition. In *principle* then—and we require no more—the effects of these fluctuations may be accommodated within the standard (perturbative) methods discussed in § I.5.1.1. Their principal effect is, then, to 'dress' the bare soft-mode frequency-squared (determined by the negative quantity $\bar{A}$) with a static self-energy correction $\sum_{\Lambda,T}(\mathbf{q}_s, \omega = 0)$, resulting in a partially renormalized (i.e. renormalized by the non-critical fluctuations, but not, as yet, by the critical modes themselves) soft-mode frequency, given by

$$\tilde{\omega}(\mathbf{q}_s = 0, \omega = 0)^2 = \bar{A} + \textstyle\sum_{\Lambda,T}(\mathbf{q}_s, \omega = 0) = \sum_{\Lambda,T} - \sum_{\Lambda,T_0}, \qquad \text{(II.4.3}\,a)$$

where the temperature T_0 is defined, implicitly, by

$$\textstyle\sum_{\Lambda,T_0} + \bar{A} = 0, \qquad \text{(II.4.3}\,b)$$

and locates the temperature at which the partially-dressed frequency (II.4.3 *a*) vanishes. (T_0 does not coincide with the true transition temperature, T_c, the difference reflecting the further, less tractable, renormalization effects arising from the interactions among the critical modes.)

At this stage of the argument the effective hamiltonian to be associated with the configurational energy (II.2.3 *c*) reads

$$\mathscr{H}_0 = \frac{\beta}{2} \sum_{|\mathbf{q}|<\Lambda} \left[\textstyle\sum_{\Lambda,T} - \sum_{\Lambda,T_0} + 2Cq^2a^2 \right] |Q(\mathbf{q})|^2 - \beta E \sqrt{N} Q(0)$$
$$+ \frac{\beta B}{4N^2} \sum_{|\mathbf{q}_1|,\ldots|\mathbf{q}_4|<\Lambda} Q(\mathbf{q}_1)Q(\mathbf{q}_2)Q(\mathbf{q}_3)Q(\mathbf{q}_4)\Delta(\mathbf{q}_1 + \mathbf{q}_2 + \mathbf{q}_3 + \mathbf{q}_4). \qquad \text{(II.4.4)}$$

The sums over the wavevectors of the critical modes are now replaced by integrals over reciprocal space. In anticipation of the usefulness of the borderline-dimensionality concept, these integrals are written in their generalized d-dimensional form, so that

$$\frac{1}{N} \sum_{|\mathbf{q}|<\Lambda} \to \left(\frac{a}{2\pi} \right)^d \int_{0<|\mathbf{q}|<\Lambda} d^d q \equiv \int_{\mathbf{q}}^{\Lambda}. \qquad \text{(II.4.5)}$$

Finally, the coordinates $Q(\mathbf{q})$ are rescaled, and new interaction parameters are defined, in such a way that the coefficient of q^2 in the dispersion of the soft mode becomes equal to unity. (This is an inessential, but helpful convention.) Thus we define parameters r_0, u_0 and E_0 by

$$r_0 = \frac{\sum_{\Lambda,T} - \sum_{\Lambda,T_0}}{2Ca^2}, \qquad \text{(II.4.6}\,a)$$

$$u_0 = \frac{B}{4\beta(2Ca^2)^2}, \tag{II.4.6 b}$$

$$E_0 \equiv \frac{\beta^{1/2}E}{(2Ca^2)^{1/2}}, \tag{II.4.6 c}$$

and make the replacement (note, particularly, the way in which a factor of $\sqrt{N}$ is subsumed into the new coordinates)

$$(2\beta Ca^2 N)^{1/2} Q(\mathbf{q}) \to Q(\mathbf{q}). \tag{II.4.6 d}$$

With these conventions the effective hamiltonian (II.4.4) assumes the form

$$\mathscr{H}_0 \equiv \mathscr{H}_0(n=1) = \frac{1}{2}\int_{\mathbf{q}}^{\Lambda} (r_0 + q^2)|Q(\mathbf{q})|^2 - E_0 \int_{\mathbf{q}}^{\Lambda} Q(\mathbf{q})\delta(\mathbf{q})$$
$$+ u_0 \int_{\mathbf{q}_1}^{\Lambda}\int_{\mathbf{q}_2}^{\Lambda}\int_{\mathbf{q}_3}^{\Lambda}\int_{\mathbf{q}_4}^{\Lambda} Q(\mathbf{q}_1)Q(\mathbf{q}_2)Q(\mathbf{q}_3)Q(\mathbf{q}_4)\delta(\mathbf{q}_1+\mathbf{q}_2+\mathbf{q}_3+\mathbf{q}_4), \tag{II.4.7 a}$$

where $\delta(\mathbf{q})$ is a d-dimensional δ function

$$\delta(\mathbf{q}) = \left(\frac{2\pi}{a}\right)^d \delta^{(d)}(\mathbf{q}). \tag{II.4.7 b}$$

For future reference we note that this effective hamiltonian may also be written in a more compact (but usually less-immediately useful) direct-space representation

$$\mathscr{H}_0(n=1) = \int_{\mathbf{x}} \left\{ \frac{1}{2}(r_0 u^2(\mathbf{x}) + (\nabla u(\mathbf{x}))^2) - E_0 u(\mathbf{x}) + u_0 u^4(\mathbf{x}) \right\}, \tag{II.4.8 a}$$

where we have introduced the displacement field

$$u(\mathbf{x}) \equiv \int_{\mathbf{q}}^{\Lambda} Q(\mathbf{q}) \exp(i\mathbf{q}\,.\,\mathbf{x}), \tag{II.4.8 b}$$

while

$$\int_{\mathbf{x}} \equiv \frac{1}{a^d}\int d^d x. \tag{II.4.8 c}$$

It will prove useful to note, also, that, except at very low temperatures (when the applicability of the classical approach we have adopted may itself be called in question: cf. §II.5.4.3), the dominant temperature-dependence of the effective hamiltonian (II.4.7) resides in the parameter r_0 (essentially the partially dressed soft-mode frequency-squared) which, at temperatures near T_0, will have the form (cf. eqn. (II.4.6 a))

$$r_0 = a_0(T - T_0). \tag{II.4.9}$$

The arguments deployed above may be applied, in an extended form, to construct the simplest effective hamiltonian capable of reproducing the universal critical behaviour of any real system. The guiding principle is that it is necessary to retain in the effective hamiltonian only the coordinates of those 'critical' modes that belong to one of the n 'soft' branches, and that have wavevectors $\mathbf{q}$ such that $\mathbf{k} \equiv \mathbf{q} - \mathbf{q}_s$ is small. (If there is more than one wavevector in the star of the soft mode, $\{\mathbf{q}_s\}$, as at a transition to an incommensurate phase, care must be taken to include the contributions of the critical modes adjoining *all* the soft-mode wavevectors $\{\mathbf{q}_s\}$. By an appropriate folding of the Brillouin zone, the resulting effective hamiltonian may still be cast in the general form indicated below (see e.g. Cowley and Bruce 1978). However the effective number of components of the order parameter is, then, the product of the soft-mode degeneracy and the number of independent wavevectors in the star $\{\mathbf{q}_s\}$, and may thus be larger than 3: cf. §I.4.4.) The hamiltonian must, moreover, correctly reflect those essential features of the real system discussed in the preceding section: other features (the contributions of non-critical phonon branches, the wavevector-dependence of anharmonic coupling constants, the effects of higher-order anharmonicity ...) which may be essential in calculations of *non*-universal quantities, may be entirely ignored. One finds then that the effective hamiltonians representative of a great spectrum of universality classes can be written in the form

$$\mathscr{H}_0 = \mathscr{H}_0(n) + \Delta\mathscr{H}_0. \qquad \text{(II.4.10}\,a)$$

Here $\mathscr{H}_0(n)$ represents the ubiquitous n-component Landau–Ginzburg–Wilson (L.G.W.) hamiltonian (a natural generalization of (II.4.7)):

$$\mathscr{H}_0(n) = \frac{1}{2}\sum_{j=1}^{n}\int_{\mathbf{k}}^{\Lambda}(r_0+k^2)|Q_j(\mathbf{k})|^2 - E_0\int_{\mathbf{k}}^{\Lambda}Q_1(\mathbf{k})\delta(\mathbf{k})$$
$$+u_0\sum_{i,j=1}^{n}\int_{\mathbf{k}_1}^{\Lambda}\int_{\mathbf{k}_2}^{\Lambda}\int_{\mathbf{k}_3}^{\Lambda}\int_{\mathbf{k}_4}^{\Lambda}Q_i(\mathbf{k}_1)Q_i(\mathbf{k}_2)Q_j(\mathbf{k}_3)Q_j(\mathbf{k}_4)\delta(\mathbf{k}_1+\mathbf{k}_2+\mathbf{k}_3+\mathbf{k}_4). \qquad \text{(II.4.10}\,b)$$

where (as in I) we adopt the convention that wavevectors denoted by $\mathbf{k}$ (rather than $\mathbf{q}$) measure 'distances' from the soft-mode wave vector as origin, while we have allowed for the presence of an ordering field (cf. §I.2.2) coupling to the $j=1$ component of the order parameter.

It is clear that $\mathscr{H}_0(n)$ conveys the first two of the E.Q.F.s of §II.4.1), namely the dimensionality d (implicit in the dimensionality of the k-space integrations) and the number n of soft branches (alias, the number of components of the order parameter). The two additional E.Q.F.s—the form of the soft mode dispersion, and the order parameter symmetry—can be controlled by the supplementary term $\Delta\mathscr{H}_0$, which will vary according to the problem of interest. One may, for example (cf. §II.5.1.3 for details) allow, through $\Delta\mathscr{H}_0$, for the Coulomb singularity present in the dispersion of a ferroelectric mode, and absent from the form of soft-mode dispersion presupposed in $\mathscr{H}_0(n)$.

Although to pursue the matter would take us beyond the scope of this article, it is appropriate at least to mention the fact that the spectrum of effective hamiltonians defined by eqns. (II.4.7 a) and (II.4.10) prove relevant in descriptions of critical phenomena, not only at structural phase transitions but also at phase transitions in magnetic systems (Wilson and Kogut 1974, Hubbard 1972; for a review see Aharony 1976) and in fluids (Hubbard and Schofield 1972; for a review see Hohenberg 1977). Indeed, this observation reflects the extent to which theories of critical phenomena in these diverse systems have, in recent years, been unified—largely through the insights of universality, firmly grounded in the renormalization group methods to which we now turn.

II.4.3. *The renormalization group method, in principle*

While its domain of applicability is, in fact, a wide one (Wilson and Kogut 1974), we shall discuss the renormalization group method solely in the context of the problems of time-independent critical behaviour—specifically, the problems associated with the evaluation of the partition function

$$Z=\int_{\{Q\}}\exp[-\mathscr{H}_0(\{Q\})], \tag{II.4.11 a}$$

where the effective hamiltonian is of the form specified in (II.4.10). While these problems are usually most tractable when posed in the reciprocal space representation employed in (II.4.11 *a*), it proves conceptually helpful, occasionally, to think in terms of a direct space representation,

$$Z=\int_{\{u\}}\exp[-\mathscr{H}_0(\{u\})]. \tag{II.4.11 b}$$

Here $\{u\}$ defines a set of local (direct-space) coordinates, and $\mathscr{H}_0(\{u\})$ gives the effective hamiltonian in this representation.

We shall subdivide our discussion of the R.G. approach to these problems into two parts. In this section we shall attempt to convey the specific aim and key concepts of the method, deferring to § II.4.4 a (brief) guide to the problems encountered in putting the principles into practice.

II.4.3.1. *The renormalization group idea*

The essential idea of the renormalization group method, devised by Wilson (1971) and described in detail by Wilson and Kogut (1974), may be stated thus:

To relate the physical properties of a system close to its critical point to the properties of a similar system that is far from criticality.

The motivation for the idea is transparent: the properties of the near-critical system are hard to determine, because the strong fluctuations invalidate the conventional approximation schemes; by contrast, in the non-critical system the fluctuations are bounded, and their effects susceptible to conventional (perturbative) methods.

To make these statements more precise we must introduce an appropriate measure of a system's 'closeness to criticality'. Instead of adopting the reduced temperature $t\equiv|T-T_c|/T_c$ (the obvious candidate), we select a quantity with a more immediate physical appeal—the ratio of the correlation length ξ to the lattice spacing a. This we shall call the *dimensionless correlation length*, denoting it by $\hat{\xi}$:

$$\hat{\xi}\equiv\xi/a. \tag{II.4.12}$$

The principal R.G. objective, prescribed above, may now be re-expressed as the task of relating the properties of a system with a *large* dimensionless correlation length, to those of a similar system with a *small* dimensionless correlation length.

The essential feature of the operation which will achieve this end is most clearly revealed by reverting, temporarily, to the direct-space, lattice, formulation of the phase-transition problem (eqn. (II.4.11 *b*)). Consider, to be specific, and for simplicity, the case of a one-dimensional system, so that the set $\{u\}$ describe the local displacements from lattice sites forming a linear chain. (Such a system can, in fact, undergo a phase transition only at $T=0$ K, but this is of no consequence in the following qualitative discussion.)

Let us subdivide the N local variables into two groups, every bth variable along the chain (b being integral) going into one group, and the remainder forming the

Fig. II.2

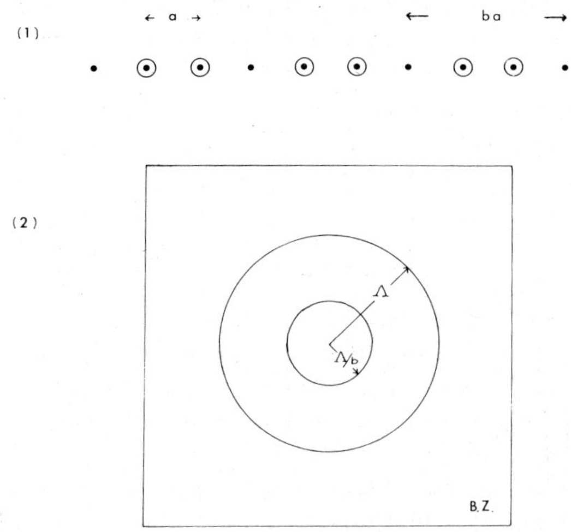

Schematic representation of the degrees of freedom eliminated in (1) a direct space realization of the R.G. with $b=3$: the coordinate associated with each circled lattice site is integrated from the partition function; (2) a reciprocal space realization of the R.G. with $b\approx 3$: the coordinates associated with the shell of wavevectors with $\Lambda/b<|\mathbf{k}|<\Lambda$ are integrated from the partition function. The 'non-critical' modes in the Brillouin zone—those with $|\mathbf{k}|>\Lambda$—are eliminated at the outset.

second group. In a notation whose significance will become clearer below we denote the first set by $\{u\}^{<}$ and the second set by $\{u\}^{>}$ (see fig. II.2(1)).

We now consider the consequences of performing that part of the configurational integral (II.4.11 *b*) associated with the coordinates $\{u\}^{>}$. Thus, formally, we write

$$
\begin{aligned}
Z &= \int_{\{u\}} \exp[-\mathscr{H}_0(\{u\})] \\
&= \int_{\{u\}^<,\{u\}^>} \exp[-\mathscr{H}_0(\{u\})] \\
&\equiv \exp[-N\Delta G(b)] \int_{\{u\}^<} \exp[-\mathscr{H}(b,\{u\}^<)]. \qquad \text{(II.4.13)}
\end{aligned}
$$

Here $\exp[-N\Delta G(b)]$ represents the contribution of that part of the integral over the $\{u\}^{>}$ coordinates that does not involve the coordinates $\{u\}^{<}$, while $\exp[-\mathscr{H}(b,\{u\}^{<})]$ represents the remaining terms which do involve these coordinates explicitly.

Now it is by no means clear that this 'partial' evaluation of Z could, in practice, prove any easier than the complete evaluation (which is, after all, *the* problem!). Leaving this question to one side, however, a little reflection establishes that the observables relevant to the critical behaviour of the hamiltonian $\mathscr{H}_0$ are implicit in (may be calculated from) the new effective hamiltonian $\mathscr{H}(b,\{u\}^{<})\equiv\mathscr{H}(b)$: this is clearly true for the order parameter, the susceptibility, and for the correlation function $\langle u(l)u(l')\rangle$, at least when $u(l)$ and $u(l')$ belong to the set $\{u\}^{<}$. (It is less evidently true for the free energy, and its temperature derivatives, to which $\Delta G(b)$ will contribute.)

Now it is clear, furthermore, that the 'true' correlation length (measured in Å) associated with the system described by $\mathscr{H}_0$ is the *same* as that associated with the effective hamiltonian $\mathscr{H}(b)$ (since each, true, correlation length reflects the distance over which, in particular, the variables $\{u\}^<$ are correlated). However, the lattice spacing of $\mathscr{H}(b)$ (the spacing between the members of the set $\{u\}^<$) is larger than that of $\mathscr{H}_0$ (the spacing a between the members of the set $\{u\}$) by a factor of b. Thus the dimensionless correlation lengths $\hat{\xi}$ and $\hat{\xi}(b)$ (associated with $\mathscr{H}_0$ and $\mathscr{H}(b)$) are related by

$$\hat{\xi}(b) = \hat{\xi}/b, \tag{II.4.14}$$

showing that the operation fulfils the cardinal objective of the renormalization group, as set out above: the act of integrating over the fraction $1-1/b$ of the degrees of freedom maps the physics of a critical system (with a large dimensionless correlation length $\hat{\xi}$) onto that of a less critical system (with a smaller dimensionless correlation length $\hat{\xi}(b) = \hat{\xi}/b$).

The R.G. operation we have described (whose origins may be traced to the seminal work of Kadanoff (1966)) proves practicable only for one-dimensional systems (cf. Nelson and Fisher 1975, Prange 1974, Bruce and Schneider 1977). Although many variants of this 'direct space' realisation of the R.G. idea have been developed (Niemeyer and van Leeuven 1973, 1974, Wilson 1975, Kadanoff and Houghton 1975, and others) the Fourier space realization, to which we now turn, has, in most respects, proved to be the more powerful and general.

II.4.3.2. *The renormalization group operation*

To recast the R.G. operation in its most general form, we return to the problem posed by the partition function Z expressed (eqn. (II.4.11 a)) as a configurational integral over the set of N Fourier coordinates $\{Q\} \equiv \{Q_j(\mathbf{k})\}$, with wavevectors $\mathbf{k}$ lying within a sphere of radius Λ, about the condensing mode. Motivated by the observation that the set of coordinates $\{u\}^<$ (left unintegrated in our direct-space renormalization group operation) contain, predominantly, information about the *long* wavelength (*small* wavevector) fluctuations, we subdivide the coordinates $\{Q\}$ into two sets, $\{Q\}^<$ and $\{Q\}^>$, consisting, respectively, of those coordinates with wavevectors less than Λ/b (where b is an arbitrary scale factor, larger than unity), and those coordinates with wavevectors *larger* than Λ/b (see fig. II.2(2)).

The first stage of the renormalization group operation then requires (cf. eqn. (II.4.13)) that we perform that part of the configurational integral associated with the larger wavevector (less critical) modes:

$$\begin{aligned} Z &= \int_{\{Q\}} \exp[-\mathscr{H}_0(\{Q\})] \\ &= \int_{\{Q\}^<, \{Q\}^>} \exp[-\mathscr{H}_0(\{Q\})] \\ &= \exp[-N\Delta G(b)] \int_{\{Q\}^<} \exp[-\bar{\mathscr{H}}(b, \{Q\}^<)]. \end{aligned} \tag{II.4.15}$$

The prefactor $\exp[N\Delta G(b)]$ represents the contribution from the integration over the $\{Q\}^>$ coordinates which does not involve the coordinates $\{Q\}^<$ of the shorter wavevector (more critical) modes which remain (as yet) unintegrated; the term $\exp[-\bar{\mathscr{H}}(b, \{Q\}^<)]$ represents the remaining contributions, which *do* involve these coordinates. (Again we shelve the question as to how the integration implied by (II.4.15) may actually be achieved.)

As it stands $\bar{\mathscr{H}}(b, \{Q\}^<)$ describes a system with a fluctuation spectrum with wavevectors in the range $0<|\mathbf{k}|<\Lambda/b$. It proves convenient to *rescale* the vectors of reciprocal space so that they again cover the range between zero and Λ. It is also helpful (and legitimate) to rescale the Fourier coordinates $\{Q\}^<$ at the same time. Thus, in this, the second, stage of the R.G. operation, we define new coordinates $Q'_j(\mathbf{k}')$ by

$$Q_j(\mathbf{k})=\zeta Q'_j(\mathbf{k}'), \quad \mathbf{k}'\equiv b\mathbf{k}. \tag{II.4.16}$$

Here ζ is a coordinate-rescaling factor, which we are free to choose as we please (a specific choice, and its rationale, will be discussed in §II.4.4). As a result of the rescaling, the effective hamiltonian $\bar{\mathscr{H}}(b, \{Q\}^<)$ may be replaced with a final effective hamiltonian $\mathscr{H}(b, \{Q'\})$, in which the wavevectors $\mathbf{k}'$ of the coordinates $Q'_j(\mathbf{k}')$ once again cover the range within a radius Λ of the condensing mode. At this stage the primes on the rescaled wavevectors and rescaled coordinates are redundant and may be dropped. The partition function (II.4.15) may then, finally, be expressed as

$$Z=\exp[-N\Delta G(b)]\zeta^{N/b^d}\int_{\{Q\}}\exp[-\mathscr{H}(b,\{Q\})], \tag{II.4.17}$$

where the factor of ζ^{N/b^d} arises from the jacobian of the transformation (II.4.16), there being a fraction $1/b^d$ of the original N coordinates $\{Q\}$ left in the set $\{Q\}^<$.

The two-stage operation by which we have (in principle—§II.4.4 includes a practical example) 'constructed' $\mathscr{H}((b), \{Q|)\equiv\mathscr{H}(b)$, given $\mathscr{H}_0$, defines the renormalization group operation $\mathfrak{R}(b)$ (Wilson and Kogut 1974). Symbolically

$$\mathfrak{R}(b)\mathscr{H}_0=\mathscr{H}(b). \tag{II.4.18}$$

It is evident, once again, that most of the interesting properties of the system described by the hamiltonian $\mathscr{H}_0$ (specifically, the properties that involve statistical averages of functions of those original coordinates with wavevectors less than Λ/b) can be expressed in terms of similar properties of $\mathscr{H}(b)$.

It is, perhaps, a little less evident that the operation does, in fact, fulfil the objectives set out in the preceding section, and that $\mathscr{H}(b)$ is, indeed, 'less critical' than $\mathscr{H}_0$. However, the first stage of the operation (the integration over the large k modes) does, in effect, produce a system with a lattice spacing ('inverse Brillouin zone size') *larger* by a factor of b, and the relabelling of all lengths (in the second stage) does not change this. Thus $\mathscr{H}_0$ and $\mathscr{H}(b)$ are, once again, characterized by dimensionless correlation lengths $\hat{\xi}$ and $\hat{\xi}(b)$, which are related by

$$\hat{\xi}(b)=\hat{\xi}/b. \tag{II.4.19}$$

Indeed, this equation constitutes the simplest of the relationships (alluded to above) between the properties of the system described by $\mathscr{H}_0$, and the properties associated with $\mathscr{H}(b)$. It is a relationship to which we shall return, once we have given more detailed consideration to the R.G. operation itself.

II.4.3.3. *Hamiltonian flow under the renormalization group*

It will now prove helpful to express the effects of the R.G. operation in a somewhat more explicit fashion. To do so we note that the effective hamiltonians defined by (II.4.10) may be written in the general form

$$\mathscr{H}_0 = \sum_i p_{i,0} 0_i. \tag{II.4.20}$$

Here the 0_i constitute a set $\{0\}$ of 'operators', whose basic members (the operators associated with the L.G.W. hamiltonian $\mathscr{H}_0(n)$) are

$$\left.\begin{aligned} 0_1 &\equiv \frac{1}{2}\int_{\mathbf{k}}^{\Lambda} \sum_{j=1}^{n} |Q_j(\mathbf{k})|^2, \\ 0_2 &\equiv \int_{\mathbf{k}}^{\Lambda} Q_1(\mathbf{k})\delta(\mathbf{k}), \\ 0_3 &\equiv \frac{1}{2}\int_{\mathbf{k}}^{\Lambda} \sum_{j=1}^{n} k^2 |Q_j(\mathbf{k})|^2, \\ 0_4 &\equiv \int_{\mathbf{k}_1}^{\Lambda}\int_{\mathbf{k}_2}^{\Lambda}\int_{\mathbf{k}_3}^{\Lambda}\int_{\mathbf{k}_4}^{\Lambda} \sum_{i,\,j=1}^{n} Q_i(\mathbf{k}_1)Q_i(\mathbf{k}_2)Q_j(\mathbf{k}_3)Q_j(\mathbf{k}_4)\delta(\mathbf{k}_1+\mathbf{k}_2+\mathbf{k}_3+\mathbf{k}_4). \end{aligned}\right\} \tag{II.4.21 a}$$

The $p_{i,0}$ constitute a set $\{p_0\}$ of parameters (or 'coupling constants') which characterize the strength with which the various operators appear in the effective hamiltonian. Thus

$$\left.\begin{aligned} p_{1,0} &\equiv r_0, \\ p_{2,0} &\equiv E_0, \\ p_{3,0} &\equiv 1, \\ p_{4,0} &\equiv u_0. \end{aligned}\right\} \tag{II.4.21 b}$$

The effective hamiltonian $\mathscr{H}_0$ (to which we shall refer as the 'physical' effective hamiltonian, since its parameters characterize directly the physical, near-critical, system of interest) may thus be associated with a point $\{p_0\}$ in a 'space' of parameters $\{p\}$. Furthermore, the rescaling of momentum space involved in the second stage of the R.G. operation ensures that the spectrum of operators appearing in $\mathscr{H}(b)$ have precisely the same form as those, typified by (II.4.21 *a*), which appear in $\mathscr{H}_0$ (cf., however, the important qualifying remarks at the end of this section). Thus we may write

$$\mathscr{H}(b) = \sum_i p_i(b) 0_i, \tag{II.4.22}$$

where the parameters $p_i(b)$ differ from the original, 'physical' parameters $p_{i,0}$ both as a result of the renormalization of the interactions among the 'more critical' fluctuations, by the 'less critical' fluctuations (integrated over during the *first* stage of $\mathfrak{R}(b)$), and as a result of the redefinition of parameters necessary to accommodate the coordinate rescaling (occurring during the *second* stage of $\mathfrak{R}(b)$, eqn. (II.4.16)). The effective hamiltonian $\mathscr{H}(b)$ may, therefore, be characterized by a different 'point' $\{p(b)\}$ in the parameter space.

Evidently, then, repeated application of the R.G. operation will generate a whole sequence of effective hamiltonians which trace out a trajectory or 'flow line' in parameter space. (Again, § II.4.4 offers a simple but explicit illustration.) Since the renormalization group operation has the semigroup property

$$\mathfrak{R}(b_2)\mathfrak{R}(b_1)=\mathfrak{R}(b_2b_1), \qquad \text{(II.4.23)}$$

repeated application of the operation generates the same spectrum of effective hamiltonians as does a single application of the operation with a spectrum of values of b: the lines of hamiltonian flow are thus parameterized by the scale factor b, increasing in the direction of the flow. Alternatively (and rather more profitably) the trajectories may be regarded as lines of decreasing (dimensionless!) correlation length—repeated application of $\mathfrak{R}(b)$ (or increasing b in $\mathfrak{R}(b)$) bringing repeated reduction in $\hat{\xi}(b)$ (cf. eqn. (II.4.19)). The flow lines thus connect a sequence of effective hamiltonians, each of which is 'less critical' than its ('upstream') predecessor, and each of which contains the information that determines the critical ('small k') properties of the physical hamiltonian $\mathscr{H}_0$.

It is important to appreciate that the lines of hamiltonian flow will not be restricted to the space defined by the non-zero parameters of the physical hamiltonian. An effective hamiltonian generated by operating on, say, the L.G.W. hamiltonian $\mathscr{H}_0(n)$, eqn. (II.4.10 b), will contain (for example) sixth-order anharmonic terms, wavevector-dependent fourth-order anharmonic terms—indeed, an infinite spectrum of terms not present in the physical hamiltonian. Thus the physical hamiltonian must be regarded as occupying a site in a special subspace of the general parameter space in which the flow occurs.

While it is not hard to convince oneself that the hamiltonian flow will, at least, conserve the E.Q.F.s of the physical hamiltonian, this aspect of the renormalization group operation appears to wipe out the simplifications made possible (§ II.4.2) by the arguments founded on the universality hypothesis. To some extent this is true: the fact that, in general, the R.G. operation generates non-zero parameters for an infinite spectrum of operators, represents the central obstacle encountered in practical R.G. calculations. These become tractable (to analytic methods, at least) only in those situations where the hamiltonian flow is (effectively) restricted to a space characterized by a *few* parameters. The principal situation where this occurs is discussed in § II.4.4.

II.4.3.4. *Recursion relations and the fixed-point concept*

We have seen that the essential effect of the renormalization group is to generate, from the physical hamiltonian $\mathscr{H}_0$, characterized by parameters $\{p_0\}$, a continuous sequence of new effective hamiltonians, $\mathscr{H}(b)$, characterized by new parameters $\{p(b)\}$. The new parameters will be functions of the physical parameters (and of the scale factor b), and we may write

$$p_i(b)=\mathfrak{R}_i(b,\{p_0\}), \qquad \text{(II.4.24)}$$

thus introducing formally a set of R.G. equations, or 'recursion relations'—one for each parameter in the space in which the hamiltonian flow occurs.

Of the variety of possibilities for the ultimate behaviour of the flow, in the limit of large b (or, equivalently, in the limit of very many applications of $\mathfrak{R}$) the simplest is that the flow lines will terminate in a *fixed point* of the R.G. operation—that is, an effective hamiltonian $\mathscr{H}^*$ satisfying the invariance condition

$$\mathscr{H}^*=\mathfrak{R}(b)\mathscr{H}^*, \qquad \text{(II.4.25)}$$

while the parameters $\{p^*\}$ with which it is associated satisfy the equations

$$p_i^* = \Re_i(b, \{p^*\}). \qquad \text{(II.4.26)}$$

That fixed-point hamiltonians fall into two categories is revealed by (II.4.19), which we may rewrite, more explicitly, as

$$\hat{\xi}^{-1}[\mathscr{H}(b)] \equiv \hat{\xi}^{-1}[\Re(b)\mathscr{H}_0] = b\hat{\xi}^{-1}[\mathscr{H}_0], \qquad \text{(II.4.27)}$$

showing that the dimensionless correlation length of a fixed point hamiltonian must satisfy

$$\hat{\xi}^{-1}[\Re(b)\mathscr{H}^*] = b\hat{\xi}^{-1}[\mathscr{H}^*], \qquad \text{(II.4.28}\,a)$$

or, in view of (II.4.25),

$$\hat{\xi}^{-1}[\mathscr{H}^*] = b\hat{\xi}^{-1}[\mathscr{H}^*]. \qquad \text{(II.4.28}\,b)$$

Thus a fixed-point hamiltonian may have either an infinite correlation length, or a zero correlation length—i.e. may be (precisely) 'critical', or (totally) 'non-critical'.

Now, since the lines of hamiltonian flow are (cf. § II.4.3.3) lines of *decreasing* correlation length, it is evident that the flow line emanating from any point in hamiltonian space (i.e. the space of parameters $\{p\}$) can terminate in a 'critical' fixed point only if the hamiltonian (associated with that point) is *itself* 'critical'; otherwise the flow line can terminate only in a 'non-critical' fixed point.

The ultimate destination of the flow line emanating from a physical hamiltonian depends, then, in a singular way upon whether or not the hamiltonian is precisely at its critical point, this singular behaviour mirroring the critical singularities exhibited by the physical system. More specifically, the closer the physical system is to its critical point, the closer does its flow line pass to a critical fixed point, and the more are its thermodynamic properties dominated (in a manner to be made explicit below) by the nature of that fixed point.

Moreover, since lines of hamiltonian flow are (we suggested in § II.4.3.3) E.Q.F.-preserving, the critical fixed point, towards which critical hamiltonian flow occurs, must itself be representative of the universality class of the physical hamiltonian. Indeed, it is here that the ideas of universality find (in principle, at least) a basis within the renormalization group. For we may suppose—and it is a conjecture largely substantiated by explicit calculation—that the fixed point which 'dominates' the critical behaviour of one member of its universality class, 'dominates' also the behaviour of all the other members: the universality of critical behaviour thus emerges as a consequence of the network of hamiltonian flow lines which 'connect' a whole spectrum of physical hamiltonians (with *different* microscopic coupling constants, but the *same* E.Q.F.s) to the *same* critical fixed point.

Evidently, to substantiate this view requires a global analysis of the flow patterns generated by the R.G. operation, to establish that there exists but one accessible critical fixed point (for any given universality class) towards which all critical flow lines are directed. There are, however, relatively few instances where it has proved possible to carry through this programme—real-space R.G. analyses of the $d=1$ Ising model (Nelson and Fisher 1975) and —to some extent—of the $d=2$ Ising model (Niemeyer and van Leeuven 1973, 1974) offer illuminating exceptions. Rather, the philosophy of most renormalization group calculations has been to look for *one* critical fixed point representative of the universality class of interest, and to *assume* that this fixed point (if it exists: cf. § II.4.3.8) is unique. The critical behaviour of the universality class is then extracted by consideration of the flow line emerging from some near-critical hamiltonian chosen (for convenience—and with universality as 'justification') to lie *near* that critical fixed point. This is the standpoint we now adopt.

II.4.3.5. *Hamiltonian flow near a fixed point: scaling and critical exponents*

Consider an effective hamiltonian $\mathscr{H}_0$ characterized by a set of parameters $\{p_0\}$, which lie close to a set $\{p^*\}$ locating a critical fixed point (to which we shall henceforth refer simply as a fixed point, since we shall have nothing to say about non-critical fixed points: see, however, Nienhuis and Nauenberg (1975) and Nelson and Fisher (1975)). 'Close to' is to be taken to signify that the deviations

$$\Delta p_i \equiv p_{i,0} - p_i^* \tag{II.4.29}$$

of the parameters $p_{i,0}$ from their fixed point values p_i^* are sufficiently small that the effects of terms $O(\Delta p_i^2)$ upon the flow line emanating from $\mathscr{H}_0$ may be neglected. The R.G. equations (II.4.24), which describe this flow, may thus be replaced by their linearized counterparts

$$\Delta p_i(b) \equiv p_i(b) - p_i^* = \sum_j \mathfrak{R}_{ij}^*(b) \Delta p_j, \tag{II.4.30}$$

where the coefficients $\mathfrak{R}_{ij}^*(b)$ go to make up a matrix which depends only upon the fixed point values of the parameters, and the scale factor b.

It is natural now to introduce those linear combinations, $\mu_i(b)$ and μ_i, of the deviations, $\Delta p_i(b)$ and Δp_i, that diagonalize (II.4.30). These new deviations, to which we shall refer as 'scaling fields', then satisfy

$$\mu_i(b) = \Lambda_i^*(b)\mu_i, \tag{II.4.31}$$

where the $\Lambda_i^*(b)$ are eigenvalues of the matrix $\mathfrak{R}^*(b)$. Almost invariably, these eigenvalues have the simple form

$$\Lambda_i^*(b) = b^{\lambda_i}, \tag{II.4.32}$$

where the exponents λ_i are independent of the scale factor b, and, while involving the values of the fixed point parameters, are independent also of the cut-off Λ (which, it will be recalled, is preserved by the hamiltonian flow). The cut-off independence of the exponents λ_i (which, we shall see below, determine the critical exponents) affords further vindication of the universality concept.

It is clear from (II.4.32) that the evolution of a scaling field μ_i under the operation $\mathfrak{R}(b)$, depends, crucially, upon the *sign* of the associated exponent λ_i. A scaling field associated with a *positive* exponent λ_i is *amplified* by the operation ($\mu_i(b) > \mu_i$), driving the line of hamiltonian flow away from the fixed point: such fields are termed *relevant*. In contrast, a scaling field associated with a *negative* exponent λ_i is *deamplified* ('renormalized away') by the operation ($\mu_i(b) < \mu_i$): such fields are termed *irrelevant*. (A scaling field whose exponent λ_i is *zero* is termed 'marginal': when such a field is present, it requires separate and careful attention. See Wegner and Riedel (1973), and further comments in §II.5.1.3.)

At this point we recall our argument that a line of hamiltonian flow may terminate in a (critical) fixed point only if its source is a hamiltonian that is, itself, 'at criticality'. In the laboratory criticality is achieved by setting to zero *two* quantities: the reduced temperature $t \equiv (T - T_c)/T_c$ and the field E conjugate to the order parameter. In consequence, the spectrum of scaling fields associated with a fixed point must contain (at least) *two* that are relevant, one vanishing (analytically) with t and the other vanishing (analytically) with E. Denoting these two fields by μ_1 and μ_2 we may write, for *small* t and E, simply

$$\left.\begin{aligned}\mu_1&=t,\\ \mu_2&=E.\end{aligned}\right\} \tag{II.4.33}$$

To investigate the consequences of these results we consider once more the fundamental equation (II.4.19) relating the dimensionless correlation lengths of effective hamiltonians connected by the operation $\mathfrak{R}(b)$. Since the effective hamiltonians $\mathscr{H}(b)$ and $\mathscr{H}_0$ may be parameterized, respectively, in terms of the scaling fields $\{\mu(b)\}$ and $\{\mu\}$ together with the fixed point parameters $\{p^*\}$, we may usefully rewrite this equation in the form

$$\hat{\xi}[\{p^*\},\{\mu(b)\}]=b^{-1}\hat{\xi}[\{p^*\},\{\mu\}]. \tag{II.4.34}$$

Suppressing, for clarity, the common dependence upon $\{p^*\}$, and making use of eqns. (II.4.31)–(II.4.33), we then have

$$\hat{\xi}(t,E,\mu_3)=b\hat{\xi}(b^{\lambda_1}t,b^{\lambda_2}E,b^{\lambda_3}\mu_3\ldots). \tag{II.4.35}$$

The 'scaling' or 'homogeneity' properties exemplified by this equation were anticipated (in a less general form) many years before Wilson's development of renormalization group methods. (See the original papers by Widom (1965) and Kadanoff (1966), and the exposition by Stanley (1971).) They may be shown, by similar means, to extend to the complete range of thermodynamic functions (or, more precisely, their singular parts). In particular, the free energy per degree of freedom (in units of k_BT) may be written as

$$G=-\frac{\ln Z}{N}\equiv b^{-d}\left(\frac{-\ln Z}{N/b^d}\right). \tag{II.4.36}$$

Making use of the representation of the partition afforded by (II.4.17) then suggests that the most singular part of this free-energy density has the homogeneity property

$$G(t,E,\mu_3,\ldots)=b^{-d}G(b^{\lambda_1}t,b^{\lambda_2}E,b^{\lambda_3}\mu_3\ldots). \tag{II.4.37}$$

Strictly, this equation follows from (II.4.17) and (II.4.36) only if the contribution to the free energy made by the less critical fluctuations ($\Delta G(b)$ in (II.4.17)) is neglected as 'less singular'—an assumption that is difficult to justify. See Ma (1974) for a careful discussion of this point.

The homogeneity properties of the order parameter Q_0, and susceptibility χ,

$$Q_0(t,E,\mu_3\ldots)=b^{\lambda_2-d}Q_0(b^{\lambda_1}t,b^{\lambda_2}E,b^{\lambda_3}\mu_3\ldots), \tag{II.4.38a}$$

$$\chi(t,E,\mu_3\ldots)=b^{2\lambda_2-d}\chi(b^{\lambda_1}t,b^{\lambda_2}E,b^{\lambda_3}\mu_3\ldots), \tag{II.4.38b}$$

follow on differentiating (II.4.37). In fact, these results stand on more solid ground than does (II.4.37) itself, since the effects of $\Delta G(b)$ may be seen, quite explicitly, to cancel from the statistical averages involved in the calculation of such (strictly small k) properties (see e.g. Ma 1974).

Let us suppose now that, on examination, the spectrum of additional scaling fields (represented by $\mu_3\ldots$ in the above equations) turns out to contain no further relevant (or marginal) members. (The meaning of this supposition will be made plain in § II.4.3.7.) The effects of the (irrelevant) fields $\mu_3 \ldots$ are then removed by the critical fluctuations, as one may see most clearly by choosing the (hitherto arbitrary) scale factor b such that

$$b^{\lambda_1}|t| = 1. \tag{II.4.39}$$

Then the homogeneity relations (II.4.38) become

$$Q_0(t, E, \ldots) = |t|^{-(\lambda_2 - d)[\lambda_1} Q_0\left(\frac{t}{|t|}, \frac{E}{|t|^{\lambda_2/\lambda_1}}, \mu_3|t|^{-\lambda_3/\lambda_1}, \ldots\right) \tag{II.4.40 a}$$

$$\approx |t|^{(d-\lambda_2)/\lambda_1} Q_0\left(\pm 1, \frac{E}{|t|^{\lambda_2/\lambda_1}}, 0, \ldots\right)$$

$$\equiv |t|^{(d-\lambda_2)/\lambda_1} \tilde{Q}_{\pm}\left(\frac{E}{|t|^{\lambda_2/\lambda_1}}\right) \tag{II.4.40 b}$$

$$\chi(t, E, \ldots) = |t|^{-(2\lambda_2 - d)/\lambda_1} \chi\left(\frac{t}{|t|}, \frac{E}{|t|^{\lambda_2/\lambda_1}}, \mu_3|t|^{-\lambda_3/\lambda_1}, \ldots\right) \tag{II.4.40 c}$$

$$\approx |t|^{-(2\lambda_2 - d)/\lambda_1} \chi\left(\pm 1, \frac{E}{|t|^{\lambda_2/\lambda_1}}, 0, \ldots\right)$$

$$\equiv |t|^{-(2\lambda_2 - d)/\lambda_1} \tilde{\chi}_{\pm}\left(\frac{E}{|t|^{\lambda_2/\lambda_1}}\right). \tag{II.4.40 d}$$

Equation (II.4.40 b) reveals the scaling form of the equation of state (see e.g. Stanley 1971). In particular, in the absence of the ordering field we have, for $t \to 0^-$,

$$Q_0 \approx D|t|^{\beta}, \tag{II.4.41 a}$$

with

$$\beta \equiv \frac{d - \lambda_2}{\lambda_1}, \tag{II.4.41 b}$$

$$D \equiv \tilde{Q}_-(0) \equiv Q_0(-1, 0, 0 \ldots). \tag{II.4.41 c}$$

Similarly (II.4.40 d) shows that, in zero field, the susceptibility diverges according to the power-law form

$$\chi \approx C_{\pm}|t|^{-\gamma}, \tag{II.4.42 a}$$

with

$$\gamma = \frac{2\lambda_2 - d}{\lambda_1} \tag{II.4.42 b}$$

$$C_{\pm} \equiv \tilde{\chi}_{\pm}(0) \equiv \chi(\pm 1, 0, 0 \ldots). \tag{II.4.42 c}$$

The foregoing results illustrate several central and general principles. In particular we see that the choice of the scale factor b, eqn. (II.4.39), enables us to fulfil the cardinal R.G. objective set out in § II.4.3.1, namely, that of relating the physics of a near critical system (i.e. one with small reduced temperature $|t|$) to that of a system far from criticality (i.e. one with reduced temperature $b^{\lambda_1}|t| = 1$). The critical singularities are then seen to manifest themselves in the singular character of the hamiltonian flow relating critical and non-critical systems: critical exponents (eqns.

(II.4.41 *b*) and (II.4.42 *b*)) are expressed in terms of flow exponents, while subsidiary quantities—e.g. the amplitudes of the critical singularities (eqns. (II.4.41 *c*) and (II.4.42 *c*))—are given in terms of the properties of the non-critical system, which may be treated within perturbation theory.

More specifically, one may derive from further similar homogeneity relations expressions analogous to (II.4.40), which reveal that the complete spectrum of critical exponents ($\alpha, \beta, \gamma \ldots$) may be expressed (cf. table II.1) in terms of the *two* 'hamiltonian-flow exponents' λ_1 and λ_2 (themselves characteristic of the fixed point involved) and the system dimensionality. Thus, subject to certain reservations (cf. the comments following (II.4.37)) the renormalization group offers a basis for 'two-exponent scaling'—the strongest of a variety of scaling theories suggested by phenomenological arguments (see Fisher and Jasnow 1980 for a review), which, however, afford no means of *calculating* the values of the exponents.

By contrast, the R.G. framework affords a prescription for the calculation not only of these exponents (which characterize the *asymptotic* critical behaviour) but also of additional quantities, describing the *approach* to the critical regime. These form the topics of the two succeeding sections.

Table II.1. The principal critical exponents (for definitions see §II.2.2 and Fisher 1967) expressed in terms of the two R.G. flow exponents λ_1 and λ_2 (§II.4.3.5). Note that an exponent characterizing a singularity as $t \to 0^+$ is taken to be equal to that characterizing the same singularity as $t \to 0^-$—an assumption which, within R.G. theory, may fail only in exceptional circumstances. See Nelson (1976).

$\alpha = 2 - d/\lambda_1$	$\delta = \lambda_2/(d - \lambda_2)$
$\beta = (d - \lambda_2)/\lambda_1$	$\nu = 1/\lambda_1$
$\gamma = (2\lambda_2 - d)/\lambda_1$	$\eta = d + 2 - 2\lambda_2$

II.4.3.6. *The approach to criticality: corrections to the leading scaling behaviour*

We have seen that, in the asymptotic critical region (defined by small $|t|$ and E), the effects of all the irrelevant scaling fields associated with a fixed point are washed away by the critical fluctuations, leaving only a homogeneous dependence of observables upon E and t (as exemplified in eqns. (II.4.40 *b*, *d*)). The *range* of $|t|$ (and E) in which this 'asymptotic' behaviour will in fact be displayed falls into the category of non-universal properties, about which few definitive statements can be made, since it will reflect the extent to which the ('irrelevant') parameters of the physical system differ from their fixed-point values. However, as suggested by Wegner (1972), one may allow for the possibly non-asymptotic character of observed critical behaviour, to some extent at least, by taking into account the residual temperature dependence associated with the leading irrelevant scaling field—i.e. the scaling field, μ_3 say, which has the least negative scaling exponent λ_3, and which therefore decays most slowly with reduced temperature.

For example, within this approximation, the zero-field behaviour of the susceptibility follows from (4.40 *c*) as

$$\chi(t, \ldots) = |t|^{-(2\lambda_2 - d)/\lambda_1} \chi(\pm 1, 0, \mu_3 |t|^{-\lambda_3/\lambda_1}) \approx C_\pm |t|^{-\gamma}(1 + C'_\pm |t|^{\bar{\omega}} + \ldots). \quad \text{(II.4.43)}$$

Here we have made use of (II.4.42 *b*) and (II.4.42 *c*) and, following Wegner (1972), have expanded the (non-critical!) function $\chi(\pm 1, \ldots)$ in powers of the irrelevant scaling field, retaining only the leading term. The correction to scaling exponent

$$\bar{\omega} \equiv -\lambda_3/\lambda_1 \tag{II.4.44}$$

is, like the basic two flow exponents λ_1 and λ_2, characteristic of the fixed point concerned and may also be calculated from explicit realizations of the R.G. operation. Naturally, the amplitudes $C'_{\pm}$ (like the leading amplitudes $C_{\pm}$) are non-universal, depending upon the details of the physical system. The principal utility of (II.4.43) is, therefore, that it provides a simple but explicit refinement of the pure power-law form (II.4.42 *a*), for fitting to experimental data. A similar form, involving the *same* exponent $\bar{\omega}$, applies to other critical observables.

II.4.3.7. *The approach to criticality: crossover behaviour*

The full richness of the critical region, and the full power of renormalization group methods, are displayed in the context of the *crossover behaviour* exhibited by systems which *almost* belong to one universality class, but actually belong to another (cf. Griffiths 1970). Such a system is characterized by an effective hamiltonian of the form

$$\mathscr{H}_0 = \mathscr{H}_0^{pr} + g_0 \Delta\mathscr{H}_0. \tag{II.4.45}$$

Here $\mathscr{H}_0^{pr}$ is some ('primary') effective hamiltonian, the representative of a universality class, while $\Delta\mathscr{H}_0$ is some perturbation, which appears with strength g_0 in the total effective hamiltonian, and which is such that the universality class of $\mathscr{H}_0$ is different from the universality class of $\mathscr{H}_0^{pr}$. As an example we cite the problem of the uniaxial ferroelectric (cf. § II.5.1.3) with *weak* dipolar forces: $\mathscr{H}_0^{pr}$ is then the Landau–Ginzburg hamiltonian $\mathscr{H}_0(1)$, eqn. (II.4.10 *b*), describing a system with short-range forces and a non-degenerate soft mode, while $\Delta\mathscr{H}_0$ represents the effects of the dipole–dipole (Coulomb) interactions upon the soft-mode dispersion (g_0 then giving a measure of the Coulomb splitting: cf. (II.5.15 *b*) below).

In the absence of the perturbation ($g_0 = 0$) the critical behaviour of the hamiltonian (II.4.45) will be governed by the 'primary' critical exponents associated with the universality class of $\mathscr{H}_0^{pr}$; with g_0 non-zero the critical singularities must reflect the 'secondary' exponents of the universality class of the perturbed system—no matter how small the perturbation g_0! The 'discontinuity' in the exponents as a function of g_0 is, however, illusory, in that one expects (and the renormalization group confirms) that the system will, on approaching criticality, first display the 'primary' critical behaviour, characteristic of the unperturbed system, exhibiting a 'crossover' to its asymptotic, 'secondary', critical behaviour only in a temperature range whose smallness reflects the smallness of g_0.

To make these ideas more precise we consider the R.G. flow in the space of hamiltonians of the form (II.4.45). For $g_0 = 0$ the line of flow emanating from a (near-critical) hamiltonian $\mathscr{H}_0$ will pass close to the fixed-point characteristic of the unperturbed system. (We shall refer to this as the 'primary' fixed point.) For g_0 non-zero, the line of flow must ultimately approach the fixed point of the universality class of the perturbed system (the 'secondary' fixed point). However, the analyticity of the R.G. equations guarantees that, if g_0 is 'small', the flow will still (en route) traverse the region close to the *primary* fixed point, where the equations may be linearized to establish, once again, homogeneity relations such as (II.4.40 *c*). The crucial new feature is that, among the scaling fields, $\mu_3 \ldots$ one would discover one new field, μ_g say, proportional to g_0 (for small g_0) and with a *positive* exponent λ_g, making the field *relevant*. Thus the spectrum of scaling fields allowed by the form of $\mathscr{H}_0$ has *three* members that are relevant with respect to the primary fixed point, rather than the minimum of *two* (supposed in the analysis of § II.4.3.5—see remarks following (II.4.38)). The primary fixed point is said to be *stable* in the space of hamiltonians $\mathscr{H}_0^{\rm pr}$ ('only-twice-unstable' would be more apt, but less succinct) and *unstable* in the full space of hamiltonians $\mathscr{H}_0$.

Explicitly, then, the homogeneity relation (II.4.40 *c*) yields (we replace t by $\tilde{t}$, for reasons soon to be apparent and choose $\tilde{t}>0$ for simplicity):

$$\chi(\tilde{t},g_0)\approx\tilde{t}^{-\frac{2\lambda_2-d}{\lambda_1}}\chi(1,0,\ldots g_0/\tilde{t}^{\lambda_g/\lambda_1}), \tag{II.4.46}$$

where we have denoted the field μ_g simply by g_0, have set the ordering field E to zero (again for simplicity) and have suppressed all the other (irrelevant) fields, which are once more renormalized away in the small $\tilde{t}$ regime.

Equation (II.4.46) has the form

$$\chi(\tilde{t},g_0)\approx\tilde{t}^{-\gamma}\hat{\chi}(g_0/\tilde{t}^{\phi}), \tag{II.4.47 a}$$

where we have introduced a *crossover scaling function* $\hat{\chi}$, and a (positive) *crossover exponent*

$$\phi\equiv\lambda_g/\lambda_1. \tag{II.4.47 b}$$

We have also made use of (II.4.42 *b*) to reintroduce the exponent γ, which is characteristic of the *primary* fixed point, and thus describes the manner in which the susceptibility *would* diverge in the *absence* of the perturbation g_0. It must likewise be recognized that the quantity $\tilde{t}$ (which enters the analysis as one of the two basic relevant fields of the *primary* fixed point—recall the argument preceding (II.4.33)) must be interpreted as the deviation from the transition temperature that the system would have in the absence of the perturbation:

$$\tilde{t}\equiv(T-T_{\rm c}(g_0=0))/T_{\rm c}(g_0=0). \tag{II.4.48}$$

(More generally the relevant fields μ_1 and μ_g will, in fact, be linear combinations of g_0 and $\tilde{t}$: see e.g. Fisher 1975.)

Equation (II.4.47) makes plain that, for reduced temperatures

$$\tilde{t}\gg\tilde{t}_g\equiv g_0^{1/\phi} \tag{II.4.49}$$

(but still within the critical region) it will, indeed, appear as if the susceptibility is to diverge (at $T_{\rm c}(g_0=0)$) with the primary exponent γ: only at reduced temperatures $\tilde{t}\sim\tilde{t}_g$ does the perturbation begin to make itself felt. Since we expect the susceptibility to diverge at some temperature $T_{\rm c}(g)\neq T_{\rm c}(g=0)$, with some secondary exponent $\dot{\gamma}$, say, it is evident that the function $\hat{\chi}$ must itself exhibit a divergence:

$$\hat{\chi}(x)\to\infty,\quad \text{as } x\to x_{\rm c},$$

with

$$x\equiv g_0/\tilde{t}^{\phi}, \tag{II.4.50 a}$$

$$x_{\rm c}\equiv g_0/\tilde{t}_{\rm c}^{\phi}, \tag{II.4.50 b}$$

$$\tilde{t}_{\rm c}\equiv\tilde{t}\Big|_{T_{\rm c}(g_0)}\equiv(T_{\rm c}(g_0)-T_{\rm c}(0))/T_{\rm c}(g_0). \tag{II.4.50 c}$$

Specifically we expect that

$$\hat{\chi}(x) \sim (1 - x/x_c)^{-\dot{\gamma}}, \tag{II.4.50 d}$$

thus bringing about the anticipated divergence of χ.

It should be made clear that the renormalization group arguments we have given so far offer little more than a basis for the 'extended scaling hypothesis', (II.4.46), previously grounded in phenomenological arguments (see e.g. Pfeuty *et al.* 1974, and references therein). There is nothing in the argument, as given, that establishes the properties we have ascribed to the function $\hat{\chi}(x)$. Nevertheless, the practical renormalization group methods we shall describe briefly in § II.4.4 may be used to determine crossover functions explicitly (as shown first by Riedel and Wegner 1974), to confirm their anticipated behaviour, and to demonstrate that, like the equation of state, they have universal features (Pfeuty *et al.* 1974).

It is appropriate to mention at this point that the manner in which the behaviour of a system evolves from its primary to its secondary form may be described, in a particularly succinct fashion, in terms of an *effective* critical exponent defined (see e.g. Riedel and Wegner 1974) for the susceptibility (say) as the logarithmic derivative

$$\gamma_{\text{eff}} \equiv -\frac{\partial \ln \chi}{\partial \ln t}, \tag{II.4.51 a}$$

where in accord with our usual notation we write

$$t \equiv (T - T_c(g_0))/T_c(g_0) \tag{II.4.51 b}$$

for the deviation from the true transition temperature. It is easily checked, with the aid of (II.4.47 *a*) and (II.4.50), that γ_{eff} evolves from the primary value γ (when $t \gg g_0^{1/\phi}$) to the secondary value $\dot{\gamma}$ (when $t \ll g_0^{1/\phi}$), its intermediate behaviour reflecting the details of the crossover function $\hat{\chi}(x)$—details which emerge from the explicit calculations alluded to in § II.4.5, and exemplified in § II.5.

Two final remarks are in order. Firstly it is to be emphasized that the crossover theory sketched above is applicable only if the universality-class-changing perturbation is 'small': if this condition is not fulfilled the system's behaviour will evolve from its 'non-critical' to its 'secondary critical' form, without displaying an intermediate primary critical regime. (In renormalization group language the line of flow emanating from a large-g_0 hamiltonian never strays into the territory of the primary fixed point.)

Secondly we remark that, while in many situations the symmetry-breaking perturbation is not subject to experimental control, in certain situations it can be adjusted, and turned off altogether! As an example (which, again, we consider explicitly below: cf. § II.5.3.3) we cite the situation where the perturbation g_0 takes the form of a symmetry-breaking stress which lifts the degeneracy of (say) a 3-fold degenerate soft mode, leaving a non-degenerate mode as the one that is 'softest'. The primary behaviour will then be that of an $n=3$-component system, and the secondary behaviour that of an $n=1$-component system, while the extent of the perturbation is controllable through the applied stress.

Evidently, in these circumstances, *fully* critical primary behaviour will be observed only in the limit in which the *three* quantities $T - T_c$, E *and* g_0 all tend to zero (the R.G. analogue of this statement being that there are three relevant fields associated with the primary fixed point). Such a critical point is termed '*multi-*

critical' (and is, thus, associated with an unstable fixed point) to distinguish it from an ordinary critical point (associated with a stable fixed point) which can be reached by adjusting only the *two* quantities, t and E. (We caution that these conventions are potentially confusing in that any 'critical' point may be regarded as a 'multicritical' point—and any 'stable' fixed point regarded as an 'unstable' fixed point—when viewed in an appropriate hamiltonian space.)

We will not chronicle the definitions of the four basic types of multicritical point—'tricritical' (see e.g. Griffiths 1973), 'bicritical' and 'tetracritical' (see Fisher and Nelson 1974, Aharony and Bruce 1974) and 'Lifshitz' (Hornreich *et al*. 1975)—but will describe their essential features in the context of the specific examples we shall meet in § II.5.3.

II.4.3.8. *First-order phase transitions and the renormalization group*

Our discussion up to this point has focused upon continuous phase transitions: we have seen that, as a result of critical fluctuations, classical Landau-type theories fail to meet the central challenge posed by such phase changes—the determination of universal quantities—and we have discussed the way in which the renormalization group theory meets this challenge.

We now turn to consider discontinuous (first-order) phase transitions where, clearly, the key task of theory is to isolate the feature, or features, of the physical system that are responsible for the discontinuous nature of the phase change, and hence to provide a means of *predicting* whether a given system will exhibit a continuous or discontinuous phase transition. The fact that a first-order phase change pre-empts the region of strongest critical fluctuations suggests that Landau theory might prove more successful in this sphere, and so we begin by recalling the features which, within Landau theory (cf. § I.4.1.1), issue in a first-order transition: such features may be usefully subdivided into two categories.

Firstly, the occurrence of a discontinuous phase transition may be a *necessary* consequence of the *symmetry* of the problem, when it allows the presence of certain types of microscopic interaction (cf. the effects of the third-order anharmonic term considered in § I.4.1).

Secondly, a first-order transition may be *contingent* upon the specific magnitudes (and *signs*) of certain microscopic interaction coefficients which, given other values, would result in a continuous phase transition (cf. the effect of the negative fourth-order anharmonic term considered in § I.4.1.1).

This subdivision of first-order phase transitions into two categories (to which we shall refer as categories one and two) remain valid when fluctuations are taken into account: however, renormalization group methods show that the Landau predictions have to be refined in two respects. Firstly, in the case of category-one phase changes, these methods reveal that the class of symmetries leading to a first-order transition has to be extended to include cases which, within Landau theory, could exhibit a continuous phase transition. Secondly, in the case of category-two phase changes, one finds that the boundaries of the domains of parameter space, within which a first-order transition occurs, are modified by critical fluctuations.

To elaborate upon these remarks we note first that a system exhibiting a first-order transition may usefully be regarded as one in which criticality cannot be reached by adjusting the usual two quantities (the temperature and the applied field); the natural renormalization group corollary is that such a system will have an effective hamiltonian which cannot, by adjusting two quantities, be set upon the critical surface of hamiltonians whose flow lines reach a (critical) fixed point. This state of affairs may come about in one of two conceptually different ways, which correspond to the two categories introduced above.

Firstly, one may find that, in the space of hamiltonians of the prescribed symmetry, there is *no stable fixed point*—i.e. that any fixed point in the space has at least *three* relevant scaling fields. A calculation which fails to reveal a stable fixed point is, then, taken as a signal that the system under investigation will exhibit a (category one) first-order phase transition (Bak *et al.* 1976)—the implied, and potentially dangerous, assumption being that *all* the fixed points have been identified in the calculation. With one or two exceptions (cf. the $d=2$ 3-state Potts model: Straley and Fisher 1973, Baxter 1973) those systems which within Landau theory exhibit a category-one first-order phase transition are indeed found to possess no stable fixed point. However, explicit renormalization group calculations reveal other systems (with potentially continuous transitions within Landau theory) which appear to have no stable fixed point, indicating that the Landau list of category-one features is incomplete. As examples we mention certain systems with multi-component ($n \geqslant 4$) order parameters (Bak *et al.* 1976) and (possibly) cubic ferroelectrics (Bruce and Aharony 1974; cf. § II.5.2.3). Further references to such 'fluctuation-driven' first-order transitions, and comparisons with experiment, may be found in §§ II.4.5.1*a*, II.5.1.2 and II.5.2.2.

Secondly, there occur situations where a stable fixed point *does* exist (within the prescribed hamiltonian space) but is inaccessible to lines of flow originating in certain domains of the space: systems with hamiltonians in such domains may thus be expected to exhibit (category-two) first-order transitions, as surmised in the seminal paper by Wilson and Fisher (1972) and exemplified in the explicit renormalization group analysis to be presented in the following section. The boundaries of the domains do not, in general, coincide with those predicted by Landau theory—reflecting the renormalization effects of critical fluctuations (cf. Aharony 1976, pp. 397, 398). However, this is an area in which the theory has not been extensively developed—largely because its relevance to any specific system can be demonstrated only if the ('non-universal') details of the system hamiltonian are known.

II.4.4. *The renormalization group method, in practice: a worked example*

In this section we shall outline the way in which the principles we have described are put into practice in the methods devised and developed by Wilson and Kogut (1974) and Wilson and Fisher (1972), and reviewed in some detail by Aharony (1976). We will describe only the most significant of the practical details—particularly the way in which the existence of a critical dimensionality d_c is exploited by the theory—and we will restrict our discussion to the simplest non-trivial application of the theory, namely that associated with the $n=1$ L.G.W. hamiltonian (eqn. (II.4.7)) in the absence of an ordering field:

$$\mathscr{H}_0 \equiv \mathscr{H}_0(\{Q\}) = \frac{1}{2}\int_{\mathbf{k}}^{\Lambda} (r_0 + k^2)|Q(\mathbf{k})|^2$$

$$+ u_0 \int_{\mathbf{k}_1}^{\Lambda}\int_{\mathbf{k}_2}^{\Lambda}\int_{\mathbf{k}_3}^{\Lambda}\int_{\mathbf{k}_4}^{\Lambda} Q(\mathbf{k}_1)Q(\mathbf{k}_2)Q(\mathbf{k}_3)Q(\mathbf{k}_4)\delta(\mathbf{k}_1+\mathbf{k}_2+\mathbf{k}_3+\mathbf{k}_4). \qquad \text{(II.4.52)}$$

It should be clear from the context in which this effective hamiltonian was first introduced (§ II.4.2) that it is expected to describe the universality class of systems for which the phase transition is associated with a non-degenerate non-ferroelectric soft mode: this universality class includes the model system defined by (II.2.1), and the Ising model in particular (so that the R.G. predictions may, in *this* case, be tested against the results established by other methods: cf. § II.2.2).

It will be recalled from § II.4.3.2 that the first stage of the R.G. operation requires that one integrate out from the partition function

$$Z=\int_{\{Q\}} \exp[-\mathscr{H}_0(\{Q\})] \tag{II.4.53}$$

the set of coordinates $Q(\mathbf{k})$ with wavevectors $k>\Lambda/b$ and thence determine the effective hamiltonian $\bar{\mathscr{H}}(b,\{Q\}^<)$, eqn. (II.4.15), in which the interactions among the $k<\Lambda/b$ modes have been renormalized through processes mediated by the $k>\Lambda/b$ modes.

It is abundantly clear that it is not possible to do this exactly—any more than it is possible to integrate directly over *all* the coordinates and hence calculate Z exactly. Evidently an approximation scheme is necessary, and, in this case, the techniques of anharmonic perturbation theory (cf. § I.5.1.1) are appropriate.

At first sight such a policy might seem doomed to failure in view of the inadequacies of such approximation schemes in the critical region (cf. § II.3.3). However, it turns out that, while a direct perturbation expansion for observables is uncontrollable in the critical region, a perturbative-realization of the renormalization group operation *can* be controlled, in a fashion to be made clear below.

Now it is clear that the effective hamiltonian $\bar{\mathscr{H}}(b,\{Q\}^<)$ must have the form

$$\bar{\mathscr{H}}(b,\{Q\}^<)=\frac{1}{2}\int_{\mathbf{k}}^{\Lambda/b}(\bar{r}(b)+\bar{f}(b)k^2)|Q(\mathbf{k})|^2$$

$$+\bar{u}(b)\int_{\mathbf{k}_1}^{\Lambda/b}\int_{\mathbf{k}_2}^{\Lambda/b}\int_{\mathbf{k}_3}^{\Lambda/b}\int_{\mathbf{k}_4}^{\Lambda/b} Q(\mathbf{k}_1)Q(\mathbf{k}_2)Q(\mathbf{k}_3)Q(\mathbf{k}_4)\delta(\mathbf{k}_1+\mathbf{k}_2+\mathbf{k}_3+\mathbf{k}_4)+\dots \tag{II.4.54}$$

In writing this equation we have allowed for the two effects of the $k>\Lambda/b$ fluctuations: firstly, that they will renormalize the parameters describing the original interactions among the $k<\Lambda/b$ modes in $\mathscr{H}_0$ – hence the new coefficients $\bar{r}(b)$, $\bar{f}(b)\dots$; secondly that they will result in *additional* interactions—e.g. a sixth-order anharmonic term—not present in $\mathscr{H}_0$, so that the hamiltonian flow would appear to depart into a space characterized by an intractably large number of parameters. However, within the perturbation scheme introduced below it proves possible to neglect these new interactions altogether, and we include the '...' in (II.4.54) merely as a reminder that they are present in principle.

The relationships between the new parameters ($\bar{r}(b),\dots$) and the old parameters ($r_0,\dots$) may be constructed using the methods of diagrammatic perturbation theory, referred to in § I.5.1.1, with certain modifications. Firstly, we work in the strictly classical and static limits: thus we require the $\hbar\to 0$ and $\omega\to 0$ limits of the general perturbation theory. Secondly, and more significantly, the operation requires that, at this stage, we take into account the renormalization effects of the $k>\Lambda/b$ modes only. Thus the intermediate 'phonon lines' in the diagrammatic representations of perturbation expansions (cf. fig. I.46) can carry only momenta in the restricted range $\Lambda/b<|\mathbf{k}|<\Lambda$.

To leading order in the coupling constant, the renormalization of the parameter r_0 (which plays the role of the soft-mode frequency squared) involves only the self-energy contribution represented diagrammatically by fig. I.46 (*b*). Explicitly one finds

$$\bar{r}(b)=r_0+12u_0 \int_{\mathbf{k}}^{\Lambda} \frac{1}{(r_0+k^2)}+O(u_0^2), \tag{II.4.55 a}$$

where we introduce the notation

$$\int_{\mathbf{k}}^{\Lambda} \equiv \left(\frac{a}{2\pi}\right)^d \int_{\Lambda/b<|\mathbf{k}|<\Lambda} d^d k. \tag{II.4.56}$$

The leading renormalization of the coupling constant arises through the process depicted in fig. I.46 (*g*), and gives

$$\bar{u}(b)=u_0-36u_0^2 \int_{\mathbf{k}}^{\Lambda} \frac{1}{(r_0+k^2)^2}+O(u_0^3). \tag{II.4.55 b}$$

Finally the coefficient of k^2 in the quadratic term is renormalized only through self-energy contributions that depend upon the external momentum: the leading contribution of this type is $O(u_0^2)$ (cf. fig. I.46 (*d*) and can be neglected, so that

$$\bar{f}(b)=1+O(u_0^2), \tag{II.4.55 c}$$

which completes the first stage of the renormalization group operation.

The second stage of the operation consists of the coordinate rescaling (cf. eqn. II.4.16):

$$Q(\mathbf{k})\equiv\zeta Q'(\mathbf{k}') \qquad \mathbf{k}'\equiv b\mathbf{k}, \tag{II.4.57}$$

where ζ is a rescaling factor, to be specified below. The terms in the hamiltonian (II.4.54) must now be re-expressed as functionals of the rescaled coordinates. Thus, for example, we write

$$\bar{u}(b)\int_{\mathbf{k}_1}^{\Lambda/b}\int_{\mathbf{k}_2}^{\Lambda/b}\int_{\mathbf{k}_3}^{\Lambda/b}\int_{\mathbf{k}_4}^{\Lambda/b} Q(\mathbf{k}_1)Q(\mathbf{k}_2)Q(\mathbf{k}_3)Q(\mathbf{k}_4)\delta(\mathbf{k}_1+\mathbf{k}_2+\mathbf{k}_3+\mathbf{k}_4)$$

$$=\bar{u}(b)\left(\frac{a}{2\pi}\right)^d \int_{0<|\mathbf{k}_1|<\Lambda/b} d^d k_1 \ldots \left(\frac{a}{2\pi}\right)^d \int_{0<|\mathbf{k}_4|<\Lambda/b} d^d k_4 Q(\mathbf{k}_1)\ldots Q(\mathbf{k}_4)\delta(\mathbf{k}_1+\ldots+\mathbf{k}_4)$$

$$=\bar{u}(b)\left(\frac{a}{2\pi}\right)^d \int_{0<|\mathbf{k}_1|<\Lambda/b} d^d(bk_1)\ldots \left(\frac{a}{2\pi}\right)^d \int_{0<|\mathbf{k}_4|<\Lambda/b} d^d(bk_4) b^{-4d}\zeta^4 Q'(b\mathbf{k}_1)$$

$$\ldots Q'(b\mathbf{k}_4)b^d\delta(b\mathbf{k}_1+\ldots+b\mathbf{k}_4)$$

$$=\zeta^4 b^{-3d}\bar{u}(b)\int_{\mathbf{k}_1'}^{\Lambda}\ldots\int_{\mathbf{k}_4'}^{\Lambda} Q'(\mathbf{k}_1')\ldots Q'(\mathbf{k}_4')\delta(\mathbf{k}_1'+\ldots+\mathbf{k}_4'). \tag{II.4.58}$$

In this way the 'intermediate' hamiltonian $\bar{\mathscr{H}}(b,\{Q\}^<)$ may be re-expressed in its final form $\mathscr{H}(b,\{Q\})$:

$$\bar{\mathscr{H}}(b,\{Q\}^<)=\mathscr{H}(b,\{Q\})=\frac{1}{2}\int_{\mathbf{k}}^{\Lambda}(r(b)+f(b)k^2)|Q(\mathbf{k})|^2$$

$$+u(b)\int_{\mathbf{k}_1}^{\Lambda}\int_{\mathbf{k}_2}^{\Lambda}\int_{\mathbf{k}_3}^{\Lambda}\int_{\mathbf{k}_4}^{\Lambda} Q(\mathbf{k}_1)Q(\mathbf{k}_2)Q(\mathbf{k}_3)Q(\mathbf{k}_4)\delta(\mathbf{k}_1+\mathbf{k}_2+\mathbf{k}_3+\mathbf{k}_4), \tag{II.4.59}$$

where we have dropped the (now-redundant) primes, and have introduced new parameters $r(b), f(b) \ldots$ into which have been subsumed the effects of the coordinate rescaling. Explicitly

$$r(b) \equiv \zeta^2 b^{-d} \bar{r}(b), \tag{II.4.60 a}$$

$$u(b) \equiv \zeta^4 b^{-3d} \bar{u}(b), \tag{II.4.60 b}$$

$$f(b) \equiv \zeta^2 b^{-d-2} \bar{f}(b). \tag{II.4.60 c}$$

To complete the operation we must make a suitable choice for the scale factor ζ, which is at our disposal. This factor controls the *overall* scale of the parameters in the hamiltonian $\mathscr{H}(b)$: it is convenient to fix its value so as to expose as clearly as possible the manner in which the parameters evolve *relative* to one another. It is therefore natural to select a value for ζ such that one of the parameters becomes an invariant of the renormalization group operation. In the case in hand it proves most convenient to fix ζ so that the coefficient of k^2 remains at its initial value of unity (cf. eqn. (II.4.52)). In some cases a different choice of ζ may prove more convenient. Equations (II.4.55 c) and (II.4.60 c) then give as the defining equation for ζ:

$$\zeta^2 = b^{d+2}[1 + O(u_0^2)]. \tag{II.4.61}$$

The R.G. equations relating the final set of parameters $r(b)$, $u(b)$ to the initial set r_0, u_0 then follow from (II.4.55), (II.4.60) and (II.4.61) as

$$r(b) = b^2\left[r_0 + 12u_0 \int_{\mathbf{k}}^{\Lambda} \frac{1}{(r_0 + k^2)} + O(u_0^2)\right], \tag{II.4.62 a}$$

$$u(b) = b^{4-d} u_0 \left[1 - 36u_0 \int_{\mathbf{k}}^{\Lambda} \frac{1}{(r_0 + k^2)^2} + O(u_0^2)\right]. \tag{II.4.62 b}$$

The fixed points of these equations consist of the sets of parameters r^*, u^* satisfying

$$r^* = b^2\left[r^* + 12u^* \int_{\mathbf{k}}^{\Lambda} \frac{1}{(r^* + k^2)}\right], \tag{II.4.63 a}$$

$$u^* = b^{4-d} u^* \left[1 - 36u^* \int_{\mathbf{k}}^{\Lambda} \frac{1}{(r^* + k^2)^2}\right]. \tag{II.4.63 b}$$

In seeking the solutions to these equations we must be guided by the two conditions we expect them to fulfil. Firstly, our approximate realization of the R.G. transformation will be self-consistent only if the fixed point is found to lie in the region of parameter space where the coupling constant is small, as assumed in the perturbation expansion. Secondly, the whole procedure will be meaningful only if it yields fixed-point parameters that are independent of the scale factor b.

The first of these conditions makes it clear that, in looking for an acceptable solution to (II.4.63), we may neglect the r^* terms that appear in the integrals, since r^* will itself be of order u^*, while the integrals are already multiplied by a coupling constant. The integrals may then be approximated as

$$\int_{\mathbf{k}}^{\Lambda} \frac{1}{(r^* + k^2)} \approx \left(\frac{a}{2\pi}\right)^d \int_{\Lambda/b < |\mathbf{k}| < \Lambda} \frac{d^d k}{k^2} = \frac{K_d a^d}{d-2}[\Lambda^{d-2} - (\Lambda/b)^{d-2}], \tag{II.4.64 a}$$

$$\int_{\mathbf{k}} \frac{1}{(r^* + k^2)^2} \approx \left(\frac{a}{2\pi}\right)^d \int_{\Lambda/b < |\mathbf{k}| < \Lambda} \frac{d^d k}{k^4} = \frac{K_d a^d}{d-4}[\Lambda^{d-4} - (\Lambda/b)^{d-4}], \tag{II.4.64 b}$$

where K_d is a dimensionality-dependent constant.

It is now evident that, in addition to the solution $u^*=0$, (II.4.63 *b*) has the non-trivial solution

$$u^* = \frac{1-b^{d-4}}{\dfrac{36K_d a^d}{d-4}[\Lambda^{d-4}-(\Lambda/b)^{d-4}]},$$

$$\equiv \frac{1-b^{-\varepsilon}}{\dfrac{36K_d a^d}{\varepsilon}[(\Lambda/b)^{-\varepsilon}-\Lambda^{-\varepsilon}]}, \tag{II.4.65}$$

where we have introduced the parameter

$$\varepsilon \equiv 4-d. \tag{II.4.66}$$

Now, physical systems obviously have integral dimensionalities d (generally $d=3$, occasionally—in effect—$d=1$ or 2); formally, however, d may be regarded as a continuous variable. In particular one may explore the 'physics' of systems characterized by small values of ε—i.e. systems with dimensionality d 'close to' 4.

The remarkable utility of ε as a small parameter, in the theory of critical phenomena, was first recognized and implemented by Wilson and Fisher (1972). The power of the approach is revealed by inspection of (II.4.65), which shows that, in the small ε limit, the solution for u^* fulfils *both* of the consistency conditions noted above. Explicitly, making use of the expansion

$$x^\varepsilon \equiv \exp(\varepsilon \ln x) = 1+\varepsilon \ln x + O(\varepsilon^2), \tag{II.4.67}$$

we find

$$u^* = \frac{\varepsilon}{36K_d a^d} + O(\varepsilon^2)$$

$$\equiv u_I^*, \tag{II.4.68 a}$$

which is, indeed, independent of b, and 'small' if ε is small. The non-trivial solution to (II.4.63 *a*) then follows, using the approximation (II.4.64 *a*), as

$$r^* = -\frac{12u^* K_d a^d b^2}{(b^2-1)(d-2)}[\Lambda^{d-2}-(\Lambda/b)^{d-2}]$$

$$= -\frac{\varepsilon\Lambda^2}{6} + O(\varepsilon^2)$$

$$\equiv r_I^*. \tag{II.4.68 b}$$

We have attached the subscript I to these fixed-point values to distinguish them from the self-evident additional solutions to (II.4.63):

$$u^* = 0 \equiv u_G^*, \tag{II.4.69 a}$$

$$r^* = 0 \equiv r_G^*. \tag{II.4.69 b}$$

The basic character of each fixed point is reflected in the nature of the hamiltonian flow lines generated by the renormalization group equations (II.4.62): a few trajectories are displayed in fig. II.3. The essential features of the flow may be

Fig. II.3

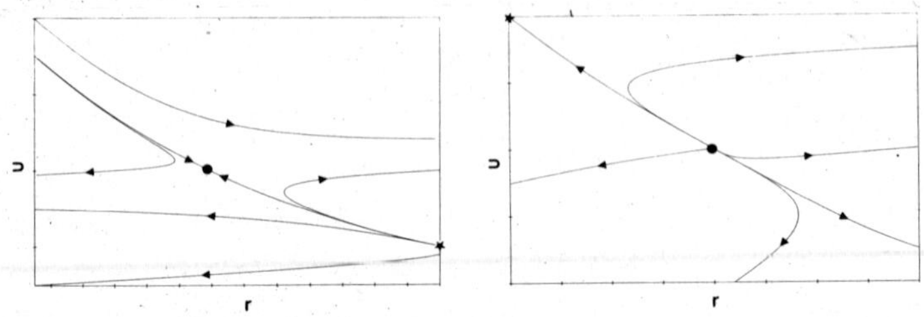

(a) (b)

Lines of hamiltonian flow generated by the R.G. equations for the Ising universality class, at leading order in $\varepsilon=4-d$, (a) for $\varepsilon=1(d=3)$, (b) for $\varepsilon=-1(d=5)$. In each case the origin $r=u=0$ is located by the gaussian fixed point (II.4.69) denoted by *, while the Ising fixed point (II.4.68) is located by the symbol ●. Note that the flow lines were actually generated from the *differential* form of the R.G. eqns. (II.4.62), described in § II.4.5.2 (b).

characterized analytically through the linearized equations (cf. eqn. (II.4.30)):

$$\begin{pmatrix}\Delta r(b)\\ \\ \Delta u(b)\end{pmatrix}=\mathfrak{R}^*(b)\begin{pmatrix}\Delta r\\ \\ \Delta u\end{pmatrix}, \tag{II.4.70}$$

which describe the evolution of the deviations (cf. eqns. (II.4.29))

$$\Delta r\equiv r_0-r^*;\quad \Delta u\equiv u_0-u^*, \tag{II.4.71}$$

assumed to be small.

Considering, first of all, the fixed point (II.4.69), one readily finds, from (II.4.62), linearized equations of the form (II.4.70) with

$$\mathfrak{R}^*(b)=\mathfrak{R}^*_G(b)\equiv\begin{pmatrix}b^2 & 12b^2I_0\\ 0 & b^\varepsilon\end{pmatrix}, \tag{II.4.72 a}$$

where

$$I_0\equiv\int_{\mathbf{k}}^{\Lambda}\frac{1}{k^2}=\frac{a^d\Lambda^2K_d}{2}(1-1/b^2)+O(\varepsilon). \tag{II.4.72 b}$$

The eigenvalues of the matrix (II.4.72 a) follow trivially as

$$\Lambda^*_{1,G}(b)=b^2, \tag{II.4.73 a}$$

$$\Lambda^*_{3,G}(b)=b^\varepsilon, \tag{II.4.73 b}$$

so that the associated scaling field exponents are (cf. eqns. (II.4.32))

$$\lambda_{1,G}=2, \tag{II.4.74 a}$$

$$\lambda_{3,G}=\varepsilon. \tag{II.4.74 b}$$

To these results we shall append the scaling field exponent for the ordering field (which we set to zero at the outset of this analysis)

$$\lambda_{2,G}=3-\varepsilon/2, \tag{II.4.74 c}$$

which follows from the R.G. equation for the ordering field, readily derivable as

$$\begin{aligned} E(b) &= \zeta E_0 \\ &= b^{(d+2)/2} E_0[1+O(u_0^2)]. \end{aligned} \tag{II.4.75}$$

It is evident from (II.4.74) that, for dimensionalities $d<4$ (i.e. $\varepsilon>0$), the fixed point (II.4.69) has (at least) three—it is in fact only three, for small ε—relevant scaling fields. This fixed point is, therefore, *unstable* and cannot describe an ordinary critical point, for $d<4$. (It can, and does, describe a *tricritical* point in three dimensions: cf. §II.5.3.1.)

Turning, then, to examine the stability properties of the other fixed point, one finds, from (II.4.62) and (II.4.68), the linearized equations (II.4.70) with

$$\mathfrak{R}^*(b) \equiv \mathfrak{R}_I^*(b) = \begin{pmatrix} b^2(1-12u^*I_1) & 12b^2 I_0 \\ 0 & b^{\varepsilon}(1-72u^*I_1) \end{pmatrix}, \tag{II.4.76 a}$$

where

$$I_1 \equiv \int_{\mathbf{k}}^{\Lambda} \frac{1}{k^4} = a^d K_d \ln b + O(\varepsilon) \tag{II.4.76 b}$$

and we have neglected terms of order ε^2. One eigenvalue of the matrix (II.4.76 *a*) is

$$\begin{aligned} \Lambda_{1,I}^*(b) &= b^2(1-12u^*I_1) \\ &= b^2\left(1-\frac{\varepsilon}{3}\ln b\right), \end{aligned} \tag{II.4.77 a}$$

which may be exponentiated to give

$$\Lambda_{1,I}^*(b) = b^{2-\varepsilon/3} + O(\varepsilon^2), \tag{II.4.77 b}$$

so that

$$\lambda_{1,I} = 2 - \varepsilon/3 + O(\varepsilon^2). \tag{II.4.77 c}$$

The other eigenvalue is

$$\begin{aligned} \Lambda_{3,I}^*(b) &= b^{\varepsilon}(1-72u^*I_1) \\ &= b^{\varepsilon}(1-2\varepsilon \ln b) && \text{(II.4.78 a)} \\ &= b^{-\varepsilon} + O(\varepsilon^2), && \text{(II.4.78 b)} \end{aligned}$$

so that

$$\lambda_{3,I} = -\varepsilon + O(\varepsilon^2). \tag{II.4.78 c}$$

The spectrum of important flow exponents is completed by making use of (II.4.75) once more, to give for the exponent governing the evolution of the ordering field

$$\lambda_{2,I} = 3 - \varepsilon/2 + O(\varepsilon^2). \tag{II.4.79}$$

Equations (II.4.77 *c*), (II.4.78 *c*) and (II.4.79) reveal that (for $\varepsilon>0$) the fixed point (II.4.68) has only *two* relevant scaling fields—the one characterized by the exponent λ_2, and associated with the ordering field, and the other characterized by the exponent λ_1 and associated, we would expect, with the reduced temperature. Indeed, this expectation is confirmed by inspection of the eigenvectors of the matrix $\mathfrak{R}_I^*(b)$

(eqn. (II.4.76 *a*)) which are readily found to be

$$\mu_{3,I}=\Delta u_I, \tag{II.4.80 a}$$

$$\begin{aligned}\mu_{1,I}&=\Delta r_I+6K_d a^d\Lambda^2\Delta u_I\\&\equiv r_0-r_I^*+6K_d a^d\Lambda^2(u_0-u_I^*)\\&=r_0+6u_0K_d a^d\Lambda^2,\end{aligned} \tag{II.4.80 b}$$

where we have made use of (II.4.76), (II.4.72 *b*), and in the last step, (II.4.68). Now it will be recalled (cf. eqn. (II.4.9) and the preceding discussion) that the temperature dependence of the physical hamiltonian enters through the parameter r_0, which vanishes at some temperature T_0—a zeroth-order approximation to the critical temperature T_c: the second term on the right-hand side of (II.4.80 *b*) may then be identified as the leading approximation to the *shift* in the transition temperature induced by the $k<\Lambda$ fluctuations.

In the spirit of the discussion in §II.4.3.4, we must now assume that, having discovered *one* (stable) fixed point in the space of Ising hamiltonians (II.4.52), we may take it to be *the* (only) fixed point of the universality class, so that the universal quantities characterizing the class are determined by the nature of this ('Ising') fixed point.

Specifically, the principal critical exponents follow from the flow exponents $\lambda_{1,I}$ and $\lambda_{2,I}$ (eqns. (II.4.77 *c*) and (II.4.79)) as described in §II.4.3.5 and summarized in table II.1. For example,

$$\gamma=\frac{2\lambda_{2,I}-d}{\lambda_{1,I}}=\frac{2}{2-\varepsilon/3}=1+\frac{\varepsilon}{6}+O(\varepsilon^2), \tag{II.4.81 a}$$

$$\beta=\frac{d-\lambda_{2,I}}{\lambda_{1,I}}=\frac{1-\varepsilon/2}{2-\varepsilon/3}=\frac{1}{2}-\frac{\varepsilon}{6}+O(\varepsilon^2). \tag{II.4.81 b}$$

Furthermore the field $\mu_{3,I}$ (eqn. (II.4.80 *a*)) turns out to be the leading irrelevant scaling field, and so the exponent $\bar{\omega}$ (eqn. (II.4.44)) characterizing the leading correction to scaling (cf. §II.4.3.6) follows from (II.4.77 *c*) and (II.4.78 *c*) as

$$\bar{\omega}=-\lambda_{3,I}/\lambda_{1,I}=\frac{\varepsilon}{2}+O(\varepsilon^2). \tag{II.4.81 c}$$

Strictly this result—indeed the entire analysis we have outlined—rests on careful consideration of the additional interactions generated by the renormalization group transformation. One finds, however (Wilson and Kogut 1974), that these interactions have fixed-point values that are at least $O(\varepsilon^2)$, and are associated with scaling fields, all of which are more strongly irrelevant (decay more rapidly) than the field μ_3.

The foregoing arguments are, therefore, consistent to leading order in ε, regarded as a small parameter. Their qualitative implications, at least, are immediate and persuasive.

Specifically we see that the calculations confirm (and, indeed, exploit the fact) that—as suggested by the arguments of §II.3.3—there exists a critical dimensionality, $d_c=4$, at which the critical behaviour of the hamiltonian (II.4.52) becomes mean-field-like. In particular, as $d\to 4^-$, the exponents (II.4.81 a, *b*) tend towards their mean-field values, as the Ising fixed point (II.4.68) moves closer to the fixed point (II.4.69) which describes a system whose critical fluctuations are effectively non-interacting, and whose critical exponents are naturally those of classical ('independent mode') theory. Moreover, (II.4.74*b*) and (II.4.78*c*) make it plain that this 'gaussian' fixed point (cf. the 'gaussian model' of Berlin and Kac 1952) becomes *the* stable fixed point above four dimensions, confirming the validity of classical theory, above the borderline dimensionality.

This state of affairs is clearly reflected in the different topologies of the hamiltonian flow in three and five dimensions, as shown in fig. II.3. Figure II.3 (*a*) also offers one simple example of the situation alluded to in § II.4.3.8, where there exists a domain of parameter space—here the regime of negative coupling constant—in which there exists no critical trajectory to the stable fixed point, reflecting the 1st-order transition exhibited by a system in this domain.

At a quantitative level one could expect little more of the leading-order results (II.4.81) than that they should correctly predict the signs of the deviations from mean-field exponent values found in three-dimensional systems (of the Ising universality class): this they clearly do (cf. table II.2). The remarkable quantitative success of calculations going to higher orders in ε, and the theoretical refinements necessary to make these calculations tractable, form part of the ground surveyed in the following sections.

Table II.2. The critical exponents γ and β for the Ising universality class in three-dimensions: (*a*) series expansions (Camp *et al.* 1976, Domb 1974): (*b*) ($O(\varepsilon)$ results at $\varepsilon = 1$ (eqns. (II.4.81)); (*c*) $O(\varepsilon^2)$ results at $\varepsilon = 1$ (Wilson 1972); (*d*) re-summed ε expansion at $\varepsilon = 1$ (Nickel 1973); (*e*) renormalization group in three dimensions (Le Guillou and Zinn-Justing 1977).

	(*a*)	(*b*)	(*c*)	(*d*)	(*e*)
γ	$1{\cdot}25^{+0{\cdot}003}_{-0{\cdot}007}$	1·17	1·24	1·250	1·2402 ± 0·0009
β	0·312 ± 0·005	0·33	0·34	0·344	0·325 ± 0·001

II.4.5. *Further calculation techniques: a summary*

The ideas set out in § II.4.3 and the techniques outlined in § II.4.4 constitute the backbone of the contemporary theory of the critical region. Thus, while the range of theoretical tools currently deployed in the field of critical phenomena is, in fact, extensive (too extensive for detailed elaboration here), most can be regarded as variants or refinements of the methods which we have outlined, and which originate in the work of Wilson (1971) and Wilson and Fisher (1972). It is in this spirit that we now attempt to chronicle the available techniques, indicating their relative merits, and the key references where the reader will find them described in greater detail. We shall see that the variety of techniques reflects both the (distinctly limited) number of quantities, such as $\varepsilon = 4 - d$, which can be used as 'small parameters' in probing the critical region, and the different ways in which the existence of a small parameter may be exploited.

II.4.5.1. *'Small parameters' for theories of critical phenomena*

The absence of exact solutions to (most) problems of critical phenomena—whether formulated within the renormalization group framework or not—has motivated the search for small parameters in terms of which one may look for the systematic expansions of universal quantities. It is evident that the range of such parameters is restricted

(i) by the spectrum of physical parameters upon which universal quantities have (or are expected to have) an explicit dependence;

(ii) by the requirement that the 'zeroth-order' theory be tractable.

The first of these criteria focuses attention on the two explicit E.Q.F.s discussed in § II.4.1—the system dimensionality d, and the order parameter dimensionality n. To date these quantities have provided three classes of small parameter meeting the second criterion.

The first (encountered explicitly on the preceding section) exploits the existence of a critical borderline dimensionality d_c *above* which fluctuations are so weak that the mean field predictions for universal quantities become exact.

The second makes use of the existence of a further borderline dimensionality—the infrared dimensionality d_i—*below* which fluctuations are so strong that the phase transition occurs only at $T_c = 0\,$K.

The third exploits the simplicity of certain universality classes in the limit of large n.

(*a*) *The small parameter* $d_c - d$

We have already seen how the existence of a critical borderline dimensionality d_c may be exploited: expansions in $d_c - d$ provide systematic fluctuation-induced corrections to the results of mean field theory (Wilson and Fisher 1972).

Now, it will be recalled from the arguments of § II.3.3 that d_c locates the dimensionality at which the fluctuation-induced divergences in certain k-space integrals are quelled by the phase–space weighting factor $d^d k$. For those universality classes in which the soft-mode dispersion has the simple form supposed in the L.G.W. hamiltonian (II.4.10 *b*), the arguments of § II.3.3 (complemented by the renormalization group analysis of § II.4.4) then show that $d_c = 4$, so that the appropriate small parameter is $\varepsilon = 4 - d$.

There are, however, universality classes in which the soft-mode dispersion differs qualitatively from that supposed in (II.4.10 *b*) and where, in consequence, the critical dimensionality is different from 4.

In fact, a generalization of the argument leading to (II.3.45 *a*) (one version of the Ginzburg (1960)–Levanyuk (1959) criterion) shows that d_c may always be located as the dimension above which the integral

$$L \equiv -\frac{d}{dt}\int_{\mathbf{k}}^{\Lambda} \sum_{i,j=1}^{n} \chi_{ij}^{\text{I.S.}}(\mathbf{k}) \tag{II.4.82}$$

is convergent in the $t \to 0$ limit; here $\chi^{\text{I.S.}}(\mathbf{k})$ is the appropriate wavevector-dependent susceptibility, within an independent-site (Landau–Ginzburg) approximation.

We shall have frequent occasion to exploit this result in later sections. For some universality classes, we shall see, the soft-mode dispersion is such that the fluctuations associated with certain sectors of k space are suppressed, with the consequence that d_c is less that 4: the uniaxial ferroelectric actually has $d_c = 3$, so that no $d_c - d$ expansion is necessary in this case (cf. § II.5.1.3). In other universality classes the dispersion leads to an enhancement of the fluctuations originating in certain regions of k space, and d_c is *larger* than 4: thus the uniaxial Lifshitz system (discussed in § II.5.3.2) has $d_c = 4{\cdot}5$, and fluctuation-induced corrections to mean field predictions may be calculated through expansions in $4{\cdot}5 - d$.

The numerical reliability of such expansion schemes, when applied to the three-dimensional systems of interest, is evidently of paramount importance. This question has been examined most thoroughly in the context of the $\varepsilon = 4 - d$ expansion—itself the most widely used approximation scheme—for the Ising universality class, where comparison with tried and tested high-temperature series expansions offer a sensitive test of the approach. The remarkable success of the $O(\varepsilon^2)$ calculations (Wilson 1972, and table II.2) is soured, somewhat, by the discovery that the discrepancies with the series results are accentuated, rather than diminished, by the $O(\varepsilon^3)$ corrections (Brézin *et al.* 1973). Indeed, as made clear by Wilson (1972) at the outset, and as confirmed by very recent work (Brézin *et al.* 1977), the ε-expansion does not produce a convergent series (though, with careful, and distinctly laborious, resummation techniques it may be made to yield very accurate results in three dimensions: see the review by Wallace 1976, and table II.2).

However, the (possibly illusory) success of the $O(\varepsilon^2)$ calculations for the Ising universality class has left, as a legacy, the expectation that calculations to similar order for other universality classes (for which no substantiating series expansion results may exist) will give a not unreasonable estimate of three-dimensional behaviour.

This view finds some further vindication in the remarkable reliability with which low-order ε-expansion methods have been shown to predict the *order* of certain classes of phase transition in three-dimensional systems (Bak *et al.* 1976): systems which *within a low order ε-expansion* are found to possess no stable fixed point do indeed exhibit the fluctuation-driven ('category one') first-order phase transitions discussed in §II.4.3.8. (A particularly convincing demonstration of the efficacy of this rule is afforded by recent molecular dynamics calculations: Mouritsen *et al.* 1977).

Naturally, expansions about critical dimensionalities larger than four can be expected to give at best a qualitative indication of the three-dimensional behaviour. Nevertheless, the $d_c - d$ expansion probably constitutes the single most versatile tool in the theory of the critical region: the two other classes of small parameters, which we now survey briefly, yield expansions that are, in comparison, numerically much less reliable.

(b) The small parameter $d - d_i$

We have already alluded (§II.2.2) to the well-known result that, in one-dimensional systems, with finite-range interactions, an ordered phase cannot be stable (against domain-producing fluctuations: cf. §III.4.2.2) at any finite temperature: a phase transition in such a system can thus occur only at 0 K. It is also a long-established result that in certain high-symmetry systems with multi-component order parameters—the L.G.W. hamiltonian (II.4.10 *b*) with $n > 2$ is the paradigm—the fluctuations in the direction of the local ordering are so strong that an ordered phase is unstable (at any finite temperature) already in two dimensions (Hohenberg 1967, Mermin and Wagner 1966).

In each case the transition temperature must therefore vanish with $\varepsilon_i = d - d_i$, where d_i is the appropriate 'infra-red' borderline dimensionality ($d_i = 1$ for most systems; $d_i = 2$ for a restricted class of high-symmetry systems).

To appreciate the use to which this result may be put we recall the way in which (§II.4.4) the realization of the renormalization group transformation as a perturbation expansion in the coupling constant is seen to be justifiable, ultimately, as an expansion in $\varepsilon=4-d$ (since u^* is $0(\varepsilon)$). By the same token a *low-temperature* expansion for the renormalization group transformation (Migdal 1975, Polyakov 1975, Brézin and Zinn Justin, 1976) is justifiable, ultimately, as an expansion in $\varepsilon_l=d-d_l$, since the fixed-point value of the temperature is itself a critical temperature and is therefore $0(\varepsilon_l)$. Universal quantities can be calculated systematically in powers of ε_l, though the series are typically much more badly behaved than the ε-expansion (Hikami and Brézin 1978).

(*c*) *The small parameter 1/n*

The quantity $1/n$ proves a useful small parameter in somewhat restricted circumstances—essentially in the context only of the isotropic L.G.W. hamiltonian (II.4.10*b*). Its utility in this sphere (Abe 1972, Suzuki 1972, Ma 1973; see also the review by Ma 1976) reflects the fact that, in the large n-limit the critical behaviour of this universality class may be determined exactly, coinciding (Stanley 1968) with that of the spherical model introduced by Berlin and Kac (1952). Expansions in $1/n$ thus provide systematic corrections to the universal quantities of the spherical model. When applied to the small n $(=1,2,3,\ldots)$ systems of physical interest the results of such expansion schemes are disappointing: nevertheless the insights that they offer often complement those of the ε expansion in a qualitatively useful way.

II.4.5.2 *Techniques in the theory of critical phenomena*

Most of the techniques currently employed in the theory of the critical region can be subdivided into two categories, reflecting the two levels at which it is possible to exploit the existence of the small parameters listed above.

Firstly, there exists a class of techniques—again originated by Wilson (1972)—which, while resting to varying degrees on R.G. insights, bypass explicit R.G. methods to obtain perturbation expansions for universal quantities directly.

Secondly, at the more refined and powerful level, there is a class of techniques—essentially variants of the R.G. method described in §§II.4.3 and II.4.4—where the small parameter is used to justify a perturbative treatment of the R.G. transformation.

Finally, there exists one further group of techniques—those in which renormalization group methods are used without the aid of a systematizing small parameter.

We now survey each of these categories in turn.

(*a*) *Direct perturbation expansion techniques*

Many of the expansions in ε and $1/n$ alluded to above were, in fact, obtained by an approach which, while lacking the transparency and, ultimately, the power of the renormalization group, wins out on efficiency. We shall illustrate the philosophy underlying the technique in the context of the ε expansion for critical exponents (Wilson 1972), since it is in this context that renormalization group insights prove most valuable. Once again we choose to consider the Ising universality class, described by the hamiltonian (II.4.52).

In contrast to renormalization group methods, the direct ('Feynman graph') approach requires that

(i) We *assume* that the critical behaviour of our system is characterized, asymptotically, by pure power laws. Thus, for example, the zero-field susceptibility is supposed to behave, in the limit as $t\to 0^+$, as

$$\chi=C_+t^{-\gamma}. \qquad \text{(II.4.83}a\text{)}$$

(ii) We *assume* that there exists an expansion for universal quantities in powers of the chosen small parameter. Thus, if the small parameter to be utilized is ε, we suppose that γ has an expansion about its mean field value

$$\gamma = 1 + \gamma_1 \varepsilon + \gamma_2 \varepsilon^2 + \ldots . \qquad \text{(II.4.83}\,b)$$

Combining (II.4.83 *a*) and (II.4.83 *b*), and making use of the expansion (II.4.67), one finds the prediction that the susceptibility itself should have the ε expansion:

$$\chi = C_+ t^{-1}(1 - C_1 \ln t + C_2 \ln^2 t \ldots) \qquad \text{(II.4.84}\,a)$$

where

$$C_1 = \gamma_1 \varepsilon + \gamma_2 \varepsilon^2 + O(\varepsilon^3), \qquad \text{(II.4.84}\,b)$$

$$C_2 = \frac{1}{2}\gamma_1^2 \varepsilon^2 + O(\varepsilon^3), \qquad \text{(II.4.84}\,c)$$

so that, in particular,

$$C_2/C_1^2 = \frac{1}{2} + O(\varepsilon). \qquad \text{(II.4.84}\,d)$$

A matching of the *expected* form (II.4.84 *a*) to that given by an *explicit* perturbation expansion for χ should then allow identification of the coefficients $\gamma_1, \gamma_2 \ldots$

In fact, a double perturbation expansion in u_0 and ε leads (in the case of the hamiltonian (II.4.52) with the cutoff Λ set equal to unity, for convenience) to a result of the form (II.4.84 *a*) with

$$C_1 = 6u_0 K_d a^d - 72(u_0 K_d a^d)^2 + O(\varepsilon, u_0)^3, \qquad \text{(II.4.85}\,a)$$

$$C_2 = -36(u_0 K_d a^d)^2 + \frac{3}{2}\varepsilon u_0 K_d a^d + O(\varepsilon, u_0)^3. \qquad \text{(II.4.85}\,b)$$

(We do not present the details of this calculation, but refer the interested reader to the review article by Wallace (1976), where the relevant techniques are thoroughly discussed.) Now it is evident, and at first sight worrying, that (II.4.85) are consistent with the expectation (II.4.84 *a*, *d*) only if

$$\frac{-36(u_0 K_d a^d)^2 + \frac{3}{2}\varepsilon u_0 K_d a^d}{36(u_0 K_d a^d)^2} = \frac{1}{2} + O(\varepsilon); \qquad \text{(II.4.86}\,a)$$

that is, only if the coupling constant has the special value

$$u_0 = u_{0c} \equiv \frac{\varepsilon}{36 K_d a^d} + O(\varepsilon^2). \qquad \text{(II.4.86}\,b)$$

Refraining from comment on this observation for a moment, we note that, with this choice of coupling constant, the coefficient γ_1 (eqn. (II.4.84 *b*)) may be identified as

$$\gamma_1 = \frac{C_1}{\varepsilon} = \frac{6u_{0c} K_d a^d}{\varepsilon} = \frac{1}{6}, \qquad \text{(II.4.87}\,a)$$

so that

$$\gamma = 1 + \frac{\varepsilon}{6} + O(\varepsilon^2) \qquad \text{(II.4.87}\,b)$$

in accord with (II.4.81 *a*).

That a special choice of coupling constant should be necessary to ensure consistency of (II.4.84) and (II.4.85) seems at first sight to conflict with the notion of universality. This inference is incorrect, however, as is made plain by the R.G. analysis of §II.4.4, which shows that the reason that a particular choice of coupling constant is necessary lies elsewhere. Specifically, we note from (II.4.81 *c*) that the exponent $\bar{\omega}$ controlling the corrections to the leading scaling behaviour (§II.4.3.6) is itself of order ε. Inspection of (II.4.43) then shows that the ε expansion of the susceptibility (or, indeed, any other observable) will *in general* contain logarithmic terms reflecting this correction to scaling term *as well as* the logarithmic terms reflecting the corrections to the exponent (γ in this case) of the leading power law: the additional logarithmic terms will be absent, and the ε expansion of the suceptibility will have the pure power-law form supposed in (II.4.83) and (II.4.84), only if the amplitude of this correction-to-scaling field is precisely zero. That this amplitude is controllable through the coupling constant is made clear by the R.G. analysis described in §II.4.4, which shows, in particular (cf. eqn. (II.4.80 *a*)) that, at leading order in ε, this amplitude reflects simply the deviation of the coupling constant from its fixed-point value, and may be set by choosing $u_0 = u_I^*$. Indeed, inspection of equations (II.4.68 *a*) and (II.4.86 *b*) reveals that the special value of the coupling constant, u_{0c}, necessary to 'expose' the pure scaling behaviour is (at this order) nothing but the Ising fixed-point value u_I^*.

This special value may be determined order-by-order in ε (without the machinery of the R.G., and, in fact, rather more efficiently than the above argument would suggest: see Wilson 1972) so that the direct Feynman graph 'matching' philosophy, outlined above, may be implemented to give a compact way of determining the systematic ε expansions (or, more generally, $d_c - d$ expansions) of universal quantities. Expansions in $1/n$ (Ma 1976) are performed in essentially the same spirit, though they are technically somewhat more demanding, have seldom been carried out beyond leading order, and are (as remarked earlier) numerically disappointing.

While these direct expansion techniques afford a particularly efficient means of determining systematic expansions for critical exponents, it is important to appreciate their limitations. Their central disadvantage lies in the fact that they yield, in the first instance, the small-parameter *expansion*(*s*) of the power law(s) characterizing the critical behaviour, rather than the power laws themselves. If the behaviour in question is *known* to be that of a pure power law then the exponent, and amplitude, of that power law can be identified by the 'matching' technique we have outlined. If, however, the behaviour in question involves more than one power law (as in the problems posed by the equation of state, by crossover phenomena, and, indeed, by the corrections-to-scaling that are present when $u_0 \neq u_{0c}$) it frequently proves impossible to identify these powers, unambiguously, from their expansions. (It is hard to make sense of the results (II.4.84 *a*) and (II.4.85) when $u_0 \neq u_{0c}$.) It is in its capacity to cope with these more complex situations (illustrated below) that the renormalization group technique displays its full power.

(*b*) *Perturbative realizations of the renormalization group*

While the techniques described in §II.4.4 constitute the single most versatile realization of the renormalization group transformation, there exists a number of alternative realizations which, for certain purposes, prove more amenable: these we now survey, emphasizing the greatly enhanced utility of the small parameters of §II.4.5.1, when employed in conjunction with the R.G. method.

As they stand, the methods of § II.4.4 have two principal drawbacks.

Firstly, (II.4.62) do not describe the hamiltonian flow in a satisfactory way, even for small ε, and u_0: for the 'small parameter' in the expansion that yields these equations is actually $\varepsilon \ln b$ (or $u_0 \ln b$) rather than ε (or u_0) itself. The limiting, large b, behaviour of the hamiltonian flow cannot, therefore, be reliably determined from these equations: this is precisely the behaviour we require to relate the physics of a critical ($\xi \to \infty$) to a non-critical ($\xi(b) \sim 1$) system—the central R.G. aim, of which we lost sight in § II.4.4 (where we focused, merely, on the determination of exponents), but which becomes crucially important in the more complex situations referred to above.

This problem is surmounted by constructing, and solving, the *differential* equations of hamiltonian flow (Wegner and Houghton 1973) describing the parameter renormalization accompanying the elimination of an infinitesimally thin shell of modes in reciprocal space.

Taking the Ising universality class as an example, once again, the differential equations (at leading order) are readily constructed from (II.4.62). We write

$$b = 1 + \delta l, \quad \delta l \ll 1 \qquad \text{(II.4.88}\,a)$$

and approximate the integrals involved by

$$\int_{\mathbf{k}}^{\Lambda} \frac{1}{(r_0 + k^2)} = \frac{K_d a^d \Lambda^d \delta l}{(r_0 + \Lambda^2)} + O(\delta l)^2, \qquad \text{(II.4.88}\,b)$$

$$\int_{\mathbf{k}}^{\Lambda} \frac{1}{(r_0 + k^2)^2} = \frac{K_d a^d \Lambda^d \delta l}{(r_0 + \Lambda^2)^2} + O(\delta l)^2. \qquad \text{(II.4.88}\,c)$$

Equations (II.4.62) may then be rewritten as

$$\frac{r(1+\delta l) - r_0}{\delta l} = 2r_0 + \frac{12 u_0 K_d a^d \Lambda^d}{r_0 + \Lambda^2}, \qquad \text{(II.4.89}\,a)$$

$$\frac{u(1+\delta l) - u_0}{\delta l} = \varepsilon u_0 - \frac{36 u_0^2 K_d a^d \Lambda^d}{(r_0 + \Lambda^2)^2}. \qquad \text{(II.4.89}\,b)$$

Successive applications of the infinitesimal process thus generate a continuous sequence of parameters $r(l)$, $u(l)$ obeying the differential flow equations (actually used to construct figs. II.3 (a) and 3 (b)):

$$\frac{dr(l)}{dl} = 2r(l) + \frac{12 u(l) K_d a^d \Lambda^d}{(r(l) + \Lambda^2)}, \qquad \text{(II.4.90}\,a)$$

$$\frac{du(l)}{dl} = \varepsilon u(l) - \frac{36 u^2(l) K_d a^d \Lambda^d}{(r(l) + \Lambda^2)^2}. \qquad \text{(II.4.90}\,b)$$

These equations must be set alongside the differential analogue of the fundamental relationship (II.4.19), namely

$$\frac{d\xi(l)}{dl} = -\xi(l), \qquad \text{(II.4.91}\,a)$$

which is trivially solvable to give

$$\xi(l) = \xi \exp(-l), \qquad \text{(II.4.91}\,b)$$

implying the correspondence between the scale factor b, and the flow parameter l (compare (II.4.19) with (II.4.91 b)):

$$b=\exp l. \tag{II.4.91 c}$$

The two fixed points of the transformation, eqns. (II.4.68) and (II.4.69), are readily recovered from (II.4.90), as the locations at which both $r(l)$ and $u(l)$ are stationary.

The merit of the differential form is revealed (in a simple way) in a study of the linearized equations, analogous to (II.4.76). Specifically, linearizing (II.4.90 b) about the Ising fixed point (II.4.68), we obtain

$$\begin{aligned}\frac{d\Delta u_I}{dl}&=(\varepsilon-72u_I^*K_d a^d)\Delta u_I(l)\\ &=-\varepsilon\Delta u_I(l),\end{aligned} \tag{II.4.92 a}$$

whose solution is immediately seen to be

$$\Delta u_I(l)=\exp(-\varepsilon l)\Delta u_I, \tag{II.4.92 b}$$

or, making use of the correspondence (II.4.91 c),

$$\Delta u_I(b)=b^{-\varepsilon}\Delta u_I. \tag{II.4.92 c}$$

Thus we recover the result (II.4.78b) for one eigenvalue of the linearized transformation,

$$\Lambda_{3,I}^*(b)=b^{-\varepsilon}. \tag{II.4.92 d}$$

However, in deriving (II.4.78b) it was necessary to *assume* that a logarithm of b (that in (II.4.78a)) arose from the expansion of a pure power law—a reflection of the uncontrollable character of the recursion relations (II.4.62) in the large b limit, and a legacy of the problems encountered in direct perturbation expansions. By contrast, in the approach sketched above, no 'exponentiation by hand' is required: in effect, the act of solving the equations of hamiltonian flow, (II.4.90), to obtain trajectories that are faithful in the large b large ($\exp l$) limit, achieves this exponentiation in a systematic way.

The capacity of the renormalization group method to deliver power laws in their exponentiated (rather than their small-parameter-expanded) form takes on a vital significance in the problems alluded to at the end of § II.4.5.2 (a) where more than one power law is involved.

Once again we take as an example the problem of the zero-field susceptibility of the Ising universality class. One may solve the differential equations (II.4.90) (Nelson and Rudnick 1975) to forge a link between the parameters of the near-critical system of physical interest, and the parameters of a non-critical system. The susceptibility of the non-critical system can be calculated with the aid of perturbation theory (specifically, eqns. (II.4.84 a), (II.4.85 a, b), which are satisfactory for a system with a *large* reduced temperature, e.g. $t=1$!) and the two susceptibilities related (through an equation of the form (II.4.38b)) to give, finally, the leading-order expression for the susceptibility of the critical system (Nelson and Rudnick 1975)

$$\chi=C_+t^{-1}\left[1+\frac{u_0}{u_I^*}(t^{-\varepsilon/2}-1)\right]^{1/3}, \tag{II.4.93 a}$$

or

$$\chi = C_+ t^{-\gamma}\left[\left(1-\frac{u_0}{u_I^*}\right)t^{\bar{\omega}}+\frac{u_0}{u_I^*}\right]^{1/3}, \qquad \text{(II.4.93}\,b)$$

where γ and $\bar{\omega}$ have their leading-order values (II.4.81 a, c).

These equations have many features worthy of comment: they provide an explicit realization of the general 'correction-to-scaling' theory outlined in § II.4.3.6; they show that, as argued in § II.4.5.2a, these corrections to scaling are absent for a system with $u_0 = u_I^*$. (In effect $\Delta u_I \equiv u_0 - u_I^*$ controls the range of temperature over which pure power-law behaviour will be observed—i.e. the width of the critical region.)

However, from a general point of view, the most salient feature of these results is their contrast with those of the direct perturbation theory, eqns. (II.4.84 a) and (II.4.85 a, b): the logarithms yielded by the perturbation theory are replaced, in the course of the full R.G. treatment, by the power laws of which they are ε expansions—it is easy to check that, when re-expanded in ε, (II.4.93 a) does indeed recover the Feynman graph result (II.4.84 a), (II.4.85 a, b).

The difference between the two results gives a measure of the difference between the power of those calculations which utilize small parameters in direct expansions, and those which use the small parameter to yield perturbative realizations of the renormalization group operation.

The second drawback of the methods outlined in § II.4.4 lies in the fact that they become somewhat intractable at higher order (see, however, Bruce *et al.* 1974) as a result of the many parameters (with fixed points $0(\varepsilon^2)$) that are generated by the R.G. operation; the differential formulation (Wegner and Houghton 1973) encounters the same problems in an even more acute form.

That these difficulties are, in some sense, illusory is made clear by the fact that it is possible to realize the R.G. operation in a rather different way—one in which these problems are not encountered. The methods in question (reviewed by Brézin *et al.* 1976) are field-theoretic in conception, and focus on the dependence of observables (vertex functions or Greens functions) on the cutoff Λ: the 'renormalizability' of hamiltonians of the form (II.4.10) (the existence of renormalized vertex functions that are finite in the $\Lambda \to \infty$ limit) leads to differential equations which, like (II.4.90), are the generators of lines of hamiltonian flow connecting systems with different length scales. The special utility of the field-theoretic approach (in particular that expounded by Zinn-Justin 1973) lies in the way in which they reveal that the flow properties can be characterized, to *any* order in ε, by (only) *two* differential equations: one describes coupling constant flow (as in (II.4.90 b)) while the other describes reduced-temperature flow (and is a close, but not precise, analogue of (II.4.90 a)).

These methods, then, probably constitute the most tractable, if not the most transparent, realization of the R.G. operation, offering simplifications that are quite essential in calculations going beyond leading order.

In summary, the full power of the renormalization group is apparent only within a differential formulation, the field theoretic variant being technically superior. Both differential formulations have been used in conjunction with expansions in ε and ε_i to tackle the wide variety of problems (equations of state—Rudnick and Nelson 1976; corrections to scaling—Rudnick and Nelson 1976, Lawrie, 1976; crossover phenomena—Nelson and Domany 1976, Bruce and Wallace 1977 ...) where the complexity of the critical behaviour makes the use of renormalization group methods essential. The results of some of these calculations will be presented in the course of § II.5.

(*c*) *Renormalization group methods without a small parameter*

We conclude this survey by citing, without detail, certain renormalization group techniques which do not employ a small parameter.

We mention first an approximate realization of the R.G. transformation—'the approximate recursion formula'—originated by Wilson (1971) and applied by him to the Ising universality class. Avoiding any use of perturbation theory (with the aid of approximations whose reasonableness it is difficult to assess), the procedure offers a qualitatively-instructive picture of the dimensionality-dependence of the physics of the Ising hamiltonian (II.4.52).

Specifically, we rewrite this effective hamiltonian in the direct space representation (eqn. (II.4.8 *a*) in zero field) as

$$\mathscr{H}_0=\mathscr{H}_0(\{u\})=\int_{\mathbf{x}}\left\{\frac{1}{2}(\nabla u(\mathbf{x}))^2+V_{s,0}(u(\mathbf{x}))\right\}, \tag{II.4.94 a}$$

with

$$V_{s,0}(u)=\frac{1}{2}r_0u^2+u_0u^4. \tag{II.4.94 b}$$

This, it will be recognized, is a continuumized version of our original lattice model (II.2.1), with $V_{s,0}$ playing the role of a partially-renormalized single-particle potential. The fixed-point form of the single particle potential, $V_s^*(u)$, (as given by the approximate recursion formula) is displayed in fig. II.4, for different dimensionalities (Wilson and Kogut 1974). We shall have occasion to comment on these results in § II.4.3.1.

Secondly we mention that there exist numerous direct space renormalization group calculations (Wilson 1975, Kadanoff and Houghton, 1975, Niemeyer and Van Leeuven 1973, 1974) close in spirit to the original Kadanoff transformation discussed in § II.4.3.1. The lack of a systematizing small parameter means that the flow equations generated by this approach have to be truncated in an essentially arbitrary way.

Fig. II.4

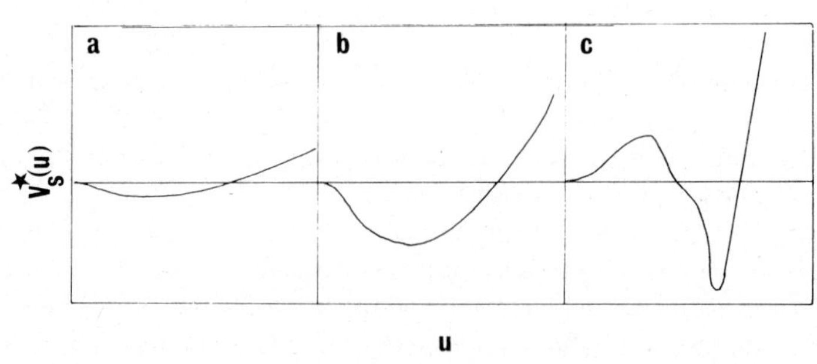

The fixed-point form of the single particle potential $V_s^*(u)$ as given by the approximate recursion formula, (*a*) for $d=3{\cdot}9$, (*b*) for $d=3$, (*c*) for $d=2$. After Wilson and Kogut (1974).

Finally we draw attention to recent studies of the field-theoretic renormalization group equations (for the Ising universality class, and its n-component generalization, eqn. (II.4.10*b*)) in *three* dimensions (Baker *et al.* 1976, Le Guillou and Zinn Justin 1977). Analysis of the contributions to these equations made by terms of *large* order in the coupling constant, coupled with Pade Borel techniques, has permitted an accurate location of the Ising fixed point in three dimensions: the associated exponents (table II.2) are largely in accord with series expansions, and, indeed, are assigned a precision that is far beyond the resolution of current experimental techniques.

§II.5. Static critical phenomena at structural phase transitions

In the course of the last few years the theoretical methods we have described in the preceding sections have been applied to a wide range of universality classes, yielding explicit predictions for their diverse critical behaviour. Physical realizations of each universality class may, in general, be found among the ranks both of systems undergoing magnetic phase transitions and of systems undergoing structural phase transitions, so that these predictions may, in principle, be subjected to experimental test in physically very different situations.

In fact, a number of key renormalization group results have been quite strikingly confirmed by careful, and impressively precise, experiments (which we shall cite in passing below) on magnetic systems. By contrast, the number of existing experiments exposing critical phenomena at structural phase transitions—and the precision of these experiments—is disappointing.

This dearth of high-precision experimental data may be attributed to a number of sources.

First of all, the coupling between the primary-order parameter and acoustic degrees of freedom is typically rather stronger in the structural phase transition case, as is reflected in the pressure dependence of critical temperatures. (For the paraelectric to ferroelectric phase transition in $BaTiO_3$, $(dT_c/dp) \approx -8{\cdot}4$ K/kbar (Samara 1971); for the paramagnetic to ferromagnetic phase transition in Eu0, $(dT_c/dp) \approx 0{\cdot}37$ K/kbar (McWhan *et al.* 1966).) As we shall see, the enhancement of this coupling may be at least partially responsible for the fact that so many structural phase changes are first order to a sufficient extent as to preclude the observation of critical phenomena.

Even when the phase transition is effectively continuous the presence of inhomogeneous strains may, through the same enhanced order-parameter–strain coupling, result in a smearing of the transition temperature that is sufficient to make exponent determination ambiguous. This is particularly true of measurements of the specific heat, which typically exhibits a *weak* singularity upon a thermal phonon background making precision experiments *very* close to T_c quite essential.

Of the structural systems that *do* exhibit continuous phase transitions, the ones that can most readily be subjected to precision experiments are, naturally, those where the order parameter is a spontaneous polarization—for then the ordering field is physically realizable and the order parameter susceptibility may be determined directly. Regrettably the *ferroelectrics* with *continuous* phase changes are generally *uniaxial*: as we shall discuss below, the deviations from mean field predictions are particularly subtle in this case, and have evaded detection until recently.

Thus it is among the class of systems exhibiting non-ferroelectric phase transitions (particularly members of the perovskite family with zone-boundary instabilities) that non-trivial critical behaviour has proved to be most readily detectable. Here, however, high-precision probes are less readily constructed.

In particular the unavailability of an ordering field means that the static response (the order parameter susceptibility) can be measured only by a scattering experiment—effectively by integrating over the appropriate frequency-dependent response (cf. § I.3.6.2). Inelastic neutron scattering techniques do not allow this to be done with great accuracy, while (somewhat surprisingly) diffuse X-ray scattering techniques do not appear to have been exploited in this way.

High-precision specific heat experiments (of great utility in the investigation of antiferromagnetic critical behaviour) are difficult, for the reasons mentioned above.

In such instances, then, it is only the order-parameter exponent β (arguably less sensitive than, say, γ to the differences between universality classes) that has been determined with the requisite accuracy, either by elastic neutron scattering, or with the aid of local (E.P.R., N.M.R.) probes. (Even the utility of these techniques may be limited by 'precursor' effects: cf. § III.4.3.2.)

In the light of these remarks it is not surprising that many of the explicit renormalization group predictions, to which this section is devoted, have yet to be substantiated in the context of structural phase transitions. It is to be hoped that, in summarizing these predictions, and in reviewing the limited extent to which they have been borne out by experiment, we shall stimulate further experiments—we shall cite specific examples!—of the precision necessary to provide adequate tests for the theory.

II.5.1. *Single-component order parameters: ordinary critical behaviour*

II.5.1.1. *Short-range interactions*

We begin with a discussion of the simplest universality class, whose representative is the generalized Ising effective hamiltonian (II.4.7), which we used to provide an illustrative example of renormalization group methods in § II.4.4.:

$$\mathscr{H}_0 = \mathscr{H}_0(n=1). \qquad \text{(II.5.1)}$$

In addition to the true Ising model, this universality class is expected to include all those systems undergoing structural phase transitions associated with a scalar, non-ferroelectric order parameter (i.e. phase transitions where the 'soft mode' is non-degenerate and non-polar). According to the principles set out in § II.4.3 and realized in the explicit calculations of § II.4.4, *all* such systems should exhibit critical behaviour characterized by the *same* universal quantities, associated with the (presumably—cf. § II.4.3.4) unique Ising fixed point of the renormalization group.

This assertion—in effect the universality hypothesis—is clearly vindicated by the results of molecular dynamics studies of the structural phase transitions exhibited by models of the form defined by (II.2.1) (Schneider and Stoll 1976 *a*, 1978). In figs. II.5 and II.6 we show the results for the temperature-dependence of the order-parameter and susceptibility in three two-dimensional models with different values of the parameter s (eqn. II.2.8) which, it will be recalled, locates the system on the order–disorder/displacive spectrum. In each case the asymptotic behaviour is seen to be consistent with the power laws (II.2.10 *b*) and (II.2.12 *a*), with $\beta = 1/8$ and $\gamma = 7/4$—the 2-*d* Ising values. The amplitudes of the power laws are, as expected, parameter-dependent, though the *ratios* of the amplitudes of the susceptibility divergence (as $t \to 0^{\pm}$) are consistent with the universal prediction (II.2.12 *c*).

Fig. II.5

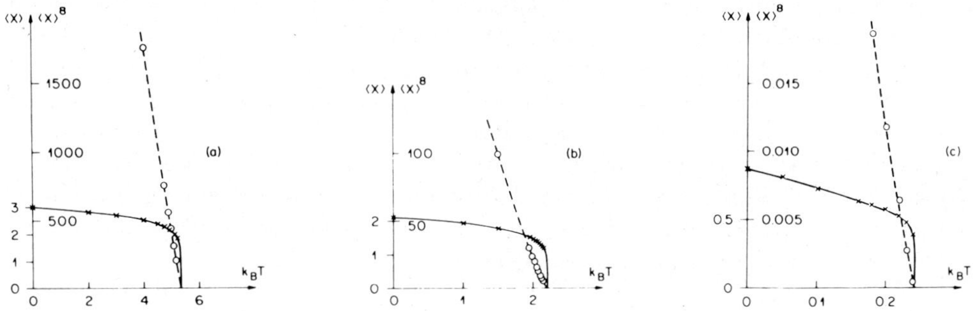

The results of molecular dynamics calculations of the order parameter for two-dimensional models of the form (II.2.1) (Schneider and Stoll 1976 *a*). (*a*) is an order–disorder model, (*b*) an intermediate model and (*c*) a displacive model.

Fig. II.6

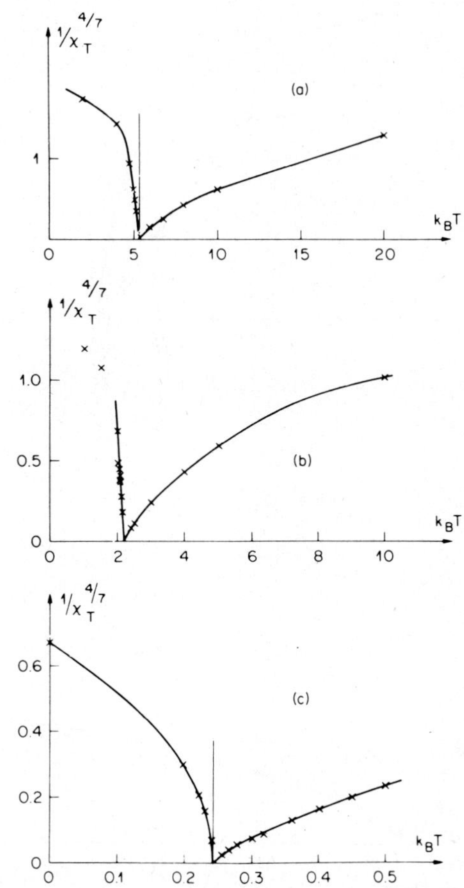

The results of molecular dynamics calculations of the order parameter susceptibility for two-dimensional models of the form (II.2.1) (Schneider and Stoll 1976 *a*). (*a*) is an order–disorder model, (*b*) an intermediate model and (*c*) a displacive model.

The s-dependence of the width of the critical region is also revealed in the susceptibility results, the region of applicability of the simple power law becoming smaller the more displacive the system (the smaller its s value), in accord with the analysis of § II.3.3. In this context it is interesting to note that, while the critical region proved to be inaccessibly small in the *susceptibility* measurements on the most displacive model (fig. II.6 (c)), pure power-law behaviour is clearly demonstrated (fig. II.5 (c)) in the corresponding *order parameter* measurements. The moral is that the ranges of reduced temperature over which pure power law behaviour is observable may vary, not only from system to system, but also from one observable to another within the same physical system.

Fig. II.7

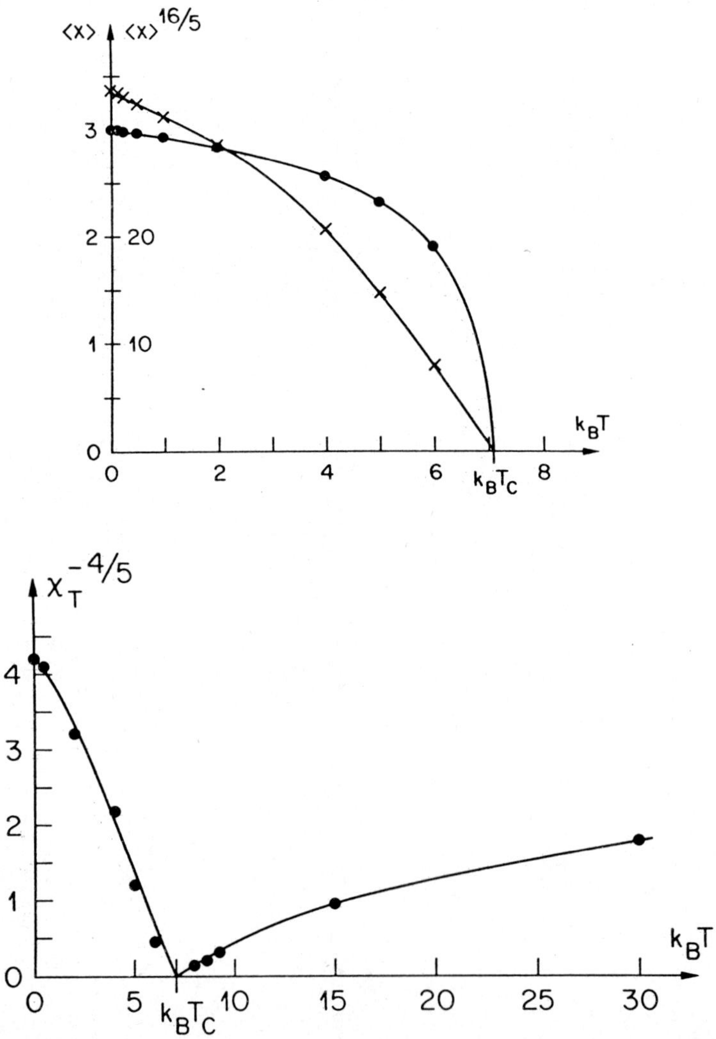

The results of molecular dynamics calculations of the order parameter and susceptibility of a three-dimensional model of the form (II.2.1) (Schneider and Stoll 1978).

The results (fig. II.7) of similar calculations on the three-dimensional form of the model (2.1) again reveal universal Ising critical behaviour, while re-emphasizing the dimensionality-dependence of universal quantities: the relevant exponents are found to be consistent with the three-dimensional Ising values (eqns. II.2.10*c*, II.2.12*b*; table II.2) $\beta \approx 5/16$, $\gamma \approx 5/4$.

There appear to be few physical realizations of this particular universality class among 'real' systems exhibiting structural phase transitions (cf., however, the discussion of monodomain-transforming perovskites in § II.5.3.3).

Although quartz (SiO_2) falls into this category its phase transition is clearly first-order. While in itself this need not necessarily preclude the existence of a region where Ising-like critical behaviour could be observed, the assignment of exponents in such instances is fraught with problems of measurement and ambiguities of interpretation (cf. § II.5.3.1). Thus the value of $\beta = 0{\cdot}34 \pm 0{\cdot}02$ obtained by Höchli and Scott (1971) through a synthesis of X-ray, ultrasonic and Raman measurements on quartz, is of doubtful significance. (We have, in any case, been unable to find any justification for the ansatz which these authors use in the fitting of their data.)

The order–disorder phase transitions in ND_4Cl and NH_4Cl, continuous at high pressures (cf. § II.5.3.1) are believed to be Ising-like, and neutron-scattering measurements (Yelon *et al.* 1974) reveal an order-parameter exponent that is close to the three-dimensional Ising value (fig. II.8).

Fig. II.8

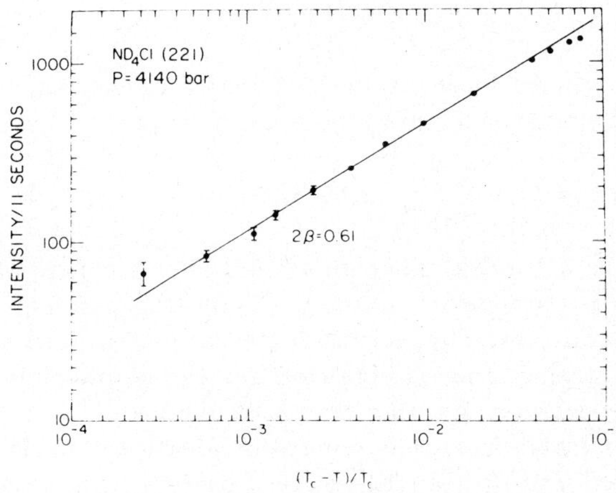

The critical behaviour of the square of the order parameter in ND_4Cl (at a pressure well in excess of its tricritical value) as measured by elastic neutron scattering (Yelon *et al.* 1974).

II.5.1.2. *Short-range interactions with order-parameter–strain coupling*

One of the many problems of critical phenomena to have become tractable, for the first time, with the development of R.G. methods, is that posed by the coupling of an order parameter to the system's elastic degrees of freedom. This problem, much studied in the context of various 'compressible Ising models' (see Bergman and Halperin 1976 for a review), assumes particular importance at structural phase transitions where, as remarked earlier, the degree of order-parameter–strain coupling tends to be more significant.

In general the problem is rather complex, and we shall outline explicitly only the analysis of the simplest case (the first to be treated by R.G. methods: Sak (1974)) which arises when the system is taken to be elastically isotropic. The appropriate effective hamiltonian is then (cf. eqn. (I.4.12)

$$\mathscr{H}_0=\mathscr{H}_0(n=1)+\mathscr{H}_0^e, \tag{II.5.2a}$$

where the term originating in the elastic interactions can be written, in the type of direct-space representation employed in (II.4.8a) as

$$\mathscr{H}_0^e=\int_{\mathbf{x}}\left\{\frac{1}{2}C_{11}^0\sum_{\alpha=1}^{d}e_{\alpha\alpha}^2(\mathbf{x})+C_{12}^0\sum_{\alpha<\beta}e_{\alpha\alpha}(\mathbf{x})e_{\beta\beta}(\mathbf{x})\right.$$
$$\left.+2C_{44}^0\sum_{\alpha<\beta}e_{\alpha\beta}^2(\mathbf{x})+e_0u^2(\mathbf{x})\sum_{\alpha=1}^{d}e_{\alpha\alpha}(\mathbf{x})+(p-p_0)\sum_{\alpha=1}^{d}e_{\alpha\alpha}(\mathbf{x})\right\}. \tag{II.5.2b}$$

Here we introduce elastic constants (into which we have subsumed a factor of k_BT) which, by virtue of the assumed elastic isotropy, are related by

$$C_{11}^0-C_{12}^0=2C_{44}^0 \tag{II.5.3a}$$

so that the elastic properties can be specified by the bulk and shear moduli

$$\left.\begin{aligned}K_0&\equiv\frac{1}{d}[C_{11}^0+(d-1)C_{12}^0],\\ \mu_0&\equiv\frac{1}{2}(C_{11}^0-C_{12}^0).\end{aligned}\right\} \tag{II.5.3b}$$

We have also introduced local strain coordinates $e_{\alpha\beta}(\mathbf{x})$, which may be usefully decomposed into homogeneous and inhomogeneous parts;

$$e_{\alpha\beta}(\mathbf{x})=\eta_{\alpha\beta}+\frac{i}{2N}\sum_{\mathbf{k}\neq 0}\{k_\beta w_\alpha(\mathbf{k})+k_\alpha w_\beta(\mathbf{k})\}\exp(-i\mathbf{k}\,.\,\mathbf{x}), \tag{II.5.3c}$$

where the sum over $\mathbf{k}$ may be taken to extend over a sphere of long-wavelength acoustic modes with wavevector cutoff Λ. Finally the parameter e_0 measures the order-parameter–strain coupling, p is the external pressure, which couples only to the homogeneous strains, and p_0 characterizes the equilibrium state, of volume $V_0\equiv Na^d$, with respect to which the strains are defined.

Since the strain variables appear at most quadratically in (II.5.2), they may be precisely integrated out of the associated partition function. However, this procedure requires careful consideration of the separate roles of the inhomogeneous and homogeneous strain coordinates, since there are d degrees of freedom (per wavevector $\mathbf{k}$) associated with the former, but $[d(d+1)]/2$ degrees of freedom associated with the latter. Thus we write

$$\mathscr{H}_0^e=\mathscr{H}_0^{e,h}+\mathscr{H}_0^{e,i}, \tag{II.5.4a}$$

with

$$\mathscr{H}_0^{e,h}=\frac{V_0}{a^d}\left\{\left(\frac{1}{2}K_0-\frac{\mu_0}{d}\right)\left(\sum_\alpha\eta_{\alpha\alpha}\right)^2\right.$$
$$\left.+\mu_0\sum_{\alpha,\beta}\eta_{\alpha\beta}^2+\left(\sum_\alpha\eta_{\alpha\alpha}\right)\left[(p-p_0)+\frac{e_0}{N}\int_{\mathbf{x}}u^2(\mathbf{x})\right]\right\} \tag{II.5.4b}$$

and

$$\mathscr{H}_0^{e,i}=\frac{1}{N}\sum_{\substack{\mathbf{k}\neq 0\\ \alpha,\beta}}\left\{\frac{1}{2}D_{\alpha\beta}(\mathbf{k})w_\alpha(\mathbf{k})w_\beta(-\mathbf{k})+i\delta_{\alpha\beta}e_0k_\alpha w_\alpha(\mathbf{k})\varepsilon(-\mathbf{k})\right\}, \qquad \text{(II.5.4}\,c\text{)}$$

where $D_{\alpha\beta}$ is the acoustic mode dynamical matrix

$$D_{\alpha\beta}(\mathbf{k})=\left(K_0+\frac{d-2}{d}\mu_0\right)k_\alpha k_\beta+\mu_0k^2\delta_{\alpha\beta}, \qquad \text{(II.5.4}\,d\text{)}$$

while $\varepsilon(\mathbf{k})$ is the Fourier transform of the local squared-order parameter fluctuation

$$\varepsilon(\mathbf{k})=\int_{\mathbf{x}}u^2(\mathbf{x})\exp(i\mathbf{k}\,.\,\mathbf{x}). \qquad \text{(II.5.4}\,e\text{)}$$

The elimination of the elastic degrees of freedom may now be accomplished by completing the appropriate squares in (II.5.4 b) and (II.5.4 c). Taking the latter as an example, we write

$$\mathscr{H}_0^{e,i}=\frac{1}{N}\sum_{\substack{\mathbf{k}\neq 0\\ \alpha,\beta}}\frac{1}{2}D_{\alpha\beta}(\mathbf{k})\tilde{w}_\alpha(\mathbf{k})\tilde{w}_\beta(-\mathbf{k})-\frac{e_0^2}{2}\cdot\frac{1}{N}\sum_{\substack{\mathbf{k}\neq 0\\ \gamma,\gamma'}}D_{\gamma\gamma'}^{-1}(\mathbf{k})k_\gamma k_{\gamma'}\varepsilon(\mathbf{k})\varepsilon(-\mathbf{k}), \qquad \text{(II.5.5}\,a\text{)}$$

where

$$\tilde{w}_\alpha(\mathbf{k})=w_\alpha(\mathbf{k})-ie_0\varepsilon(\mathbf{k})\sum_\gamma k_\gamma D_{\alpha\gamma}^{-1}(\mathbf{k}). \qquad \text{(II.5.5}\,b\text{)}$$

The integration over the inhomogeneous strain coordinates (in effect over the 'shifted' coordinates $\tilde{\omega}_\alpha(\mathbf{k})$) is now trivial: their elimination leaves only the second term in (II.5.5 a):

$$\Delta\mathscr{H}_0^{e,i}=-\frac{e_0^2}{2}\cdot\frac{1}{N}\sum_{\mathbf{k}\neq 0}\varepsilon(\mathbf{k})\varepsilon(-\mathbf{k})\sum_{\gamma,\gamma'}D_{\gamma\gamma'}^{-1}(\mathbf{k})k_\gamma k_{\gamma'}. \qquad \text{(II.5.6)}$$

As a result of the assumed elastic isotropy, the last factor in the right-hand side of this equation is a constant (independent of $\mathbf{k}$):

$$\sum_{\gamma,\gamma'}D_{\gamma\gamma'}^{-1}(\mathbf{k})k_\gamma k_{\gamma'}=\frac{1}{2\left[\frac{1}{2}K_0+(d-1/d)\mu_0\right]}, \qquad \text{(II.5.7)}$$

and (II.5.6) becomes

$$\Delta\mathscr{H}_0^{e,i}=-\frac{e_0^2}{4\left[\frac{1}{2}K_0+(d-1/d)\mu_0\right]}\cdot\int_{\mathbf{x}}u^4(\mathbf{x})+\frac{Ne_0^2}{4\left[\frac{1}{2}K_0+(d-1/d)\mu_0\right]}\left[\frac{1}{N}\int_{\mathbf{x}}u^2(\mathbf{x})\right]^2 \qquad \text{(II.5.8}\,a\text{)}$$

where we have made use of (II.5.4 e).

The macroscopic strain coordinates can be treated in a similar way (note that we are assuming conditions of constant pressure) and yield the net contribution

$$\Delta\mathscr{H}_0^{e,h}=-\frac{(p-p_0)e_0}{K_0}\int_{\mathbf{x}}u^2(\mathbf{x})-\frac{Ne_0^2}{2K_0}\left(\frac{1}{N}\int_{\mathbf{x}}u^2(\mathbf{x})\right)^2. \qquad \text{(II.5.8}\,b\text{)}$$

Combining (II.5.8 a) and (II.5.8 b) with the basic Ising hamiltonian yields a final effective hamiltonian

$$\mathscr{H}_0 = \mathscr{H}_0(n=1) + \Delta\mathscr{H}_0^{e,i} + \Delta\mathscr{H}_0^{e,h} = \tilde{\mathscr{H}}_0(n=1) + \Delta\mathscr{H}_0, \tag{II.5.9 a}$$

where $\tilde{\mathscr{H}}_0$ $(n=1)$ has the original Ising form (eqns. (II.4.7 a) and (II.4.8 a)), with renormalized parameters

$$\tilde{r}_0 = r_0 - \frac{2e_0}{K_0}(p-p_0), \tag{II.5.9 b}$$

$$\tilde{u}_0 = u_0 - \frac{e_0^2}{4\left[\frac{1}{2}K_0 + \frac{d-1}{d}\mu_0\right]}. \tag{II.5.9 c}$$

The new term in (II.5.9 a) is

$$\Delta\mathscr{H}_0 = \tilde{v}_0 N\left(\frac{1}{N}\int_{\mathbf{x}} u^2(\mathbf{x})\right)^2 = \frac{\tilde{v}_0}{N}\left(\int_{\mathbf{k}}^{\Lambda} |Q(\mathbf{k})|^2\right)^2, \tag{II.5.9 d}$$

where

$$\tilde{v}_0 \equiv \frac{e_0^2}{4}\left[\frac{1}{\frac{1}{2}K_0 + \frac{d-1}{d}\mu_0} - \frac{1}{\frac{1}{2}K_0}\right]. \tag{II.5.9 e}$$

However, before proceeding to discuss the crucial role of this term, one or two general observations may usefully be made.

First of all we note that, according to (II.5.9 b), the parameter e_0 characterizes the pressure-dependent shift in the 'Landau' approximation to the transition temperature (cf. eqn. (II.4.9))

$$T_0(p) - T_0(p_0) = \frac{2e_0}{a_0 K_0}(p-p_0), \tag{II.5.10}$$

a result which gives a zeroth-order estimate to the pressure dependence of the actual transition temperature. Thus we see that the scale of the effects derived from the order-parameter–strain coupling is indeed set by the quantity $[dT_c(p)]/dp$ (or, perhaps more appropriately, $\{1/[T_c(p)]\}\{[dT_c(p)]/dp\}$, to which we referred earlier.

Secondly we note the tendency (already alluded to in §I.4.1.2) for the order-parameter–strain interaction to renormalize the coupling constant (eqn. II.5.9 c) towards smaller and, ultimately, possibly negative values, raising the *possibility* of a first-order phase transition, and a tricritical point (cf. §II.5.3.1)

Next, we observe that, from (II.5.8 b), the strain at the equilibrium pressure $p=p_0$ satisfies

$$\sum_{\alpha=1}^{d} \langle \eta_{\alpha\alpha} \rangle = -\frac{e_0}{K_0}\langle u^2(\mathbf{x})\rangle$$

$$= -\frac{e_0}{K_0}\cdot\frac{1}{N}\int_{\mathbf{k}}^{\Lambda} \langle |Q(\mathbf{k})|^2 \rangle. \tag{II.5.11 a}$$

Finally, with the aid of this result, and (II.5.8 *b*) again, it is easily shown that the bulk modulus is fluctuation-renormalized from its bare value K_0, to a value

$$K = K_0\left\{1 + \frac{e_0^2}{K_0}\frac{1}{N}\int_{\mathbf{k}_1}^{\Lambda}\int_{\mathbf{k}_2}^{\Lambda}(\langle Q(\mathbf{k}_1)Q(-\mathbf{k}_1)Q(\mathbf{k}_2)Q(-\mathbf{k}_2)\rangle - \langle Q(\mathbf{k}_1)Q(-\mathbf{k}_1)\rangle\langle Q(\mathbf{k}_2)Q(-\mathbf{k}_2)\rangle)\right\}^{-1}. \quad \text{(II.5.11 }b\text{)}$$

We now turn to consider the qualitatively new term in the effective hamiltonian (II.5.9 *a*)—defined explicitly in (II.5.9 *d*)—which represents an interaction between the order-parameter degrees of freedom that is *long-range* in character, being mediated by long-wavelength (acoustic) modes.

The role of this term is elucidated by constructing the appropriate R.G. equations (Sak 1974). The flow equation for $\tilde{v}$ has the leading order form

$$\tilde{v}(b) = b^{\varepsilon}\tilde{v}_0\left[1 - 24\tilde{u}\int_{\mathbf{k}}^{\Lambda}\frac{1}{(\tilde{r}_0 + k^2)^2}\right]. \quad \text{(II.5.12)}$$

It is easily checked that the normal Ising fixed point (eqn. (II.4.68)), which, of course, has $\tilde{v}^* = 0$, is then *unstable* against the parameter $\tilde{v}$, which constitutes a relevant scaling field with flow exponent.

$$\begin{aligned}\lambda_{\tilde{v}} &= \varepsilon - 24u^* \,.\, K_d a^d \\ &= \varepsilon/3 \\ &= \alpha/\nu + O(\varepsilon^2),\end{aligned} \quad \text{(II.5.13)}$$

where α and ν are the usual Ising exponents: the latter identification can be shown to be true to all orders in ε (Sak 1974).

The normal Ising fixed point cannot, therefore, describe (asymptotic) critical behaviour in the presence of acoustic-mode coupling. Furthermore, a full analysis of the R.G. equations for the hamiltonian (II.5.9 *a*) reveals *no* other fixed point that is both 'stable' and accessible to a line of hamiltonian flow emanating from 'physical' values of $\tilde{r}_0$, $\tilde{u}_0$ and $\tilde{v}_0$. (Note that, according to (II.5.9 *e*) $\tilde{v}_0$ must be *negative* for a system with the positive shear modulus required for stability.)

Recalling the discussion of § II.4.3.8, the implication is clear: the existence of order-parameter–strain coupling necessitates a fluctuation-driven first-order phase transition under conditions of constant stress.

There are several points which must now be made.

First of all we remark that the underlying character of the first-order instability, left somewhat obscure in the approach we have sketched, is beautifully revealed in the more elaborate R.G. analysis of Bergman and Halperin (1976). In this approach R.G. equations are set up for the full effective hamiltonian (II.5.2 *a*), the elastic degrees of freedom being explicitly retained. This analysis shows that the continuous phase transition is preempted by an elastic instability, caused by the renormalization of the bulk modulus K.

Our discussion so far has been restricted to an elastically-isotropic solid. Bergman and Halperin (1976) and de Moura *et al.* (1976) have also studied the more realistic, and more complex, problem that ensues when elastic anisotropy is taken into account. Their conclusions are that, again, the phase change is first-order, through in this case the instability may be traced to the order parameter degrees of freedom themselves.

While this mechanism may, then, always be invoked to account for the character of a phase transition that is observed to be first-order, it is likely that the effect is, in fact, small (in part because the crossover exponent characterizing the growth in $\tilde{v}$ as $t \to 0$ is small: eqn. (II.5.13)) and that the real reasons for observed discontinuities lie elsewhere—perhaps merely in the sign of the renormalized coupling coefficient (II.5.9 *c*) (cf. § II.5.3.1). Indeed, it is clear that the effects *must* frequently be 'small' in view of the existence of many structural phase transitions that are 'effectively' continuous. In such instances the system will still exhibit normal Ising critical behaviour (under conditions of constant stress—under conditions of constant strain the behaviour is more subtle: see, again, Bergman and Halperin 1976), before the lines of hamiltonian flow are driven from the Ising fixed point.

In this range of reduced temperature the normal Ising behaviour will be reflected in the behaviour of the elastic degrees of freedom. In particular, since the local mean-square distortion $\langle u^2(\mathbf{x})\rangle$ should display the critical behaviour of the *energy* (cf. § II.5.4.1) we expect a pseudocritical behaviour for the strain (eqn. II.5.11 *a*):

$$\sum_{\alpha=1}^{d} \langle \eta_{\alpha\alpha} \rangle \sim E_1 + E_2 t + E_3^{\pm} |t|^{1-\alpha}, \tag{II.5.14 a}$$

where the E_i are constants. (Notice that, now that fluctuations have been taken into account, the strain does *not* vanish with the order parameter—in sharp contrast to the Landau prediction of § I.4.1.2.)

Similarly, it is readily shown that the 'energy–energy' correlation function appearing in (II.5.11 *b*) follows the specific heat, so that

$$K \approx \frac{K_0}{1 + E_4^{\pm} |t|^{-\alpha}} \sim |t|^{\alpha}, \tag{II.5.14 b}$$

although the pure power-law form will be evident only over a range of t which itself reflects the smallness of e_0.

These results are rather more general than our development would suggest. They are relevant to any system in which

(i) the order parameter is a single-component quantity. (The behaviour of the elastic degrees of freedom in systems with multi-component order parameters is more exotic: cf. § II.5.2.2.)

(ii) the strain couples to the square of the ordering coordinate and constitutes, therefore, a truly 'secondary' order parameter: cf. § I.3.2. (In systems where the strain couples linearly to 'the' ordering coordinate—e.g. piezoelectrics—it has, itself, the status of a primary order parameter: the appropriate components of the strain and the elastic constants then display the usual strong singularities, characterized by exponents β and γ, and the theory of acoustic instabilities, discussed in § II.5.1.4, is applicable.)

In these circumstances, then, the strain coordinates involved will carry an energy-type singularity, while the appropriate elastic constants will vanish with a specific-heat singularity

$$C_{ij} \sim |t|^{\alpha}. \tag{II.5.14 c}$$

Since the singularities involved are typically very weak, the 'vanishing' of the elastic constants (like the 'divergence' of the specific heat) will generally be masked by impurity-induced smearing of T_c. Indeed we are aware of no experiments offering a

Fig. II.9

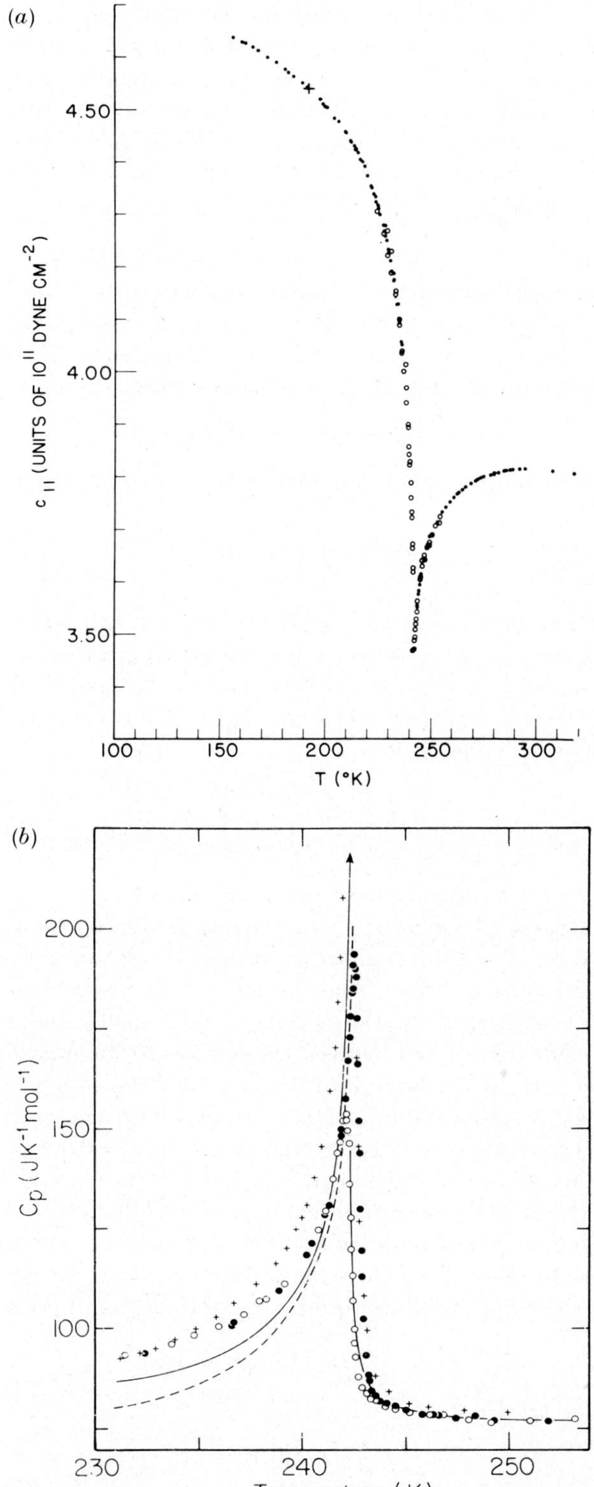

The temperature dependence of (*a*) the elastic constant C_{11} (Garland 1970) and (*b*) the specific heat (Garland and Baloga 1977) in NH_4Cl at atmospheric pressure.

quantitative verification of the prediction (II.5.14 c). Qualitatively, however, the way in which the critical behaviour of the specific heat is mirrored in elastic properties is afforded by measurements on the Ising-like system NH_4Cl (Garland 1970, Garland and Baloga 1977; fig. II.9). In this case the proximity of the tricritical point (the phase transition is actually first-order at atmospheric pressure: cf. § II.5.3.1) results in a relatively strong anomaly in the specific heat ($\alpha=1/2$ for true gaussian tricritical behaviour: cf. § II.5.3.1) making the aforementioned problems less acute.

II.5.1.3. *Short-range and dipolar interactions* (*uniaxial ferroelectrics*)

As discussed in some detail in § I.4.1.4, the dispersion of a *ferroelectric* soft mode in a uniaxial system has a distinctive Coulomb singularity which must (cf. § II.4.1) be incorporated into the effective hamiltonian for such systems, to give (we now ignore coupling to acoustic modes, but bear in mind its implications!)

$$\mathscr{H}_0=\mathscr{H}_0^{\cdot}(n=1)+\mathscr{H}_0^{g,u}, \tag{II.5.15 a}$$

where the term originating in the dipolar interactions is

$$\mathscr{H}_0^{g,u}=\frac{g_0}{2}\int_{\mathbf{k}}^{\Lambda}\frac{k_z^2}{k^2}|Q(\mathbf{k})|^2. \tag{II.5.15 b}$$

As discussed in I, the modified soft-mode dispersion results in a strong suppression of the critical fluctuations with wavevectors along the z-direction (the incipient ferroelectric axis), with a consequent *lowering* of the critical dimensionality d_c. Specifically, the Levanyuk integral (II.4.82) takes the form (recall that, to the extent that Landau theory is consistent, $r_0 \sim t$)

$$L=\int_{\mathbf{k}}^{\Lambda}(t+k^2+g_0k_z^2/k^2)^{-2}\sim\int_0^{\Lambda}dk\int_0^{\pi}d\theta\frac{k^{d-1}\sin^{d-2}\theta}{(t+k^2+g_0\cos^2\theta)^2}, \tag{II.5.16}$$

which is convergent for dimensionalities $d>d_c=3$.

Thus, in the uniaxial ferroelectric, the central predictions of mean field theories become correct above $d_c=3$ dimensions. However, the behaviour in a real (precisely three-dimensional) system differs from the mean field predictions in a subtle way, that mirrors the behaviour of a system with purely short-range interactions at *its* borderline dimensionality, $d_c=4$. Indeed the physics of the uniaxial ferroelectric in $d=3$ can largely be mapped onto that of the short range system in $d=4$. In the latter case it is immediately clear that difficulties will arise from the scaling field associated with the coupling constant, which is *marginal* in $d=4$ ($\lambda_{3,I}=0$ for $\varepsilon=0$, eqn. II.4.78 c). Such situations have been discussed in a general fashion by Wegner and Riedel (1973): the slow decay of the marginal scaling field results in logarithmic corrections to the pure mean-field power laws. In fact this behaviour is already implicit in the expression (II.4.93 a) describing the susceptibility of the short-range Ising system: taking the limit $\varepsilon\to 0$ and making use of eqns. (II.4.67) and (II.4.68 a) we find

$$\chi\approx C_+t^{-1}(1-18u_0K_da^d\ln t)^{1/3} \tag{II.5.17 a}$$

whose asymptotic form is

$$\chi\sim t^{-1}|\ln t|^{1/3}. \tag{II.5.17 b}$$

Indeed, this is precisely the behaviour predicted for the uniaxial ferroelectric—

originally, in the pre-renormalization group era, by Larkin and Khmel'nitskii (1969), and subsequently, using renormalization group methods, by Aharony (1973*a*).

Within the renormalization group framework the importance of the dipolar coupling is revealed by the fact that it is strongly relevant with respect to the short-range Ising fixed point, the crossover exponent ϕ being equal to the Ising exponent γ. There is, however, an accessible stable fixed point (effectively gaussian in character) with $g^* = \infty$, a result which graphically re-emphasizes the fact that, as $t \to 0$, the fluctuation spectrum is totally dominated by fluctuations with $\mathbf{k} \| \hat{z}$.

Two limiting situations may usefully be distinguished. If the dipole–dipole interaction is *large* in comparison with the short-range interaction (a situation typical of ferroelectrics), there will be no range of t in which normal Ising critical behaviour is observed: as $t \to 0$ the behaviour will evolve directly from its classical (non-critical) form to its uniaxial dipolar critical form.

In particular the susceptibility behaves as (Larkin and Khmel'nitskii 1969)

$$\chi \approx C_{\pm}|t|^{-1}(1 - G_1 \ln |t|)^{1/3} \qquad \text{(II.5.18}a)$$

$$\sim |t|^{-1}|\ln t|^{1/3}, \qquad \text{(II.5.18}b)$$

where G_1 depends upon the strength of the dipolar interactions, vanishing logarithmically with g_0, while the amplitudes $C_{\pm}$ obey the 'law of 2' (cf. eqn. (II.3.18*d*)):

$$C_+/C_- = 2. \qquad \text{(II.5.18}c)$$

The specific heat has a similar form:

$$\tilde{C} \approx A_{\pm}(1 - G_1 \ln |t|)^{1/3} \qquad \text{(II.5.19}a)$$

$$\sim |\ln |t||^{1/3}, \qquad \text{(II.5.19}b)$$

with

$$A_+/A_- = 1/4. \qquad \text{(II.5.19}c)$$

Logarithmic singularities also appear in the equation of state as has recently been emphasized by Binder *et al.* (1976). Writing, for this equation, the expansion

$$E = Q_0[f_0(t) + f_2(t)Q_0^2 + \ldots], \qquad \text{(II.5.20}a)$$

one has

$$f_0(t) \sim t|\ln |t||^{-1/3}, \qquad \text{(II.5.20}b)$$

$$f_2(t) \sim (\ln |t|)^{-1}. \qquad \text{(II.5.20}c)$$

Deferring discussion of the experimental situation we consider now the second limiting possibility for the hamiltonian (II.5.15), when the dipolar coupling g_0 is *small*. There will then exist a range of t (*inside* the critical region, but *outside* the dipole-dominated part of this region) where normal short-range Ising critical behaviour will be exhibited with a *crossover* (cf. §II.4.3.7) to dipole-dominated behaviour as $t \to 0$.

As discussed in §II.4.3.7, such crossover behaviour may be described (less satisfactorily on this occasion, in view of the absence of *pure* asymptotic power-law behaviour) in terms of effective critical exponents. The behaviour of the susceptibility exponent γ_{eff} (eqn. II.4.51*a*)) has been determined by $O(\varepsilon)$ R.G. calculations (Bailin and Love 1977), with the results shown in fig. II.10.

Fig. II.10

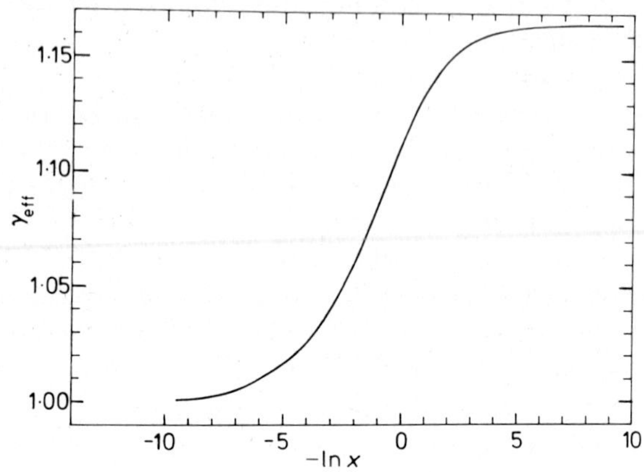

The effective susceptibility exponent for a uniaxial ferroelectric (Bailin and Love 1977) showing the crossover from 'short range' Ising behaviour ($\gamma_{\mathrm{eff}} \approx 1{\cdot}16$ at $O(\varepsilon)$) to a dipole-dominated, near mean-field, behaviour ($\gamma_{\mathrm{eff}} \approx 1$). The variable x is a measure of the reduced temperature introduced in the text. Note the *natural* logarithmic scale.

In interpreting these results one must appreciate that the temperature scale is set, not by the *true* reduced temperature t, but by a reduced temperature $\bar{t} \sim t/g_0^{1/\phi}$. Thus, in a system with a very small value of g_0, the dipole-dominated region may not be accessible (requiring an unrealizably small value for t) while, in a system with a large value of g_0 (as discussed above) *only* the dipolar critical behaviour will be observable, the crossover to short-range critical behaviour at larger $\bar{t}$ being pre-empted by a crossover to 'mean-field' behaviour (not shown in fig. II.10).

Figure II.10 also illustrates the fact that, in general, crossover does not occur abruptly but may take several *decades* of reduced temperature to complete (the smaller the crossover exponent, the wider the range). Thus it is unlikely that an experiment on a single system (with a given g_0) will be able to probe both primary and secondary regimes.

In fact the results shown probably overestimate quite considerably the width of the crossover region in this case, since they underestimate the Ising susceptibility exponent (and hence the crossover exponent) as a result of the $O(\varepsilon)$ approximations. Furthermore, a general consequence of the use of the ε expansion (necessary to cope with the short-range behaviour) is that the calculation does not capture the logarithmic correction to the asymptotic power law—a correction which would manifest itself in a *very* slow approach of γ_{eff} to its asymptotic value of 1. Although these two failings may be mutually compensatory, a more quantitatively satisfactory calculation would prove valuable.

In view of the subtlety of the logarithmic corrections it is not surprising that, until recently, experiments on uniaxial ferroelectrics have revealed only mean field behaviour. (See e.g. the review by Mitsui *et al.* 1973.) A number of more recent experiments have, however, supplied some evidence to substantiate the theory sketched above, though none are as convincing as the vindication of (II.5.19 *b*, *c*) through measurements on the uniaxial ferro*magnet* $LiTbF_4$ (Ahlers *et al.* 1975).

Among the more strongly dipolar ferromagnets, TGS has perhaps the most studied continuous phase transition. Measurements of its specific heat (and elastic properties), severely rounded near T_c, show some evidence (Strukov 1976; fig. II.11) for the *intermediate* behaviour suggested by (II.5.19 *a*), namely

$$\tilde{C} \approx A_+ \left[1 - \frac{1}{3} G_1 \ln|t| \ldots \right]. \qquad \text{(II.5.21)}$$

Further evidence for the logarithmic corrections is presented by Binder *et al.* (1976), who point out that measurements of the coefficient f_2 (eqn. (II.5.20 *a*)) for TGS are, indeed, consistent with the prediction (II.5.20 *c*) (fig. II.12).

However, more detailed experiments are clearly necessary in this area.

Among the uniaxial ferroelectrics where the dipolar coupling is weaker (see Lajzerowicz and Legrand 1978 for a quantitative classification) so that crossover behaviour is to be expected, DSP apparently shows near-Ising behaviour (Deguchi and Nakamura 1976), while experiments on Tanane (Lajzerowicz and Legrand 1978)

Fig. II.11

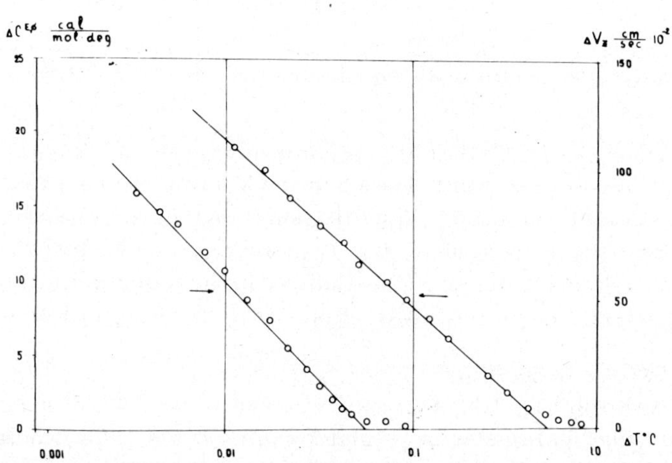

The temperature dependence of the specific heat, and of the elastic constant C_{33}, in TGS, plotted against $\ln t$ to compare with the prediction (II.5.21) (Strukov 1976).

Fig. II.12

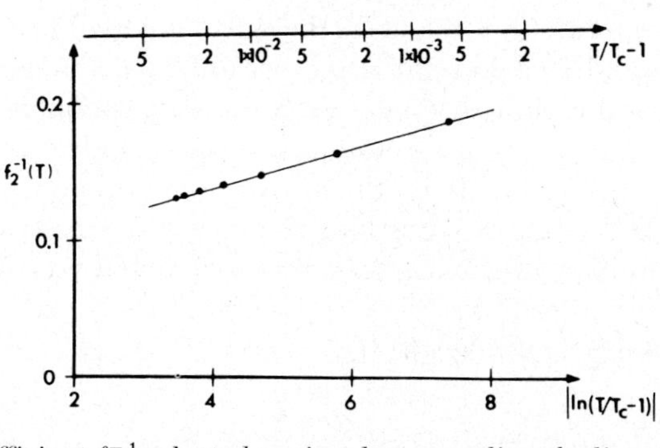

TGS data for the coefficient f_2^{-1}, plotted against $\ln t$, revealing the linear behaviour predicted in (II.5.20 *c*) (Binder *et al.* 1976).

Fig. II.13

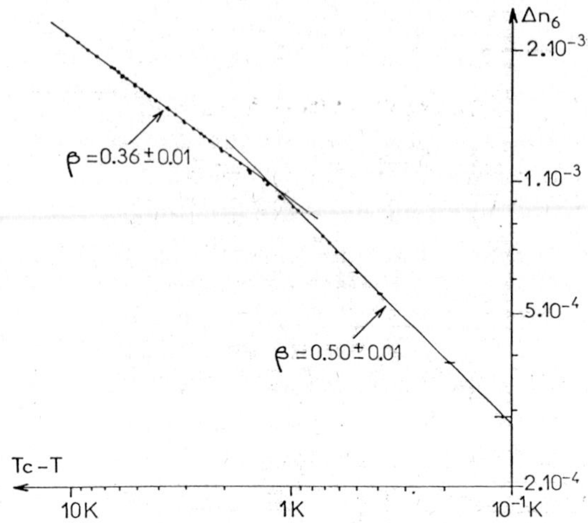

The critical behaviour of the order parameter in the uniaxial ferroelectric Tanane, as measured by birefringence experiments (Lajzerowicz and Legrand 1978).

show behaviour (fig. II.13) that has been interpreted as a crossover from the Ising to the uniaxial dipolar fixed point. The abruptness of this 'crossover' casts considerable doubt upon this interpretation. The situation is further complicated by the fact that Tanane is actually a ferroelastic (i.e. is piezoelectric even for $T>T_c$). As argued originally by Villain (1970), and as we shall see in the course of the next section, mean field theory should (asymptotically) hold *exactly* in these situations.

II.5.1.4. *Acoustic instabilities*

As discussed in §I.4.1.5, there exists a class of structural phase transitions for which the order parameter is a macroscopic strain, and whose occurrence is accompanied by the vanishing of the *velocity(ies)* of acoustic modes in certain sectors of **k** space. The spectrum of different instabilities permitted in the different crystal classes, and their elucidation by renormalization group methods, have been the subject of extensive analyses by Cowley (1976) and by Folk *et al.* (1976). Here we shall discuss, in a simple way, those situations where (as supposed throughout §II.5.1) the order parameter is a scalar.

When the order parameter is a scalar (the 'soft' irreducible representation of the elastic constant is one-dimensional) the sectors of **k** space in which mode softening occurs are themselves one-dimensional: i.e. the acoustic phonon velocity is low only along certain *lines* in reciprocal space, and, furthermore, only for a polarization of the acoustic wave prescribed uniquely by the propagation direction. (See §I.4.1.5 for an explicit example.) With this recognition, the simplest effective hamiltonian representative of this universality class may be constructed as (Folk *et al.* 1976)

$$\mathscr{H}_0=\frac{1}{2}\int_{\mathbf{k}}^{\Lambda}(r_0p^2+g_0q^2+p^2k^2)|w(\mathbf{k})|^2$$

$$+u_0\int_{\mathbf{k}_1}^{\Lambda}\int_{\mathbf{k}_2}^{\Lambda}\int_{\mathbf{k}_3}^{\Lambda}\int_{\mathbf{k}_4}^{\Lambda}p_1p_2p_3p_4w(\mathbf{k}_1)w(\mathbf{k}_2)w(\mathbf{k}_3)w(\mathbf{k}_4)\delta(\mathbf{k}_1+\mathbf{k}_2+\mathbf{k}_3+\mathbf{k}_4), \qquad \text{(II.5.22)}$$

where we have decomposed the wavevector $\mathbf{k}$ into a 'soft' component p, and a $(d-1)$-dimensional 'hard' component $\mathbf{q}$. Thus r_0 measures the *partially*-renormalized elastic constant with the strong temperature dependence (again given by eqn. (II.4.9)), while g_0 may be taken to be a constant. Because the (renormalized) coefficient of p^2 vanishes at the transition it is necessary to include the p^2k^2 (or p^4) term in the dispersion. The coordinates $w(\mathbf{k})$ are the appropriate linear combinations of the coordinates $w_\alpha(\mathbf{k})$, introduced in (II.5.3*c*). Finally, notice that the 4-phonon interaction term vanishes with the soft components p_i of the participating modes: those interactions involving 'hard-components' can be shown to be irrelevant, and have therefore been discarded.

In writing eqn. (II.5.22) we have allowed for the existence of only a *single* soft direction, while in practice there may be several (cf. eqn. (I.4.23)). It would be necessary to consider the different sectors of soft degrees of freedom, and the coupling between them, if the fluctuation-effects associated with a single sector proved to be important. That such fluctuation effects are, in fact, unimportant, one may show either by applying the R.G. method directly to (II.5.22) or—the approach we adopt—by applying the Ginzburg criterion, having transformed (II.5.22) to a more familiar form. Thus we set $Q(\mathbf{k}) \equiv pw(\mathbf{k})$, when (II.5.22) assumes the form

$$\mathscr{H}_0 = \mathscr{H}_0(n=1) + \mathscr{H}_0^g, \tag{II.5.23a}$$

where

$$\mathscr{H}_0^g = \frac{g_0}{2} \int_{\mathbf{k}}^{\Lambda} \frac{q^2}{p^2} |Q(\mathbf{k})|^2. \tag{II.5.23b}$$

The analogy with the uniaxial dipolar system (eqn. (II.5.15)) is now apparent. However, in contrast to (II.5.15*b*) the perturbation (II.5.23*b*) now (strongly!) suppresses all fluctuations *except* those with wavevectors close to the singled-out direction. This difference is reflected in the borderline dimensionality which may, again, be determined from the Levanyuk integral (II.4.82):

$$L = \int_{\mathbf{k}}^{\Lambda} (t + k^2 + g_0 q^2/p^2)^{-2} \sim \int_0^{\Lambda} dk \int_0^{\pi} d\theta \frac{k^{d-1} \sin^{d-2}\theta}{(t + k^2 + g_0 \tan^2\theta)^2} \tag{II.5.24}$$

which remains convergent in the $t \to 0$ limit, provided $d > d_c = 2{\cdot}5$.

In three dimensions, therefore, Landau theory is entirely consistent, and we conclude that the critical behaviour exhibited at elastic phase transitions, with a scalar order parameter, should be characterized by mean field exponents, with *no* logarithmic corrections. This conclusion extends to those systems where the order parameter is some linear combination of optic mode (or pseudo-spin mode) and strain coordinates—the uniaxial ferroelectrics that are piezoelectric in their high-temperature phase (e.g. KH_2PO_4) being prominent examples (Villain 1970).

These conclusions are substantially borne out by experiment. Strictly classical behaviour is shown, for example, at the 151 K Jahn–Teller phase transition in $PrAlO_3$ (fig. II.14) where the order parameter is a linear combination of electronic mode, optic mode and strain coordinates (Birgeneau *et al.* 1974). Birefringence measurements (fig. II.15) of the order parameter for the Jahn–Teller phase transition in $TbVO_4$ also reveal classical behaviour (Harley and Macfarlane 1975). However, non-classical (Ising-like) behaviour is apparent at the similar phase transition occurring in $DyVO_4$ (fig. II.15). This unexpected result has been attributed (Harley and MacFarlane 1975) to the chance cancellation (Sandercock *et al.* 1972) of the effects of dipolar interactions and elastic interactions in the suppression of fluctuations. Presumably, as in Tanane (fig. II.13), a crossover to mean field behaviour will occur closer to T_c.

Fig. II.14

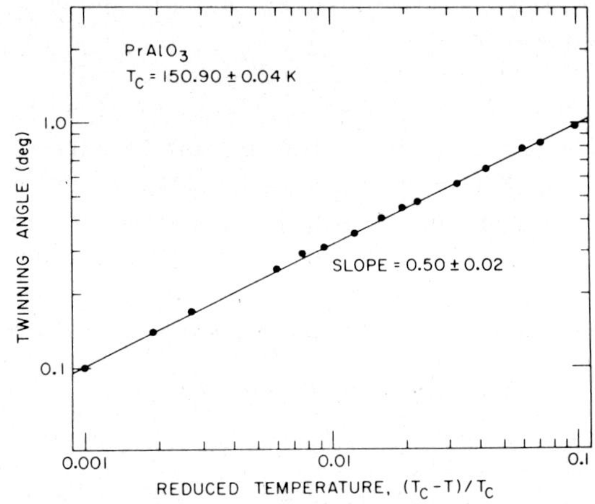

The critical behaviour of a twinning angle, mirroring the strain order parameter, in $PrAlO_3$ (Birgeneau *et al.* 1974).

Fig. II.15

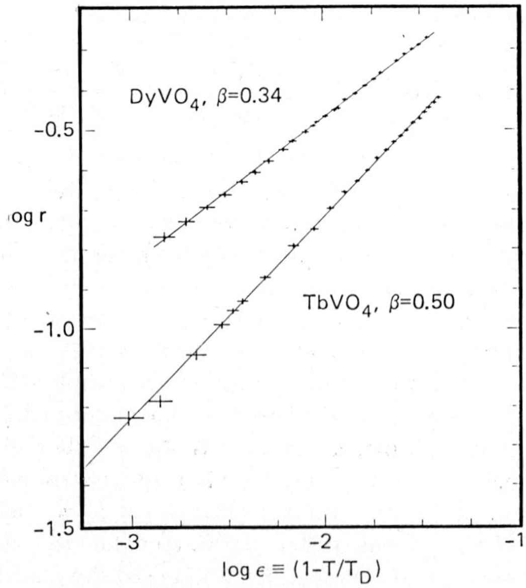

The critical behaviour of the birefringence, and thence the order parameter, in $TbVO_4$ and $DyVO_4$ (Harley and MacFarlane 1975).

II.5.2. *Multiple-component order parameters: ordinary critical behaviour*

II.5.2.1. *Short-range interactions*

The basic effective hamiltonian appropriate for structural phase transitions associated with an n-fold degenerate soft mode is the n-component L.G.W. model, introduced in (II.4.10 b). This effective hamiltonian has a stable fixed point, referred to frequently as the (isotropic) n-component Heisenberg fixed point, since it characterizes the universality class to which the Heisenberg model of ferromagnetism belongs. The fixed-point parameters, and the associated critical exponents (table II.3), show the explicit (but relatively weak) n-dependence anticipated in the discussion of § II.4.1, reducing, for $n=1$, to the Ising model results.

However, the appropriateness of the hamiltonian (II.4.10 b), and the relevance of the results shown in table II.3 to any real structural phase transition, requires more extensive justification than is necessary in the $n=1$-component problem, the reason being that, for $n>1$, the order parameter symmetry will generally permit various additional interactions which do not have the isotropy of (II.4.10b). Such symmetry-breaking interactions are potentially universality-class changing, and their effects must be carefully considered.

Table II.3. The critical exponents γ and β of the isotropic n-component system $\mathscr{H}_0(n)$, as given by the second-order ε expansion (Wilson 1972) and (in parentheses) series expansions, as summarized by Fisher and Jasnow (1980). The final column gives the exponents for an ($n=3$)-component *ferroelectric* system, when interactions of cubic symmetry are ignored (Bruce and Aharony 1974).

	$n=1$	$n=2$	$n=3$	$n=3$ (ferroelectric)
γ	1·24 (1·25)	1·30 (1·32)	1·36 (1·375)	1·37
β	0·34 (0·31)	0·36 (0·34)	0·38 (0·38)	0·38

The perturbations whose effects have been most extensively studied are those of cubic symmetry. (See Aharony 1976, for an extensive review.) These are of two types.

First of all the cubic symmetry is reflected in the additional (4th-order) invariant allowed in a Landau free-energy expansion (cf. eqn. (I.4.33)), which manifests itself as the additional interaction

$$\mathscr{H}_0^v = v_0 \sum_\alpha \int_{\mathbf{k}_1}^\Lambda \int_{\mathbf{k}_2}^\Lambda \int_{\mathbf{k}_3}^\Lambda \int_{\mathbf{k}_4}^\Lambda Q_\alpha(\mathbf{k}_1) Q_\alpha(\mathbf{k}_2) Q_\alpha(\mathbf{k}_3) Q_\alpha(\mathbf{k}_4) \delta(\mathbf{k}_1+\mathbf{k}_2+\mathbf{k}_3+\mathbf{k}_4). \quad \text{(II.5.25)}$$

Secondly, the cubic symmetry may be reflected in the soft-mode dispersion, through an additional term of the form

$$\mathscr{H}_0^f = \frac{f_0}{2} \sum_\alpha \int_{\mathbf{k}}^\Lambda k_\alpha^2 |Q_\alpha(\mathbf{k})|^2. \quad \text{(II.5.26)}$$

We remark, parenthetically, that while anisotropy in the dispersion of a *singlet* soft mode is, of course, also possible, it is not usually important (i.e. universality-class-changing)—hence the isotropy assumed throughout our discussion of the $n=1$ system. (See, however, § II.5.3.2.)

Together with the basic term, (II.4.10b), these interactions yield the much-studied effective hamiltonian

$$\mathscr{H}_0 = \mathscr{H}_0(n=d) + \mathscr{H}_0^f + \mathscr{H}_0^v, \qquad \text{(II.5.27)}$$

where the ($n=d$)-component case has been singled out since the form of the perturbation (II.5.26) imposes the condition $n=d$. The consensus of a number of independent analyses—see Aharony (1976), and references therein—is that in $d=3$ the stable fixed point of the effective hamiltonian (II.5.27) remains the isotropic fixed point discussed above: asymptotically (cf. fig. II.16) the cubic perturbations (II.5.25) and (II.5.26) will be renormalized to zero fixed-point values by the critical fluctuations, and the appropriate critical exponents will be those of the ($n=3$)-component Heisenberg model (table II.3).

The significance of this result is, however, tempered by the smallness of the exponents, λ_v and λ_f, which characterize the 'rate' at which, as a function of t, the anisotropies decay: λ_f, in particular, is so small that the associated anisotropy will not change appreciably over any reasonable interval of reduced temperature (Nattermann 1976). In physical situations where the anisotropy is large, the observed critical behaviour will then be characterized, not by the isotropic exponents, but by effective critical exponents, which reflect the anisotropy. Calculations to leading order in ε, which presumably give a qualitatively correct indication of the effects of the anisotropy, suggest (Nattermann and Trimper 1975) that, in those situations where the parameter f_0 is close to -1 (so that the αth branch is flat along the αth cube axis), the effective exponents $\beta_{\text{eff}}(f_0)$ and $\gamma_{\text{eff}}(f_0)$ will be somewhat *larger* than their asymptotic counterparts.

However, a large anisotropy ($f_0 \sim -1$) may have rather more drastic effects: as conjectured by Bruce (1974) and demonstrated by Nattermann (1976), a system with f_0 lying close enough to -1 actually lies outside the domain of attraction of the

Fig. II.16

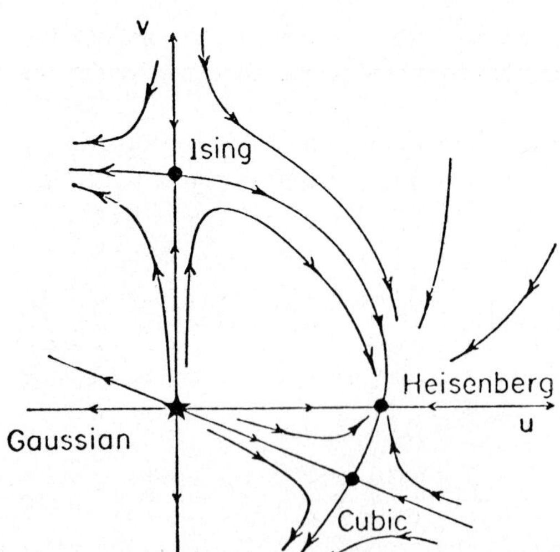

Lines of *critical* hamiltonian flow for the $n=d=3$ cubic system (II.5.27), with $f_0=0$. The Heisenberg fixed point is stable, and accessible provided $u_0>0$ and $v_0>v_c(u_0)$, where $v_c(u)$ describes the one trajectory leading to the cubic fixed point (Aharony 1976).

stable Heisenberg fixed point, the f_0-induced anisotropy of the critical fluctuations driving the renormalized coefficient of quartic anisotropy, v_0, to large negative values, thus promoting a first-order phase transition.

In systems with order parameter dimensionalities n larger than 3 (recall the discussion following (II.4.9), and §I.4.2.1), the situation does not even have the virtue of being simple in principle: additional interactions of new symmetries are, in general, permitted and the Heisenberg fixed point is no longer stable. Two possibilities then arise.

First of all the behaviour near the phase transition may be characterized by a different fixed point, of lower symmetry, e.g. the 'cubic' fixed point discussed by Aharony (1973 b), or the 'tetragonal' fixed point discussed by Mukamel (1975); experience suggests, however, that the associated critical exponents would not then differ appreciably from their Heisenberg counterparts.

Secondly the effective hamiltonian may (and frequently does) turn out to have no stable fixed point. The absence of stable fixed points in many $n \geqslant 4$-component systems has been shown to correlate very well with the occurrence of first-order phase transitions in the magnetic realisations of these universality classes (Bak *et al.* 1976).

A discussion of the experimental situation begins, of necessity, with the prototype members of the universality class described by (II.5.27)—those members of the perovskite family exhibiting phase transitions associated with the ($n=3$)-component soft mode at the R-point on the Brillouin-zone boundary. The critical region in $SrTiO_3$ was among the first to attract detailed experimental study (Berre *et al.* 1969). Measurements of the order parameter exponent in $SrTiO_3$ and $LaAlO_3$ (fig. II.17) by E.P.R. techniques (Müller and Berlinger 1971) and by elastic neutron scattering (Riste *et al.* 1971) yield $\beta = 0{\cdot}33 \pm 0{\cdot}02$, rather low in comparison with the

Fig. II.17

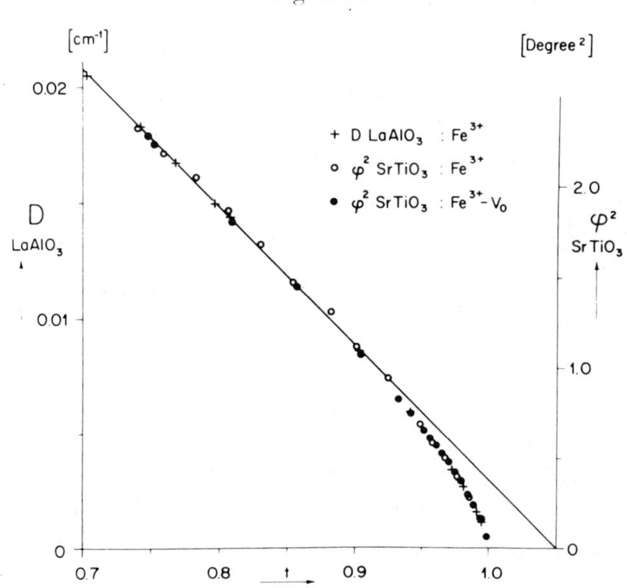

E.P.R. measurements of the square of the order parameter in $SrTiO_3$ ('ϕ^2') and $LaAlO_3$ ('D'), as a function of $t \equiv T/T_c$, showing the deviations from mean-field theory (Müller and Berlinger 1971).

Heisenberg exponent which, we have seen, should be appropriate asymptotically. It seems unlikely that this discrepancy can be attributed to the slow decay of the anisotropy parameter f_0, since this anisotropy, while pronounced in $SrTiO_3$ ($f_0 \approx -0{\cdot}97$: Stirling 1972: cf. eqn. (I.4.40)) is apparently not significant in $LaAlO_3$ (Kjems *et al.* 1973): in any case, the effective exponent $\beta_{\text{eff}}(f_0)$ is, it will be recalled, expected to be *larger* than the asymptotic exponent! An alternative source for the discrepancy will be discussed in § II.5.3.3.

None of the other static critical exponents, for this group, has been determined with comparable accuracy. Neutron inelastic scattering measurements on $SrTiO_3$ indicate a curiously large value for γ (Shapiro *et al.* 1972, Töpler *et al.* 1977); an accurate X-ray diffuse scattering measurement would be useful here. Further E.P.R. experiments give $\nu = 0{\cdot}65 \pm 0{\cdot}05$ (von Waldkirch *et al.* 1972). The problems alluded to at the outset of § II.5 seem to preclude an accurate determination of α (Hatta *et al.* 1977).

At a qualitative level, however, the fact that $KMnF_3$, in which the f_0-anisotropy is particularly pronounced ($f_0 \approx -0{\cdot}99$; Shapiro *et al.* 1972), has a first-order phase transition, while the transition in $SrTiO_3$ is not discernibly discontinuous, lends support to the renormalization group and analyses of the effects of highly anisotropic fluctuations. (See also Aharony and Bruce 1979, Buzaré *et al.* 1979).

The only ($n > 3$)-component structural phase transition to have been studied in detail is NbO_2, which has a 4-component order parameter (cf. § I.4.5.1). The results of neutron scattering measurements (Pynn and Axe 1976) are shown in fig. II.18. The 'asymptotic' value $\beta \approx 0{\cdot}4$ is close to the tetragonal fixed point value ($\beta \approx 0{\cdot}39$) which, Mukamel (1975) has argued, should be appropriate for this system. The low value of β found further from T_c (≈ 810 K) is suggestive of an effectively low-dimensional behaviour in this regime (Pynn and Axe 1976; see also § II.5.4.2).

It will be clear from the foregoing discussion that it is proving considerably more difficult to match critical exponent theory with experiment, in a quantitatively satisfactory way, in systems with $n > 1$: the theoretical predictions are less clear cut, and the usefulness of asymptotic exponents limited by irrelevant perturbations that decay only slowly with t. These difficulties may be met in two ways.

First of all, more detailed calculations of effective critical exponents would be helpful—particularly those appropriate for the system with cubic perturbations (eqn. (II.5.27)).

Fig. II.18

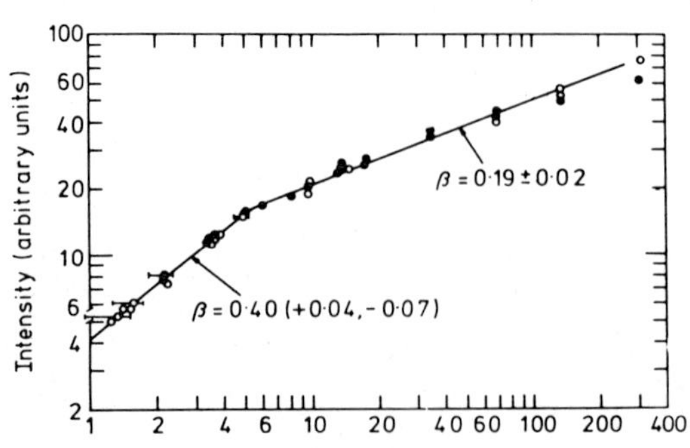

The critical behaviour of the intensity of a superlattice reflection, measuring the square of the order parameter, in NbO_2 (Pynn and Axe 1976).

Secondly it would prove useful to perform careful experiments on the one class of ($n>1$)-component structural phase transitions for which the isotropy presupposed in $\mathscr{H}_0(n)$ is *not* broken by interactions of lower symmetry. We have in mind the simple phase transition to an *incommensurate* phase (discussed in § I.4.4.1) where, to the extent that the phason mode is unpinned, the ($n=2$)-component isotropic hamiltonian is appropriate without modification. The only relevant experimental work on such a system, of which we are aware, is the measurement of the order parameter exponent in K_2SeO_4 (Iizumi *et al.* 1977); but the result of their preliminary measurements, $\beta=0{\cdot}4\pm0{\cdot}05$, is of insufficient accuracy for a comparison with the ($n=2$)-component result ($\beta\approx0{\cdot}34$) to be meaningful.

II.5.2.2. *Short-range interactions with order-parameter–strain coupling*

In § II.5.1.2 we discussed the effects of order-parameter–strain coupling on the critical behaviour of the Ising system $\mathscr{H}_0(n=1)$. The analogous, though somewhat more complex, effects occurring in ($n>1$)-component systems have been comprehensively analysed by Murata (1977), Nattermann (1977) and Bender (1976). We shall state only the essential conclusion of these analyses, which is that a stable fixed point is accessible to the interacting strain-order parameter system only if the strain couples purely to the squared magnitude of the local ordering variable. This condition will not usually be fulfilled (it would require, for example, $e_2=e_3$ in eqn. (I.4.37)), and so the associated phase transitions may be presumed to be first-order. However, although the crossover exponents characterizing the growth of the instability-inducing perturbations are larger in this case than in the $n=1$ problem (cf. eqn. (II.5.13)), the first-order component can be (and, in view of the experimental situation, frequently is!) so small that the phase transition is effectively continuous, with a critical region dominated by the fixed point appropriate in the absence of the coupling to strain coordinates.

The critical behaviour of the ordering degrees of freedom will, then, once more be reflected in the behaviour of the elastic coordinates. Now, however, one must be careful to treat separately the different irreducible representations of the elastic coordinates, which couple to correspondingly different combinations of the ordering variables, and (may) as a result display different critical singularities. (See Murata 1976 for a careful discussion.)

One finds in particular that the volume strain behaves as (we continue to suppose that $n=d$)

$$\sum_{\alpha=1}^{d}\eta_{\alpha\alpha}\sim\int_{\mathbf{k}}^{\Lambda}\sum_{\alpha=1}^{d}\langle|Q_\alpha(\mathbf{k})|^2\rangle. \tag{II.5.28}$$

The operator on the right-hand side of this equation represents the energy density for this problem, and (as in eqn. (II.5.14a)) has the critical behaviour

$$\int_{\mathbf{k}}^{\Lambda}\sum_{\alpha=1}^{d}\langle|Q_\alpha(\mathbf{k})|^2\rangle\sim E_1+E_2t+E_3^{\pm}|t|^{1-\alpha}+\dots \tag{II.5.29}$$

However the tetragonal strain accompanying the phase transition (and acting as a secondary order parameter) involves a different combination of ordering variables:

$$\eta_{11}-\frac{1}{d-1}\sum_{\alpha\neq1}^{d}\eta_{\alpha\alpha}\sim\int_{\mathbf{k}}^{\Lambda}\left\{\langle|Q_1(\mathbf{k})|^2\rangle-\frac{1}{d-1}\sum_{\alpha\neq1}^{d}\langle|Q_\alpha(\mathbf{k})|^2\rangle\right\}. \tag{II.5.30}$$

Within a Landau theory the operator on the right-hand side of this expression would vanish as $|t|^{2\beta}$ as $t \to 0^-$; a more careful analysis (to be discussed, with a different emphasis, in §II.5.3.3) shows that, in fact (Fisher and Nelson 1974, Aharony and Bruce 1974)

$$\int_{\mathbf{k}}^{\Lambda} \left\{ \langle |Q_1(\mathbf{k})|^2 \rangle - \frac{1}{d-1} \sum_{\alpha \neq 1}^{d} \langle |Q_\alpha(\mathbf{k})|^2 \rangle \right\} \sim |t|^{\tilde{\beta}} \qquad \text{(II.5.31 } a\text{)}$$

with

$$\tilde{\beta} = 2 - \alpha - \phi. \qquad \text{(II.5.31 } b\text{)}$$

Here $\phi = \phi(n)$ is a crossover exponent reflecting the instability of the Heisenberg fixed point (assumed to characterize the critical behaviour) against a 'perturbation' (the operator in (II.5.30)) of uniaxial symmetry. For $n = d = 3$, one finds

$$\tilde{\beta} \approx 0{\cdot}85, \qquad \text{(II.5.31 } c\text{)}$$

which differs appreciably from the exponent

$$2\beta \approx 0{\cdot}74, \qquad \text{(II.5.31 } d\text{)}$$

which has frequently, and incorrectly, been assumed to characterize the operator (II.5.30). Indeed, more generally, secondary order parameters which, within a naïvely-modified Landau theory, would appear to vanish as $|t|^{2\beta}$, require careful attention: the secondary Fourier components induced in an incommensurate phase, which determine the intensities of secondary Bragg peaks, offer a further cautionary example (Cowley and Bruce 1978).

The elastic constants, too, carry a singularity involving ϕ (Aharony and Bruce 1974):

$$C_{ij} \sim |t|^{\tilde{\gamma}}, \qquad \text{(II.5.32 } a\text{)}$$

with

$$\tilde{\gamma} = 2\phi + \alpha - 2, \qquad \text{(II.5.32 } b\text{)}$$

where, for $n = d = 3$,

$$\tilde{\gamma} \approx 0{\cdot}4. \qquad \text{(II.5.32 } c\text{)}$$

We remark, however, that if the critical singularities were governed by the cubic fixed point or, as seems likely, by effective exponents reflecting (weakly irrelevant) cubic perturbations, different components of the elastic constant matrix will carry different singularities (Murata 1976). In any event the anticipated singularities are much stronger than the specific-heat-like behaviour expected for ($n = 1$)-component systems (cf. eqn. (II.5.14 c)).

Experiments on perovskites offer qualitative support for these predictions. Measurements by Courtens (1972) show that $SrTiO_3$ has an optical birefringence (which should follow the behaviour of the operator (II.5.30): cf. Gehring 1977) characterized by the power law (II.5.31 a) with $\tilde{\beta} = 0{\cdot}72 \pm 0{\cdot}02$ (fig. II.19) which is suggestively larger than the experimental value of $2\beta = 0{\cdot}66 \pm 0{\cdot}04$ (cf. also the experiments on squaric acid cited in §II.5.4.2). Further, ultrasonic measurements on $SrTiO_3$ (Rehwald 1977) suggest that the elastic constant C_{11} vanishes according to (II.5.32 a) with $\tilde{\gamma} \approx 0{\cdot}66$. However a comparison with the prediction (II.5.32 c) is not really meaningful in view of the very limited softening displayed in the temperature region to which attenuation effectively restricts the pulse-echo techniques used.

Fig. II.19

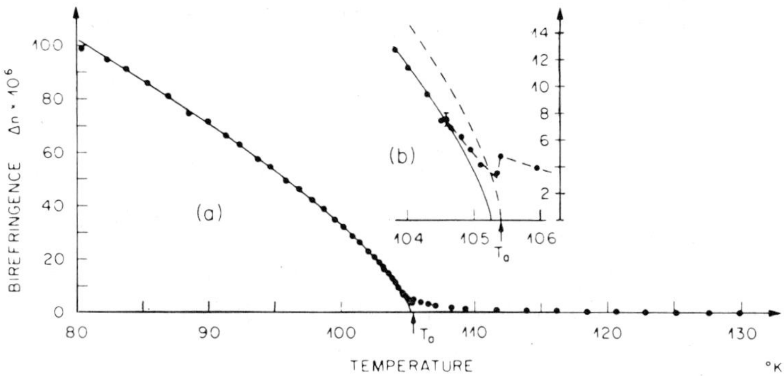

The critical behaviour of the birefringence in $SrTiO_3$ (Courtens 1972). The solid line is a fit to a pure power-law behaviour (II.5.31 *a*). The inset shows, on an extended scale, the crossover from ideal behaviour, dominated by the *spontaneous* symmetry breaking, to a behaviour dominated by some inbuilt 'orienting mechanism' (§ II.5.3.3). The dash–dot line represents the order-parameter data of fig. II.17, and T_a is the transition temperature.

II.5.2.3. *Short-range and dipolar interactions (cubic ferroelectrics)*

In systems with $n>1$, the nature of the Coulomb singularity in the dispersion of a ferroelectric soft mode (§I.4.2.5), and its effects upon the critical behaviour, are quite different from the case of the $n=1$, uniaxial, ferroelectric discussed in § II.5.1.3. Taking the case of the cubic ferroelectric, for which $n=d=3$ (but ignoring, for the moment, the effects of interactions of cubic symmetry), the appropriate effective hamiltonian is

$$\mathscr{H}_0=\mathscr{H}_0(n=d)+\mathscr{H}_0^{g,c}, \tag{II.5.33 a}$$

where

$$\mathscr{H}_0^{g,c}=\frac{g_0}{2}\int_{\mathbf{k}}^{\Lambda}\sum_{\alpha,\beta}\frac{k_\alpha k_\beta}{k^2}Q_\alpha(\mathbf{k})Q_\beta(-\mathbf{k}) \tag{II.5.33 b}$$

is the dipolar term that is responsible for the familiar L.O. mode/T.O. mode splitting in polar crystals.

Within the Landau–Ginzburg approximation the wavevector-dependent susceptibility in the small t, small k, limit has the form

$$\chi_{\alpha\beta}^{\text{I.S.}}(\mathbf{k})=\frac{1}{t+k^2}\left(\delta_{\alpha\beta}-\frac{k_\alpha k_\beta}{k^2}\right), \tag{II.5.34}$$

a result which reveals the strong suppression of fluctuations in the displacement *parallel* to the propagation vector **k**. (The 'longitudinal' correlation function

$$\sum_{\alpha,\beta}\langle k_\alpha Q_\alpha(\mathbf{k})k_\beta Q_\beta(-\mathbf{k})\rangle$$

actually vanishes in the limit presupposed in (II.5.34)). In *contrast* to the uniaxial ferroelectric, however, this partial suppression of critical fluctuations is *not* sufficient to shift the critical dimensionality. The transverse fluctuations ensure that the Levanyuk integral

$$L = \sum_{\alpha,\beta=1}^{d} \int_{\mathbf{k}}^{\Lambda} \frac{1}{(t+k^2)^2} \left(\delta_{\alpha\beta} - \frac{k_\alpha k_\beta}{k^2} \right) \tag{II.5.35}$$

is still divergent for $d < d_c = 4$, as $t \to 0$.

Thus the isotropic dipolar system described by (II.5.33) has non-classical behaviour in three dimensions. More explicitly, renormalization group analysis shows that the short range Heisenberg fixed point is strongly unstable against the dipolar term (II.5.33 *b*) (the crossover exponent is $\phi = \gamma(n=d)$), and that the critical behaviour is characterized by a different, dipolar, fixed point with $g^* = \infty$ (Aharony and Fisher 1973). The associated critical exponents, calculated to $O(\varepsilon^2)$, are certainly non-classical, but are practically indistinguishable from their short-range counterparts (Bruce and Aharony 1974; table II.3).

However, calculations show that, in its crossover region, such a system may be characterized by an effective susceptibility exponent that is quite different from the limiting primary (Heisenberg) and secondary (dipolar) values (Nattermann and Trimper 1976, Bruce 1977). The remarkable dip in the effective exponent (fig. II.20), which may account for the anomalously low value of γ observed in the cubic *ferromagnet* EuO, can be simply understood in the following way.

Deviations from mean-field behaviour reflect both the 'weight' of critical fluctuations (the 'density' of soft modes) and the strength with which they interact (measured by the renormalized coupling constant). As $t \to 0$, the growth in the effective strength g of the dipolar perturbation is rapid, and the weight of the (longitudinal) fluctuations correspondingly rapidly suppressed. Asymptotically this effect is compensated by an increase in the renormalized coupling constant (hence the close equivalence of Heisenberg and dipolar exponents)—but the coupling

Fig. II.20

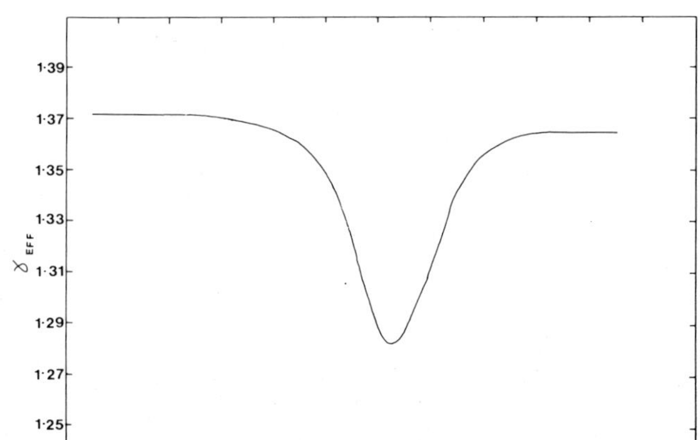

The effective susceptibility exponent for an ($n=3$)-component ferroelectric (when perturbations of cubic symmetry are ignored), showing the crossover from short-range to dipolar behaviour. The reduced temperature $\bar{t}$ is defined in § II.5.1.3. (Bruce 1977).

constant renormalization takes place more gradually as a function of t. Thus there is an intermediate regime where the weight of the fluctuations is reduced with no compensating increase in their interaction strength: this, one surmises, is the region where γ_{eff} is reduced towards its mean field value.

However, while apparently offering a reasonable account of the critical behaviour of cubic ferromagnets, these calculations probably represent no more than the first stage of a theoretical treatment of the cubic ferroelectrics, where interactions of cubic symmetry seem to play a much more significant role. The appropriate effective hamiltonian is then (cf. eqns. (II.5.25) and (II.5.26))

$$\mathscr{H}_0 = \mathscr{H}_0(n=d) + \mathscr{H}_0^{g,c} + \mathscr{H}_0^{v} + \mathscr{H}_0^{f}. \qquad \text{(II.5.36)}$$

The critical behaviour of this hamiltonian has not yet been clearly established, the combination of dipolar and anisotropic terms in the dispersion complicating R.G. analysis. It is clear that the cubic perturbations are strongly relevant with respect to the isotropic dipolar fixed point (Bruce and Aharony 1974). It is not clear whether an alternative stable ('cubic dipolar') fixed point exists. A recent numerical study of approximate R.G. equations, in the strong dipolar limit (Sokolov and Tagantsev 1978), revealed a cubic dipolar fixed point that is unstable. It is possible that no stable fixed point exists, so that such phase transitions are necessarily first-order (even in the absence of coupling to strain!). This possibility must, however, be viewed cautiously in the light of the experimental situation to which we now turn.

In cubic ferroelectrics the ratio of dipolar to short range coupling is typically *much* larger than it is in cubic ferromagnets. Thus it seems unlikely that crossover phenomena of the type discussed above will be seen in ferroelectrics: any critical behaviour observed will, almost certainly, be dominated by the dipolar forces. It is, however, unlikely that such behaviour would be characterized by the isotropic fixed point discussed above, in view of the strong cubic anisotropy observed in the dispersion of ferroelectric soft modes in perovskites (Yamada *et al*. 1969, Comès and Shirane 1972): in the notation of (II.5.26), this anisotropy is characterized by a large *positive* value of the parameter f_0, expressing the fact that the dominant fluctuations are transverse in character *and* polarized along a cube axis.

Indeed, it is at first sight tempting to surmise that such an anisotropic system does not have access to a fixed point, thereby explaining the fact that the cubic ferroelectrics seem invariably to have markedly first-order phase transitions. However, it appears that these phase transitions may, in fact, become continuous at high hydrostatic pressures (Samara 1971): $BaTiO_3$, in particular, is reported to have a tricritical point (cf. §II.5.3.1) at a pressure of 34 ± 2 kbar (Clarke and Benguigi 1977). This observation suggests that the source of the first-order character of the (normal pressure) phase transitions may lie elsewhere (§II.5.3.1) and that a stable fixed point for the cubic ferroelectrics does, in fact, exist.

Further experiments on $BaTiO_3$ and its isomorphs, probing the (potentially non-classical) critical behaviour beyond the tricritical point, might motivate the further theoretical effort necessary to elucidate the critical properties of the hamiltonian (II.5.36).

II.5.2.4. *Acoustic instabilities*

Acoustic instabilities for which the soft irreducible representation of the strain has a dimensionality $n > 1$ are, again, more complex than their $n = 1$ counterparts, discussed in §II.5.1.4, and their character has not been fully elucidated. In many situations a third-order invariant in the free energy is allowed by symmetry (Anderson and Blount 1965, Cowley 1976) so that the associated phase transition is expected to be first-order: however, the phase transition in Nb_3Sn falls into this category, and is not discernibly discontinuous (Shirane and Axe 1971).

In those situations where the third-order invariant is absent by symmetry, the soft sectors of reciprocal space are invariably two-dimensional, and there is a close analogy with the uniaxial ferroelectric, the critical dimensionality being, apparently, $d_c = 3$ (Cowley 1976, Folk *et al.* 1976). Although the details of the critical behaviour are, once again, obscured by cubic perturbations (Cowley 1976), it seems likely that it will be classical, with, at most, logarithmic corrections.

Indeed, we are aware of no reports of non-classical behaviour at such structural phase transitions.

II.5.3. *Multicritical phenomena*

Our discussion so far has focused on *ordinary* critical points, which are reached by adjusting *two* quantities (t and E). We now turn to discuss a variety of *multicritical* points (cf. §II.4.3.7) which are characterized by special values of at least *three* quantities.

II.5.3.1. *The tricritical point*

Consider once again the Ising system with order parameter strain coupling, discussed in §II.5.1.2. Equation (II.5.10) makes it plain that the transition temperature will be a function of pressure, sweeping out a line of phase transitions, as depicted in fig. II.21. Now the coupling constant u_0 (eqn. (II.5.9 *c*)) will in general have some weak temperature dependence (over and above the linear temperature dependence that would be displayed by *both* terms on the right-hand side of (II.5.9 *c*), had we kept careful trace of factors of $k_B T$!) so that the relative size of the two terms in (II.5.9 *c*), determining the effective coupling constant $\tilde{u}_0$, will be temperature-

Fig. II.21

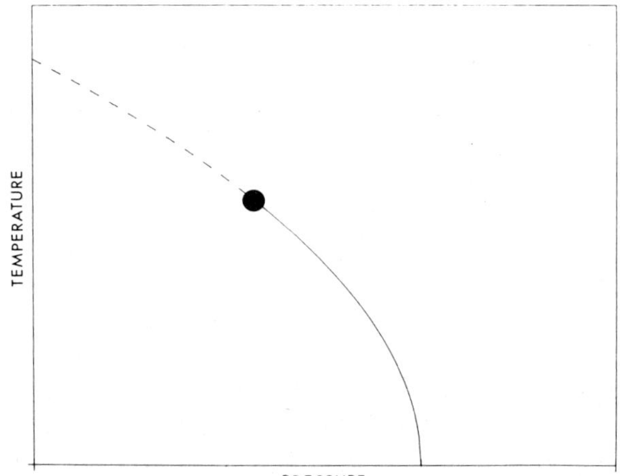

Schematic representation of the pressure dependence of the transition temperature for a structural phase transition with order-parameter–strain coupling. The solid line locates continuous phase changes, and the dashed line first-order phase changes: the two sections are separated by a tricritical point.

dependent. In particular, the *sign* of the effective coupling constant may change at some point on the line $T_c(p)$—a state of affairs which may also come about (without fluctuation renormalization of u_0) through terms in the effective hamiltonian of higher order in the strain: see Benguigi (1973).

When the coupling constant is negative, Landau theory shows (§I.4.1.1; cf. also the ensuing R.G. hamiltonian flow—fig. II.3) that the phase transition is of first-order. When the coupling constant is positive, the phase transition will be continuous (to the extent that one can ignore the more subtle strain-induced effects discussed in §II.5.1.2). The point p_{tr}, $T_{tr}=T_c(p_{tr})$ separating the first- and second-order sections of the line $T_c(p)$ is a *tricritical* point: it is reached by the adjustment of *three* quantities (though this is *not* the motivation for the prefix 'tri') namely, in this case, T, p and (implicitly) the conjugate field E.

The behaviour near a tricritical point was discussed, within Landau theory, in §I.4.3.1. To examine the validity of Landau theory in this context we must generalize the Levanyuk–Ginzburg criterion, embodied in the convergence properties of the integral (II.4.82). It is easily shown that, according to Landau theory, at a generalized critical point at which the first non-vanishing coefficient in the free-energy expansion is of order $2m$($m=2$ for an ordinary critical point; $m=3$ for the tricritical point under discussion), the ordered phase specific heat diverges as $|t|^{-(m-2)/(m-1)}$. The argument which led, via (II.3.45 a), to (II.4.82), then shows that a necessary condition for the consistency of the classical theory is the convergence of the integral

$$L(m)\equiv -t^{(m-2)/(m-1)}\frac{d}{dt}\int_{\mathbf{k}}^{\Lambda}\sum_{i,j=1}^{n}\chi_{ij}^{\mathrm{I.S.}}(\mathbf{k}). \tag{II.5.37}$$

For the systems with short-range forces this gives a critical dimensionality

$$d_c(m)=4-2\left(\frac{m-2}{m-1}\right), \tag{II.5.38}$$

so that, for a tricritical point (Bausch 1972),

$$d_c=3. \tag{II.5.39}$$

We expect, therefore, that the classical theory of tricritical phenomena should be essentially correct in $d=3$. Indeed, one can recover the classical tricritical results immediately from the R.G. analysis of §II.4.4, recognising that, since the tricriticality under discussion is achieved by 'switching off' the coupling constant, it will be characterized by the *gaussian* fixed point (II.4.69). (Riedel and Wegner 1972). Equations (II.4.74 a,c), together with table II.1, then immediately yield the tricritical exponents

$$\left.\begin{aligned}\gamma&=1,\\ \beta&=\tfrac{1}{4}.\end{aligned}\right\} \tag{II.5.40}$$

However, as is customary at borderline dimensionalities, careful analysis reveals the existence of logarithmic corrections to certain power laws. On this occasion these corrections originate in the marginal character of the scaling field associated (loosely) with the sixth-order interaction term, whose presence is required for the stability of the effective hamiltonian when the fourth-order term is negative. Specifically,

Wegner and Riedel (1973) find that, while the susceptibility has the pure power law behaviour of the Landau theory:

$$\chi \sim |t|^{-1}, \qquad \text{(II.5.41 } a\text{)}$$

the $\beta = \frac{1}{4}$ power law yielded by the classical tricritical theory becomes

$$Q_0 \sim |t|^{1/4} |\ln|t||^{1/4}. \qquad \text{(II.5.41 } b\text{)}$$

These results prompt several cautionary remarks.

First of all, if (in the language of the example cited above) the applied pressure is somewhat greater than the tricritical pressure, so that the system actually undergoes a phase transition associated with an ordinary critical point, the behaviour predicted by (II.5.41) will be displayed only over some range of $|t|$, with a slow crossover to the appropriate critical behaviour for small $|t|$. (See Rudnick and Nelson 1976 for a discussion of the crossover theory.)

Secondly, it is worth mentioning that not all tricritical points are of the 'gaussian' type we have discussed. Consider, for example, the properties of the cubic system (eqn. (II.5.27) with $f_0 = 0$) summarized in fig. II.16 (Aharony 1976). For $v_0 > v_c(u_0)$ there exist flow lines to the Heisenberg fixed point, and the phase transition will be continuous; for $v_0 < v_c(u_0)$ there is none, indicating a first-order phase change for systems with parameters in this region. For $v_0 = v_c(u_0)$ there is a line of (critical) hamiltonian flow reaching the *cubic* fixed point. Thus the $v_0 - T$ phase diagram for this system will resemble the $p - T$ phase diagram of fig. II.22, with $v_0 = v_c(u_0)$ locating a *tricritical* point—but one which, on this occasion, has the *non-classical* critical exponents of the cubic fixed point.

The ambiguities inherent in fitting power laws to data—particularly order parameter data—characterizing behaviour near a first-order phase transition should also now be apparent. A 'low value of β' may indeed be an indication of critical-point fluctuations promoting deviations from classical behaviour: a system close to the 'cubic' tricritical point offers one example where this is the case; the fluctuation-driven first-order phase changes referred to in §II.4.3.8 offer others. However, a low value of β may simply indicate the proximity of a gaussian tricritical point, where the fluctuations are entirely classical.

Fig. II.22

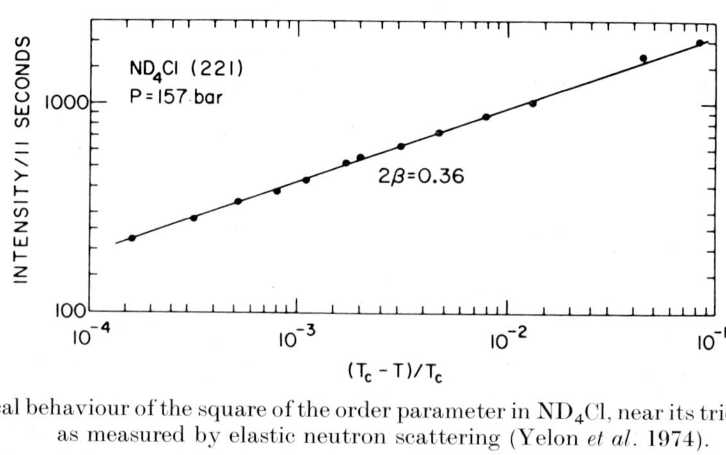

The critical behaviour of the square of the order parameter in ND_4Cl, near its tricritical point, as measured by elastic neutron scattering (Yelon *et al.* 1974).

Turning now to the experimental situation, it is pleasing to find that structural phase transitions may prove to offer a particularly rich vein of tricritical behaviour, as a result of the relative sensitivity of the order of many such phase changes to hydrostatic pressure.

The most detailed measurements of tricritical exponents at a structural phase transition are on the Ising-like system ND_4Cl. Experiments by Yelon *et al.* (1974) locate a tricritical point at $T_{tr} = 250{\cdot}2 \pm 0{\cdot}01$ K, $p_{tr} = 128 \pm 10$ bar. The temperature dependence of the order parameter, near the tricritical point, is well described by a simple power law with exponent $\beta \approx 0{\cdot}18$ (fig. II.22)—the low value possibly reflecting the influence of the logarithmic corrections (eqn. (II.5.41 *b*)). Gaussian tricritical behaviour is also suggested by specific heat measurements on ND_4Cl, which display a singularity with an exponent close to the tricritical value, $\alpha = \frac{1}{2}$ (Garland and Baloga 1977).

Pressure induced tricritical points in SbSI (Peercy 1975) and $BaTiO_3$ (Clarke and Benguigi 1977) have also been located by Raman scattering and dielectric studies, respectively: no exponent measurements have yet been reported.

Very recent measurements by Bastie *et al.* (1978) provide evidence for an order parameter exponent close to $\frac{1}{4}$ near the tricritical point occurring in KDP at pressures of about 2 kbar (Schmidt *et al.* 1976). The latter authors make a tentative proposal (partially substantiated by experiment) for the full E–T–p phase diagram (fig. II.23). This figure displays the surfaces of first-order phase changes ('tricritical wings') that burgeon from the first-order line in the T–p plane, as a function of the ordering field. (See e.g. Griffiths 1973). The edges of these surfaces are formed by lines of critical points, at which the system undergoes a continuous phase transition, with no symmetry change (cf. §I.4.3.2). Thus the tricritical point is seen to be the

Fig. II.23

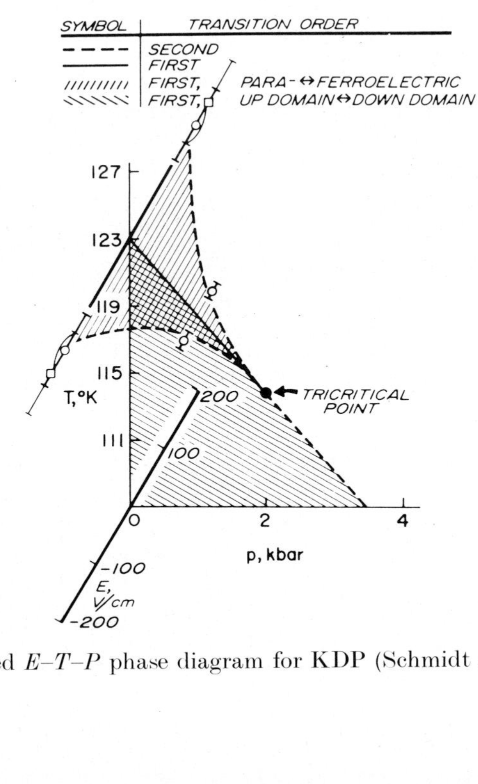

Proposed E–T–P phase diagram for KDP (Schmidt *et al.* 1976).

intersection of *3 lines of critical points*—the original motivation for the name *tri*critical.

In fact, in the other realizations of gaussian tricritical points studied to date (e.g. certain antiferromagnets, ^{3}He–^{4}He mixtures: see Riedel 1973 for a review), the ordering field is not controllable in the laboratory: thus experiments on appropriate ferroelectrics may well turn out to offer the best opportunity for detailed exploration of tricritical phase diagrams.

II.5.3.2. *The uniaxial Lifshitz point*

We have seen that the critical behaviour of systems with multi-component order parameters is frequently complicated by anisotropies in the soft-mode dispersion: the effects of such anisotropies are particularly difficult to elucidate because they tend to couple k-space and soft-mode coordinate space in an awkward fashion (cf. the discussion of the f_0-anisotropy in §II.5.2.1).

In fact, crystal symmetry may also permit anisotropies in the dispersion that are 'tied' to the lattice: thus, for example, the effective hamiltonian for an $(n=1)$-component system may be written, more generally, as

$$\mathscr{H}_0=\mathscr{H}_0(n=1)-\sum_{\alpha=1}^{d} h_{0,\alpha}\int_{\mathbf{k}}^{\Lambda} k_\alpha^2|Q(\mathbf{k})|^2. \tag{II.5.42}$$

Now it is readily seen that *this* form of anisotropy will not, in general, affect the asymptotic critical behaviour: with the aid of a simple rescaling of reciprocal space it may be 'transferred' from the soft-mode dispersion to the shape of the Brillouin zone, when its unimportance becomes apparent (Bruce 1974).

This argument breaks down, however, if one (or more) of the anisotropy coefficients $h_{0,\alpha}$ is close to unity, so that the soft-mode dispersion is nearly flat in one (or more) direction(s). Specifically, we consider the case of uniaxial anisotropy

$$\left.\begin{aligned} h_{0,\alpha}&=h_0 \quad (\alpha=z),\\ h_{0,\alpha}&=0 \quad (\alpha\neq z). \end{aligned}\right\} \tag{II.5.43}$$

The width of the asymptotic region of normal, d-dimensional Ising, critical behaviour then vanishes with $1-h_0$, and the special value $h_0=1$ locates a multi-critical point, which has been termed a (uniaxial) Lifshitz point (Hornreich *et al.* 1975).

In characterizing the behaviour associated with this multi-critical point it is necessary, in view of the vanishing of the k_z^2 term in the dispersion, to include the hitherto unimportant term proportional to k_z^4—or, more generally, k_z^{2p}, where $2p$ is the smallest power of k_z with non-vanishing coefficient. The effective hamiltonian to be studied is then

$$\mathscr{H}_0=\mathscr{H}_0(n=1)+\int_{\mathbf{k}}^{\Lambda}(-h_0k_z^2+k_z^{2p})|Q(\mathbf{k})|^2. \tag{II.5.44}$$

To explore the multi-critical point behaviour we consider the Levanyuk integral (II.4.82):

$$L(p)=\int_{\mathbf{k}}^{\Lambda}(t+k^2-k_z^2+k_z^{2p})^{-2}\sim\int_0^{\Lambda}dk\int_0^{\pi}d\theta\frac{k^{d-1}\sin^{d-2}\theta}{(t+k^2\sin^2\theta+k^{2p}\cos^{2p}\theta)^2}, \tag{II.5.45}$$

which is convergent in the $t\to 0$ limit, provided $d>d_c(p)$, where

$$d_c(p)=5-1/p. \tag{II.5.46}$$

Thus fluctuation corrections to the mean-field critical exponents may be calculated (cf. §II.4.5.1 (*a*)) systematically in powers of $\varepsilon_L(p) \equiv d_c(p) - d$. These calculations are less tractable (and, presumably, less reliable) than conventional $(4-d)$ expansions, and have been carried out only to leading order (Hornreich *et al*. 1975, Nicoll *et al*. 1976) with the results

$$\left.\begin{aligned} \gamma &= 1 + \frac{1}{6}\varepsilon_L(p) + O(\varepsilon_L^2(p)). \\ \beta &= \frac{1}{2} - \frac{1}{6}\varepsilon_L(p) + O(\varepsilon_L^2(p)). \end{aligned}\right\} \qquad \text{(II.5.47)}$$

In the simplest case (the 'ordinary' Lifshitz point) where the coefficient of k_z^4 is non-zero, so that $p=2$, (II.5.46) and (II.5.47) give

$$\left.\begin{aligned} \gamma &= 1{\cdot}25 + O(\varepsilon_L^2). \\ \beta &= 0{\cdot}25 + O(\varepsilon_L^2). \end{aligned}\right\} \qquad \text{(II.5.48)}$$

While of dubious accuracy, these results at least display the trends towards two-dimensional behaviour (cf. (II.2.10 *c*), (II.2.12 *b*), and §II.5.4.2) that one might expect in view of the weakness of the coupling between planes perpendicular to the z-direction, expressed in the flatness of the dispersion in the hamiltonian (II.5.44). Indeed, the critical points of systems for which the coefficients of $k_z^4 \ldots k_z^{2(p-1)}$ also vanish (Lifshitz points of 'high order') correspond ever more closely to an ordinary critical point of a $(d-1)$-dimensional system: thus the $\varepsilon_L(p)$-expansion results (II.5.47) coincide, in the $p \to \infty$ limit, with the leading order ε-expansion results for a $(d-1)$-dimensional system (cf. eqns. (II.4.81)).

In principle, then, the spectrum of Lifshitz critical points form the necessary conceptual link between d- and $(d-1)$-dimensional critical behaviour. In practice, however, it may seldom prove possible to dinstinguish simple Lifshitz-point behaviour ($p=2$) from genuinely low-dimensional behaviour ($p \to \infty$), especially in view of the uncertainties in the theoretical estimates for Lifshitz-point exponents.

Nevertheless it is tempting to associate the low value of the order parameter exponent observed in $BaMnF_4$, $\beta = 0{\cdot}225 \pm 0{\cdot}025$ (Shapiro *et al*. 1976) with Liftshitz-type behaviour. Here the soft-mode dispersion is flat along the $[\zeta, \frac{1}{2}, \frac{1}{2}]$ direction, in the region *between* the soft-mode wavevector $\mathbf{q}_s$, at $\zeta = 0{\cdot}39$ (the low-temperature phase is actually incommensurate) and the zone boundary, and rises more steeply on the *other* side of $\mathbf{q}_s$, suggesting a Lifshitz form of critical behaviour (Cowley and Bruce 1978).

II.5.3.3. *Bicritical and tetracritical points*

The discussion of §II.5.3.1 revealed how, by virtue of order-parameter–strain coupling, the order of a phase transition may be changed by an *isotropic* stress. In this section we shall consider the interesting range of multi-critical behaviour that may ensue when an *anisotropic* stress, or some other *symmetry-breaking* perturbation, is applied to a system with a multiple-component order parameter.

We shall focus, in particular, on the problems discussed within Landau theory in §I.4.3.3—the behaviour of perovskite crystals, undergoing R_{25}-mode instabilities, subjected to [100] or [111]-stress (Aharony and Bruce 1974, Bruce and Aharony 1975). We shall simplify the discussion by ignoring the anisotropy of the fluctuations (represented in $\mathscr{H}_0^f$, eqn. (II.5.26)): this neglect may invalidate the finer details, but not the essential spirit of the analysis. We shall also ignore the long-range interactions (among the soft coordinates) mediated by the acoustic modes (§ II.5.2.2): these, we have seen, should invariably make phase transitions (slightly) first-order.

(*a*) [*100*]-*stress*

Following the methods outlined in § II.5.1.2, the elastic degrees of freedom may be eliminated from the coupled order-parameter–strain system, leaving the effective hamiltonian (a natural generalization of the Landau free energy (I.4.46))

$$\mathscr{H}_0 = \mathscr{H}_0(n=d=3) + \mathscr{H}_0^v + \mathscr{H}_0^p, \tag{II.5.49 a}$$

where $\mathscr{H}_0(n)$ and $\mathscr{H}_0^v$ are defined by (II.4.10 *b*) and (II.5.25), while $\mathscr{H}_0^p$ is given by

$$\mathscr{H}_0^p = p_0 \int_{\mathbf{k}}^{\Lambda} \left\{ |Q_1(\mathbf{k})|^2 - \frac{1}{d-1} \sum_{\alpha \neq 1}^{d} |Q_\alpha(\mathbf{k})|^2 \right\}. \tag{II.5.49 b}$$

Here, the coefficient p_0 is proportional to the anisotropic stress along [100], and, without significant loss of generality, we have set to zero the coefficient characterizing the coupling of the order parameter to a volume strain.

As discussed in § I.4.3.3, the perturbation p_0 lifts the degeneracy of the R_{25} soft mode, making fluctuations about the [100] axis softer/harder, for p_0 negative/positive, than the fluctuations associated with the remaining coordinates. The universality principle then implies that the asymptotic critical behaviour of the stressed system will be ($n=1$)-component or ($n=2$)-component, according to the sign of p_0. In the language of the renormalization group, the stress term (II.5.49 *b*) constitutes a *relevant* perturbation, which promotes a *crossover* from the critical behaviour of the stress free system (which, we shall assume, is characterized by the ($n=3$)-component *Heisenberg* fixed point), the appropriate crossover exponent—that appearing in eqns. (II.5.31 *b*) and (II.5.32 *b*)—having the value (Pfeuty *et al.* 1974)

$$\phi(n=3) = 1{\cdot}25 \pm 0{\cdot}015. \tag{II.5.50}$$

Several key results now follow immediately from a crossover scaling ansatz for the free energy (cf. the corresponding crossover ansatz for the susceptibility, eqn. (II.4.47 *a*),

$$G(\tilde{t}, p_0) \approx |\tilde{t}|^{2-\alpha} \hat{G}_\pm (p_0/|\tilde{t}|^\phi), \tag{II.5.51}$$

where α is the Heisenberg specific heat exponent, and $\tilde{t}$ (II.4.48) measures the deviation from the zero-stress transition temperature:

$$\tilde{t} \equiv (T - T_c(p_0=0))/T_c(p_0=0). \tag{II.5.52}$$

In particular, (II.5.49) and (II.5.51) show that, in the zero-stress limit,

$$\int_{\mathbf{k}}^{\Lambda} \left\{ \langle |Q_1(\mathbf{k})|^2 \rangle - \frac{1}{d-1} \sum_{\alpha \neq 1}^{d} \langle |Q_\alpha(\mathbf{k})|^2 \rangle \right\} = \frac{\partial}{\partial p_0} G(\tilde{t}, p_0) \bigg|_{p_0=0} = |\tilde{t}|^{2-\alpha-\phi} \hat{G}'_\pm(0), \tag{II.5.53}$$

confirming that, as $t \to 0^-$, in the zero-stress limit, the tetragonal strain (II.5.30) vanishes with the exponent $\tilde{\beta}$ (II.5.31 *b*). The stress-free behaviour of the elastic constants (II.5.32 *a*) also follows readily.

We turn now to the critical behaviour in the presence of the anisotropic stress—in particular, to the nature of the p_0–T phase diagrams. The general scaling result (II.5.51) shows that the mean-field predictions for these phase diagrams (cf. fig. I.35,

§ I.4.3.3) are qualitatively incorrect in one crucial respect: according to (II.5.51) (see also the crossover theory of § II.4.3.7, and (II.4.50 *b*) in particular) the transition temperature depends upon the stress in a *non-analytic* fashion:

$$T_c(p_0) - T_c(0) \sim p_0^{1/\phi}. \qquad \text{(II.5.54)}$$

Thus the lines of ($n=1$)-component and ($n=2$)-component phase transitions are expected to approach the zero-stress axis *tangentially*, as shown schematically in figs. II.24 (a, b), which are appropriate, respectively, to the $v_0<0$ and $v_0>0$ cases exemplified by $SrTiO_3$ and $LaAlO_3$. (The two lines of continuous phase transitions bounding the intermediate phase, present in the latter case, exhibit more subtle fluctuation-induced warping: Bruce and Aharony 1975.)

With the perspectives of fig. II.24, the zero-stress critical points may be thought of as multi-critical: according to whether v_0 is negative (or positive) the point $p_0=0$, $T=T_c$ is the intersection of *two* (or *four*) lines of continuous phase transitions, and has been termed a *bi*critical (or *tetra*critical) point, following the analysis by Fisher and Nelson (1974) of the analogous phenomena exhibited by anisotropic antiferromagnets in a uniform magnetic field. (The $p_0=0$ line in fig. II.24 (a) locates a line of first-order phase transitions closely analogous to the 'spin–flop' phase transition exhibited by these magnetic systems.)

Indeed, turning to the experimental situation, it is in the context of magnetic systems that the foregoing theory has been most extensively tested. In particular, predictions for the phase-boundary shapes near the multi-critical points have been beautifully verified in experiments on MnF_2 and $GdAlO_3$ (Rohrer 1975, Rohrer and Gerber, 1977); in the context of [100]-stressed perovskites, there exist no experiments with a remotely comparable accuracy.

There is, nevertheless, experimental evidence offering qualitative support for some aspects of the foregoing theory. Ultrasonic measurements by Rehwald (1977) show that the exponent characterizing the vanishing of the elastic constant C_{11} in $SrTiO_3$ *decreases* from its stress-free value of 0·66 (§ II.5.2.2) to a value of 0·55, under

Fig. II.24

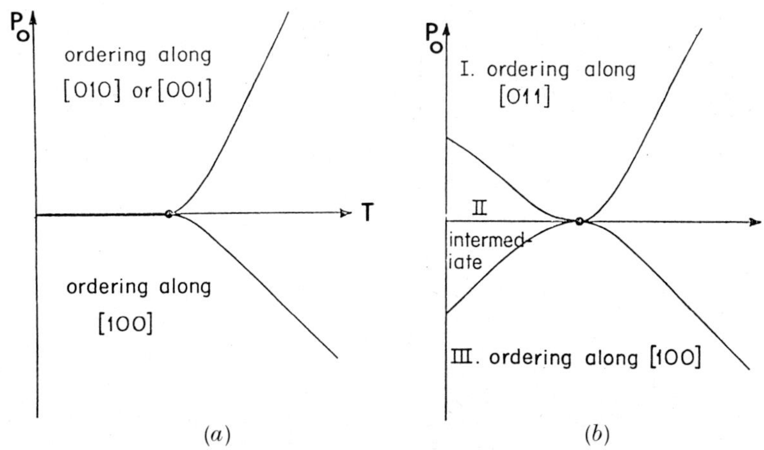

Schematic phase diagrams for [100]-stressed perovskites, showing the fluctuation-induced warping of the phase boundaries (Bruce and Aharony 1975). (a) is the case $v_0<0$ (e.g. $SrTiO_3$) and (b) is the case $v_0>0$ (e.g. $LaAlO_3$).

a biaxial compression of $\sim 0{\cdot}2$ kbar. This is in qualitative accord with the theory, which suggests a stress-induced crossover in this exponent, as $T \to T_c$ with $p_0 \neq 0$, from the value $\tilde{\gamma}$ (II.5.32) to the value $\alpha(n=1) \approx 0{\cdot}12$ (II.5.14*c*). Presumably the applied stress was insufficient to expose fully the $(n=1)$-component character of the asymptotic behaviour (obscured, very close to T_c, by strong attenuation).

Fig. II.25

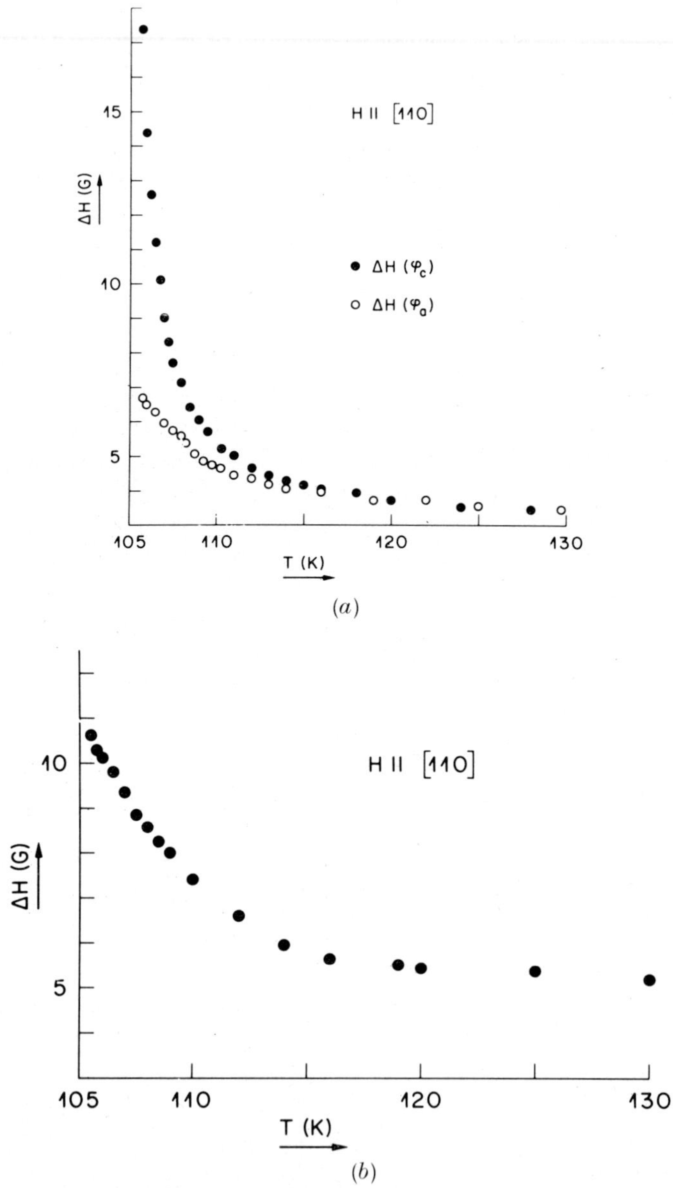

Electron paramagnetic resonance line widths measuring the strength of local fluctuations in $SrTiO_3$ (Müller and Berlinger, private communication), (*a*) in a monodomain-transforming sample, showing the difference between fluctuations about (●) and perpendicular to (○) the monodomain axis; (*b*) in a polydomain-transforming sample, showing the degeneracy of these fluctuations.

Further, more oblique, support for the theory is to be found in the experiments of Müller and Berlinger (1971) and Courtens (1972) cited in § II.5.2.1. According to the former experiments the value of the order-parameter exponent in $SrTiO_3$ is closer to the Ising value, than to the (anticipated) ($n=3$)-component Heisenberg value (table II.3), while the latter experiments (which probe the operator (II.5.49 *b*): cf. § II.5.2.2) also reveal a behaviour reminiscent of an Ising system, in the region close to T_c (see the inset of fig. II.19). Taken together with the fact that both experiments were performed upon a $SrTiO_3$ sample prepared in such a way as to transform into a *monodomain* low-temperature phase, these results prompted the suggestion (Courtens 1972, Aharony and Bruce 1974) of the existence of an 'orienting mechanism'—presumed to be a strain field—which singles out the monodomain axis, and promotes Ising-like critical behaviour.

This suggestion has since been partially substantiated in further E.P.R. experiments by Müller and Berlinger (1975), which show that the strengths of fluctuations of oxygen octahedra about, and perpendicular to, the monodomain axis, different in the monodomain-transforming samples, become the same when the sample is annealed so as to transform into a polydomain low-temperature phase (fig. II.25).

These authors also report a reanalysis of the $SrTiO_3$ order parameter data of fig. II.17 in which the results were fitted to the form anticipated for an Ising system when account is taken of the leading correction to pure power-law behaviour (cf. eqn. (II.4.43)):

$$Q_0 = D|t|^{\beta}(1 + D'|t|^{\bar{\omega}}). \tag{II.5.55}$$

With a *presupposition* of Ising behaviour $\beta = 0{\cdot}31$, this analysis led to a correction-to-scaling exponent $\bar{\omega}$ close to 0·5; the predicted Ising value is $0{\cdot}493 \pm 0{\cdot}007$ (Le Guillou and Zinn-Justin 1977). The close accord, though pleasing, may be somewhat illusory, since (in the light of fig. II.19) it is unlikely that the inbuilt strain field is sufficiently strong to establish the Ising behaviour (II.5.55) over the wide range of reduced temperature employed in the analysis.

(*b*) [*111*]-*stress*

We shall discuss the more complex behaviour of [111]-stressed perovskites only qualitatively, referring the reader to Bruce and Aharony (1975) for details. We shall focus on the $v_0 < 0$ case where the natural (tetragonal) ordering of the system conflicts with that favoured by the stress, to produce an intermediate phase.

Again R.G. arguments suggest a fluctuation-induced warping of the phase boundaries near the bicritical point (compare fig. II.26 with the mean field prediction, fig. I.36 (*a*)). Furthermore, fluctuations also turn out to play an important role at the first-order phase boundary (separating [111]-ordered, and intermediate-ordered phases) near the bicritical point (Aharony *et al.* 1977). This phase transition is characterized by the ordering of the coordinates spanning the two-dimensional order-parameter space perpendicular to the [111]-direction, and it proves useful to introduce the Fourier coordinates

$$\left.\begin{aligned} \tilde{Q}_1(\mathbf{k}) &\equiv \frac{1}{\sqrt{2}}(Q_1(\mathbf{k}) - Q_2(\mathbf{k})), \\ \tilde{Q}_2(\mathbf{k}) &\equiv \frac{1}{\sqrt{6}}(Q_1(\mathbf{k}) + Q_2(\mathbf{k}) - 2Q_3(\mathbf{k})). \end{aligned}\right\} \tag{II.5.56}$$

Fig. II.26

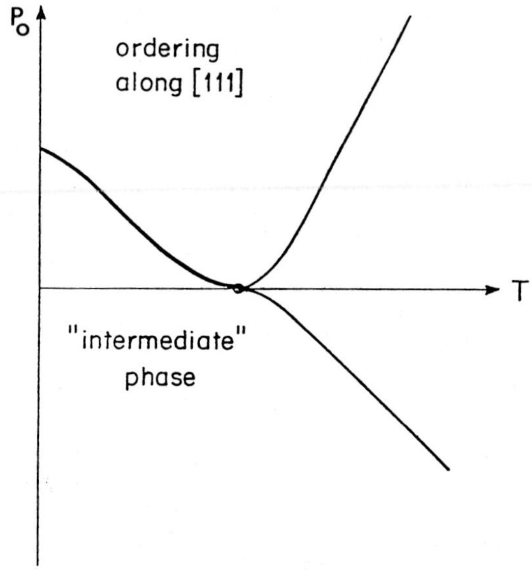

Schematic phase diagram for a [111]-stressed perovskite with $v_0 < 0$ (e.g. $SrTiO_3$) (Bruce and Aharony 1975). The boundary separating [111]-ordered, and 'intermediate' ordered phases is a line of Potts-model phase transitions (Aharony *et al.* 1977).

The appropriate effective hamiltonian may then be shown to have the form

$$\mathscr{H}_0 = \mathscr{H}_0(n=2) + \mathscr{H}_0^w, \tag{II.5.57 a}$$

where $\mathscr{H}_0(n=2)$ is the $n=2$ component L.G.W. functional (II.4.10b) of the coordinates $\tilde{Q}_1$, $\tilde{Q}_2$, while $\mathscr{H}_0^w$ is defined by

$$\mathscr{H}_0^w = w_0 \int_{\mathbf{k}_1}^{\Lambda} \int_{\mathbf{k}_2}^{\Lambda} \int_{\mathbf{k}_3}^{\Lambda} \tilde{Q}_2(\mathbf{k}_1) \left[\tilde{Q}_1(\mathbf{k}_2)\tilde{Q}_1(\mathbf{k}_3) - \frac{1}{3}\tilde{Q}_2(\mathbf{k}_2)\tilde{Q}_2(\mathbf{k}_3) \right] \delta(\mathbf{k}_1 + \mathbf{k}_2 + \mathbf{k}_3). \tag{II.5.57 b}$$

This term originates in the [111]-ordering, and the coupling 'constant' w_0 varies with the order parameter Q_0 for this (trigonal) phase:

$$w_0 \approx 2\sqrt{2 v_0} Q_0. \tag{II.5.57 c}$$

The effective hamiltonian (II.5.57a) (the 'continuous spin' form of the three-state Potts model of ferromagnetism) has been the subject of a number of renormalization group analyses (see e.g. Golner 1973, Rudnick 1975). It transpires that the w_0 interaction does indeed promote the first-order instability suggested by Landau theory; nevertheless, for small values of w_0, critical fluctuations are sufficiently strong to invalidate certain of the Landau predictions. Specifically, the order-parameter discontinuity at the phase transition is expected to vary *non-analytically* (rather than *linearly*) with w_0, and thence with Q_0 (the magnitude of the [111]-ordering):

$$\Delta\tilde{Q}_0 \sim Q_0^{\delta^*}, \tag{II.5.58}$$

where $\delta^* \approx 0{\cdot}6$ (Rudnick 1975).

On the experimental front, these predictions have met with mixed success. Experiments by Müller *et al.* (1970) and Currat *et al.* (1978) on [111]-stressed $SrTiO_3$ (see fig. II.27) seem to show an essentially mean-field form of phase diagram (cf. fig. I.36). It is, however, pertinent to note that the fluctuation-induced warping of the phase boundaries near the bicritical points of magnetic systems—e.g. MnF_2—is made (convincingly!) evident only by the remarkable precision achieved by King and Roher (1979), whose experiments probed the range of reduced temperature within 10^{-3}, and with a reduced temperature resolution of the order of 10^{-5}: the precision achieved in the stressed-perovskite experiments is not remotely comparable.

Nevertheless these experiments do offer a partial vindication of the theory of the 'Potts' transition between the ordered phases (Aharony *et al.* 1977), the order-parameter discontinuity exhibiting a behaviour of the form (II.5.58), with $\delta^* = 0{\cdot}62 \pm 0{\cdot}1$, in close accord with the theoretical value, and irreconcilable with the mean-field prediction, $\delta^* = 1$.

Fig. II.27

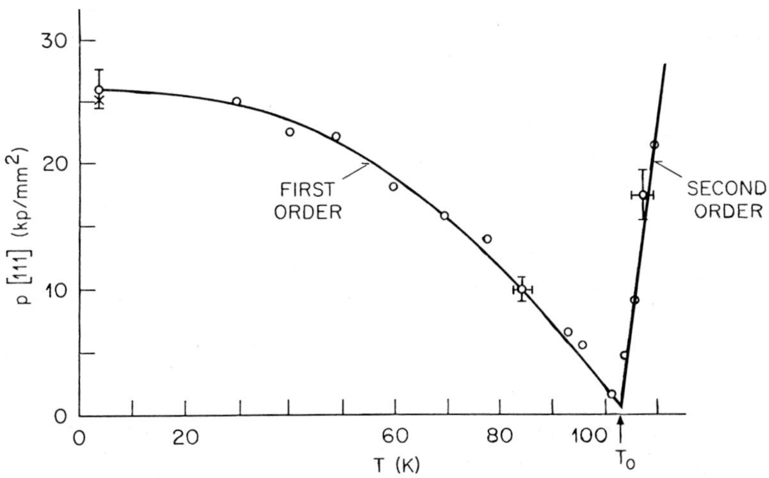

Experimentally-determined phase diagram for [111]-stressed $SrTiO_3$ (Müller *et al.* 1970).

II.5.4. *Miscellaneous problems*

While we have attempted to cover those areas exhibiting the greatest overlap of theoretical and experimental activity, the foregoing survey is not comprehensive: we now we make brief reference to a number of the principal topics we have omitted.

II.5.4.1. *The Debye–Waller factor*

The theory of the Debye–Waller factor of a system undergoing a structural phase transition merits brief mention, if only to dispel some of the misconceptions generated by the surprising number of incorrect treatments appearing in the literature: a short survey is to be found in the paper by Meissner and Binder (1975) who present what we believe to be a correct treatment of the problem, based upon scaling theory.

The essential conclusion of their analysis may be illustrated by taking, once again, the example of the Ising universality class, described by the hamiltonian $\mathscr{H}_0(n=1)$, in its direct space representation (II.4.8*a*).

Writing the temperature factor of an atom at position $\mathbf{x}$ as

$$d_{\mathbf{x}}=\exp(-W_{\mathbf{x}}), \tag{II.5.59}$$

the contribution to the exponent $W_{\mathbf{x}}$ that is *quadratic* in the momentum transfer $\mathbf{K}$ has the form

$$W_{\mathbf{x}} \approx \frac{K^2}{2}\left\{\langle u^2(\mathbf{x})\rangle - \langle u(\mathbf{x})\rangle^2\right\} \equiv \frac{K^2}{2}\langle \delta u^2(\mathbf{x})\rangle. \tag{II.5.60}$$

The behaviour of the thermal averages $\langle u^2(\mathbf{x})\rangle$ and $\langle \delta u^2(\mathbf{x})\rangle$ is displayed in fig. II.28, which is taken from the molecular dynamics calculations of Schneider and Stoll (1976*a*) on the $n=1$, $d=2$ universality class: the sharp drop (below T_c) in the local mean-square fluctuation $\langle \delta u^2(\mathbf{x})\rangle$ is noteworthy, since a cursory glance might suggest that this quantity is divergent (as, indeed, some authors have incorrectly predicted). The *local* mean-square fluctuation is, however, finite at T_c.

Fig. II.28

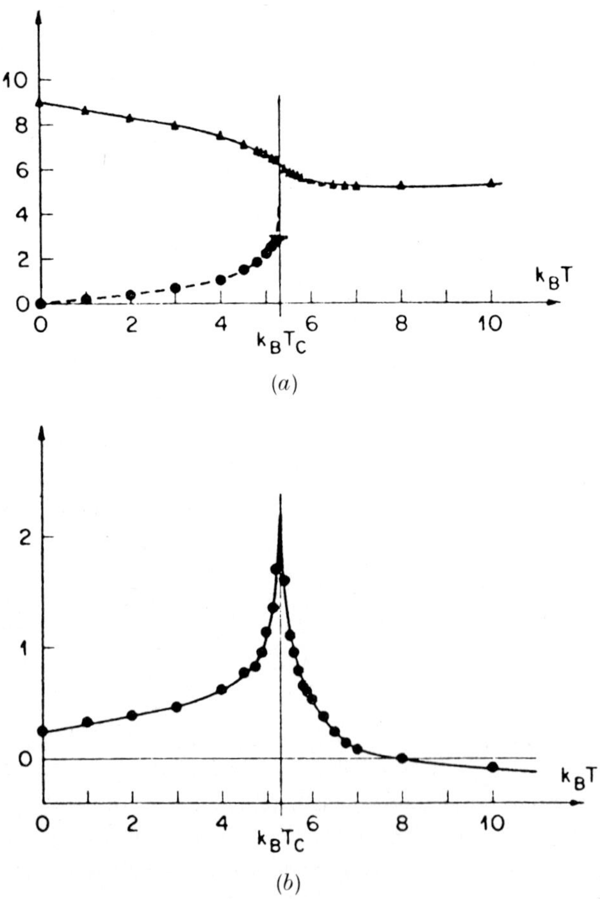

The results of molecular dynamics calculations on the $d=2$ Ising universality class (Schneider and Stoll 1976 a). (*a*) shows the temperature dependence of $\langle u^2\rangle$ and $\langle \delta u^2\rangle$ (dashed line, below T_c), and (*b*) shows the temperature dependence of $-(d/dt)\langle u^2\rangle$, displaying the behaviour of the specific heat.

More explicitly it follows from (II.4.8a) and (II.4.11b) that

$$\langle u^2(\mathbf{x})\rangle = -\frac{2}{N}\frac{\partial}{\partial r_0}\ln Z, \tag{II.5.61}$$

so that, recalling that the temperature dependence of $\mathscr{H}_0$ resides in the parameter r_0, the quantity $\langle u^2(\mathbf{x})\rangle$ is seen to possess the critical behaviour of the energy density

$$\langle u^2(\mathbf{x})\rangle \sim E_1 + E_2 t + E_3^{\pm}|t|^{1-\alpha}, \tag{II.5.62}$$

where α is the specific heat exponent. The exponent $(1-\alpha)$ of the non-analytic piece contrasts with that (2β or $\gamma/2$ if the exponent η is neglected: Meyer and Cowley 1976; see also eqn. (I.4.20)) suggested by a calculation of $\langle u^2(\mathbf{x})\rangle$ as the integral over the wavevector-dependent susceptibility—the difference reflecting a subtle cancellation of terms missed by the latter approach. We anticipate that the form of this result (with the appropriate value of α) will be quite generally applicable, one exception being the uniaxial ferroelectric where logarithmic singularities appear (Binder *et al.* 1976).

While the recognition that the local mean-square amplitude follows the energy (and its temperature derivative follows the specific heat) may prove qualitatively helpful in interpreting a number of different types of experiment (e.g. measurements of spin-lattice relaxation rates: cf. fig. I.24), the relative weakness of the critical singularities involved will presumably make it difficult to achieve a quantitative match between theory and experiment.

II.5.4.2. *Quasi-two-dimensional systems*

There exists a class of systems undergoing structural phase transitions, where the dominant interactions among the (local) ordering variables are confined to *planes* of the lattice. It will be clear from the theory of crossover described in §II.4.3.7 that such systems should exhibit a ('primary') critical behaviour that is two-dimensional in character, prior to a small $|t|$ crossover (to full three-dimensional critical behaviour) induced by the inter-planar interactions, which will always be present in practice, and which ensure three-dimensional ordering.

In fact, systems of this type may be described, typically, by a hamiltonian with the strong 'lattice' anisotropy discussed in §II.5.3.2. However, the crossover itself proves more tractable when the hamiltonian is reformulated so as to make its crossover form, eqn. (II.4.45), more explicit (Grover 1973)—though there exist no detailed calculations of crossover behaviour of this type.

As regards the range of possible primary, two-dimensional, behaviour, only the 'short range' Ising ($n=1$) universality class is well understood (cf. the Onsager results touched on in §§II.2.2 and II.5.1.1. A number of two-dimensional models, which include the Ising system as a special case, have also been solved exactly, largely through the remarkable work of Baxter (1971, 1978); however, their relevance to any real physical system has yet to be demonstrated). While renormalization group analyses (José *et al.* 1977) have established the importance of interactions of square, or lower, symmetry in the $n=2$ universality classes, the details of their critical behaviour are, as yet, unresolved—a situation which, in part, reflects the futility of $\varepsilon=4-d$ expansions for two-dimensional systems.

In any case, the universality classes of the layered compounds studied so far themselves remain somewhat uncertain.

An $n=2$, $d=2$ model with 'square' anisotropy is probably appropriate for the 'perovskite-like' layered compound $(CH_2NH_3)_2MnCl_4$, where experiment gives $\beta=0{\cdot}25\pm0{\cdot}005$ (Kind and Roos 1976), a value which suggests behaviour intermediate between that of two and three dimensions.

Feder (1976) has argued that the layered compound 'squaric acid' also falls into this universality class. Here, neutron scattering and N.M.R. measurements give $\beta\approx0{\cdot}14$ (Samuelsen and Semmingsen 1975, Becker *et al.* 1978) interestingly close to the $d=2$ Ising value (cf. also the molecular dynamics investigation of this universality class by Schneider and Stoll 1976 b). Birefringence measurements which, presumably, determine the exponent $\bar{\beta}$, and *not* β (cf. eqn. (II.5.31 *a*), and concluding comments in § II.5.2.2), give $\bar{\beta}=0{\cdot}34\pm0{\cdot}02$ (Semmingsen and Feder 1974)—significantly larger than 2β, a trait which, we have seen, is exhibited by experiment and theory in three-dimensional systems.

II.5.4.3. *Quantum-mechanical effects*

Our entire discussion of structural phase transitions has been predicated on the assumption that quantum-mechanical effects are unimportant in the critical region. Physically, the justification for this assumption lies in the fact that, near T_c, the dominant fluctuations of necessity have energies small in comparison with k_BT. More formally, the absence of quantum effects reflects the fact that, in the sums over Matsubara frequencies $\omega_m=(2\pi mk_BT)\hbar$ encountered in perturbative expansions for an anharmonic crystal (see e.g. Cowley 1963), asymptotically the classical, $m=0$, terms dominate; the $m\neq0$ terms act in much the same way as do non-critical branches of the phonon dispersion (Young 1975, Oppermann and Thomas 1975).

It is evident that the regime of temperature in which these arguments are valid will become smaller the lower the transition temperature, and will vanish altogether for a phase transition occurring precisely at 0 K. In such a case each sum over Matsubara frequencies must be transformed to an integral over a continuous variable, which acts in much the same way as an additional component (dimension) of wavevector space. Thus the zero-point fluctuation spectrum of a real (quantum) d-dimensional system exhibiting a structural phase transition is, in effect, that of a $(d+1)$-dimensional classical system. As a result, in systems whose critical temperature T_c vanishes (cf. fig. II.29) for a critical value y_c of some external field y—the pressure, for example—the critical singularities exhibited as $y\to y_c$ at 0 K will be mean-field like (with logarithmic corrections) in three dimensions, as described by Schneider *et al.* (1976), and Oppermann and Thomas (1975), who also show that the transition temperature will typically vanish with $y-y_c$ as

$$T_c(y)\sim(y-y_c)^{1/\psi}, \qquad \text{(II.5.63 a)}$$

with

$$\psi=d-1(d>2). \qquad \text{(II.5.63 b)}$$

The latter simple and, it would appear, rather general result is nicely borne out by experiments on KDP (Samara 1974; fig. II.30), which show that the transition temperature vanishes at a critical pressure $p_c\approx17$ kbar, with an exponent $\psi\approx2$. A similar result has been reported by Höchli *et al.* (1977), in experiments on the mixed system $KTa_{1-y}Nb_yO_3$, where the 'field' y is now the concentration of the Nb atoms; however, the critical behaviour of mixed systems is not well understood at present (§ II.5.4.4) and the applicability of the foregoing theory remains to be justified.

Fig. II.29

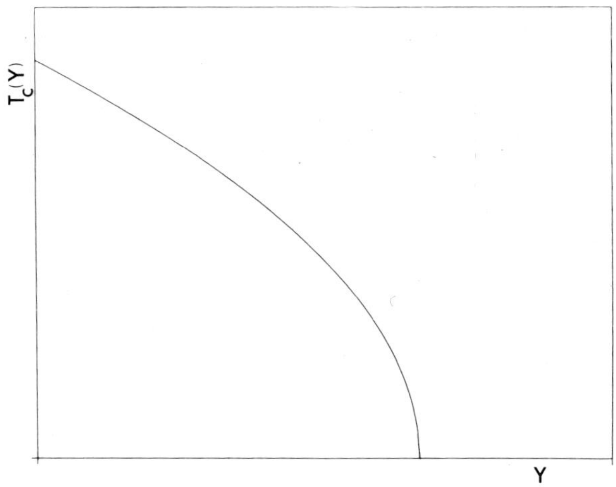

Schematic phase diagram for a system whose critical temperature T_c vanishes at a special value y_c of some external field y.

Fig. II.30

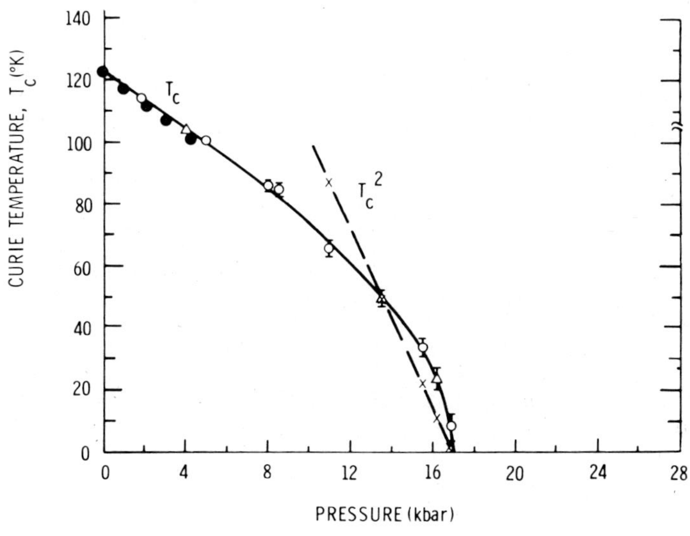

The transition temperature in KDP, as a function of pressure (Samara 1974).

Finally, at a more qualitative level, we remark that quantum-mechanical (zero-point) fluctuations compound the tendency of classical critical fluctuations to suppress the critical temperature (Schneider *et al.* 1976)—a mechanism which, Müller and Burkard (1979) have argued, is responsible for the absence of a ferroelectric phase transition in $SrTiO_3$, at normal pressures.

II.5.4.4. *The effects of crystal imperfections*

The nature of the critical behaviour in the presence of crystal imperfections has attracted considerable attention in recent years. While the greatest effort has been devoted to the elucidation of static properties (for a review see Aharony 1978) it is actually in the dynamic behaviour near the critical point that crystal defects manifest themselves most clearly: for this reason we defer discussion of these issues—which necessarily involves a classification of defects—to the concluding article in this series (III).

§II.6. Conclusions

Two general points should, we trust, have emerged in the course of this review.

Firstly, the theoretical survey constituting the initial portion of this work should have conveyed the extent of the theoretical armoury currently available to deal with the problems of the critical region.

Secondly, our survey of static critical phenomena at structural phase transitions will, we hope, have drawn attention to the wide range of critical behaviour displayed (actually or potentially) by such systems—a range which matches and, in some respects, surpasses the spectrum encountered in the context of magnetic systems, where much of the theory was originally developed.

It will be clear, furthermore, that the synthesis of experiment and theory is far from complete. While there remain unresolved theoretical issues (the nature of the critical behaviour in cubic ferroelectrics, for example) the outstanding need is for appropriate experiments (on uniaxial ferroelectrics, tricritical points, cubic ferroelectrics at high pressures, transitions to incommensurate phases ...) analysed with due awareness of the relevant theory.

Acknowledgment

I am happy to acknowledge much illuminating collaboration with Amnon Aharony.

References

Abe, R., 1972, *Prog. theor. Phys.*, **48,** 1414.

Aharony, A., 1973 a, *Phys. Rev.* B, **8,** 3363; 1973 b, *Ibid.*, **8,** 4270; 1976, *Phase Transitions and Critical Phenomena*, Vol. 6, edited by C. Domb and M. S. Green (New York: Academic Press), p. 357; 1978, *J. Magn. magn. Mater.*, **7,** 198.

Aharony, A., and Bruce, A. D., 1974, *Phys. Rev. Lett.*, **33,** 427; 1979, *Ibid.*, **42,** 462.

Aharony, A., and Fisher, M. E., 1973, *Phys. Rev.* B, **8,** 3323.

Aharony, A., Müller, K. A., and Berlinger, W., 1977, *Phys. Rev. Lett.*, **38,** 33.

Ahlers, G., Kornblit, A., and Gugenheim, H. J., 1975, *Phys. Rev. Lett.*, **34,** 1227.

Anderson, P. W., and Blount, E. I., 1965, *Phys. Rev. Lett.*, **14,** 217.

Aubry, S., 1975, *J. chem. Phys.*, **62,** 3217; 1976, *Ibid.*, **64,** 3392.

Bailin, D., and Love, A., 1977, *J. Phys.* C, **10,** 3633.

Bak, P., Krinsky, S., and Mukamel, D., 1976, *Phys. Rev. Lett.*, **36,** 52.

Baker, G. A., Nickel, B. G., Green, M. S., and Meiron, P. I., 1976, *Phys. Rev. Lett.*, **36,** 1351.

Bastie, P., Vallade, M., Vettier, C., and Zeyen, C. M. E., 1978, *Phys. Rev. Lett.*, **40,** 337.

Bausch, R., 1972, *Z. Phys.*, **254,** 81.

Baxter, R. J., 1971, *Phys. Rev. Lett.*, **26,** 832; 1973, *J. Phys.*, C, **6,** L445; 1978, *Proceedings of the International Conference on Statistical Mechanics, Annals of Israel Physical Society*, Vol. 2 (Bristol: Adam Hilger), p. 37.

Becker, J. D., Suwelack, D., and Mehring, M., 1978, *Solid St. Commun.*, **25,** 1145.

Bender, E., 1976, *Z. Phys.* B, **23,** 285.

Benguigi, L., 1973, *Phys. Stat. Sol.* (b), **60,** 835; 1975, *Phys. Rev.* B, **11,** 4547.

Bergman, D. J., and Halperin, B. I., 1976, *Phys. Rev.* B, **13,** 2145.

Berlin, T. H., and Kac, M., 1952, *Phys. Rev.*, **86,** 821.

Berre, B., Fossheim, K., and Müller, K. A., 1969, *Phys. Rev. Lett.*, **23,** 589.

BINDER, K., MEISSNER, G., and MAIS, H., 1976, *Phys. Rev.* B, **13,** 4890.

BIRGENEAU, R. J., KJEMS, J. K., SHIRANE, G., and VAN UITERT, L. G., 1974, *Phys. Rev.* B, **10,** 2512.

BOCCARA, N., and SARMA, G., 1965, *Physics*, **1,** 219.

BRÉZIN, E., LE GUILLOU, J. C., and ZINN-JUSTIN, J., 1974, *Phys. Lett.* A, **47,** 285; 1976, *Phase Transitions and Critical Phenomena*, Vol. 6, edited by C. Domb and M. S. Green (New York: Academic Press), p. 125; 1977, *Phys. Rev.* D, **15,** 1544.

BRÉZIN, E. LE GUILLOU, J. C., ZINN'JUSTIN, J., and NICKEL, B. G., 1973, *Phys. Lett.* A, **44,** 227.

BRÉZIN, E., and ZINN-JUSTIN, J., 1976, *Phys. Rev. Lett.*, **36,** 691.

BRUCE, A. D., 1974, *J. Phys.* C, **7,** 2089; 1977, *Ibid.*, **10,** 419.

BRUCE, A. D., and AHARONY, A., 1974, *Phys. Rev.* B, **10,** 2078; 1975, *Ibid.*, **11,** 478.

BRUCE, A. D., DROZ, M., and AHARONY, A., 1974, *J. Phys.* C, **7,** 3673.

BRUCE, A. D., and SCHNEIDER, T., 1977, *Phys. Rev.* B, **16,** 3991.

BRUCE, A. D., and WALLACE, D. J., 1977, *J. Phys.* A, **9,** 1117.

BUNDE, A., and LAM, F., 1977, *Solid St. Commun.*, **21,** 71.

BUZARÉ, J. Y., FAYET, J. C., BERLINGER, W., and MÜLLER, K. A., 1979, *Phys. Rev. Lett.*, **42,** 465.

CAMP, W. J., SAUL, D. M., VAN DYKE, J. P., and WORTIS, M., 1976, *Phys. Rev.* B, **14,** 3990.

CLARKE, R., and BENGUIGI, L., 1977, *J. Phys.* C, **10,** 1963.

COMÈS, R., and SHIRANE, G., 1972, *Phys. Rev.* B, **5,** 1886.

COURTENS, E., 1972, *Phys. Rev. Lett.*, **29,** 1380.

COWLEY, R. A., 1963, *Adv. Phys.*, **12,** 421; 1976, *Phys. Rev.* B, **13,** 4877.

COWLEY, R. A., and BRUCE, A. D., 1978, *J. Phys.* C, **11,** 3577.

CURRAT, R., MÜLLER, K. A., BERLINGER, W., and DENOYER, F., 1978, *Phys. Rev.* B, **17,** 2937.

DEGUCHI, K., and NAKAMURA, E., 1976, *Phys. Rev. Lett.*, **37,** 1642.

DOMB, C., 1974, *Phase Transitions and Critical Phenomena*, Vol. 3, edited by C. Domb and M. S. Green (New York: Academic Press), p. 357.

EISENRIEGLER, 1974, *Phys. Rev.* B, **9,** 1029.

FEDER, J., 1976, *Ferroelectrics*, **12,** 71.

FISHER, M. E., 1967, *Rep. Prog. Phys.*, **30,** 615; 1972, *Essays in Physics*, Vol. 4 (London: Academic Press), pp. 43–89; 1975, *Phys. Rev. Lett.*, **34,** 1634.

FISHER, M. E., and JASNOW, D., 1980, *Phase Transitions and Critical Phenomena*, Vol. 4, edited by C. Domb and M. S. Green, (New York: Academic Press) (to be published).

FISHER, M. E., and NELSON, D. R., 1974, *Phys. Rev. Lett.*, **32,** 1350.

FOLK, R., IRO, H., and SCHWABL, F., 1976, *Z. Phys.*, B, **25,** 69.

FROEHLICH, J., SIMON, B., and SPENCER, T., 1976, *Commun. Maths Phys.*, **50,** 79.

GARLAND, C. W., 1970, *Physical Acoustics*, Vol. VII (New York: Academic Press), p. 52.

GARLAND, C. W., and BALOGA, J. D., 1977, *Phys. Rev.* B, **16,** 331.

GEHRING, G. A., 1977, *J. Phys.* C, **10,** 531.

GILLIS, N. S., and KOEHLER, T. R., 1971, *Phys. Rev.* B, **4,** 3971; 1972, *Phys. Rev. Lett.*, **29,** 369.

GINZBURG, V. L., 1960, *Fiz. Tverd Tela*, **2,** 2031; 1960, *Sov. Phys. solid St.*, **2,** 1824.

GOLNER, G. R., 1973, *Phys. Rev.* B, **8,** 3419.

GRIFFITHS, R. B., 1970, *Phys. Rev. Lett.*, **24,** 1479; 1973, *Phys. Rev.* B, **7,** 545; 1974, *Ferroelectrics*, **7,** 71.

GROVER, M. K., 1973, *Phys. Lett.* A, **44,** 253.

HARLEY, R. T., and MACFARLANE, R. M., 1975, *J. Phys.* C, **8,** L451.

HATTA, I., SHIROISHI, Y., MÜLLER, K. A., and BERLINGER, W., 1977, *Phys. Rev.* B, **16,** 1138.

HIKAMI, S., and BRÉZIN, E., 1978, *J. Phys.* A, **11,** 1131.

HÖCHLI, U. T., and SCOTT, J. F., 1971, *Phys. Rev. Lett.*, **26,** 1627.

HÖCHLI, U. T., WEIBEL, H. E., and BOATNER, L. A., 1977, *Phys. Rev. Lett.*, **39,** 1158.

HOHENBERG, P. C., 1967, *Phys. Rev.*, **158,** 383; 1978, *Microscopic Structure and Dynamics of Liquids*, edited by J. Dupuy and A. J. Dianoux (New York: Plenum Press).

HORNREICH, R. M., LUBAN, M., and SHTRIKMAN, S., 1975, *Phys. Rev. Lett.*, **35,** 1678.

HUBBARD, J., 1972, *Phys. Lett.* A, **39,** 365.

HUBBARD, J., and SCHOFIELD, S., 1972, *Phys. Lett.* A, **40,** 245.

IIZUMI, M., AXE, J. D., SHIRANE, G., and SHIMAODA, K., 1977, *Phys. Rev.* B, **15,** 4392.

JOSÉ, J. V., KADANOFF, L. P., KIRKPATRICK, S., and NELSON, D. R., 1977, *Phys. Rev.* B, **16,** 1217.

KADANOFF, L. P., 1966, *Physics*, **2,** 263; 1971, *Critical Phenomena, Proc. Int. School of Physics 'Enrico Fermi'*, edited by M. S. Green (New York: Academic Press), p. 100.
KADANOFF, L. P., GÖTZE, W., HAMBLEN, D., HECHT, R., LEWIS, E. A. S., PALCIAUSKAS, V. V., RAYL, M., and SWIFT, J., 1967, *Rev. mod. Phys.*, **39,** 395.
KADANOFF, L. P., and HOUGHTON, A., 1975, *Phys. Rev.* B, **11,** 377.
KIND, R., and ROOS, J., 1976, *Phys. Rev.* B, **13,** 45.
KING, A. R., and ROHRER, H., 1979, *Phys. Rev.* B, **19,** 5864.
KJEMS, J. K., SHIRANE, G., MÜLLER, K. A., and SCHEEL, H. J., 1973, *Phys. Rev.* B, **8,** 1119.
KOSTERLITZ, J. M., 1974, *J. Phys.* C, **7,** 1046.
KRUMHANSL, J. A., and SCHRIEFFER, J. R., 1975, *Phys. Rev.* B**11,** 3535.
LARKIN, A. I., and KHMEL'NITSKII, D. E., 1969, *Zh. éksp. teor. Fiz.*, **56,** 2087; 1969, *Sov. Phys., JETP*, **29,** 1123.
LAJZEROWICZ, J., and LEGRAND, J. F., 1978, *Phys. Rev.* B, **17,** 1438.
LAWRIE, I. D., 1976, *J. Phys.* A, **9,** 961.
LE GUILLOU, J. C., and ZINN-JUSTIN, J., 1977, *Phys. Rev. Lett.*, **39,** 95.
LEVANYUK, A. P., 1959, *Zh. éksp. teor. Fiz.*, **36,** 810; 1959, *Soviet Phys. JETP*, **9,** 571.
LINES, M. E., 1974, *Phys. Rev.* B, **9,** 950.
MA, S. K., 1973, *Phys. Rev.* A, **7,** 2172; 1974, *J. math. Phys.*, **15,** 1866; 1976, *Phase Transitions and Critical Phenomena*, Vol. 6, edited by C. Domb and M. S. Green (New York: Academic Press), p. 249.
MCWHAN, D. B., SOUERS, P. C., and JURA, G., 1966, *Phys. Rev.*, **143,** 385.
MEISSNER, G., and BINDER, K., 1975, *Phys. Rev.* B, **12,** 3948.
MERMIN, N. D., and WAGNER, H., 1966, *Phys. Rev. Lett.*, **17,** 113.
MEYER, G. M., and COWLEY, R. A., 1976, *Ferroelectrics*, **11,** 479.
MIGDAL, A. A., 1975, *Zh. éksp. teor. Fiz.*, **69,** 810, 1457; 1976, *Sov. Phys. JETP*, **42,** 413, 743.
MITSUI, T., NAKAMURA, E., and TOKUNAGA, M., 1973, *Ferroelectrics*, **5,** 185.
DE MOURA, M. A., LUBENSKY, T. C., IMRY, Y., and AHARONY, A., 1976, *Phys. Rev.* B, **13,** 2177.
MOURITSEN, O. G., KNAK-JENSEN, S. J., and BAK, P., 1977, *Phys. Rev. Lett.*, **39,** 629.
MUKAMEL, D., 1975, *Phys. Rev. Lett.*, **34,** 481.
MÜLLER, K. A., and BERLINGER, W., 1971, *Phys. Rev. Lett.*, **26,** 13; 1975, *Ibid.*, **35,** 1547.
MÜLLER, K. A., BERLINGER, W., and SLONCZEWSKI, J. C., 1970, *Phys. Rev. Lett.*, **25,** 734.
MÜLLER, K. A., and BURKARD, H., 1979, *Phys. Rev.* B, **19**, 3593.
MURATA, K. K., 1976, *Phys. Rev.* B, **13,** 4015; 1977, *Ibid.*, **15,** 4328.
NATTERMANN, T., 1976, *J. Phys.* C, **9,** 3337; 1977, *J. Phys.* A, **10,** 1757.
NATTERMANN, T., and TRIMPER, S., 1975, *J. Phys.* A, **8,** 2000; 1976, *Ibid.* C, **9,** 2589.
NELSON, D. R., 1976, *Phys. Rev.* B, **13,** 2222.
NELSON, D. R., and DOMANY, E., 1976, *Phys. Rev.* B, **13,** 236.
NELSON, D. R., and FISHER, M. E., 1975, *Ann. Phys.*, **91,** 226.
NELSON, D. R., and RUDNICK, J., 1975, *Phys. Rev. Lett.*, **35,** 178.
NICKEL, B. G., 1973, Oxford preprint (unpublished).
NICOLL, J. F., CHANG, T. S., and STANLEY, H. E., 1976, *Phys. Rev.* A, **13,** 1251.
NIEMEYER, TH., and VAN LEEUWEN, J. M. J., 1973, *Phys. Rev. Lett.*, **31,** 1411; 1974, *Physica*, **71,** 17.
NIENHUIS, B., and NAUENBERG, M., 1975, *Phys. Rev. Lett.*, **35,** 477.
ONODERA, Y., 1970, *Prog. theor. Phys.*, **44,** 1477.
ONSAGER, L., 1944, *Phys. Rev.*, **65,** 117.
OPPERMANN, R., and THOMAS, H., 1975, *Z. Phys.* B, **22,** 387.
PEERCY, P. S., 1975, *Phys. Rev. Lett.*, **35,** 1581.
PFEUTY, P., JASNOW, D., and FISHER, M. E., 1974, *Phys. Rev.* B, **10,** 2088.
PIETRASS, B., 1972, *Phys. Stat. Sol.* (b), **53,** 279.
POLYAKOV, A. M., 1975, *Phys. Lett.* B, **59,** 79.
PRANGE, R. E., 1974, *Phys. Rev.* A, **9,** 1711.
PYNN, R., and AXE, J. D., 1976, *J. Phys.* C, **9,** L199.
PYTTE, E., 1972, *Phys. Rev. Lett.*, **28,** 895.
REHWALD, W., 1977, *Solid St. Commun.*, **21,** 667.
RIEDEL, E. K., 1973, *A.I.P. Conf. Proc.*, **18,** 849.
RIEDEL, E. K., and WEGNER, F. J., 1972, *Phys. Rev. Lett.*, **29,** 349; 1974, *Phys. Rev.* B, **9,** 294.
RISTE, T., SAMUELSEN, E. J., OTNES, K., and FEDER, J., 1971, *Solid St. Commun.*, **9,** 1455.
ROHRER, H., 1975, *Phys. Rev. Lett.*, **34,** 1638.

ROHRER, H., and GERBER, CH., 1977, *Phys. Rev. Lett.*, **38,** 909.
RUDNICK, J., 1975, *J. Phys.* A, **8,** 1125.
RUDNICK, J., and NELSON, D. R., 1976, *Phys. Rev.* B, **13,** 2208.
SAK, J., 1974, *Phys. Rev.* B, **10,** 3957.
SAMARA, G. A., 1971, *Ferroelectrics*, **2,** 277; 1974, *Ibid.*, **7,** 221.
SAMUELSEN, E. J., and SEMMINGSEN, D., 1975, *Solid St. Commun.*, **17,** 217.
SANDERCOCK, J. R., PALMER, S. B., ELLIOTT, R. J., HAYES, W., and YOUNG, A. P., 1972, *J. Phys.* C, **5,** 3126.
SARBACH, S., 1977, *Phys. Rev.* B, **15,** 2694.
SCHMIDT, V. H., WESTERN, A. B., and BAKER, A. G., 1976, *Phys. Rev. Lett.*, **37,** 839; erratum, *ibid.*, **37,** 1312.
SCHNEIDER, T., BECK, H., and STOLL, E., 1976, *Phys. Rev.* B, **13,** 1123.
SCHNEIDER, T., and STOLL, E., 1973, *Phys. Rev. Lett.*, **31,** 1254; 1975, *Ibid.*, **35,** 296; 1976 a, *Phys. Rev.* B, **13,** 1216; 1976 b, *Phys. Rev. Lett.*, **36,** 1501; 1978, *Phys. Rev.* B, **17,** 1302.
SEMMINGSEN, D., and FEDER, J., 1974, *Solid St. Commun.*, **15,** 1369.
SHAPIRO, S. M., AXE, J. D., SHIRANE, G., and RISTE, T., 1972, *Phys. Rev.* B, **6,** 4332.
SHAPIRO, S. M., COWLEY, R. A., COX, D. E., and EIBSCHUTZ, M., 1976, Oak Ridge Laboratory Report CONF 760601-Pl, 399.
SHIRANE, G., and AXE, J. D., 1971, *Phys. Rev.* B, **4,** 2957.
SOKOLOV, A. I., and TAGANTSEV, A. K., 1978, *Ferroelectrics*, **20,** 141–3.
SOKOLOV, A. I., and VENDIK, O. G., 1974, *Ferroelectrics*, **7,** 391.
STANLEY, H. E., 1968, *Phys. Rev.*, **176,** 718; 1971, *Introduction to Phase Transitions and Critical Phenomena* (Oxford: Clarendon Press).
STIRLING, W. G., 1972, *J. Phys.* C, **5,** 2711.
STRALEY, J. P., and FISHER, M. E., 1973, *J. Phys.*, A, **6,** 1310.
STRUKOV, B. A., 1976, *Ferroelectrics*, **12,** 97.
SUZUKI, M., 1972, *Phys. Lett.* A, **42,** 5.
TARKO, H. B., and FISHER, M. E., 1975, *Phys. Rev.* B, **11,** 1217.
THOMAS, H., 1971, *Structural Phase Transitions and Soft Modes*, edited by E. J. Samuelsen, E. Anderson and J. Feder (Oslo: Universitetsförlaget), p. 15.
TÖPLER, J., ALEFELD, B., and HEIDMANN, A., 1977, *J. Phys.* C, **10,** 635.
VARMA, C. M., 1976, *Phys. Rev.* B, **14,** 244.
VILLAIN, J., 1970, *Solid St. Commun.*, **8,** 295.
VON WALDKIRCH, TH., MÜLLER, K. A., BERLINGER, W., and THOMAS, H., 1972, *Phys. Rev. Lett.*, **28,** 503.
WALLACE, D. J., 1976, in *Phase Transitions and Critical Phenomena*, Vol. 6, edited by C. Domb and M. S. Green (New York: Academic Press), p. 293.
WEGNER, F. J., 1972, *Phys. Rev.* B, **5,** 4529.
WEGNER, F. J., and HOUGHTON, A., 1973, *Phys. Rev.* A, **8,** 401.
WEGNER, F. J., and RIEDEL, E. K., 1973, *Phys. Rev.* B, **7,** 248.
WIDOM, B., 1965, *J. chem. Phys.*, **43,** 3892; 3898.
WILSON, K. G., 1971, *Phys. Rev.* B, **4,** 3174; 3184; 1972, *Phys. Rev. Lett.*, **28,** 548; 1975, *Rev. mod. Phys.*, **47,** 773.
WILSON, K. G., and FISHER, M. E., 1972, *Phys. Rev. Lett.*, **28,** 248.
WILSON, K. G., and KOGUT, J., 1974, *Phys. Rep.* C, **12,** 77.
YAMADA, Y., SHIRANE, G., and LINZ, A., 1969, *Phys. Rev.*, **177,** 848.
YELON, W. B., COX, D. E., KORTMAN, P. J., and DANIELS, W. B., 1974, *Phys. Rev.* B, **9,** 4843.
YOUNG, A. P., 1975, *J. Phys.* C, **8,** L309.
ZINN-JUSTIN, J., 1973, *Cargese Lectures in Field Theory and Critical Phenomena* (unpublished).

III

Critical Dynamics and Quasi-Elastic Scattering

By

A. D. BRUCE and R. A. COWLEY

SUMMARY

The dynamical behaviour of systems exhibiting structural phase transitions is reviewed, with particular emphasis on the problems posed by the simultaneous existence of quasi-elastic and phonon features in the dynamic response of the soft coordinates. The results of scattering experiments revealing this phenomenon are surveyed in detail, and are used in an assessment of current theories, subdivided into three categories: those which seek to describe the phenomenon within an extended phonon-based theory: those which attempt to trace the effect to the formation of long-lived clusters of precursor order; and those which associate the quasi-elastic scattering with crystal imperfections.

§III.1. INTRODUCTION

In parts I and II we developed the classical theory of structural phase transitions, outlined its failures in the context of static critical phenomena, and surveyed the extent to which its more sophisticated successor, the theory of the renormalization group (R.G.), accounts for equilibrium critical properties.

In this review we turn to consider the respects in which the classical theory fails to account for the *dynamic* behaviour of systems undergoing structural phase transitions. We shall be concerned not so much with the theory's (unsurprising) failure to yield quantitatively correct predictions for the exponents characterizing time-dependent behaviour (cf. however, §III.3.2) as with its more qualitative, conceptual inadequacies—latent in its failures with equilibrium properties, but more clearly exposed in the context of dynamic phenomena. These failures, and the consequent need for a more refined theory, have proved most acute in the case of displacive systems (§I.3.5; §II.2.2) to which, accordingly, we devote the bulk of this article.

According to the classical theory of displacive systems—the soft-mode phenomenology of Cochran (1960) and Anderson (1960)—the frequency-dependent response of the order-parameter degrees of freedom is that of a classically-damped simple harmonic oscillator (cf. eqn. (I.3.22))

$$\chi^{-1}(\mathbf{q}j,\omega)=\omega_0(\mathbf{q}j)^2-\omega^2-2i\omega\gamma_0. \qquad \text{(III.1.1.}a\text{)}$$

Here, within the low-order perturbation theory in which the soft-mode phenomenology is grounded (cf. §I.5.1.1) γ_0 is a weakly temperature-dependent quantity, while the renormalized frequency $\omega_0(\mathbf{q}_s j)$, which determines the static response for

$\mathbf{q}=\mathbf{q}_s$ (cf. §I.3.5) has the critical temperature dependence

$$\omega_0(\mathbf{q}_s j)^2 = a(T-T_c)^\gamma, \qquad \text{(III.1.1}\,b\text{)}$$

with a susceptibility exponent $\gamma=1$.

That the critical order-parameter response function for a displacive system may be qualitatively different from the form implied by these equations was first clearly demonstrated by inelastic neutron-scattering measurements near the 105 K phase transition in $SrTiO_3$ (Riste *et al.* 1971). Subsequently, higher resolution measurements on $SrTiO_3$, and its isomorph $KMnF_3$, were performed by Shapiro *et al.* (1972), whose results are illustrated in fig. III.1. Instead of the *single* softening resonant (or overdamped) region of response suggested by the classical theory, the observed high-temperature phase response exhibits *two* components: a phonon-like response, initially softening in accord with the soft-mode theory (and with the results displayed in fig. I.10), but ultimately saturating a few degrees above T_c, together with a narrow component, centred on zero frequency transfer, first appearing some 70 K above the phase transition, and exhibiting a critical divergence with the approach to T_c.

Fig. III.1

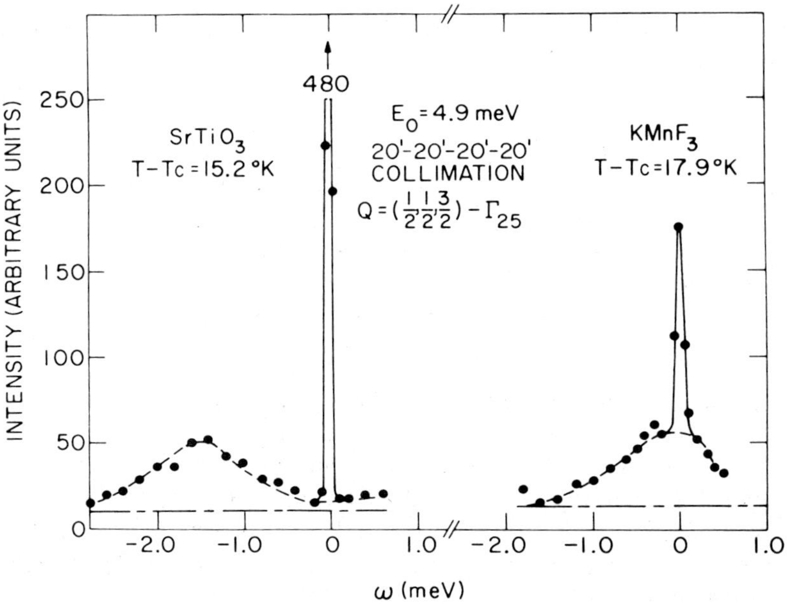

The R_{25} soft-mode spectral function in $SrTiO_3$ and in $KMnF_3$, as determined by neutron scattering (Shapiro *et al.* 1972).

The essential features of these results are captured by a phenomenological theory (Schwabl 1972, Shapiro *et al.* 1972) which supposes that the response of the ordering coordinates is that of a damped simple harmonic oscillator coupled, with interaction strength δ, to some (unspecified) internal degree of freedom with a Debye relaxation time τ_0. The response function then has the form (see Feder 1976, for a simple derivation)

$$\chi^{-1}(\mathbf{q}j,\omega) = \omega_0(\mathbf{q}j)^2 - \omega^2 - 2i\omega\gamma_0 - \frac{i\omega\tau_0\delta^2}{1-i\omega\tau_0} \qquad \text{(III.1.2)}$$

$$\equiv \omega_\infty(\mathbf{q}j)^2 - \omega^2 - 2i\omega\gamma_0 + \sum(\mathbf{q}j, \omega), \tag{III.1.3}$$

where

$$\omega_\infty(\mathbf{q}j)^2 \equiv \omega_0(\mathbf{q}j)^2 + \delta^2, \tag{III.1.4}$$

and

$$\sum(\mathbf{q}j, \omega) \equiv \frac{-\delta^2}{1 - i\omega\tau_0}. \tag{III.1.5}$$

The associated scattering properties are characterized by the spectral density (eqn. (I.2.40) in the classical approximation)

$$\mathscr{S}(\mathbf{q}, \omega) = \frac{k_B T}{\pi\omega} \chi^I(\mathbf{q}, \omega). \tag{III.1.6}$$

For sufficiently small values of the damping constant, the spectral density may be written in the form (we shall frequently suppress the labels $\mathbf{q}j$)

$$\mathscr{S}(\mathbf{q}, \omega) = \frac{k_B T}{\pi} \left\{ \frac{2\gamma_0}{(\omega_\infty^2 - \omega^2)^2 + 4\omega^2\gamma_0^2} + \frac{\delta^2}{\omega_0^2\omega_\infty^2} \cdot \frac{\omega_Q}{(\omega_Q^2 + \omega^2)} \right\} \tag{III.1.7}$$

revealing resonances at $\omega \approx \pm\omega_\infty$, and a central component of halfwidth ω_Q, where

$$\omega_Q = \tau_0^{-1} \left(\frac{\omega_0}{\omega_\infty} \right)^2. \tag{III.1.8}$$

As long as the central and side-band regions of response are well separated (i.e. as long as the approximate form (III.1.7) is appropriate), it is possible to associate a unique scattering intensity with each component: the ratio of the central-component intensity to that contained in the two side bands then follows as

$$\frac{I_C}{2I_{SB}} = \frac{\delta^2}{\omega_0^2}. \tag{III.1.9}$$

In the experiments of Shapiro *et al.* (1972), the ω width of the central component was found to lie within the instrumental resolution, so that it proved possible to assign only an upper bound of 5×10^9 Hz. The behaviour of the quantities ω_0^2, ω_∞^2 and δ^2 was determined by fitting the data to the scattering implied by (III.1.2), convoluted with the experimental resolution function, the wave-vector dependence of the coupling strength δ (and, less restrictively, of γ_0 and τ_0) being supposed small. The results of this analysis are shown in fig. III.2. The statically-renormalized soft-mode frequency ω_0 evidently vanished in accord with eqn. (III.1.1 *b*), though with a non-classical exponent γ (as is to be expected: cf. § II.5.2.1) to which, however, these measurements do not permit the assignment of a precise value. In contrast, the dynamically-renormalized soft-mode frequency ω_∞ retains a finite value ($\omega_\infty^c \approx 0{\cdot}13 \times 10^{12}$ Hz in $SrTiO_3$) at criticality.

The essential implication of these results is that the near-critical behaviour of the order-parameter degrees of freedom, in these systems, is characterized by *two* distinct time scales, rather than the *single* time scale suggested by the soft-mode theory: in addition to the relatively short (non-critical) time scale set by the soft-phonon side band, there is a longer, critical, time scale associated with the critically-narrowing (eqn. (III.1.8)) central peak.

Fig. III.2

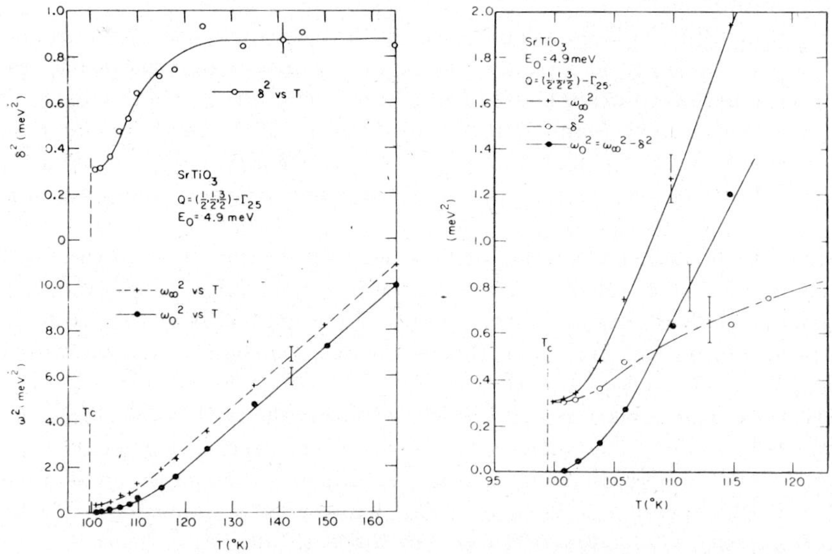

The parameters of the R_{25} soft-mode spectral function in $SrTiO_3$, as functions of temperature (Shapiro *et al.* 1972).

It should be made clear at the outset that the existence of two time scales in the order-parameter response function is problematic only to the extent that the systems concerned—particularly $SrTiO_3$—have been regarded as offering the paradigm examples of *displacive* phase transitions, where the single time scale, quasi-harmonic, soft-mode picture is at its most plausible. By contrast, the existence of two time scales in the dynamics of *order–disorder* systems, typified by $NaNO_2$ (§I.1) is experimentally well-established and, qualitatively at least, easily understood. In this case, the relatively short time scale is set by collective, small (quasi-harmonic) oscillations of the atoms about the positions of equilibrium set by the instantaneous configuration of the NO_2 groups: the associated phonon resonances, observed by scattering techniques, show no appreciable temperature dependence, even close to the phase transition (cf. fig. I.12). The second, long, time scale is set by the collective large-amplitude (strongly anharmonic) motion of NO_2 groups, between their two quasi-equilibrium configurations; the associated quasi-elastic response exhibits critical narrowing with the approach to T_c (cf. fig.I.13). The distinction between the two types of motion has not always been clearly appreciated: cf. Andrade *et al.* (1973).

Clearly, there are many physical systems that do not fall unambiguously into either of these two categories, which are to be regarded as the two extremes of a continuous spectrum: in the case of KH_2PO_4, for example, (§I.5.2.4) it is not easy to establish whether the spectral function of the ordering coordinates (a single region of central response) is better regarded as that of a displacive system with an overdamped soft mode, or that of an order–disorder system with only the tunnelling ('pseudo-spin') mode identifiable. However, the behaviour of the critical spectral function in $SrTiO_3$ raises even more fundamental questions as to the validity of the order–disorder/displacive classification: this is a theme to which we shall return later.

The observations we have discussed pose problems on both experimental and theoretical fronts.

The experimental challenge they present is the determination of the frequency and wavevector width of the central components—a task which has prompted the development of high-resolution neutron and Raman scattering techniques: to these we shall refer further in the course of a detailed survey of the relevant experiments (§ III.2), focusing on those materials—strontium titanate, lead germanate and potassium dihydrogen phosphate—in which the effect has been most extensively studied.

The theoretical challenge is to understand the origins of the second time scale in the dynamics of a displacive system—or, more specifically, to understand the coexistence of a central peak and a (partially) softening phonon. This is a challenge that has been met with a barrage of theories, which we shall subdivide into three categories.

In § III.3 we outline those theories which, adopting the traditional standpoint of a renormalized phonon basis, attempt to capture the second time scale through anharmonic phonon perturbation theories, of varying degrees of sophistication (Silberglitt 1972, Cowley and Coombs 1973, Murata 1975, Bausch and Halperin 1978). This approach is analytically explicit, but conceptually opaque.

We devote § III.4 to a survey of those theories which correlate the two time scale dynamics with the existence of precursor clusters of local order (Schneider and Stoll 1973) which introduce an order–disorder component into the collective motion (Krumhansl and Schrieffer 1975, Aubry 1975, 1976 a, b, Varma 1976). This approach—orthogonal in character if not, ultimately, in content, to that based on refined phonon perturbation theory—is conceptually appealing, but as yet without firm analytic foundation.

In § III.5 we assess the arguments which associate the central-peak phenomenon with crystal inperfections (Halperin and Varma 1976). Indeed, in the course of this article, we shall see that there is overwhelming evidence that much of the quasi-elastic scattering observed should, in fact, be attributed to such extrinsic mechanisms. We shall argue, however, that there remains both theoretical and experimental evidence for the occurrence of central peaks in ideal crystal behaviour.

We remark, finally, that in systems for which there exists a characteristic relaxation rate of the electronic degrees of freedom, which is not short in comparison with inverse phonon frequencies, that relaxation rate can manifest itself as a second time scale in the dynamical behaviour of the crystal, and give rise to quasi-elastic scattering. The central peak observed in Nb_3Sn has been attributed to a process of this type (Sinha 1978; see, however, § I.5.3.5.), while Pytte (1973) has developed a detailed theory of such a mechanism in the case of Jahn–Teller phase transitions. Our survey, however, will be restricted to systems for which the adiabatic approximation to the crystal dynamics is acceptable, since it is in this context that the existence of a second time scale has proved most problematic.

§ III.2. Scattering experiments: a survey

In this section we shall survey the results of experiments investigating the dynamic behaviour of systems undergoing structural phase transitions, with particular reference to the central-peak phenomenon. The literature devoted to such studies is extensive and we shall restrict our survey in two principal respects.

First of all we shall focus almost exclusively on the results of scattering experiments employing neutrons, light or X-rays, since these afford the most direct probes of the collective dynamic behaviour: dielectric measurements are mentioned briefly: highly selective reference to E.P.R. and N.M.R. experiments (surveyed recently, in this context, by Blinc 1978, and Müller, 1979) is deferred to §III.4.3.2, since their relationship to central-peak phenomena requires preparatory discussion; we make no menton of ultrasonic attenuation measurements, although the relevant theory is a byproduct of §III.3.2.2.

Secondly, while making brief mention of other systems—including computer 'experiments' on a simple model of a structural phase transition—we shall restrict detailed discussion to three materials: $SrTiO_3$, $Pb_5Ge_3O_{11}$ and KH_2PO_4.

The particular attention we devote to three systems reflects both the fact that they have been studied more extensively than any other, and the fact that they exemplify the three principal classes of structural phase transition, when subdivided according to the extent to which one expects fluctuations to promote deviations from 'classical' behaviour. Thus, critical singularities may be non-classical ($SrTiO_3$), classical with logarithmic corrections ($Pb_5Ge_3O_{11}$), or strictly classical (KH_2PO_4).

We caution at this stage, however, that 'classical' critical behaviour is *not* to be regarded as being synonymous with 'soft-mode' behaviour: a system whose critical singularities are classical *may* be expected to exhibit simple soft-mode behaviour *if* it is displacive in character (even here some refinements of the theory are necessary: §III.3.1.1): it cannot be expected to exhibit a softening phonon if it is order–disorder in character.

The task of surveying even the scattering experiments on these three materials is sufficiently extensive that the key results can all too readily be lost amid a sea of detail: with this in mind we summarize the central features of the experimental situation at the conclusion of this section.

III.2.1. *Strontium titanate*

The phase transition occurring in strontium titanate at approximately 105 K is characterized by a three-component order parameter, associated with the wavevector $\mathbf{q}_s = (0{\cdot}5, 0{\cdot}5, 0{\cdot}5)2\pi/a$, and describing rotations of the oxygen octahedra about the three cubic axes (cf. fig. I.2). This system thus belongs to the universality class with $n=3$, cubic symmetry and short-range interactions, for which non-classical critical behaviour is expected (cf. §II.5.2.1).

The central peak exhibited by $SrTiO_3$ has been studied more extensively than that occurring in any other material. The most readily-interpreted information has proved to be that supplied by inelastic neutron-scattering techniques which can probe the one-phonon response at the critical wavevector $\mathbf{q}_s$, both above and below T_c. The first measurements were made by Riste *et al.* (1971), and were soon repeated, under conditions of better resolution, by Shapiro *et al.* (1972). The latter experiment was performed using a conventional triple-axis spectrometer, with incident neutrons of energy 4.9 meV, giving a frequency resolution of 5×10^9 Hz, and a wavevector resolution of about 10^{-2} Å^{-1}.

Measurements were made on crystals from two sources: a clear crystal from the National Lead Company, grown by the flame fusion method, and a brownish-coloured crystal from Sanders Associates, grown by a top-seeded method. The latter crystal was less strained than the former, and was therefore used for the majority of the measurements. In fact the neutron-scattering experiments showed no significant differences between the two crystals, except for their different transition temperatures, the Sanders crystal having the lower T_c by some 5 K. (However, a Raman-scattering investigation by Steigmeier and Auderset (1975) revealed a temperature-independent defect mode in a Sanders Associate crystal, at a frequency close to that at which the soft-mode side band was observed to saturate, in the experiments of Shapiro *et al.* (1972)).

The results of these experiments were outlined in the preceding section (cf. fig. III.1): for convenience we summarize the essential conclusion here.

A central peak and a softening phonon side band were observed over a range of about 70 K above the phase transition, the latter making the dominant contribution at temperatures higher than T_c+10 K, approximately. The frequency width ('full width at half maximum') of the central component was less than the instrumental resolution of 5×10^9 Hz. The wavevector width at T_c+9 K was determined as 0·02 $\pm$0·01 Å^{-1}—in rough accord with estimates based on the phenomenological form (III.1.7) which (if q-independent values of δ^2 and γ_0 are assumed) implies a central-peak q width uniquely determined by the soft-phonon dispersion. However, we do not feel that these measurements settle this issue conclusively. Fitting the scattering data to the phenomenological form yielded the results displayed in fig. III.2, according to which the limiting frequency for the soft-phonon side band (as $T\to T_c^+$) is given by $\omega_\infty^c=\delta=0{\cdot}13\pm0{\cdot}02\times10^{12}$ Hz, while the behaviour of ω_0^2 suggests a value for the exponent γ (eqn. (III.1.1 *b*)) lying between 1·5 and 2·5. We re-emphasize, however, that the significance of these results is strongly correlated with the applicability of the phenomenological form (III.1.2), and of the additional assumptions (implicit in the analysis) of the wavevector independence of the parameters δ and τ_0.

The failure of these experiments to resolve the width of the central component in $SrTiO_3$ motivated subsequent inelastic neutron-scattering experiments (focusing on the central peak itself) employing a spin–echo spectrometer(Mezei and Heyter 1974) and a back-scattering spectrometer (Töpler *et al.* 1975, 1977) to yield still higher resolution.

In the most recent and precise of these experiments, Töpler *et al.* (1977) examined two crystals, both supplied by the National Lead Company, one of which had previously been cooled through the phase transition, while the other had not. In their experiment the frequency resolution was $2{\cdot}1\times10^8$ Hz, and the wave vector resolution about 10^{-2} Å^{-1}. Again, no frequency broadening was exhibited by the central peaks observed (in both crystals) above T_c, and it was concluded that (at least within 12 K of the phase transition) the central component width is less than 2×10^7 Hz. Since the soft-phonon side band was not investigated, no results for the limiting behaviour of ω_∞^2 are reported, but values of γ of $1{\cdot}53\pm0{\cdot}1$ (for the crystal previously taken through the phase transition) and $1{\cdot}2\pm0{\cdot}1$ (for the 'virgin' crystal) were derived from the growth in the intensity of the central component with the approach to T_c (fig. III.3); again, these values should be viewed as giving at best a qualitative indication of non-classical behaviour.

Still more recently, in the course of a study of the effects of [111]-stress on $SrTiO_3$, Currat *et al.* (1978) have used conventional inelastic neutron-scattering techniques to reinvestigate the soft-mode spectral function in a crystal grown by the Verneuil process at the National Lead Company. Their (zero-stress) results, analysed using the phenomenological form (III.1.2), are displayed in fig. III.4, where they are compared with those of Shapiro *et al.* (1972). Evidently there is substantial

Fig. III.3

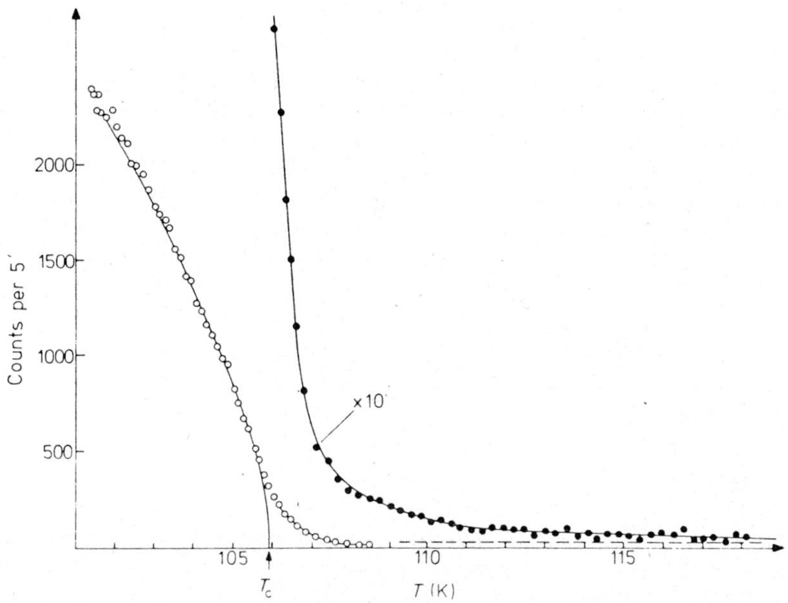

The intensity of the R-point quasi-elastic neutron scattering in $SrTiO_3$, as function of temperature (Töpler *et al.* 1977).

agreement as regards the behaviour of the phonon side band (and, therefore, the *limiting* $T \to T_c$ value of δ^2); however, this agreement does not extend to the behaviour of δ^2 at higher temperatures—a point to which we return at the conclusion of this section.

The role of crystal defects in the quasi-elastic scattering in $SrTiO_3$ has also been investigated in neutron-scattering experiments (Hastings *et al.* 1978) on crystals with defects introduced by reduction with hydrogen, to produce free carriers. Measurements were made on crystals of nominally pure $SrTiO_3$, and crystals with carrier concentrations of $1{\cdot}8 \times 10^{24}\,m^{-3}$, $8 \times 10^{25}\,m^{-3}$, $1{\cdot}2 \times 10^{26}\,m^{-3}$ and $3{\cdot}2 \times 10^{26}\,m^{-3}$, the transition temperature varying linearly from 104 K for the 'pure' samples, to 70 K for the specimen with the most carriers. The results (fig. III.5) show that, for a given $T - T_c$, the central-peak intensity does increase with carrier concentration: the effect is, however, a modest one—for $T - T_c \sim 20$ K an increase in the carrier concentration of more than 200 is accompanied by an increase in the intensity of only a factor of 3.

The scattering at the critical wavevector in $SrTiO_3$ has also been probed with X-ray and γ-ray techniques (Darlington *et al.* 1975, Darlington and O'Connor, 1976, 1978) using a colourless crystal obtained from the National Lead Company. The γ-ray experiment employed the 14·4 keV Mössbauer γ-rays emitted from a ^{57}Co source, and analysed using a matched absorber of iron diffused into rhodium. This technique gives excellent energy resolution ($3{\cdot}6 \times 10^6$ Hz), but the experiments gave no indication of frequency broadening in the scattering associated with the central component.

The inferences to be drawn from this result are, however, obscured by the fact that, quite unexpectedly, the intensity of the quasi-elastic scattering observed above

Fig. III.4

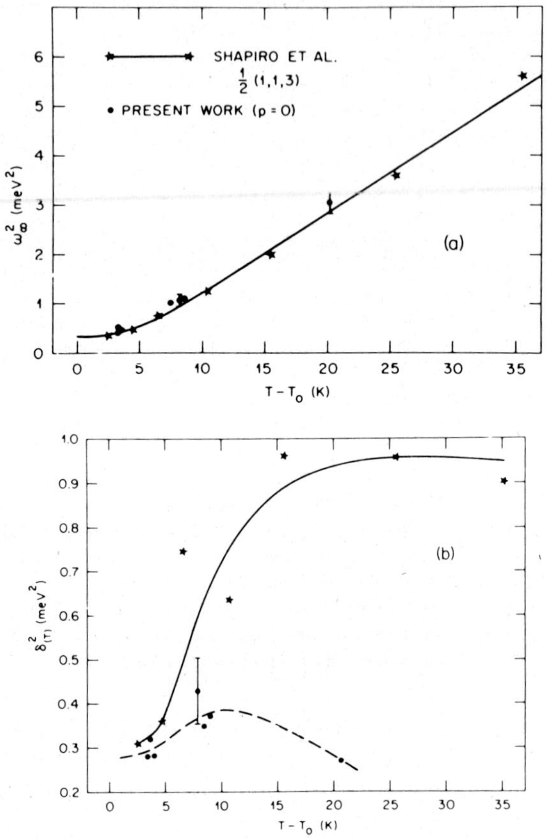

The parameters of the R_{25} soft-mode spectral function in $SrTiO_3$ as determined by Currat *et al.* (1978), and compared with the results of Shapiro *et al.* (1972).

the macroscopic transition temperature was quite comparable with the (presumably) truly elastic, Bragg scattering observed at the superlattice points below T_c—in marked contrast to the results of the inelastic neutron-scattering investigations where the quasi-elastic scattering is small on the scale of the superlattice Bragg scattering. This is also a feature of the results of the X-ray scattering measurements (fig. III.6) reported by Darlington and O'Connor (1976), which revealed, furthermore, a marked dependence of the central-peak intensity upon the X-ray energy, the intensity being larger for X-rays of lower energy. Subsequent experiments (Darlington and O'Connor 1978) also showed that the central-peak intensity was significantly lower after the crystal had been annealed for 24 h at 1300 K.

While it is the purpose of this section to summarize, rather than to interpret, the relevant experimental results, these observations merit some immediate comment. The plausible interpretation proposed by Darlington and O'Connor, is that at about 170 K (the temperature at which the quasi-elastic scattering is first in evidence) a portion of the crystal close to the surface assumes the distorted ('low-temperature') structure, the extent of the distortion, and the thickness of the surface layer, growing with the approach to the bulk transition temperature: about 5 K above T_c they

Fig. III.5

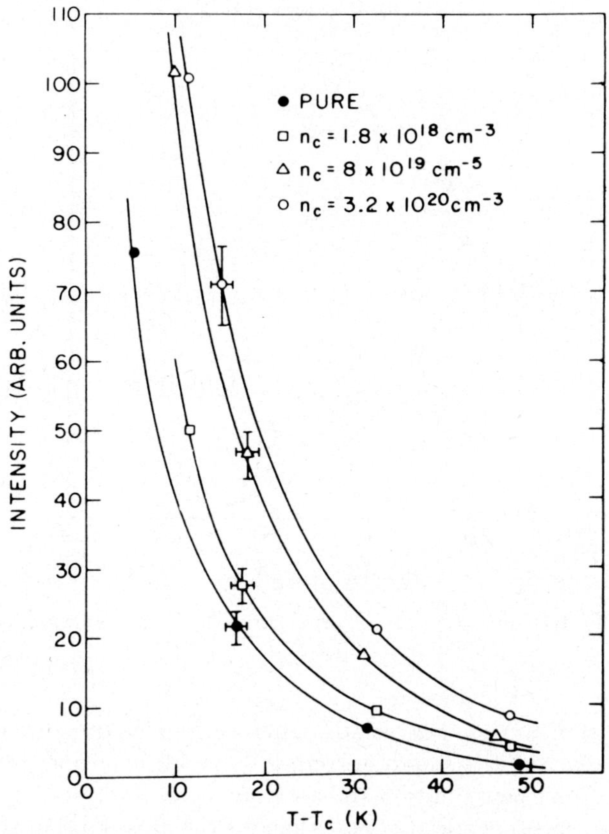

The intensity of the central peak in the soft-mode spectral function in hydrogen-reduced $SrTiO_3$, as function of temperature and carrier concentration (Hastings *et al.* 1978).

estimate that the surface layer is almost wholly distorted to a depth of at least 20 μm. Within this interpretation the 'quasi-elastic' scattering observed in these experiments is nothing but Bragg scattering from the (effectively ordered) surface layer, the intensity of the scattering being larger for the X-rays of lowest energy, since they have the smallest penetration depths and are therefore most sensitive to the state of the surface layer. The difference between the neutron and X-ray results also follows plausibly, since the neutron-scattering techniques probe the bulk crystal properties.

However, the volume of the surface layer is actually sufficient to account for the intensity of the central peak observed in the neutron-scattering experiments (Cowley and Shirane 1978). To investigate whether the whole of the central peak exhibited in the neutron-scattering results in fact originates in the crystal surface, a further neutron-scattering experiment was performed (Cowley and Shirane 1978): by an appropriate masking of the neutron beams, it was established that there is no significant difference between the scattering from the whole crystal and the scattering from the crystal interior alone.

We turn now to survey the results of Raman-scattering studies of the phase transition in $SrTiO_3$, prefacing our discussion with a brief review of the relevant theory, discussed in §I.3.6.2.

Fig. III.6

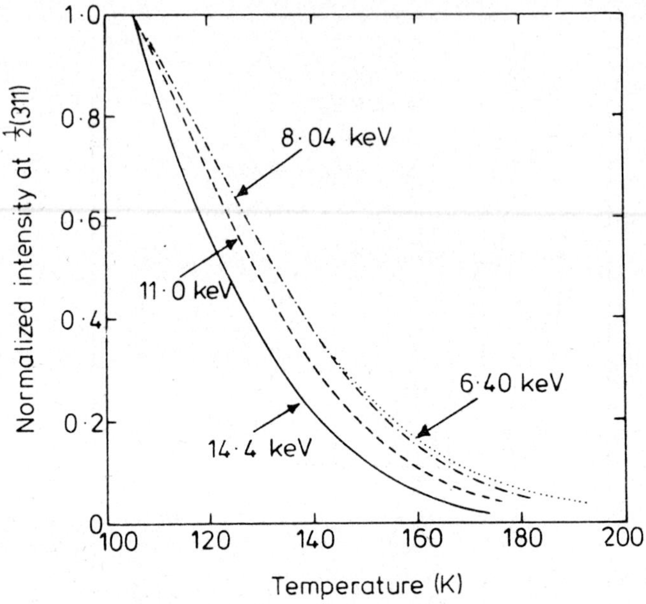

The intensity of the R-point X-ray scattering in $SrTiO_3$ as function of temperature (Darlington and O'Conner 1976).

For crystals with the cubic perovskite structure, the leading order terms in the expansion of the polarizability in powers of normal mode coordinates are absent by symmetry. The leading contribution to the Raman tensor thus originates in the second-order terms: the appropriate expression for the cross section is given in eqn. (I.3.31). In the high-temperature phase, the cross section consists entirely of two-phonon Raman-scattering processes, first-order scattering being symmetry-forbidden. Below the phase transition there exists, in addition to a two-phonon background, a set of one-phonon contributions from modes with $\mathbf{q}=\mathbf{q}_s$ (invariably including the soft mode itself), which appear in the cross section with a strength proportional to the square of the order parameter. Naturally, in such situations it is the Raman spectrum of the low-temperature phase that is the focus of attention. However, as remarked in §I.3.6.2 (see also §III.3.2.2) the interpretation of low-temperature phase Raman spectra is less straightforward than has frequently been supposed by authors who assume the dominance of the one-phonon scattering over the two-phonon background—a mean-field approximation!—throughout the critical region.

The earliest Raman-scattering investigation of the phase transition in $SrTiO_3$ was that of Fleury *et al.* (1968), which revealed well-defined ordered-phase soft-phonon resonances of A and E symmetry (fig. I.23). The temperature-dependence of the fully symmetric A_1 mode has subsequently been investigated by Steigmeier and Auderset (1973) and by Yacoby *et al.* (1976). For reduced temperatures $|t|\gtrsim 0{\cdot}03$, the resonance is sufficiently well-defined to be parameterized by the characteristic frequency and damping constant of an anharmonic oscillator, with the results displayed in fig. III.7 (Steigmeier and Auderset 1973). The damping constant increases weakly with the approach to criticality (fig. III.7 (b)). The square of the

Fig. III.7

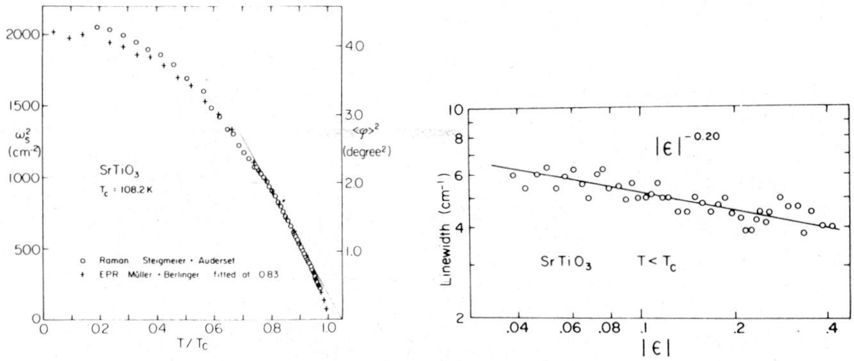

The parameters of the soft low-temperature phase A_{1g} phonon in $SrTiO_3$, within a damped simple harmonic oscillator representation. (*a*) The square of the soft-mode frequency. (*b*) The mode line width (Steigmeier and Auderset 1973).

frequency clearly follows the square of the order parameter (fig. III.7 (*a*)) even within the region where the latter displays non-classical behaviour. However, the more recent experiments of Yacoby *et al.* (1976) are consistent with an asymptotic saturation of the A-mode frequency at the limiting value assigned to the high-temperature phase soft mode on the basis of the neutron-scattering results discussed above.

Raman-scattering techniques have also revealed quasi-elastic response near the phase transition in $SrTiO_3$—usually orders of magnitude more intense than the inelastic light scattering discussed above. In an experiment without spectral analysis, Steigmeier *et al.* (1973, 1974) found quasi-elastic scattering whose intensity increased critically with the approach to T_c (fig. III.8). The dependence of the intensity of this central component upon the scattering angle is a clear indication that it originates in scattering centres which are of a size comparable with the inverse of the wavevector transfer (i.e. of order 10^3 Å) and which are therefore large compared with the value of the correlation length (10^1–10^2 Å at comparable temperatures) deduced from the soft-phonon dispersion (cf. Levanyuk 1976). It appears, however (Lockwood and Steigmeier, private communications), that the (relative) intensity of this scattering varies from one sample to another and even depends upon the sample history.

More recently Lyons and Fleury (1976) have made use of a light-scattering technique which allows the examination of spectral features lying as close as 5×10^8 Hz from the exciting line. The method uses a molecular iodine resonant absorption cell to remove the light scattered closer to the exciting line (thus eliminating any scattering as narrow as that suggested by the X-ray and neutron-scattering experiments discussed above), together with a tandem Fabry–Perot interferometer to analyse the spectrum. Fleury and Lyons (1978) have reported observations of both the diagonal and the off-diagonal spectra of $SrTiO_3$, obtained with this technique.

The diagonal spectrum (Fleury and Lyons 1978; fig. III.9) reveals evidence for two dynamic central components: a narrow peak, with a width ($\sim 7{\cdot}2 \times 10^9$ Hz at T_c) which varies with the square of the wavevector transfer, and a broader peak with a width ($\sim 7{\cdot}6 \times 10^{10}$ Hz at T_c) essentially independent of the wavevector transfer.

Fig. III.8

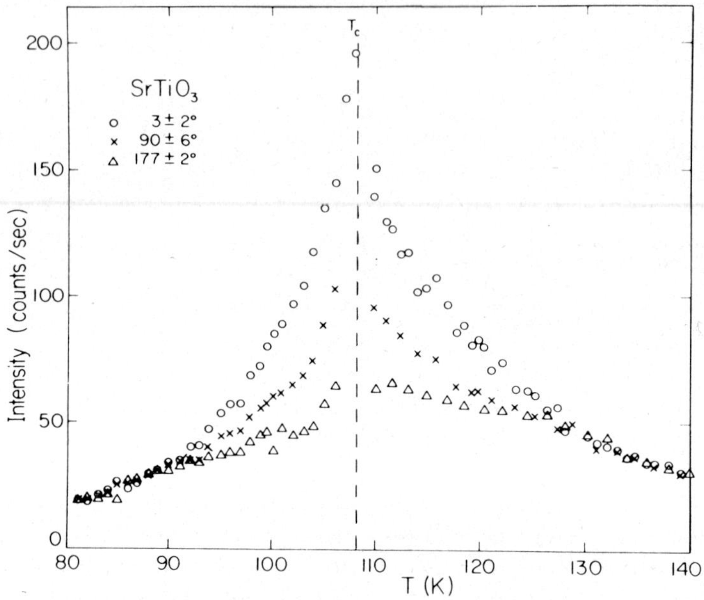

The intensity of quasi-electric light scattering in $SrTiO_3$, as a function of temperature, for three different scattering angles (Steigmeier *et al.* 1974).

While both the line widths and the intensities of these features were observed to vary considerably with temperature, neither component was found to exhibit a detectable critical anomaly: indeed, their behaviour was found to be, in all essential respects, identical to that of similar features observed in the diagonal light-scattering spectrum of $KTaO_3$ (Lyons and Fleury 1976) which does not undergo a phase transition. It is evident, therefore, that these dynamic central components are not to be associated with the occurrence of a phase transition.

In contrast, the off-diagonal spectrum (Lyons and Fleury 1977) (fig. III.10) displays a weak central component (of width 2×10^{10} Hz) visible only in the immediate vicinity of the phase transition, and exhibiting a definite maximum close to T_c. (We view with reserve the authors' conclusion that the maximum actually occurs some 1·5 K below the phase transition, since their determination of $T_c = 107$ K rests upon a mean-field analysis of the Brillouin spectrum.) The interpretation of the spectrum is complicated not only by doubts as to the appropriateness of the 'one-phonon' approximation, but also by the interaction of the soft E phonon with a transverse acoustic mode: nevertheless the coupled-mode formalism employed by Lyons and Fleury (1977, 1978) in the analysis of their data suggests that, as $T \to T_c^-$, the soft E phonon saturates at a frequency close to that at which, according to the inelastic neutron-scattering experiments discussed above, the soft-high-temperature phase phonon saturates as $T \to T_c^+$.

We conclude with a brief survey of the dynamical behaviour displayed by other members of the perovskite family, exhibiting R-point mode instabilities.

$KMnF_3$ exhibits a phase transition of this type at 187 K. Inelastic neutron-scattering experiments (Shapiro *et al.* 1972) have revealed an unresolved central mode, coexisting with a softening phonon (fig. III.1) which, in this case, is

Fig. III.9

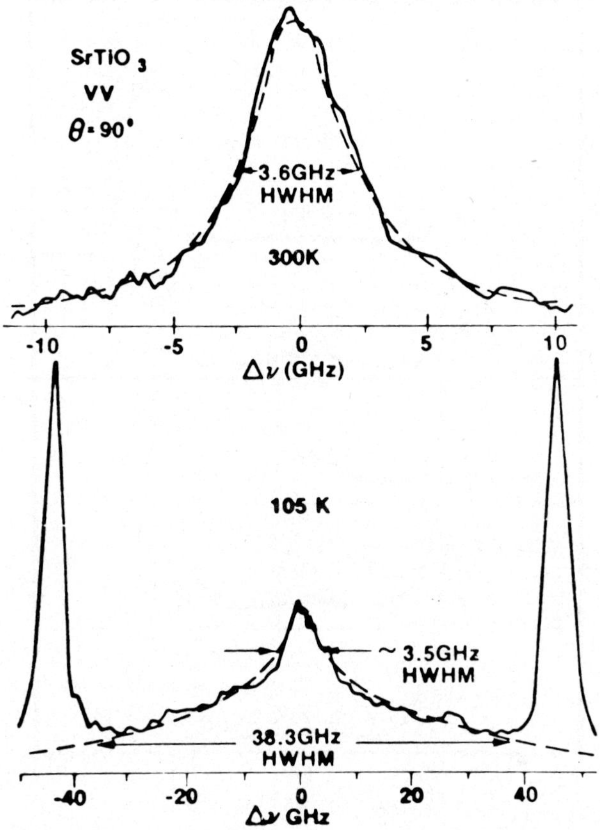

The diagonal light-scattering spectrum of $SrTiO_3$, in the presence of an iodine filter, at 300 K and 105 K. The dashed line is the sum of two lorentzians with the halfwidths indicated (Fleury and Lyons 1978).

overdamped even far from T_c. The Raman spectrum was investigated by Lockwood and Torrie (1974 a, b) who found that the low-temperature phase A_1 soft-mode response could not be adequately characterized with the simple damped harmonic oscillator formalism. These authors also studied the intensity of the 'hard' (227 cm^{-1})E_g mode which, like the soft mode itself, is Raman active in the low-temperature phase, and has a Raman inactive R-corner mode progenitor above T_c. Their observation that the scattering persists into the high-temperature phase has recently been confirmed in more detailed studies by Bruce *et al.* (1980; fig. III.11) who also develop the connection between this phenomenon and the existence of a central peak in the soft-mode response function (cf. § III.4.3.2).

$LaAlO_3$ and $RbCaF_3$ also exhibit similar phase transitions near 767 K and 193 K, respectively: again neutron-scattering experiments (Kjems *et al.* 1973, Almairac *et al.* 1977) have established the coexistence of unresolved central component and softening phonon. It is to be noted, however, that the results of analyses employing the phenomenological form (III.1.2) display little uniformity in the behaviour of the parameter δ^2. In $LaAlO_3$ this parameter is virtually temperature-independent (Kjems *et al.* 1973) in $RbCaF_3$ it decreases with increasing $T-T_c$ (Almairac *et al.*

Fig. III.10

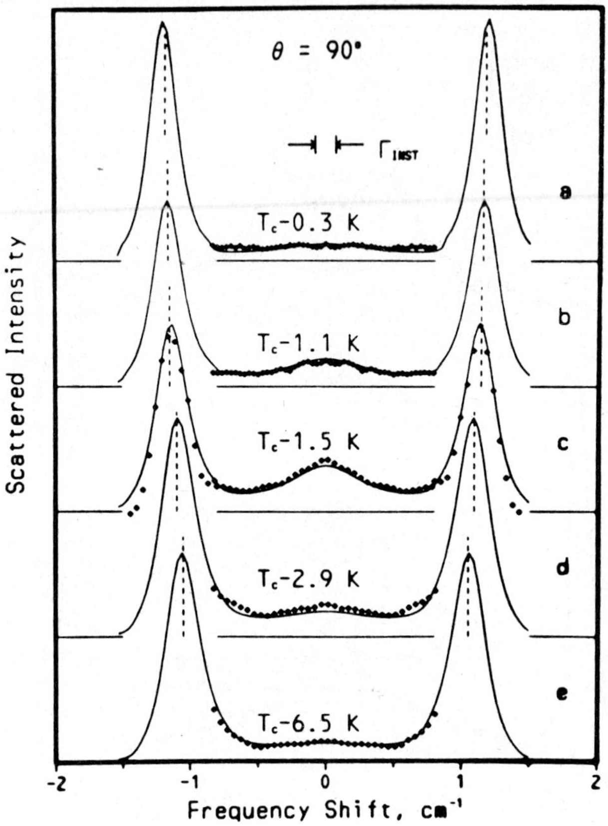

The off-diagonal light-scattering spectrum of $SrTiO_3$, in the presence of an iodine filter, at a variety of temperatures close the the phase transition (Fleury and Lyons 1978).

Fig. III.11

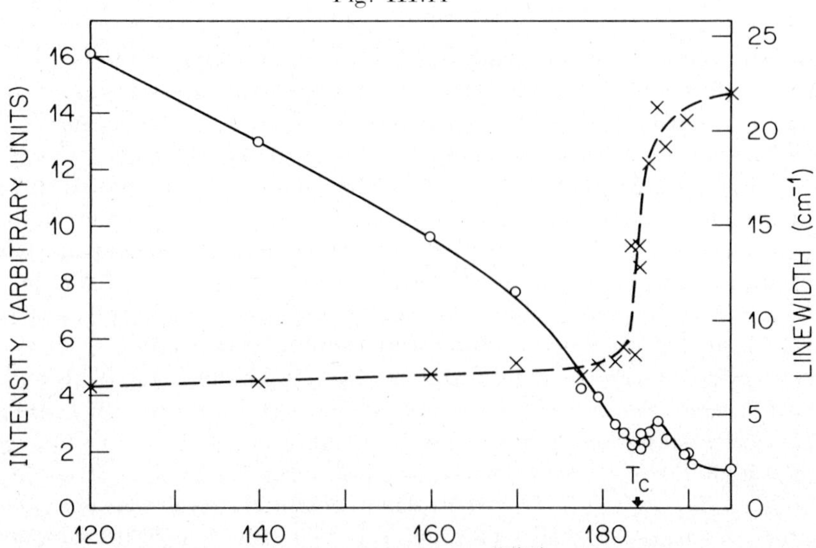

The temperature-dependence of the intensity and the *effective* line width of the *hard* 227 cm^{-1} E_g-mode feature in the light-scattering spectrum of $KMnF_3$ (Bruce *et al.* 1980).

Fig. III.12

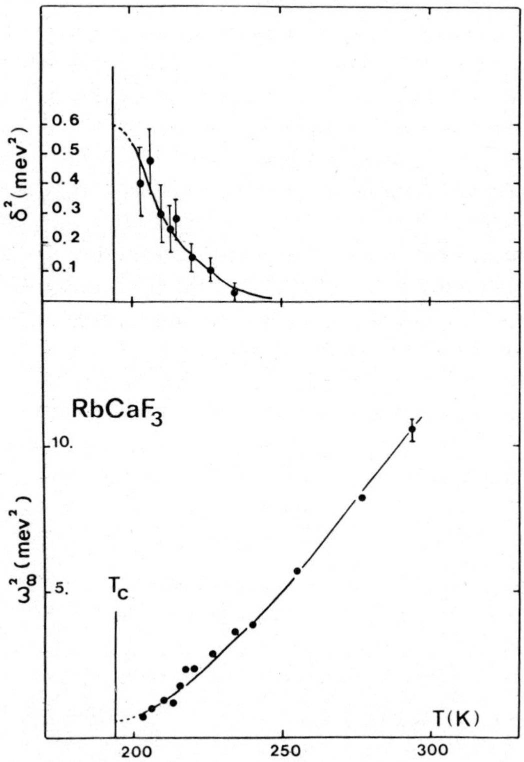

The parameters of the R_{25} soft-mode spectral function in $RbCaF_3$, as functions of temperature (Almairac *et al.* 1977).

1977; fig. III.12), while in $SrTiO_3$, as noted earlier (fig. III.4), the results of Shapiro *et al.* (1972) and Currat *et al.* (1978) are in conflict except at T_c. It is conceivable that the differences derive from the use of a (possibly) inadequate phenomenology to analyse experiments performed under different conditions. If, however, one takes the view that the results reflect real differences in the behaviour of these systems, it seems important to appreciate that these differences must reside principally in the behaviour of the central-peak intensity (effectively parameterized by δ^2), rather than in that of the soft-phonon side band: the latter, as we have seen in the case of $SrTiO_3$ (cf. fig. III.4), exhibits an apparently sample-independent (and experiment-independent!) incomplete softening, qualitatively similar to that observed in the other perovskites cited above.

III.2.2. *Lead germanate*

Lead germanate ($Pb_5Ge_3O_{11}$) has three formula units in each unit cell, and undergoes a ferroelectric transition at 451 K. Above T_c the point group is $\bar{6}$, and below T_c it is 3, the polarization lying along the unique trigonal axis. This system thus belongs to the universality class with $n=1$ and with dipolar interactions ('uniaxial ferroelectrics'), expected to exhibit classical behaviour (with logarithmic corrections: cf. §II.5.1.3)—an expectation borne out in measurements of the equilibrium properties (Iwasaki *et al.* 1972).

The spectral response of the ferroelectric fluctuations has been investigated with neutron-scattering techniques (Cowley *et al.* 1976). In the high-temperature phase, the experiments revealed a strong quasi-elastic component (fig. III.13) in addition to the scattering characteristic of an overdamped ferroelectric mode. Again, the energy width of the central component was less than the instrumental resolution (2.7×10^{10} Hz). Unlike the case of $SrTiO_3$, however, the central component was found to be observable for a *range* of wavevectors around each reciprocal lattice point, the range being restricted to the plane perpendicular to the ferroelectric axis, but extending as much as half way across the Brillouin zone. The intensity of the quasi-elastic scattering, first observable some 100 K above the phase transition, increased as the temperature was reduced, reaching a maximum at T_c. On further cooling the intensity fell off sharply, before increasing once again towards a maximum at low temperatures. This behaviour is shown in fig. III.14. Further experiments on a different specimen, subjected to an electric field to produce a nearly monodomain low-temperature phase, showed similar behaviour above and immediately below T_c but did not display the same strong low-temperature central component, suggesting that this portion of the scattering observed in the first

Fig. III.13

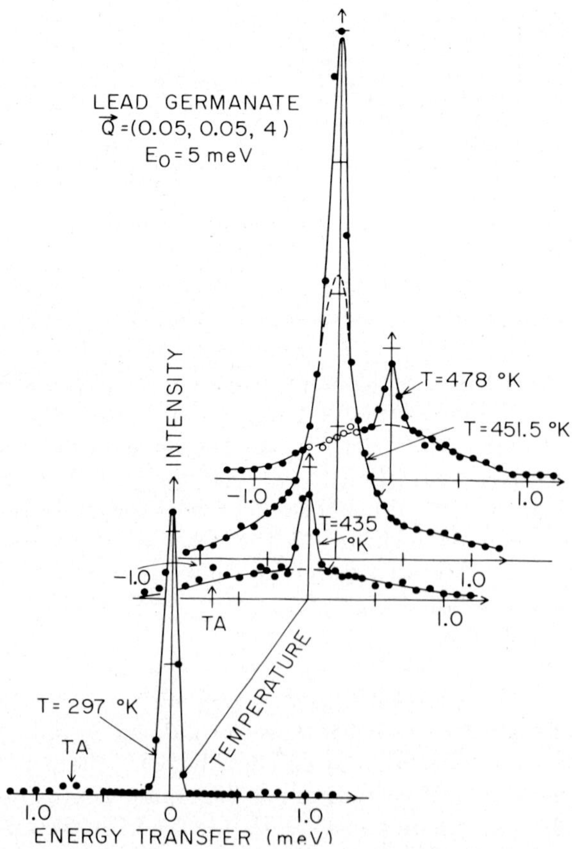

The neutron-scattering spectrum of $Pb_5Ge_3O_{11}$ at a variety of temperatures above and below T_c (Cowley *et al.* 1976).

Fig. III.14

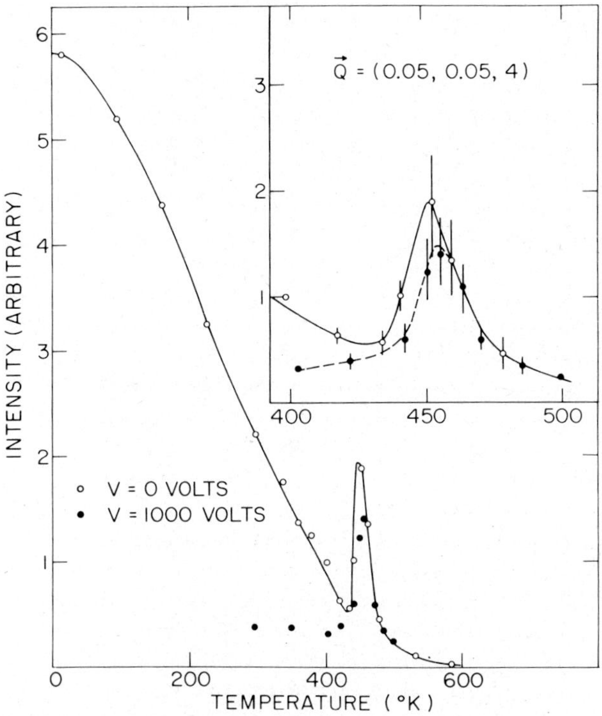

The intensity of the quasi-elastic neutron scattering in $Pb_5Ge_3O_{11}$ as a function of temperature, in zero field, and in a field sufficiently strong to produce a largely monodomain low-temperature phase (Cowley *et al.* 1976).

experiment might be attributable to a ferroelectric domain structure. Indeed, a model calculation supposing a random distribution of domain walls, with planes parallel to the ferroelectric axis, was found to give a good account of the wavevector-dependence of the low-temperature quasi-elastic scattering (fig. III.15) when the domain-wall thickness was taken to be a very reasonable 10 Å.

The wavevector-dependence of the quasi-elastic scattering observed above T_c was found to be qualitatively similar to that of the low-temperature scattering, and could also be well represented by the domain-wall model calculation (fig. III.15). The only significant differences were exhibited in scans with momentum transfer of the form $q = (\zeta, \zeta, \eta)$ with η varying and ζ fixed. At low temperatures the width in η lay within the wavevector resolution of the experiment, corresponding to scattering confined to a narrow sheet of reciprocal space perpendicular to the ferroelectric axis. The high-temperature phase scattering exhibited a discernible width in η, which increased with ζ in a qualitatively similar fashion to the behaviour predicted by the classical theory of uniaxial ferroelectrics (§ I.4.1.4.).

As in strontium titanate, the critical fluctuations in lead germanate give rise to one-phonon Raman scattering only in the low-temperature phase, which has been the subject of a number of studies (Burns and Scott 1972, Ryan and Hisano 1973, Hosea *et al.* 1979). The spectra (fig. III.16) are complicated by the large number of additional Raman active modes interacting with the soft mode. Nevertheless, they do reveal a mode whose frequency decreases with increasing temperature, becoming

Fig. III.15

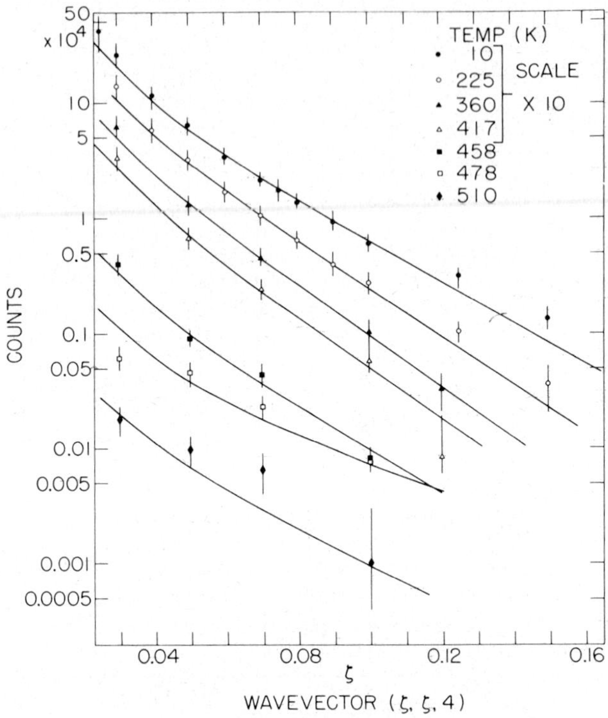

The intensity of the quasi-elastic neutron scattering in $Pb_5Ge_3O_{11}$ as a function of wavevector, for a variety of temperatures. The solid lines represent the results of calculations based on the domain-wall model described in the text (Cowley *et al.* 1976).

overdamped at about 400 K, whereafter the critical slowing down is manifested in a continued narrowing of the overdamped peak. It was found, however, that close to T_c the associated scattering could not be described very satisfactorily with the damped harmonic oscillator formalism (fig. III.16). Moreover the parameters obtained from such a fitting procedure (fig. III.17) showed clear departures from the behaviour suggested by the simple anharmonic theory of the one-phonon response.

A more precise characterization of the dynamic behaviour neat T_c has emerged from the experiments reported by Fleury and Lyons (1976), and described more extensively in a recent paper (Lyons and Fleury 1978). Employing an iodine filter (§ III.2.1) these authors were able to detect a dynamic central component (fig. III.18) in addition to that associated with the overdamped soft mode. The appearance of this central component at $T_c - T \approx 6$ K was found to coincide with the saturation of the width of the soft-mode wing at a value of approximately $1{\cdot}5 \times 10^{11}$ Hz. The width of the central component was observed to decrease with the approach to T_c, attaining a limiting value (compatible with a smearing of T_c by ≈ 1 K) of 8×10^9 Hz. Its intensity showed a very weakly singular increase with the approach to T_c, vanishing a few degrees above the phase transition.

Spectra taken without the iodine filter revealed, in addition to this dynamic central peak, a quasi-elastic component with a much narrower width (less than 2×10^6 Hz) and an intensity exhibiting a considerably stronger critical divergence, compatible with a power law with an exponent of order unity. These observations are

Fig. III.16

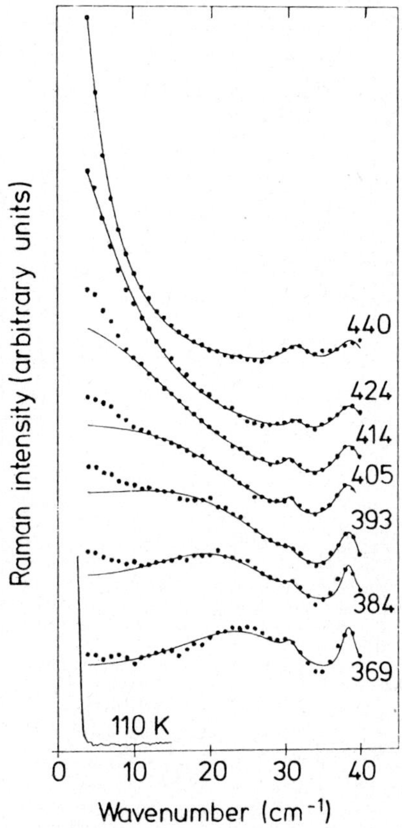

The light-scattering spectrum of the low-temperature phase of $Pb_5Ge_3O_{11}$, at a variety of temperatures. The solid lines represent optimal damped harmonic oscillator representations (Hosea *et al.* 1979).

in accord with earlier measurements (Lockwood *et al.* 1976; fig. III.19) revealing a strongly divergent central component whose width (less than 10 Hz) could not be resolved even with the aid of photon correlation techniques. The observation of a laser speckle pattern is a strong indication that this scattering should be associated with static crystal imperfections.

III.2.3. *Potassium dihydrogen phosphate*

Potassium dihydrogen phosphate (KH_2PO_4) is one of a class of uniaxial ferroelectric materials in which the ferroelectric fluctuations along the c-axis are piezoelectrically coupled to the xy shear modes, as described in §I.4.1.6. This coupling reduces the effect of fluctuations to such an extent that the universal features of the critical region should be those predicted by classical theory (§II.5.1.4).

Unfortunately, the incoherent scattering from the H atoms prevents a detailed measurement of the low-energy ferroelectric fluctuations using neutron-scattering techniques. Investigations of the deuterated material, using these techniques, have yielded interesting results (§I.4.1.4), but have failed to resolve the details of the dynamical response.

Fig. III.17

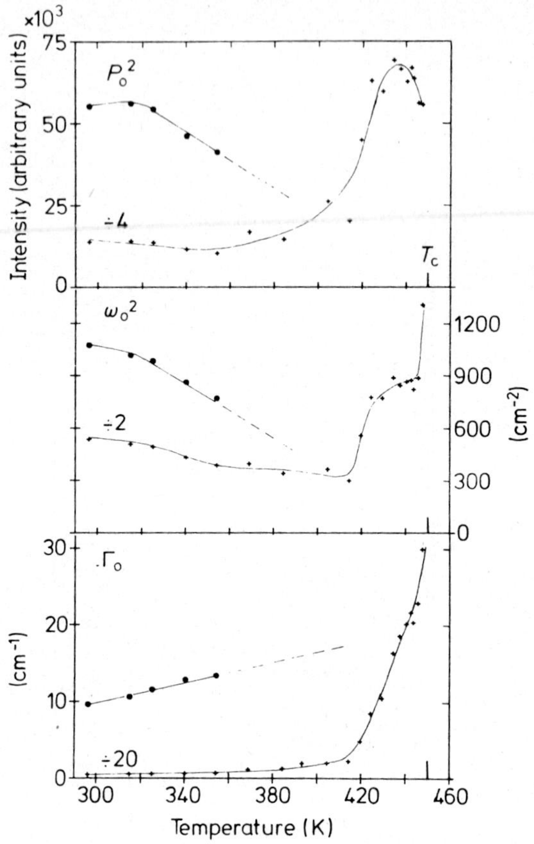

The temperature-dependence of the parameters defining the damped simple harmonic oscillator representations shown in fig. III. 16 (Hosea *et al.* 1979).

These difficulties are, to some extent, offset by the particular utility of light-scattering studies of KH_2PO_4 (and its isomorphs), where, in contrast to the cases of $SrTiO_3$ and $Pb_5Ge_3O_{11}$, the critical fluctuations give rise to one-phonon Raman scattering *above* (as well as below) the phase transition.

The first light-scattering experiments on KH_2PO_4 were performed by Kaminow and Damen (1968): their results are shown in fig. I.22. The spectra have since been remeasured by a number of authors (Wilson and Cummins 1971, She *et al.* 1972, Lagakos and Cummins 1974, Peercy 1975, Courtens 1978, Scarparo *et al.* 1978).

The analysis of the spectra is complicated by the fact that the soft ferroelectric mode is overdamped (cf. fig. I.22) and interacts both with another optic mode and with an acoustic mode. Nevertheless, the Raman scattering can be reasonably well described with the response function of two coupled classically-damped oscillators, corresponding to the two optic modes (Lagakos and Cummins 1974; fig. III.20). Moreover, the parameters derived from this analysis are quite consistent with the weakly anharmonic soft-mode theory: the damping constant of the ferroelectric mode is essentially independent of temperature, while the square of its frequency (fig. III.20) varies linearly with temperature. However, the mode frequency does not vanish at the (free) Curie temperature T_c at which (strictly speaking *close* to which:

Fig. III.18

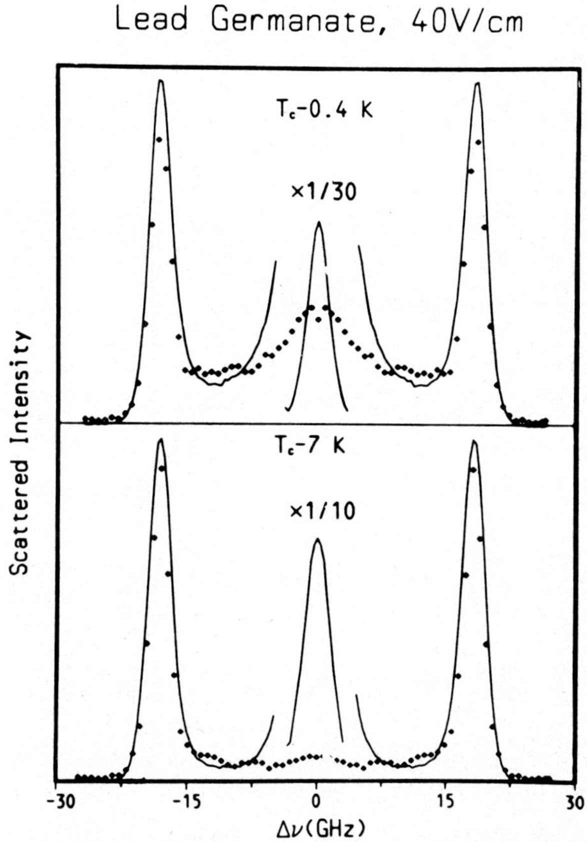

The diagonal light-scattering spectrum of $Pb_5Ge_3O_{11}$, in the presence of an iodine filter, at temperatures close to T_c (Fleury and Lyons 1978).

the transition is actually slightly first-order at atmospheric pressure) the phase transition occurs: rather, its extrapolated value vanishes at a temperature T_0 some 4 K lower than T_c. Though reminiscent of the saturation of the soft-mode frequency observed in $SrTiO_3$, this effect may, in contrast, be readily understood as a consequence of the interaction between the ferroelectric and acoustic modes (cf. § I.4.1.6): thus T_0 is, within experimental uncertainties, equal to the *clamped* Curie temperature T_c^x, while complete softening is observed only in the response of the acoustic mode. Indeed, very recently Courtens (1978) has shown that the Brillouin spectrum of an annealed sample of KH_2PO_4 can be fitted perfectly (with statistical deviations only) using the same classical coupled-mode formalism to describe the interacting ferroelectric and acoustic modes, the frequency of the latter vanishing with the approach to the free Curie temperature.

The Raman spectra of the related compounds KH_2AsO_4, CsH_2AsO_4 and RbH_2AsO_4, and their deuterated isomorphs, have also been the subject of a number of studies. Experiments by Lowndes *et al.* (1974), analysed with the coupled-mode formalism, revealed behaviour similar to that described in the case of KH_2PO_4. However, the difference between the temperatures T_c and T_0 was found to be quite large in certain cases (some 16 K in CsH_2AsO_4) though not as large as was suggested

Fig. III.19

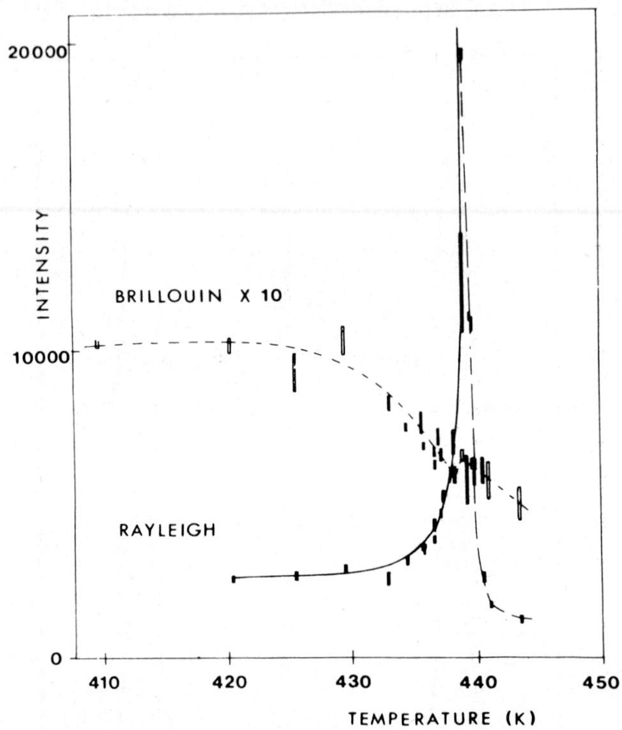

The intensities of the elastic ('Rayleigh') and acoustic-mode ('Brillouin') light scattering in $Pb_5Ge_3O_{11}$ as functions of temperature (Lockwood *et al.* 1976).

by a number of earlier investigations (see e.g. Cowley *et al.* 1971, Katiyar *et al.* 1971) in which it was incorrectly assumed that the coupling between ferroelectric and optic modes could consistently be taken to be purely real (see Lagakos and Cummins (1974) for a careful discussion of this point, which still appears to be the source of some confusion: Scarparo *et al.* 1978). Nevertheless it has since been established that these differences can be accounted for by unusually large disparities between clamped and free Curie temperatures (Lagakos and Cummins 1975, Pollina and Garland 1975).

The dynamic behaviour of the KH_2PO_4 family above T_c, as exhibited in light-scattering experiments, is then well described by a simple (coupled) soft-mode theory: there is, as yet, no evidence for an additional dynamic central peak. Below T_c, however, high-resolution Brillouin-scattering measurements on KH_2PO_4 (Mermelstein and Cummins 1977) have provided some evidence for the existence of such a component, observable only in the range $T_c - T < 0{\cdot}1$ K, and with a width (exhibiting no observable critical narrowing) of $4{\cdot}7 \times 10^7$ Hz.

This dynamic central peak coexists with the much narrower central component observed earlier (Lagakos and Cummins 1974; fig. III.21). This component is present in both phases, its intensity exhibits a marked critical anomaly, and it has been correlated with the observation of laser speckle (Durvasula and Gammon 1977), indicating that it arises from static defects. This conclusion is supported by the experiments of Courtens (1978), who found that this component could be strongly suppressed by annealing.

Fig. III.20

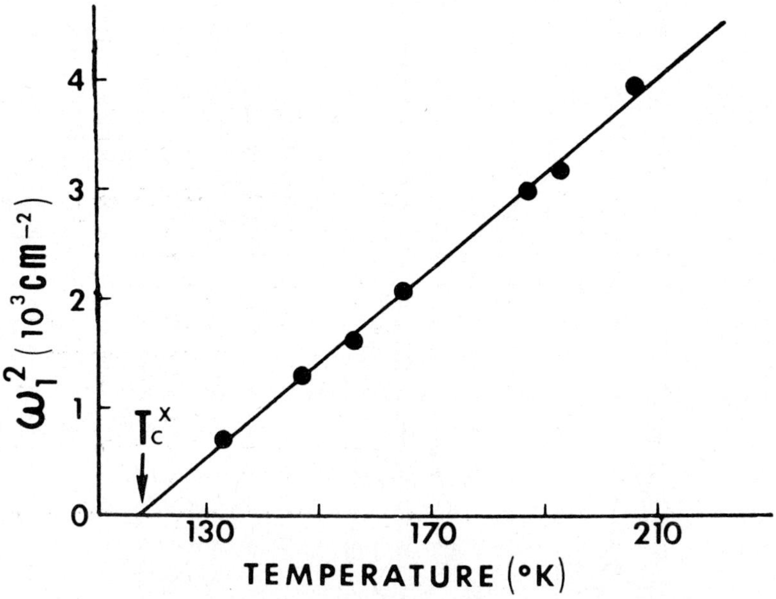

The square of the frequency of the soft ferroelectric mode in KH_2PO_4 as a function of temperature (Lagakos and Cummins 1974).

Finally, we mention that the collective low-frequency dynamical behaviour of the KH_2PO_4 family may also be conveniently probed with dielectric techniques. Measurements on the paraelectric phase of CsH_2AsO_4 and KH_2AsO_4, reported by Blinc (1978; fig. III.22), show significant dielectric loss in the vicinity of T_c at very low frequencies ($\sim 10^2$ Hz). Measurements on the ferroelectric phase of CsH_2AsO_4 and KH_2PO_4 (Tornberg and Lowndes 1977; fig. III.23) exhibit loss in a frequency range ($\sim 10^7$ Hz) which, as pointed out by Widom *et al.* (1978), is not dissimilar to that characteristic of the dynamical central peak observed in the light-scattering spectrum of KH_2PO_4.

III.2.4. *Other materials*

Quasi-elastic scattering has now been observed in experimental studies of many other structural phase transitions. In no case has the effect been studied as comprehensively as in the systems discussed above: the following survey is, correspondingly, both selective and brief.

One of the earliest experimental investigations of a central component is to be found in a neutron-scattering study of Nb_3Sn (Axe and Shirane 1974) which undergoes a near-continuous structural phase transition associated with the softening of the elastic constant $C_{11}-C_{12}$, the critical temperature being 46 K. Quasi-elastic scattering was observed for temperatures below 55 K at wavevectors for which there exist modes belonging to the soft irreducible representation. The scattering displayed a detectable q width, persisting to one-tenth of the distance across the Brillouin zone; its ω width however, was again found to be unresolvably small. Similar results have been observed near the elastic instability in $PrAlO_3$ (Birgeneau *et al.* 1974).

Fig. III.21

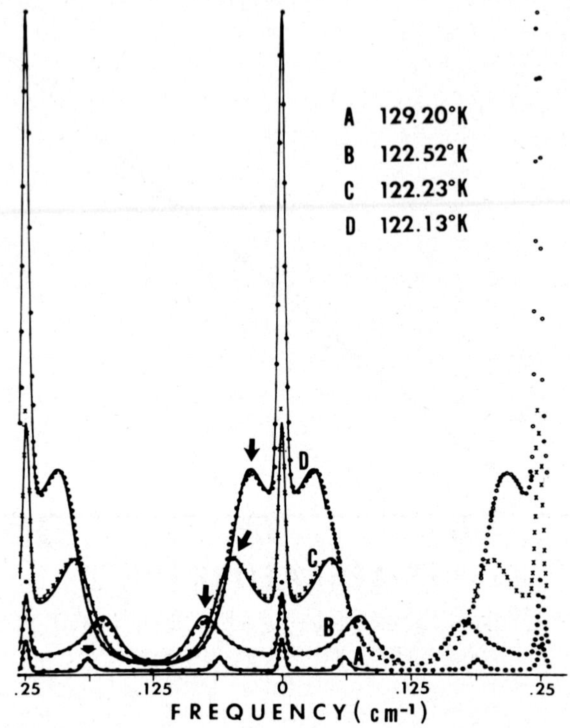

The low-frequency light-scattering spectrum of KH_2PO_4 at a variety of temperatures above T_c (Lagakos and Cummins 1974).

Certain perovskite crystals exhibit structural instabilities associated with an M-point zone boundary mode (cf. the R-point instability in $SrTiO_3$), prominent examples being the 90 K phase transition in $KMnF_3$, and the 913 K phase transition in $NaNbO_3$. A neutron-scattering study of the former (Shapiro *et al.* 1974) revealed quasi-elastic scattering with a finite wavevector spread, persisting along the direction $[\zeta, \frac{1}{2}, \frac{1}{2}]$ for values of $|\zeta| < 0{\cdot}21$. A similar study of $NaNbO_3$ (Denoyer *et al.* 1976) failed to separate a central component from an overdamped phonon background, but subsequent observations made with Mössbauer techniques (Darlington and O'Connor 1978) exhibited a well-defined quasi-elastic peak, persisting about 20 K above the phase transition—considerably less (in terms of reduced temperatures) than observed in the analogous experiments on $SrTiO_3$.

Neutron-scattering studies have shown the existence of a central component above the transition to the incommensurate phase in $BaMnF_4$ (Cox *et al.* 1979) and above the ω-phase transition exhibited by b.c.c. Zr–Nb alloys (Moss *et al.* 1973, Axe *et al.* 1975). In each case the scattering was found to occur at the wavevectors locating the Bragg superlattice points of the low-temperature phase; in neither case was the energy width of the scattering resolved.

Finally, we make mention of two light-scattering studies. Quasi-elastic scattering has been observed in the low-temperature phase of SbSI (Steigmeier *et al.* 1975) which undergoes a ferroelectric phase transition at 292·6 K, above which temperature the soft mode is Raman inactive: we are unconvinced that this scattering is dynamic in character.

Fig. III.22

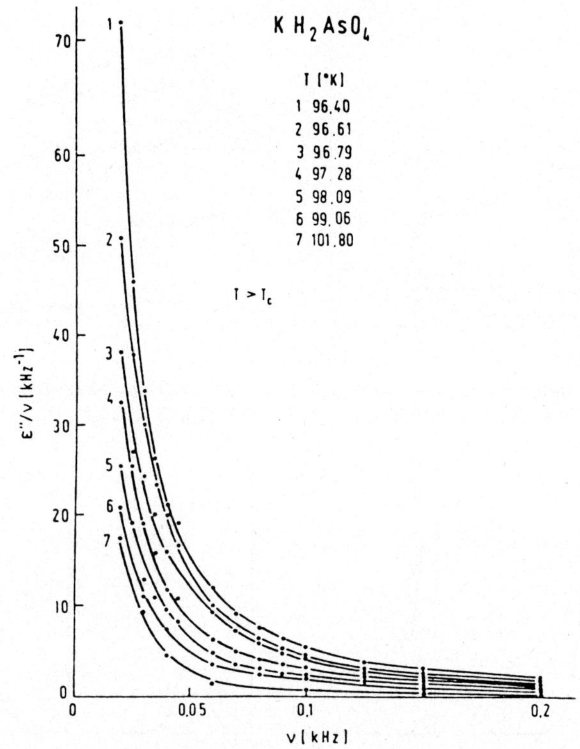

The low-frequency dielectric dispersion in KH_2AsO_4 at a variety of temperatures above T_c, as reported by Blinc (1978).

Experiments by Yagi *et al.* (1977) showed that the central component observed in the Brillouin spectrum of $KH_3(SeO_3)_2$, near its elastic instability at 201·6 K, could be significantly reduced in intensity by recycling through the phase transition.

III.2.5. *Computer experiments on a model system*

To complete our survey of experiments probing critical dynamical behaviour, we make brief mention of a sequence of molecular-dynamics studies of a simple model undergoing a structural transition (Schneider and Stoll 1973, 1975, 1976, Aubry 1975, 1976 a, b). These investigations straddle the regime between experiment and theory. We summarize at this stage the key results for the spectral function of the ordering coordinates, since these results constitute further 'experimental' data (on a defect-free system!) against which the predictions of analytic theory (§ III.3) may be assessed. We defer to § III.4 the specific *interpretation* of these results (and of the central-peak phenomenon generally) which has emerged from the attempt to correlate the features of the spectral function with configurations of the ordering coordinates (uniquely accessible to the molecular-dynamics technique).

The model system that has been studied most frequently is essentially that described in some detail in § II.2. It has a scalar order parameter, with short-range interactions, and belongs to the same universality class as the Ising model: for dimensionalities $d \leqslant 4$, its static critical behaviour is correspondingly non-classical (§ II.5.1.1).

Fig. III.23

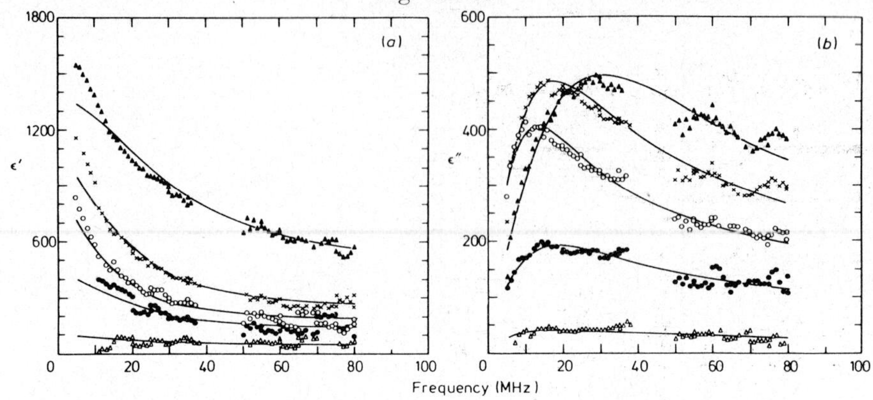

The low-frequency dielectric behaviour of KH_2PO_4 at a variety of temperatures below T_c: Δ, 91·12 K; ●, 98·47 K; ○, 103·36 K; ×, 110·46 K; ▲, 122·25 K (Tornberg and Lowndes 1977).

Fig. III.24

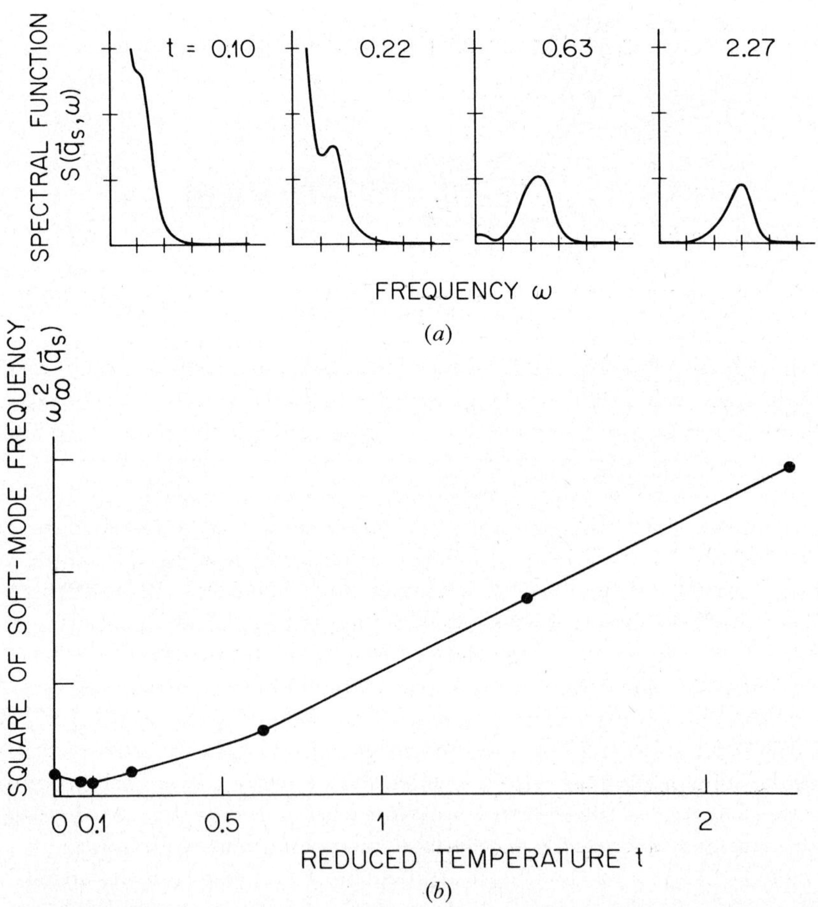

The dynamic behaviour of the $d=2$ model defined by eqn. (III. 4.1) (with $g \approx 0{\cdot}19$; eqn. (III. 4.3)) at a variety of temperatures $t \equiv (T-T_c)/T_c$. (*a*) is the spectral function for $\mathbf{q}=\mathbf{q}_s$, and (*b*) the associated soft-phonon (peak) frequencies (Schneider and Stoll 1976, 1980).

This model has been studied with different dimensionalities d and with model parameters chosen variously to give realizations of both order–disorder and displacive systems (cf. the classification schemes of § II.2.2 and § III.4.1, below). In order–disorder systems, as we have frequently emphasized, the occurrence of a two-time-scale dynamic is immediately plausible, and is inherent in the simplest (appropriate) theories: consequently, we focus attention on the properties of the model constituting the best realization of a displacive system with a true phase transition—the $d=2$ 'model III' discussed by Schneider and Stoll (1976).

At temperatures high in comparison with T_c, the spectral function of the ordering coordinates exhibits a single well-defined phonon resonance (fig. III.24 (*a*)) which softens with reducing temperature (figs. III.24 (*a*), (*b*)) in accord with simple theory. Below a certain characteristic temperature T_I (corresponding to a reduced temperature $t_I=(T_I-T_c)/T_c \approx 0{\cdot}5$ in this case), the phonon frequency begins to saturate, and there appears a central component whose width is not small on the scale of that of the soft mode. As the transition temperature is approached the central-component intensity grows, ultimately carrying all the weight of the spectral function; the soft-phonon frequency appears to remain finite at the phase transition. Independent molecular-dynamics studies (Bartolome and Kerr 1977; Kerr 1978) of a $d=2$ model, belonging to the same universality class, seem to corroborate these observations.

Finally, we remark that there have been a number of molecular-dynamics studies of the $d=1$ form of this model (Aubry 1975, 1976 a, b; Schneider and Stoll 1975) which, with some reservations (cf. § III.4.2.2), may be thought of as exhibiting a phase transition at $T_c=0$ K (§ II.2.2). The results presented in fig. III.25 (Aubry 1976 a, b) reveal behaviour that bears more than a passing resemblance to that of the $d=2$ system. Above a characteristic temperature T_I, the spectral function for momentum transfer k exhibits a well-defined softening phonon resonance; below T_I, and in a regime $k \leqslant k_I$, where k_I is a characteristic wavevector ($\approx$ one-tenth of the Brillouin-zone size in this case), the spectrum is more complex but ultimately (as $T \rightarrow 0$ K) resolves itself into a hardening phonon resonance, and an additional low-frequency component which is 'central' for small k, but which appears to be shifted to finite frequency for larger wavevector transfers (before vanishing for wavevectors $k \gtrsim k_I$); the $k=0$ central component grows in intensity and narrows as $T \rightarrow 0$ K.

III.2.6. *Summary*

We conclude the section with a rather subjective attempt to distil the key features of the results we have encountered, taking our three prototype systems in turn.

(1) Neutron-scattering studies of the phase transitions exhibited by $SrTiO_3$ and its isomorphs show a saturation of the phonon side band as $T \rightarrow T_c^+$. The strongly-critical quasi-elastic neutron scattering observed in the high-temperature phase soft-mode spectral function is narrow (less than 2×10^7 Hz within 12 K of T_c in $SrTiO_3$). There is evidence to suggest that at least some of it (though not all) resides in Bragg scattering from a surface layer, and that some originates in the influence of impurity centres. Light-scattering studies provide evidence for two diagonal spectrum dynamic central components, of widths $\sim 10^9$ and 10^{10} Hz, unrelated to the phase transition, and a weakly critical central component, of width $\sim 10^{10}$ Hz, in the depolarized spectrum.

Fig. III.25

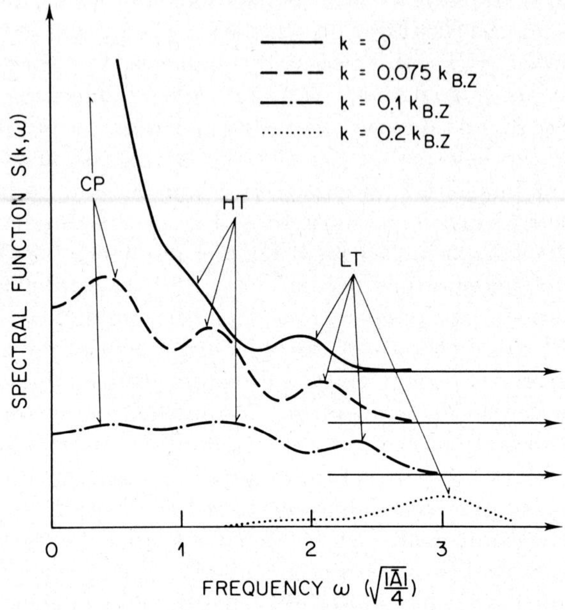

(*a*)

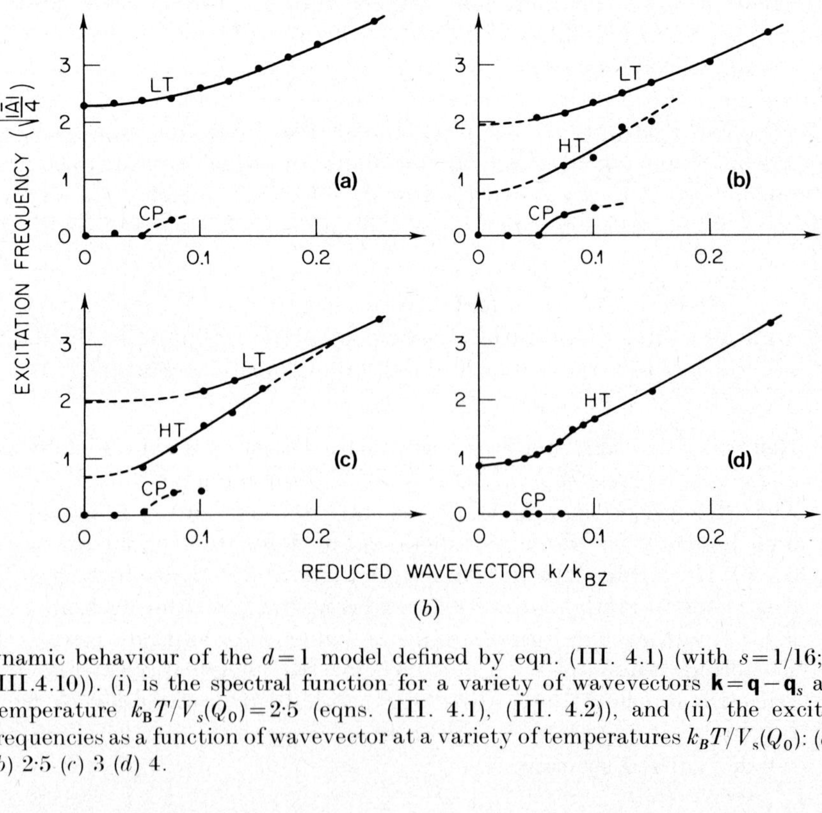

(*b*)

The dynamic behaviour of the $d=1$ model defined by eqn. (III. 4.1) (with $s=1/16$; eqn. (III.4.10)). (i) is the spectral function for a variety of wavevectors $\mathbf{k}=\mathbf{q}-\mathbf{q}_s$ at the temperature $k_B T/V_s(Q_0)=2{\cdot}5$ (eqns. (III. 4.1), (III. 4.2)), and (ii) the excitation frequencies as a function of wavevector at a variety of temperatures $k_B T/V_s(Q_0)$: (*a*) 1·5 (*b*) 2·5 (*c*) 3 (*d*) 4.

(2) Light-scattering studies of $Pb_5Ge_3O_{11}$ reveal the simultaneous existence of strongly critical scattering with a static origin (reflected also in the results of neutron-scattering studies) and a weakly critical dynamic contribution, observable close to T_c in both phases, and with a width of the order of 10^{10} Hz. This dynamic component coexists with the response of the overdamped phonon, which stabilizes as $T \to T_c^-$.

(3) Light-scattering studies of KH_2PO_4 reveal critical scattering that clearly resides in static crystal imperfections. The behaviour of the soft ferroelectric and acoustic modes in the high-temperature phase is entirely consistent with the simplest anharmonic soft-mode theory. There is some evidence for a dynamic central component immediately below T_c, with a width of the order of 10^7 Hz.

§III.3. Phonon density fluctuations

III.3.1. *The non-critical region*

III.3.1.1. *The one-phonon response*

The experimental results assembled in the preceding section reveal basic deficiencies in the soft-mode phenomenology. In attempting to understand these deficiencies, it is natural to begin with an examination of the low-order perturbation theory, upon which the phenomenology is based. Within this perturbation theory, described in §I.5.1, the character of the frequency-dependent soft-mode response function is determined by a process (depicted in fig. III.26 (a)) of virtual decay of the soft phonon into pairs of phonons with wavevector and branch labels ($\mathbf{q}_1 j_1$) and ($\mathbf{q}_2 j_2$), the process being subject to the usual conservation laws of Matsubara frequency (energy) and wavevector: the explicit form of the associated self-energy contributions—formally the result of second-order perturbation theory, or the 'golden rule'—are given in eqns. (I.5.4).

Fig. III.26

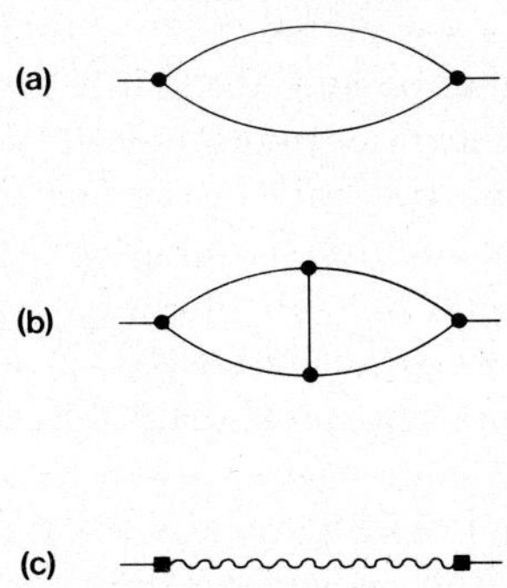

Diagrammatic representation of processes contributing to the soft-mode self-energy. (a) shows the leading-order frequency-dependent contribution, (b) one of the infinite sequence of additional processes which must be included when the soft mode can couple to pairs of modes ($\mathbf{q}_1 j_1$) and ($-\mathbf{q}_1 j_1$), and (c) is a schematic representation of the sum of the series of processes (b).

There are, in fact, two reasons why these results may fail to give an adequate account of the dynamical behaviour of a displacive system. Firstly, and most obviously, a perturbation theory of *dynamic* behaviour is necessarily as prone to failure in the critical region as the corresponding treatment of the static behaviour (to which, of course, it must reduce in the appropriate static limit): thus it is clear that the perturbation expansion for the frequency-dependent soft-mode self-energy will fail to converge in the critical region of any system where critical fluctuations are sufficiently strong (i.e. systems whose borderline dimensionality $d_c \geqslant 3$: cf. § II.4.5.1).

We shall address these difficulties, partially at least, in § III.3.2. Our immediate concern, however, lies in the fact that the perturbation theory may fail to converge even in the *absence* of critical fluctuations (indeed, even in the absence of a phase transition!). Specifically, the low-order perturbation theory fails in any system whose symmetry permits a coupling between the soft-phonon coordinate and pairs of phonons $(\mathbf{q}_1 j_1)$, $(\mathbf{q}_2 j_2)$ belonging to the same phonon branch, and with equal but opposite wavevectors $(\mathbf{q}_1 = -\mathbf{q}_2)$. In the absence of such a coupling, the response of the soft-phonon coordinate should (outside the fluctuation-dominated regime, at least) be well-approximated, over a reasonable range of probe frequency ω, by the response of the classically-damped harmonic oscillator, eqn. (III.1.1). In the presence of such processes, the inadequacy of the low-order perturbation theory is signalled by the appearance of contributions to the imaginary part of the self-energy having a singular (δ-function) dependence upon the frequency ω: the low-frequency character of the additional response reflects the fact that the complete subset of processes concerned contribute at zero frequency, while its unphysically-singular nature is a clear indication that it is necessary to take more careful account of the lifetimes of the participating modes.

It is to a review of the extensions of the anharmonic theory necessary to accommodate these effects (Cowley 1970, Wehner and Klein 1971, Coombs and Cowley 1973) that this section is devoted. We shall first discuss the circumstances in which the effects are symmetry-permitted, and then sketch the way in which they may be calculated both in displacive systems, for which the weakly anharmonic crystal model is appropriate, and, briefly, in the case of more strongly anharmonic systems. Throughout the section we shall be guided by two approximations, both of which we shall subsequently relax: firstly, as we have already indicated, we shall for the moment skirt the problems associated with critical fluctuations, assuming in particular that the fluctuating component of the soft coordinate may be taken to be negligibly small on the scale of the order parameter of the distorted phase; secondly, we shall discuss scattering properties only in the one-phonon approximation.

We begin by noting that the difficulties highlighted by our preliminary discussion may be traced to the effects, upon the soft phonon, of processes during which one phonon is created, and one phonon is destroyed, in a typical mode $(\mathbf{q}_1 j_1)$. The product of the associated phonon creation and annihilation operation is the number operator $n(\mathbf{q}_1 j_1)$: thus, the problem may be viewed as a consequence of the (linear) coupling of the soft mode to fluctuations in the numbers of other phonons, or, in other words, to fluctuations in the *phonon density*.

These fluctuations can be conveniently classified according to the irreducible representations of the point group of the crystal. By way of an example we consider the case of the point group $2/m$, which has the symmetry elements E, 2_z, m_z and I. The wavevector $\mathbf{q}_1 = (q_x, q_y, q_z)$ is related by these symmetry elements to the wavevectors $\mathbf{q}_2 = (q_x, q_y, -q_z)$, $\mathbf{q}_3 = (-q_x, -q_y, -q_z)$ and $\mathbf{q}_4 = (-q_x, -q_y, q_z)$. We will denote by n_i $(i=1,4)$ the occupation number of the mode (of some branch index j_1) of wavevector $\mathbf{q}_i$. Assuming that the interatomic forces reflect only the separations of the atoms involved, it follows that a mode can couple linearly only to combinations of the phonon numbers of the form n_1+n_3 and n_2+n_4. The representation defined by these two combinations may itself be decomposed into two irreducible representations of A_g and B_g symmetry: it follows immediately that the soft mode will be able to couple linearly to the phonon-density fluctuations only if it also has either A_g or B_g symmetry.

The above example suggests that there are phonon-density fluctuations corresponding to all those irreducible representations of the point group which are even under the inversion operator I, and none which are odd. This is, indeed, a completely general result: if the point group does not contain I then there are phonon-density fluctuations corresponding to *all* the irreducible representations; whereas if it does contain I, there are phonon-density fluctuations corresponding only to the even representations.

In the high-temperature phase there will be a linear coupling between the soft mode and the phonon-density fluctuations only if they transform according to the same representation of the space group. It follows that there is no linear coupling unless $\mathbf{q}_s=0$. Furthermore, even if $\mathbf{q}_s=0$, there will be no coupling in a centrosymmetric crystal while the soft mode is odd under inversion. This is the situation above all ferroelectric phase transitions except those in which the ferroelectric fluctuations are coupled piezoelectrically to the strains above T_c: an acoustic mode is always coupled linearly to the phonon-density fluctuations (Cowley 1967). Even when the coupling is permitted in the high-temperature phase, it can only involve the phonon-density fluctuations which do not transform according to the identity representation, since (in all but exceptional circumstances: cf. §I.4.3.2) the condensation of the soft phonon breaks the symmetry of the high-temperature phase.

In contrast, in the low-temperature phase the soft mode (or one of the soft modes, if the high-temperature-phase mode is degenerate) will *always* transform according to the identity representation, and can then couple to the phonon-density fluctuations which have this symmetry and which correspond, as we shall see below, to fluctuations in the local temperature, or 'thermodynamic' fluctuations. Evidently this linear coupling must vanish as $T\to T_c^-$ (cf. however, subsequent discussion in §III.3.2 and §III.4.3), whereas the coupling to the other ('non-thermodynamic') phonon-density fluctuations need not vanish if the associated coupling is permitted in the high-temperature phase.

We shall find further justification for certain of these assertions in the explicit forms of the self-energy contributions made by phonon-density fluctuations: to these we now turn our attention.

The central difficulties encountered in the anharmonic perturbation theory of phonon-density fluctuations originate in the fact that a strictly harmonic treatment of the phonons intermediate in the process depicted in fig. III.26 (a) gives propagators whose energy denominators vanish with the probe frequency ω. The unphysically singular behaviour that ensues is suppressed when correct account is taken of the finite lifetimes (damping) of the intermediate phonons. It then becomes apparent, however, that a consistent perturbation theory must actually include not only the process depicted in fig. III.26 (a), but also the infinite series of multiple-scattering processes, exemplified in fig. III.26 (b): although these processes include large numbers of anharmonic vertices ('small' for the weakly anharmonic crystal) they also include a corresponding number of propagators which become large ($0(1/\gamma_1)$, where γ_1 is a typical phonon damping constant) at low probe frequencies, so that each term in the series makes a similar contribution to the low-frequency part of the one-phonon self-energy. Thus, even in the absence of critical fluctuations, the anharmonic perturbation theory does not converge when symmetry permits a coupling to the phonon-density fluctuations.

The problems posed by the need to carry the perturbation theory to infinite order have been addressed by a number of authors (Klein and Wehner 1968, 1969, Niklasson 1970, Coombs and Cowley 1973). We shall not enter into the details of their analysis, but will show how (with the benefits of hindsight) the essential results may be constructed with the aid of arguments whose virtue is simplicity rather than rigour.

We consider first of all the coupling of a B_g zone centre ($\mathbf{q}_s=0$) mode with the B_g phonon-density fluctuations. A distortion characterized by the amplitude $Q(0j)$ of the B_g mode gives rise to a change in the frequency of the modes with which it interacts. Within simple perturbation theory this change in frequency is given, in terms of the anharmonic coefficients defined in eqn. (I.2.7), as

$$\delta\omega_1=\frac{3}{\omega_1}V\begin{pmatrix}0 & \mathbf{q}_1 & -\mathbf{q}_1\\ j & j_1 & j_1\end{pmatrix}Q\begin{pmatrix}0\\ j\end{pmatrix}, \tag{III.3.1}$$

where the suffix 1 stands for $(\mathbf{q}_1 j_1)$, the labels of a typical mode. If thermodynamic equilibrium at temperature T is to be maintained, the occupation number of the modes $(\mathbf{q}_1 j_1)$ must change by amounts

$$\delta n_1=\frac{\partial n_1}{\partial\omega_1}\delta\omega_1=-n_1(n_1+1)\beta\hbar\delta\omega_1. \tag{III.3.2}$$

The changes in the phonon occupation numbers promote corresponding changes in the B_g-mode self-energy, expressed in an additional contribution

$$\Sigma^{NT}(\mathbf{q}_s=0,j=B_g)=\sum_1\frac{3\hbar}{Q\begin{pmatrix}0\\ j\end{pmatrix}\omega_1}V\begin{pmatrix}0 & \mathbf{q}_1 & -\mathbf{q}_1\\ j & j_1 & j_1\end{pmatrix}\delta n_1 \tag{III.3.3 a}$$

$$=-9\hbar^2\beta\sum_1\left|V\begin{pmatrix}0 & \mathbf{q}_1 & -\mathbf{q}_1\\ j & j_1 & j_1\end{pmatrix}\right|^2\frac{n_1(n_1+1)}{\omega_1^2}. \tag{III.3.3 b}$$

This result presupposes that the phonon occupation numbers have time to relax to the local equilibrium defined by the instantaneous value of the B_g soft-mode coordinate, an assumption which is valid only if the relaxation time of the phonon-density fluctuations is short in comparison with the fluctuation time of the B_g phonon. A study of the leading term in fig. III.26 (*a*) establishes that the phonon-density fluctuations associated with the mode $(\mathbf{q}_1 j_1)$ decay with a Debye relaxation time

$$\tau_1=1/2\gamma_1, \tag{III.3.4}$$

where $2\gamma_1$ is the full width of the response of the mode concerned. Equation (III.3.3) must thus be replaced with its more general frequency-dependent counterpart

$$\Sigma^{NT}(0,B_g,\omega)=-9\hbar^2\beta\sum_1\left|V\begin{pmatrix}0 & \mathbf{q}_1 & -\mathbf{q}_1\\ j & j_1 & j_1\end{pmatrix}\right|^2\frac{n_1(n_1+1)}{\omega_1^2}\frac{1}{1-i\omega\tau_1}. \tag{III.3.5}$$

It is instructive to compare this result with the phenomenological self-energy introduced in eqn. (III.1.5): the similarities are apparent. The principal difference lies in the fact that (III.3.5) involves many relaxation processes, each with its own particular coupling constant and relaxation time. Nevertheless, one might expect that the form (III.1.5) would offer a not unreasonable representation of (III.3.5), with τ_0 and δ^2 taken as appropriate averages. Equation (III.3.5) also suggests that the parameters τ_0 and δ^2 thus obtained, would in this case also characterize, to a good approximation, the additional response at finite wavevector transfer $\mathbf{k}$, close to the soft-mode wavevector $\mathbf{q}_s = 0$.

We consider now, in a similar spirit, the contributions of phonon-density fluctuations to the self-energy of an A_g mode. A distortion $Q(0j)$ associated with such a mode not only gives rise to a change in frequency, $\delta\omega_1$, of the modes $(\mathbf{q}_1 j_1)$ with which it interacts (eqn. (III.3.1)), but will, in contrast to a distortion of B_g symmetry, also promote a change in the temperature, δT (to be specified below). The equilibrium values of the occupation numbers of the modes $(\mathbf{q}_1 j_1)$ will then change by (cf. eqn. (III.3.2)),

$$\delta n_1 = -n_1(n_1+1)\beta\hbar\left(\delta\omega_1 - \omega_1\frac{\delta T}{T}\right). \tag{III.3.6}$$

Provided the distortion takes place adiabatically, the associated change in temperature may be determined by combining (III.3.6) with the adiabatic condition $\sum_1 \omega_1 \delta n_1 = 0$, to give

$$\frac{\delta T}{T} = \frac{W}{C} Q\begin{pmatrix} 0 \\ j \end{pmatrix}, \tag{III.3.7}$$

where

$$W = 3\hbar^2\beta\sum_1 n_1(n_1+1)V\begin{pmatrix} 0 & \mathbf{q}_1 & -\mathbf{q}_1 \\ j & j_1 & j_1 \end{pmatrix} \tag{III.3.8a}$$

and

$$C = \hbar^2\beta\sum_1 n_1(n_1+1)\omega_1^2. \tag{III.3.8b}$$

The coefficient W, eqn. (III.3.8a), vanishes unless the mode $(\mathbf{q}_s = 0, j)$ transforms as the identity representation of the crystal point group, thus justifying the neglect of temperature fluctuations in our treatment of the B_g-mode self-energy; for the A_g mode under consideration, the coefficient W is non-zero. Combining (III.3.3a), (III.3.6) and (III.3.7), and taking account once more of the finite time τ_1 required for the changes (III.3.6) to be accomplished, we obtain the self-energy contribution

$$\sum\nolimits^{NT}(0, A_g, \omega) = -9\hbar^2\beta\sum_1 n_1(n_1+1)$$

$$\times\left\{\frac{\left|V\begin{pmatrix} 0 & \mathbf{q}_1 & -\mathbf{q}_1 \\ j & j_1 & j_1 \end{pmatrix}\right|^2}{\omega_1^2} - \frac{W\,V\begin{pmatrix} 0 & \mathbf{q}_1 & -\mathbf{q}_1 \\ j & j_1 & j_1 \end{pmatrix}}{3C}\right\}\frac{1}{1-i\omega\tau_1} \tag{III.3.9}$$

which has the same form as (III.3.5).

One must now allow for the fact that the fluctuation in temperature, induced by the distortion $Q(0j)$, will itself diffuse away. At this point a less cavalier treatment of the wavevector-dependence of the response is required: a distortion $Q(\mathbf{q}j)$, with $\mathbf{q}\approx\mathbf{q}_s=0$, will induce correspondingly wavevector-dependent fluctuations $\delta n_1(\mathbf{q})$ and $\delta T(\mathbf{q})$ in the phonon occupation numbers and the temperature. According to the theory of heat transport, such a temperature fluctuation will diffuse away with a relaxation time

$$\tau_T(\mathbf{q})=1/(D_T q^2), \tag{III.3.10}$$

where D_T is the thermal-diffusion constant; the occupation numbers of the modes $(\mathbf{q}_1 j_1)$ will follow this relaxation (cf. eqn. (III.3.6)), and the self-energy of the A_g mode will acquire the additional contribution

$$\begin{aligned}\Sigma^T(\mathbf{q}, A_g, \omega) &= -\frac{W^2}{C}\cdot\frac{1}{1-i\omega\tau_T(\mathbf{q})}\\ &= -\frac{W^2}{C}\cdot\frac{1}{1-i\omega/(D_T q^2)}\end{aligned} \tag{III.3.11}$$

In contrast to (III.3.5) and (III.3.9), which we now identify as self-energy contributions arising from fluctuations in the phonon numbers about the values appropriate to the local thermodynamic equilibrium ('non-thermodynamic fluctuations'), the contribution (III.3.11) reflects fluctuations in the phonon density caused by changes in the local thermodynamic equilibrium ('thermodynamic fluctuations'). While structurally similar to the phenomenological form (III.1.5), the self-energy (III.3.11) is characterized by a relaxation time which (in contrast to the supposition frequently made when employing (III.1.5)) depends strongly upon the wavevector transfer. The resulting singular structure in the phonon self-energy reflects the well-known difference between the adiabatic and isothermal response functions

$$\begin{aligned}\chi_T &\equiv \lim_{q\to 0}\lim_{\omega\to 0}\chi(qj,\omega)\\ \chi_S &\equiv \lim_{\omega\to 0}\lim_{q\to 0}\chi(qj,\omega).\end{aligned} \tag{III.3.12}$$

Specifically, (III.1.3) and (III.3.11) yield the result

$$\chi_S^{-1}-\chi_T^{-1}=\frac{W^2}{C}, \tag{III.3.13}$$

which, with the aid of (III.3.7) and (III.3.8), may be shown to be in accord with the results of thermodynamics (see e.g. Thomas 1974):

$$\chi_S^{-1}-\chi_T^{-1}=\frac{\tilde{C}_{Q_0}}{T}\left(\frac{\partial T}{\partial Q_0}\right)_S^2 \tag{III.3.14 a}$$

$$\chi_T-\chi_S=\frac{T}{\tilde{C}_E}\left(\frac{\partial Q_0}{\partial T}\right)_E^2, \tag{III.3.14 b}$$

where $\tilde{C}_Q$ and $\tilde{C}_E$ denote, respectively, the specific heats (per unit cell) at constant distortion $Q_0\equiv N^{-1/2}\,\langle Q(0A_g)\rangle$, and at constant conjugate field E. Similarly, from (III.1.9) and (III.3.11) one obtains for the ratio of the thermodynamic ('thermal-diffusion') central-peak intensity to that of the phonon side bands the generalized Landau–Placzek result

$$\frac{I_C}{2I_{\mathrm{SB}}}=\frac{\chi_T-\chi_S}{\chi_S}=\frac{T}{\chi_S\tilde{C}_E}\left(\frac{\partial Q_0}{\partial T}\right)_E^2=R^0_{\mathrm{LP}}. \tag{III.3.15}$$

This equation demonstrates explicitly that, in accord with our earlier symmetry arguments, a thermal-diffusion central peak can appear in a soft-mode spectral function only in the low-temperature phase. (Phase transitions which do not break the symmetry provide exceptions to this rule: cf. § I.4.3.2, and Courtens *et al.* 1979.) More generally one may deduce from (III.3.15) that a thermal-diffusion central peak always reflects thermal fluctuations in some coordinate with macroscopic order; as such it can appear only in conjunction with a Bragg peak (cf. Heller 1970, Schulhof *et al.* 1971).

We complete our analysis by outlining how, in microscopic terms, the symmetry-breaking occurring at T_c permits a coupling to phonon-density fluctuations in the low-temperature phase, even when it is absent above T_c. Consider, to be specific, a case (e.g. $SrTiO_3$) in which the high-temperature-phase soft mode lies at the Brillouin-zone boundary: wavevector conservation then prohibits a coupling, through third-order anharmonic coefficients, between the high-temperature-phase soft mode and pairs of phonons with equal and opposite wavevectors; the effects we have described are thus absent (at least within mean-field theory). Below T_c, however, there emerges from the anharmonic terms of fourth order an effectively third-order term with the requisite properties:

$$\begin{aligned}&\sum_{1,2,3,4} V\begin{pmatrix}\mathbf{q}_1 & \mathbf{q}_2 & \mathbf{q}_3 & \mathbf{q}_4\\ j_1 & j_2 & j_3 & j_4\end{pmatrix} Q\begin{pmatrix}\mathbf{q}_1\\ j_1\end{pmatrix} Q\begin{pmatrix}\mathbf{q}_2\\ j_2\end{pmatrix} Q\begin{pmatrix}\mathbf{q}_3\\ j_3\end{pmatrix} Q\begin{pmatrix}\mathbf{q}_4\\ j_4\end{pmatrix}\\ &\approx 4\left\langle Q\begin{pmatrix}\mathbf{q}_s\\ j\end{pmatrix}\right\rangle \sum_{1,2,3} V\begin{pmatrix}\mathbf{q}_1 & \mathbf{q}_2 & \mathbf{q}_3 & \mathbf{q}_s\\ j_1 & j_2 & j_3 & j\end{pmatrix} Q\begin{pmatrix}\mathbf{q}_1\\ j_1\end{pmatrix} Q\begin{pmatrix}\mathbf{q}_2\\ j_2\end{pmatrix} Q\begin{pmatrix}\mathbf{q}_3\\ j_3\end{pmatrix}\\ &\equiv \sum_{1,2,3} \tilde{V}\begin{pmatrix}\mathbf{q}_1 & \mathbf{q}_2 & \mathbf{q}_3\\ j_1 & j_2 & j_3\end{pmatrix} Q\begin{pmatrix}\mathbf{q}_1\\ j_1\end{pmatrix} Q\begin{pmatrix}\mathbf{q}_2\\ j_2\end{pmatrix} Q\begin{pmatrix}\mathbf{q}_3\\ j_3\end{pmatrix},\end{aligned} \tag{III.3.16 a}$$

where, in particular, the coefficient

$$\tilde{V}\begin{pmatrix}\mathbf{q}_s=0 & \mathbf{q}_1 & -\mathbf{q}_1\\ j & j_1 & j_1\end{pmatrix}=4Q_0\,N^{1/2}\,V\begin{pmatrix}\mathbf{q}_s & -\mathbf{q}_s & \mathbf{q}_1 & \mathbf{q}_1\\ j & j & j_1 & j_1\end{pmatrix} \tag{III.3.16 b}$$

couples the soft mode (at the zone centre, below the phase transition) to the phonon-density fluctuations. Evidently this coupling coefficient vanishes with the order parameter Q_0, its square (which enters the expressions of interest) varying as T_c-T. within classical theory.

While we have developed the foregoing theory in the framework of the weakly anharmonic crystal model, qualitatively similar results have been derived for the strongly anharmonic pseudo-spin model discussed in § I.5.2.

The effects of thermodynamic fluctuations were incorporated into the model by Blinc and Zeks (1974), using the formalism employed in the theory of magnetic resonance (cf. also Silvermann 1970). The time evolution of the 'spin' coordinates is then described by Bloch equations, incorporating two phenomenological relaxation times τ_1 and τ_2, which characterize the rates at which the components of the spin parallel to, and perpendicular to, the instantaneous directions of the local molecular field relax towards the quasi-equilibrium defined by that field. The inverse relaxation time $1/\tau_2$ plays essentially the same role as the normal soft-mode damping (γ_0 in eqn. (III.1.3)), while τ_1 characterizes the relaxation of energy, or temperature, fluctuations, which manifest themselves, in the low-temperature phase, in the by now familiar additional response of the ordering coordinates, at frequencies $\omega \lesssim 1/\tau_1$. In this case, however, the relaxation time τ_1 does not display the wavevector-dependence characteristic of the thermodynamic fluctuations encountered above: since the pseudo-spin coordinates are assumed to be in contact with a bath of thermal phonons, it is not appropriate to impose the energy conservation condition responsible for this form of $\tau_T(\mathbf{q})$ (eqn. (III.3.10)).

The coupling of the pseudo-spin degrees of freedom to non-thermodynamic fluctuations in the phonon density has been investigated by Young and Elliott (1974) with results of essentially the same form as those derived in the weakly anharmonic limit.

In summary, the arguments we have presented in this section show that, even in the absence of critical fluctuations, the simple soft-phonon picture will fail to describe the dynamical behaviour when the ordering degrees of freedom are coupled linearly to phonon-density fluctuations. Two cases can be distinguished.

Firstly, if (as in a piezoelectric) the coupling to the non-thermodynamic fluctuations is symmetry-allowed in the high-temperature phase (where the coupling to thermodynamic fluctuations is generally forbidden), the softening of both high- and low-temperature-phase phonons will be incomplete: the saturation frequency ($\omega_\infty^c \equiv \delta(T_c)$ in the notation of the phenomenological theory) is given by the difference between the appropriate high-frequency ($\omega \gg 1/\tau$) and low-frequency ($\omega \ll 1/\tau$) self-energies. In both phases the soft-mode spectral function will exhibit additional low-frequency scattering, whose intensity diverges critically. This situation is represented schematically in fig. III.27. (In practice, the analysis of such situations will be complicated by the interaction of the soft mode with an acoustic mode, which will itself have the same anomalous self-energy structure: see below.)

Secondly, if the coupling to non-thermodynamic fluctuations is symmetry-forbidden in the high-temperature phase (as when $\mathbf{q}_s$ lies at the zone boundary), the theory developed thus far predicts that the high-temperature-phase spectral function will display complete mode softening, with no anomalous low-frequency scattering. While, deep in the low-temperature phase, the soft-mode spectral function will again display quasi-elastic response, the theory predicts that its intensity relative to the phonon side band will not diverge as $T \to T_c^-$, the two contributions becoming asymptotically indistinguishable. This situation is depicted in fig. III.28.

A number of final observations must now be made.

Firstly, we emphasize that, in the spirit that pervades this paper, the language we have used above, and the forms of figs. III.27 and III.28, are appropriate to the weakly anharmonic crystal, where the prediction of a dynamics characterized by two distinct time scales is in sharpest contrast with the simple ('soft-mode') theory. In the case of strongly anharmonic (order–disorder) systems, the implications of the predicted 'additional low-frequency response' are less clear cut: here it is clear *a priori* (cf. § I) that the spectral response of the critical degrees of freedom will exhibit (in addition to a high-frequency resonance) a central 'tunnelling-mode' contribution (the overdamped 'soft mode' of the pseudo-spin model) which it may be difficult in

Fig. III.27

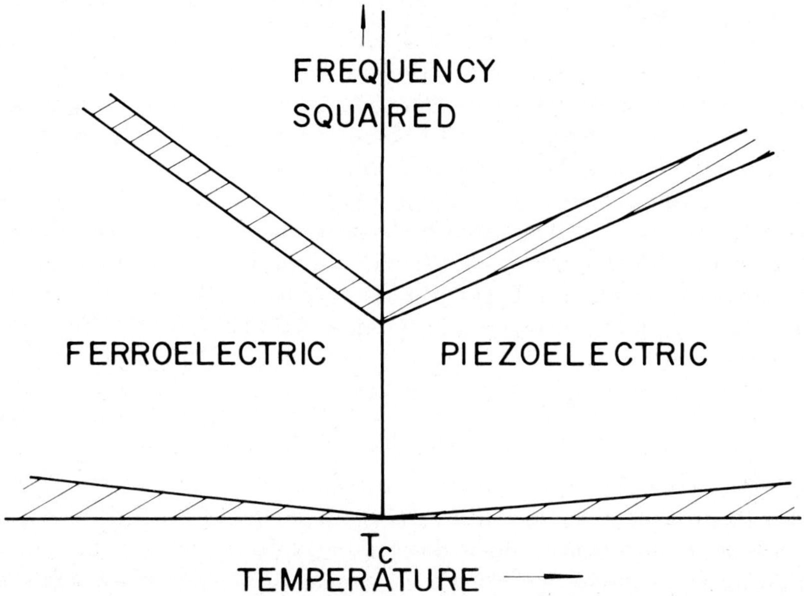

Schematic representation of the frequency-distribution of the one-phonon scattering, close to a (ferroelectric) phase transition in a system that is piezoelectric above T_c.

Fig. III.28

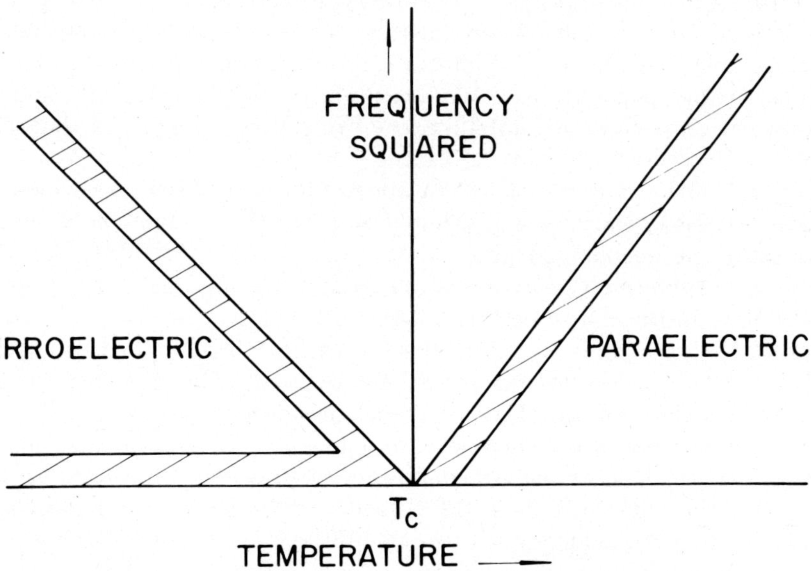

Schematic representation of the frequency-distribution of the one-phonon scattering, close to a (ferroelectric) phase transition in a system that is *not* piezoelectric above T_c.

practice (and perhaps even conceptually misleading) to distinguish from the additional scattering we have discussed.

Secondly, we remark that, while the explicit expressions for low-frequency self-energy contributions derived in this section are strictly appropriate only in the case of an optic mode, a wholly analogous development is possible in the case of an acoustic mode, as described in detail by Wehner and Klein (1972). The self-energy contribution arising from thermodynamic fluctuations (the analogue of (III.3.11)) then results in a difference between the adiabatic and isothermal elastic constants (the analogue of (III.3.13)). The associated low-frequency scattering is the Rayleigh peak familiar in the light-scattering spectrum of fluids (see e.g. Mountain 1966*a*), while the ratio of central (Rayleigh) peak to side-band (Brillouin) intensity is the Landau–Placzek result (cf. III.3.15) in its more familiar form (Wehner and Klein 1972)

$$\frac{I_{\mathrm{C}}}{2I_{\mathrm{SB}}}=\frac{C_S-C_T}{C_T}=\frac{C_S T}{\tilde{C}_\sigma}\left(\frac{\partial\eta}{\partial T}\right)_\sigma^2\equiv R^A_{\mathrm{LP}}, \tag{III.3.17}$$

where η and σ label the relevant strain (acoustic mode) and its conjugate stress. Symmetry arguments establish that (III.3.17) is zero for a strictly transverse acoustic mode: a pure shear mode does not couple linearly to thermodynamic fluctuations. Such a mode does, however, couple to non-thermodynamic fluctuations. The associated self-energy contributions (the generalization of eqn. (III.3.5)) result in a central component strongly analogous to that predicted for fluids in which there is a coupling between density fluctuations and certain internal degrees of freedom (Mountain 1966*b*).

Thirdly, we re-emphasize that the occurrence of the effects we have discussed in this section is in no respect restricted to crystals undergoing phase transitions: phonon-density fluctuations will contribute low-frequency structure to the self-energy of *any* mode of the appropriate symmetry. Nevertheless, our emphasis on the phase-transition context reflects not merely our central concern, but also the particular significance which these effects may assume in the vicinity of a phase transition where (at least in the case of the piezoelectric: cf. fig. III.27) the additional low-frequency component ultimately carries the whole weight of the spectral function.

Finally, we consider the question of the magnitude of the effects described in this section. The extent of the effects associated with the thermodynamic fluctuations may, naturally, be determined given sufficiently detailed knowledge of strictly thermodynamic quantities (cf. for example, (III.3.15) and (III.3.17)): some examples are cited in the course of §III.3.3.1.

In contrast, the calculation of effects associated with the non-thermodynamic fluctuations requires a knowledge of microscopic coupling constants (e.g. the anharmonic coefficients in eqn. (III.3.5)), which manifest themselves only indirectly in thermodynamic quantities. The results of one detailed model calculation (Bruce and Cowley 1973)—the only one of which we are aware—are shown in fig. III.29, which displays the difference (δ^2 in the notation of the phenomenological theory) between high- and low-frequency self-energies of the E and A soft modes in the low-temperature phase of $SrTiO_3$: the E-mode contribution is entirely attributable to the non-thermodynamic fluctuations, and, at 70 K, constitutes about 10% of the total self-energy (the square of the renormalized mode frequency); the A mode picks up contributions from both thermodynamic and non-thermodynamic fluctuations. All

Fig.III.29

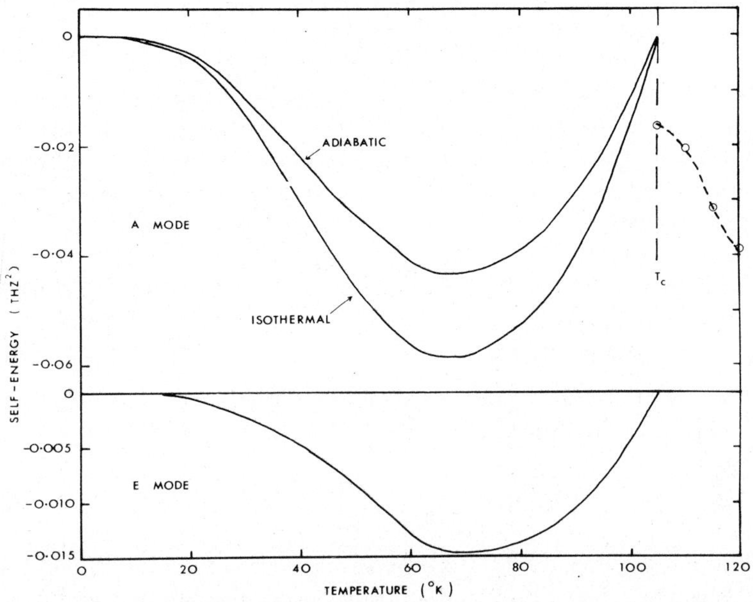

Calculations of the additional low-frequency contributions (eqns. (III.3.5), (III.3.9) and (III.3.11)) to the self-energies of the A and E low-temperature-phase soft modes in $SrTiO_3$. The calculations are based upon the mean-field approximation (III.3.16) (Bruce and Cowley 1973).

the low-frequency self-energy contributions vanish as $T \to T_c^-$, reflecting the mean-field approximation for the coupling to the phonon-density fluctuations (eqn. (III.3.16))—an assumption we shall relax in §III.3.2. We turn first, however, to assess our other major approximation (implicit, for example, in (III.3.17))—the neglect of multi-phonon contributions to the scattering cross section.

III.3.1.2. *The two-phonon response*

In the preceding section we discussed one mechanism through which phonon-density fluctuations gives rise to quasi-elastic scattering: both thermodynamic and non-thermodynamic fluctuations contribute low-frequency structure to the self-energy of any phonon mode of the appropriate symmetry; these additional self-energy contributions then manifest themselves in the scattering properties of that phonon. There is, however, a second, more direct, way in which phonon-density fluctuations may influence scattering properties: a scattering probe can, in general, couple to *pairs* of phonon modes, with the result that phonon-density fluctuations may make a *direct* contribution to the scattering.

The relative importance of direct and indirect phonon-density contributions to the scattering cross section depends upon the nature of the scattering probe. The problem has been examined in detail only in the case of light scattering (Wehner and Klein 1972) where, since symmetry often precludes first-order Raman activity, the second-order coupling frequently plays a particularly important role. It is, therefore, in this context that we discuss the effect briefly here.

The Raman-scattering cross section for wavevector transfer $\mathbf{q}$ is defined by the tensor (§ I.2.3)

$$I_{\alpha\beta\gamma\delta}(\mathbf{q},\omega)=\frac{1}{2\pi}\int_{-\infty}^{\infty}\langle\delta P_{\alpha\beta}(\mathbf{q},\tau)\delta P_{\gamma\delta}(-\mathbf{q},0)\rangle\exp(i\omega\tau)d\tau \qquad \text{(III.3.18)}$$

and may be related to the crystal dynamics through an expansion for the polarizability fluctuations in powers of the normal mode coordinates (eqn. (I.2.46)). The two-phonon contribution to the cross section derives from the second-order terms in this expansion:

$$\delta P_{\alpha\beta}(\mathbf{q})=\frac{1}{N}\sum_{\mathbf{q}_1\mathbf{q}_2 j_1 j_2} P_{\alpha\beta}\begin{pmatrix}\mathbf{q}_1 & \mathbf{q}_2\\ j_1 & j_2\end{pmatrix}\left\{Q\begin{pmatrix}\mathbf{q}_1\\ j_1\end{pmatrix}Q\begin{pmatrix}\mathbf{q}_2\\ j_2\end{pmatrix}-\left\langle Q\begin{pmatrix}\mathbf{q}_1\\ j_1\end{pmatrix}Q(\mathbf{q}_2 j_2)\right\rangle\right\}$$
$$\times\Delta(\mathbf{q}+\mathbf{q}_1+\mathbf{q}_2). \qquad \text{(III.3.19)}$$

Retaining those contributions that are diagonal in the branch indices j_1 and j_2, one obtains for (III.3.19) the approximation

$$\delta P^{(2)}_{\alpha\beta}(\mathbf{q})\approx\sum_{1}\frac{\hbar}{\omega_1}P_{\alpha\beta}\begin{pmatrix}\mathbf{q}_1 & -\mathbf{q}_1\\ j_1 & j_1\end{pmatrix}\delta n_1(\mathbf{q}), \qquad \text{(III.3.20)}$$

revealing that phonon-density fluctuations result in direct modulation of the crystal polarizability and hence in direct contributions to the Raman tensor (III.3.18). The contributions made by thermodynamic fluctuations in the phonon density have been analysed by Wehner and Klein (1972); those made by non-thermodynamic fluctuations have been considered by Coombs and Cowley (1973). In each case the fluctuations are found to give rise to quasi-elastic scattering distributed over the same frequency range as the corresponding indirect contributions which they make through the one-phonon self-energy. Again the magnitude of the non-thermodynamic contributions can be estimated only with the aid of a microscopic model. while the magnitude of the thermodynamic contributions may be related to purely macroscopic parameters. Specifically it may be shown, either by microscopic arguments, making approximations consistent with those underlying our results for the one-phonon self-energy (Wehner and Klein 1972), or by thermodynamic fluctuation theory (Landau and Lifshitz 1958), that the integrated intensity associated with the direct contributions of the thermodynamic fluctuations satisfies

$$\int_{-\infty}^{\infty}I^{(2),T}_{\alpha\beta\gamma\delta}(\mathbf{q}=0,\omega)d\omega=\left(\frac{\partial P_{\alpha\beta}}{\partial S}\right)\left(\frac{\partial P_{\gamma\delta}}{\partial S}\right)\langle\delta S^2\rangle=\left(\frac{\partial P_{\alpha\beta}}{\partial S}\right)\left(\frac{\partial P_{\gamma\delta}}{\partial S}\right)k_{\mathrm{B}}\tilde{C}. \qquad \text{(III.3.21)}$$

(Strictly, the partial derivatives and the specific heat $\tilde{C}$ appearing in this expression should bear subscripts indicating that they are to be taken with appropriate *fields* held constant.) This result underlines the fact that, as emphasized by Wehner and Klein (1972), the thermodynamic fluctuations are, in many respects, better characterized with entropy, rather than temperature chosen as the independent variable.

In assessing the importance of the direct ('two-phonon') scattering contributions of the thermodynamic fluctuations, one must also take careful account of their interference with indirect ('one-phonon') contributions. Within the framework of the thermodynamic theory (i.e. to the extent that one can ignore non-thermodynamic fluctuations), the net influence of the direct coupling upon the light-scattering spectrum may then be characterized by a renormalized Landau–Placzek ratio.

When the one-phonon scattering is dominated by an acoustic mode (cf. (III.3.17)), one finds for the diagonal ($\alpha\alpha$) spectrum (Wehner and Klein 1972):

$$\frac{I_C}{2I_{SB}} \approx R_{LP}^{A}\left[1+\frac{(\partial P_{\alpha\alpha}/\partial T)_\eta}{(\partial\eta/\partial T)_\sigma(\partial P_{\alpha\alpha}/\partial\eta)_T}\right]^2, \qquad \text{(III.3.22)}$$

while, when the one-phonon cross section is dominated by an optic mode, one finds

$$\frac{I_C}{2I_{SB}} \approx R_{LP}^{0}\left[1+\frac{(\partial P_{\alpha\alpha}/\partial T)_{Q_0}}{(\partial Q_0/\partial T)_E(\partial P_{\alpha\alpha}/\partial Q_0)_T}\right]^2 . \qquad \text{(III.3.23)}$$

Evidently the extent of the direct thermodynamic contribution to the quasi-elastic scattering is controlled by the derivative $(\partial P_{\alpha\alpha}/\partial T)$ at constant ordering coordinate. This term is generally negligible in liquids, and the Landau–Placzek approximation to (III.3.22) is then not unreasonable. In crystals, however, this term is frequently far from negligible and, indeed, can lead to a considerable enhancement (e.g. by a predicted factor of 16 in the Brillouin spectrum of $SrTiO_3$ at 300 K) of the quasi-elastic scattering (Wehner and Klein 1972).

The analysis of this section (like that of §III.3.1.1) has made no essential reference to the occurrence of a phase transition. In fact, in the immediate vicinity of a critical point, the results of both sections have to be viewed with caution. Those results, characterizing thermodynamic phonon-density fluctuations, that are expressions of purely thermodynamic relationships (e.g. (III.3.14) and (III.3.21)) are not invalidated by critical fluctuations. However, inferences drawn from them (e.g. (III.3.15) and (III.3.23)) regarding the intensity of quasi-elastic scattering are (or can be) undermined by the inadequacy (within the critical region) of our perturbation results for non-thermodynamic fluctuations (e.g. (III.3.5))—results which do not have an independent basis in thermodynamics. It is to these problems that we now turn our attention.

III.3.2. *The critical region*

III.3.2.1. *The one-phonon response*

The theory we have outlined in the preceding sections accounts for only a limited number of the experimental results surveyed in §III.2. In particular—we defer a fuller comparison of theory and experiment to §III.3.3—it fails to account for the existence of a quasi-elastic component in the $T > T_c$ soft-mode spectral function in systems, typified by $SrTiO_3$, where there is no linear coupling between the soft coordinate and phonon-density fluctuations. It is tempting to attribute this deficiency to the theory's failure to allow for the effects of critical fluctuations. This possibility has been the subject of a number of investigations which have tended to fall into one or other of two categories: we subdivide this section accordingly.

We discuss, firstly, a group of theories whose common motivation is the same physically plausible picture of the key consequences of critical fluctuations, which they accommodate (with varying degrees of sophistication) in an essentially *ad hoc* fashion. We then turn to summarize the relevant results of analyses employing dynamic renormalization-group methods, which accommodate fluctuation-effects in a more systematic way.

The essential idea linking together the first group of theories is one which will recur at various stages of this article—particularly in §III.4—where it assumes a central role. In the present context it consists of the realization that, just as a homogeneous distortion induces a non-zero coupling between soft-mode and phonon-density fluctuations in the ordered phase (eqn. (III.3.16)) so, in the high-temperature phase, the superfluity of soft phonons near T_c (the critical fluctuations!) will produce microdomains of distorted crystal within which (for the life time of the microdomains) the coupling is again allowed.

This idea has been expressed in a number of different guises. It was first proposed by Feder (1971) who argued that, in the fluctuation-dominated regime, the thermodynamic result (III.3.14 a) should be replaced by

$$\chi_S^{-1} - \chi_T^{-1} = \left\langle \frac{\tilde{C}_Q(\lambda)}{T(\lambda)} \left(\frac{\partial T(\lambda)}{\partial Q(\lambda)} \right)_S^2 \right\rangle, \tag{III.3.24}$$

where the function of strictly macroscopic coordinates appearing on the right-hand side of (III.3.14 a) has been replaced by an ensemble average over corresponding quantities characterizing the behaviour within a 'cell' of linear dimension λ. In contrast to the thermodymanic result (III.3.14 a), the difference (III.3.24) does not, in general, vanish about T_c: provided the cell size λ is less than the correlation length ξ,

$$\xi/\lambda \gtrsim 1 \tag{III.3.25}$$

the ensemble average will include significant contributions from configurations in which the instantaneous cell coordinate $Q(\lambda)$ [and, therefore, $(\partial T(\lambda)/\partial Q(\lambda))_S$] is non-vanishing. Accordingly a 'thermodynamic' central peak is predicted for the high-temperature phase also.

This argument is potentially misleading: there is no reason to suppose that strictly thermodynamic relationships such as (III.3.14) lose their validity in the critical region (cf. eqn. (III.3.33) below, and accompanying discussion), although, with the approach to criticality, it becomes increasingly difficult to measure 'strictly thermodynamic' quantities with microscopic probes. Thus, the difference between the 'true' adiabatic and isothermal susceptibilities *does* vanish identically above T_c—in apparent conflict with Feder's result (III.3.24), and with the result of an approximate treatment of a microscopic realization of Feder's scheme, described by Enz (1974).

This conflict may, however, be partially dispelled if one argues that the cell size λ (taken as the zero-temperature correlation length in Feder's treatment) should actually be taken as the effective wavelength of the scattering probe employed to determine the response, so that

$$1/\lambda \approx k, \tag{III.3.26}$$

where $\mathbf{k}=\mathbf{q}-\mathbf{q}_s$ and $\mathbf{q}$ is the corresponding wavevector transfer. Recalling the condition (III.3.25), it is then apparent that the difference (III.3.24) *does* vanish in the strict $\mathbf{k}\to 0$ (specifically, $k\xi \ll 1$) limit implicit (cf. eqn. (III.3.12)) in the microscopic identification of the true thermodynamic response functions; the right-hand side of (III.3.24) is non-vanishing only in the regime $k\xi \gtrsim 1$.

In effect, then, while leaving the hydrodynamic behaviour obscure, (III.3.24) predicts a fluctuation-induced 'thermodynamic' central peak in the one-phonon response above T_c, in the regime $k\xi \gtrsim 1$. This prediction is quite consistent with the results of dynamic renormalization-group arguments: these, we shall see, establish that fluctuation-induced coupling to the thermodynamic mode can, in specified

Fig. III.30

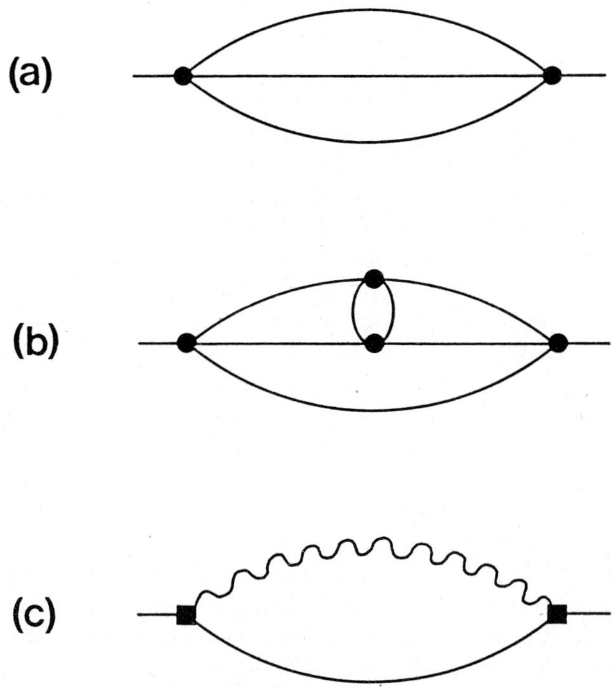

Diagrammatic representation of important processes contributing to the soft-mode self-energy in the critical region above T_c. (*a*) shows the leading frequency-dependent contribution carrying critical singularities. This process is the dynamic precursor of that depicted in fig. III.26 (*a*). (*b*) The dynamic precursor of the process shown in fig. III.26 (*b*). (*c*) The dynamic precursor of the sum of processes represented by fig. III.26 (*c*).

circumstances, give rise to a central component which, in the regime $k\xi \gtrsim 1$, has the characteristics of a thermodynamic central mode. However, in the light of the foregoing discussion, the difference between 'high-frequency' and 'low-frequency' response, underlying this central peak, should not be identified with a difference between 'adiabatic' and 'isothermal' response. Correspondingly, the arguments underlying (III.3.24) reside more comfortably, and can be explored more fully, within the strictly microscopic framework offered by phonon perturbation theory.

Developments along these lines were initiated by Silberglitt (1972), who proposed that, in the systems concerned, it is essential to include in the one-phonon self-energy the type of process depicted in fig. III.30 (*a*). The principal contributions associated with this diagram arise through those processes in which one of the intermediate phonon lines is associated with a mode belonging to a soft branch, and with $\mathbf{q} \approx \mathbf{q}_s$; the remaining lines then carry (nearly) equal and opposite wavevectors. It is readily seen that this process is the natural *dynamic* precursor of that depicted in fig. III.26 (*a*), where the third-order vertex functions are forged (below T_c) from their fourth-order counterparts through the non-zero *equilibrium* value of the soft coordinate (eqn. (III.3.16)): again the fluctuations in the ordering coordinate (the intermediate soft-phonon line in fig. III.30 (*a*)) mediate a coupling between the soft mode and phonon-density fluctuations.

This approximation captures the effects only of the (non-linear) coupling to the *non*-thermodynamic fluctuations, and the thermodynamic fluctuations in lowest order. The effects of the thermodynamic fluctuations can be more adequately incorporated by including self-energy contributions (fig. III.30 (*b*)) that are the precursors of those depicted in fig. III.26 (*b*) (in the same sense as the process shown in fig. III.30 (*a*) is the precursor of that in fig. III.26 (*a*)). Calculations along these lines have been reported by Murata (1975) and by Ohnari and Takada (1979). Both calculations showed a spectral function with phonon and central-peak features, above T_c. In the work of Ohnari and Takada the thermodynamic effects were introduced in a semi-phenomenological way, making it difficult to assess the significance of their 'very sharp' central peak. They do not discuss the relative strengths of the two contributions to the spectrum. In Murata's study of a simple microscopic model, treating the soft-phonon propagator self-consistently, the central peak was found to be much broader than the phonon contribution, on its first appearance. The widths were found to become comparable only sufficiently close to criticality that the central-peak feature carried the bulk of the spectral weight.

The intuitive appeal of the analyses we have discussed thus far is offset by their lack of rigour: their treatment of fluctuations is plausible but in no way systematic. In recent years a systematic treatment of time-dependent critical behaviour has become possible with the development of dynamic renormalization-group (R.G.) techniques (Halperin *et al.* 1972, 1974, 1976)—natural extensions of the static R.G. methods pioneered by Wilson (1971) and extensively surveyed in II.

Like its static counterpart the dynamic R.G. method constitutes, in essence, an explicit basis for scaling theories previously based upon phenomenological arguments (Ferrell *et al.* 1967, Halperin and Hohenberg 1967; see Stanley 1971 for a general introduction). The objectives of the methods (and the spirit in which we shall survey its predictions) can best be indicated by considering explicitly the central scaling ansatz for the dynamic susceptibility

$$\chi(t, \mathbf{k}, \omega) \approx b^{2\lambda_2 - d} \chi(b^{\lambda_1} t, bk, b^z \omega) \equiv b^{\gamma/\nu} \chi(b^{1/\nu} t, bk, b^z \omega). \qquad \text{(III.3.27)}$$

This equation represents the natural extension, to finite wavevector $\mathbf{k}(\equiv \mathbf{q} - \mathbf{q}_s)$ and frequency ω, of the scaling ansatz for the static susceptibility given in (II.4.38 *b*). The exponents γ and ν characterize the divergences (as function of reduced temperature $t \equiv (T - T_c)/T_c$) of, respectively, the static susceptibility and the correlation length (cf. (II.2.13 *a*) and (II.2.12 *a*)), while z is the principal dynamic critical exponent. Choosing for the arbitrary scale factor b the value $b = |t|^{-\nu}$, (III.3.27) may be rewritten in the form

$$\chi(t, \mathbf{k}, \omega) \approx |t|^{-\gamma} \chi(\pm 1, k/|t|^{\nu}, \omega/|t|^{z\nu}) \equiv |t|^{-\gamma} \tilde{\chi}_{\pm}(k/|t|^{\nu}, \omega/|t|^{z\nu}). \qquad \text{(III.3.28 } a)$$

It follows immediately that the spectral function of the ordering coordinates, related to the dynamic susceptibility (III.3.28 *a*) though (III.1.6), itself has the scaling form

$$\mathcal{S}(t, \mathbf{k}, \omega) = |t|^{-\gamma - z\nu} \mathcal{S}_{\pm}(k/|t|^{\nu}, \omega/|t|^{z\nu}). \qquad \text{(III.3.28 } b)$$

It proves useful to introduce a characteristic frequency ω_Q through the relation

$$\int_{-\omega_Q}^{\omega_Q} \mathcal{S}(t, \mathbf{k}, \omega) d\omega = \frac{1}{2} \int_{-\infty}^{\infty} \mathcal{S}(t, \mathbf{k}, \omega) d\omega. \qquad \text{(III.3.29)}$$

Evidently, ω_Q characterizes the spectral range over which the scattering associated with (III.3.28*b*) is distributed. It is easily seen that, in a regime in which the scattering is dominated by a well-defined soft-phonon resonance, ω_Q coincides with the soft-mode frequency; in a regime in which the scattering is dominated by a central resonance ω_Q characterizes the width of the component. However, if the scattering resides in more than one distinct frequency regime, ω_Q does not have such an immediate physical interpretation. In any event, (III.3.28*b*) and (III.3.29) now imply that the characteristic frequency ω_Q has the scaling form

$$\omega_Q(t, k) \approx |t|^{z\nu} \tilde{\omega}_{Q,\pm}(k/|t|^{\nu}). \tag{III.3.30}$$

Equations (III.3.28) and (III.3.30) define the principal goals of the dynamic R.G. method: the calculation of the dynamic critical exponent z and the explicit determination of scaling functions for the characteristic frequency $\tilde{\omega}_Q$ and the spectral density $\mathscr{S}$, or (equivalently) the 'shape function' f defined by

$$\mathscr{S}(t, \mathbf{k}, \omega) \equiv \frac{\chi(t, \mathbf{k}, \omega = 0)}{\omega_Q(t, k)} f_x(y) \tag{III.3.31}$$

with $x \equiv k\xi$ and $y \equiv \omega/\omega_Q(t, k)$.

The method itself, which we shall not discuss in detail, is an immediate generalization of its static counterpart. The basic ingredient is an equation of motion (taking the place of the 'effective hamiltonian', see §II.4.2), characterizing the time evolution of the critical fluctuations (with wavevector $k < \Lambda$), and embodying the system's 'essential qualitative features' (cf. below): the prototype equation describes a set of coupled harmonic oscillators (the phonons of the high-temperature phase) which are damped, and subject to a random noise, to simulate the effects of the non-critical ($k > \Lambda$) fluctuations. The renormalization of the parameters governing the time evolution of the 'more critical' ($k < \Lambda/b$) fluctuations by the 'less critical' ($k > \Lambda/b$) fluctuations is calculated within a perturbation theory in the phonon coupling; the results are embodied in a set of R.G. equations which are truncated by invoking the 'smallness' of some parameter (cf. §II.4.5.1)—principally $\varepsilon = 4 - d$. The dynamic exponent and scaling functions then follow from a study of the truncated equations, near their appropriate fixed point (static critical quantities being recovered in the process). The results of extensive calculations of this type, and an account of the techniques themselves, are to be found in the recent review by Hohenberg and Halperin (1977).

Regrettably, the gulf between theory and experiment remains wide—much more so than in the field of static critical phenomena: this is particularly true in the context of structural phase transitions, where the dearth of appropriate experimental results is acute. In the discussion which follows we shall, therefore, endeavour to emphasize those aspects of dynamic R.G. calculations which illuminate the *qualitative* character of the dynamic response near criticality.

Like static critical behaviour, dynamic behaviour near a critical point is characterized by 'universal' quantities (the dynamic exponent z being the prime example). However, dynamic R.G. arguments establish that the 'essential qualitative features' (spatial dimensionality d, order-parameter dimensionality n . . ., cf. §II.4.1), which define a system's universality class as regards its static critical behaviour, do not uniquely determine the dynamic universality class to which the system belongs: dynamic critical behaviour reflects, in addition, the fundamental *conservation laws* characterizing the motion. The importance of quantities conserved by the motion (the total energy is the principal example) derives from the slow long-wavelength modes (in particular heat diffusion) whose existence they imply: such a mode, if coupled to the ordering coordinates, can influence the manner in which critical slowing down occurs.

Exploiting universality in this wider sense, dynamic R.G. methods can be used to investigate the simplest model embodying the 'symmetry properties' (the E.Q.F.s discussed in II) *and* the conservation laws appropriate to the system of physical interest. In the case of the displacive phase transitions of primary interest here, the only potentially important conserved quantity is the energy. (The imposition of the energy conservation condition in the case of a strongly anharmonic pseudo-spin system may not be appropriate: cf. remarks in §III.3.1.1, and the discussion by Halperin *et al.* (1974) of the relative merits of their models A to C.) The influence of this conservation law may be established by R.G. studies of the equations of motion of the (n-component) ordering coordinate $Q_\alpha(\mathbf{x})$, and the energy density $\varepsilon(\mathbf{x})$, introduced as an auxiliary field (Halperin *et al.* 1974)—a procedure justified in considerable detail by Bausch and Halperin (1978). The two equations of motion are coupled through the leading (non-linear) interaction between ordering coordinate and energy density allowed in the high-temperature phase of a system displaying a phase transition at which the symmetry is broken:

$$\mathscr{H}_{Q.\varepsilon}=g_T\int_{\mathbf{x}}\sum_{\alpha=1}^{n}Q_\alpha^2(\mathbf{x})\varepsilon(\mathbf{x}). \tag{III.3.32}$$

It is helpful to note that the leading contribution to the soft-phonon self-energy, resulting from a perturbation treatment of this interaction (fig. III.30 (c)) is essentially a compact representation of the series analysed by Murata (1975) (fig. III.30 (b)), itself a refinement of the basic process singled out by Silberglitt (1972) (fig. III.30 (a)).

In contrast to such direct perturbation expansions, however, the R.G. method permits one to assess the importance of the energy-density/ordering coordinate coupling in a *systematic* way. Specifically, R.G. arguments establish (Halperin *et al.* 1974. Simanek and Ward 1976) that the fixed point value g_T^* of the interaction (III.3.32) is non-zero (and the coupling to thermodynamic fluctuations consequently of importance near the critical point) only if the specific heat of the system concerned diverges as $T\to T_c$.

This result can be made plausible in a very simple fashion. We recall the thermodynamic result (III.3.14b) in the form

$$\frac{\chi_S}{\chi_T}=1-\frac{T}{\tilde{C}_E\chi_T}\left(\frac{\partial Q_0}{\partial T}\right)_E^2\equiv 1-r_{\rm LP}, \tag{III.3.33}$$

where χ_S and χ_T are the 'true' adiabatic and isothermal susceptibilities (cf. (III.3.24) *et seq.*), so that $r_{\rm LP}\equiv 0$ for $T\to T_c$. Consider now, however, the limiting behaviour of the quantity $r_{\rm LP}$ (and therefore the ratio χ_S/χ_T) as $T\to T_c^-$. Assigning the quantities in (III.3.33) their usual limiting forms (§II.2.2):

$$\left.\begin{aligned}&\chi_T\sim|t|^{-\gamma}\\&\left(\frac{\partial Q_0}{\partial T}\right)_E^2\sim|t|^{2(\beta-1)}\\&\tilde{C}_E\sim\begin{cases}|t|^{-\alpha}, & \alpha>0\\|t|^{-\alpha}+\text{constant}, & \alpha<0\end{cases}\end{aligned}\right\} \tag{III.3.34}$$

we find

$$r_{\mathrm{LP}} = r^*_{\mathrm{LP}}|t|^{2\beta+\gamma+\tilde{\alpha}-2}, \tag{III.3.35a}$$

where

$$\tilde{\alpha} = \begin{Bmatrix} \alpha, \alpha > 0 \\ 0, \alpha < 0 \end{Bmatrix}. \tag{III.3.35b}$$

Assuming the validity of the scaling result (cf. table II.1)

$$2\beta + \gamma + \alpha - 2 = 0, \tag{III.3.36}$$

eqn. (III.3.34*a*) yields, in the $T \to T_c^-$ limit,

$$r_{\mathrm{LP}} = r^*_{\mathrm{LP}}|t|^{\tilde{\alpha}-\alpha} = \begin{Bmatrix} r^*_{\mathrm{LP}}, \alpha > 0 \\ 0, \quad \alpha < 0 \end{Bmatrix}, \tag{III.3.37a}$$

so that

$$\lim_{T \to T_c^-} \frac{\chi_S}{\chi_T} = \begin{Bmatrix} 1 - r^*_{\mathrm{LP}}, \alpha > 0 \\ 1 \qquad , \alpha < 0 \end{Bmatrix}. \tag{III.3.37b}$$

One may show (Bruce 1980, Aharony and Hohenberg 1976) that the quantity r^*_{LP} is universal (i.e. is a constant specific to each universality class). These results thus demonstrate that the thermodynamic central peak exhibited quite generally (cf. §III.3.1) by the spectral function of a low-temperature-phase soft mode will, asymptotically, carry a universal fraction r^*_{LP} of the total spectral weight. In systems characterized by a divergent specific heat ($\alpha > 0$), the intensity of this central component itself diverges as $T \to T_c^-$ with the exponent γ of the static (isothermal) susceptibility. Equations (III.3.37) do not, however, give any indication whether the remaining portion of the spectral weight (also divergent with exponent γ) resides in a phonon mode completely softening as $T \to T_c^-$ or in a saturated phonon and an additional (ultimately dominant) central peak.

In any event this simple argument establishes that, whenever the specific heat diverges, the influence of the thermal-diffusion mode upon the order-parameter response persists as $T \to T_c^-$; its persistent influence in the critical region above T_c, revealed by the R.G., is then not surprising.

In fact, the majority of explicit R.G. analyses have been restricted to this high-temperature-phase critical region: it is, therefore, to this regime that the remainder of this section is devoted. In accord with the insights yielded by the foregoing argument, we shall subdivide our discussion of the dynamic universality classes of interest according to whether they are characterized by a divergent specific heat (e.g. systems with $n = 1$ and short-range forces, §II. 5.1.1.) or by a specific heat that remains finite at T_c (e.g. systems with ($n = 3$)-component order parameters, §II.5.2). We shall focus particular attention on the $n = 1$, short-range-force ('Ising') universality class—partly because (being the simplest!) it has been the subject of the most detailed dynamic R.G. analyses (see, especially, Bausch and Halperin 1978), and partly because it is the universality class to which the models studied by Murata (1975; see above) and by Schneider and Stoll (1973, 1975; see §III.2.5 and §III.4.2.1) belong.

The key result of dynamic R.G. analyses of this universality class is that the spectral response (III.3.28*b*) is dominated, asymptotically as $T \to T_c^+$, by a 'central component' whose width ω_Q satisfies the general scaling result (III.3.30), with the critical exponent z given by (Halperin *et al.* 1974, 1976, Bausch and Halperin 1978)

$$z = 2 + \alpha/\nu. \tag{III.3.38}$$

In the limit $k\xi \ll 1$ (i.e. as $k \to 0$ at fixed t), the width ω_Q may be written in the form (see, particularly, Bausch and Halperin 1978)

$$\omega_Q \approx t^{z\nu}\tilde{\omega}_Q(0) \approx D_T\kappa^2, \tag{III.3.39 a}$$

where $\kappa \equiv \xi^{-1} \sim t^{\nu}$ is the inverse correlation length, while $D_T \sim t^{\alpha}$ is the thermal diffusivity. This result shows clearly how the rate of critical slowing down of the ordering degrees of freedom is tied to the thermal-diffusion mode.

In the limit $k\xi \gg 1$ (i.e. $t \to 0$ at fixed k), the central component has the k-dependent width characteristic of the thermodynamic mode

$$\omega_Q \sim k^z = k^{2+\alpha/\nu} \tag{III.3.39 b}$$

in accord with the qualitative expectations following from (III.3.24).

The case of the uniaxial ferroelectric, also characterized by a divergent specific heat ($\tilde{C} \sim |\ln t|^{1/3}$; eqn. (II.5.19 *b*)), has been examined by Siggia (1975), who finds that the asymptotic behaviour of the width of the $\mathbf{k} = 0$ spectral response displays the temperature-dependence

$$\omega_Q \sim \frac{|\ln t|^{-1/3}}{\chi(k=0, \omega=0)} \sim t|\ln t|^{-2/3}. \tag{III.3.40}$$

In systems whose specific heat is non-divergent (typified by the ($n = 3$)-component isotropic Heisenberg model), the non-linear coupling (III 3.32) between ordering coordinate and energy density is, as we have discussed, 'irrelevant': the heat-diffusion mode does not influence the critical slowing down in this case. The asymptotic behaviour of the spectral function is characterized by a width obeying (III.3.30) with an exponent (Halperin *et al.* 1972, 1974, 1976):

$$z = 2 + c\eta, \tag{III.3.41}$$

where η is the (typically small) exponent characterizing the wavevector-dependence of the susceptibility at criticality (cf. eqn. (I.3.20)) and c is a universality-class-dependent constant of order unity. The exponent z has been evaluated (within an ε expansion) for the isotropic n-component short-range model (Halperin *et al.* 1972), for the cubic n-component short-range model (Yamazaki 1976), and for the isotropic ($n = d$)-component dipolar model (Teitelbaum 1975). In the case of acoustic-mode instabilities, the borderline dimensionality is three or less (§§ II.5.1.4 and 5.2.4) and critical fluctuations give at most logarithmic corrections to the classical dynamical theory (Folk *et al.* 1978).

Such details are, however, of less immediate importance to us than the more qualitative inferences regarding the dynamic response that can be drawn from the R.G. analyses we have surveyed: these we now examine.

For all the universality classes, dynamic R.G. arguments establish that the asymptotic form of the spectral function is dominated, as $T \to T_c^+$, by a 'central response' (probably showing some deviations from lorentzian behaviour: cf. Hohenberg and Halperin 1977) with a width ω_Q whose asymptotic $\mathbf{k}$- and t-dependence is specified (e.g. eqns. (III.3.39 *a*) and (III.3.39 *b*)) modulo generally unknown (non-universal) amplitudes.

In systems for which $\alpha > 0$, this central response may be traced to the non-linear interaction between the ordering coordinates and the thermodynamic phonon-density fluctuations, bearing out the conjecture of Feder (1971) and the calculations of Murata (1975).

In systems with $\alpha < 0$, the coupling to the thermodynamic (heat-diffusion) mode may be neglected and the central response (to the extent that it is distinguishable from the phonon 'side band': see below) arises from the non-linear coupling to the non-thermodynamic phonon-density fluctuations, partially vindicating the original analysis of Silberglitt (1972) for this case. (Note that the process singled out by Silberglitt, fig. III.30 (*a*), represents for this class of system the leading contribution to the self-energy within an ε expansion.)

However, dynamic R.G. analysis does not readily establish the actual magnitude of the central-component width, characterized, as remarked above, by non-universal (system-specific) amplitudes; nor does it readily yield the range of reduced temperature in which the central component should be visible. Consequently the manner in which the spectral function evolves from its form far from T_c, where a single resonant soft-mode response dominates, to its asymptotic form dominated by the central resonance, is left unclear.

In particular, R.G. analysis does not easily distinguish between two possible extreme situations: firstly, the central component forms at sufficiently large values of $|t|$ and with a sufficiently small width that it is distinguishable from the spectrum of the softening phonon, which remains non-critical at T_c; secondly, the central component forms only at sufficiently small $|t|$, and with a sufficiently large width that it is indistinguishable (experimentally, and, perhaps, conceptually) from the spectrum of the by-then-overdamped soft phonon.

Nevertheless, after an extensive R.G. study of a simple Ising-like model (belonging to the same universality class as that defined by eqn. (II.2.1)), Bausch and Halperin (1978) conclude that the latter picture is likely to be the more appropriate, at least when the system is highly displacive. In any event, as Halperin and Varma (1976) have argued, it seems unlikely that there will emerge from any R.G. calculation a central peak with a time scale markedly longer than what one might expect for an overdamped soft phonon.

Two caveats must, however, be made. We remark, firstly, that dynamical R.G. studies of multicomponent ($n > 1$) systems reveal certain pathological features which have not, as yet, been resolved (Halperin *et al.* 1974, 1976, Murata 1976 *a*) and which have been linked, speculatively, with an anomalously narrow central component (Murata 1976 *a*). Secondly, we note that the dynamic R.G. has not yet been employed to analyse a system with as highly anisotropic a soft-mode dispersion as that characteristic of the perovskites with R_{25}-mode instabilities (§ II.5.2.1); again there is some evidence to suggest (Aubry 1977) that such anisotropy might result in a particularly narrow central peak.

Finally we remark that there exist a number of alternative theories for a high-temperature-phase central peak (Enz 1972, Beck and Meier 1973, Schwabl 1973 *a*) which, while based on phonon perturbation theory, do not fall into the mainstream of the two groups of theories we have discussed, lacking both the aesthetic appeal of the former and the rigour of the latter. For a review of these theories we refer the reader to earlier surveys of this problem (Cowley 1974, Schwabl 1974); we remark here, simply, that none provides a convincing explanation of an anomalously narrow central component.

III.3.2.2. *The two-phonon response*

While the results of inelastic neutron-scattering experiments reflect predominantly the one-phonon response discussed in detail in the preceding section, for symmetry reasons light-scattering experiments frequently probe, primarily, the two-phonon response. In § III.3.1.2 we discussed the two-phonon contribution to the light-scattering spectrum outside the critical region, emphasizing the contributions made by the thermodynamic fluctuations in the phonon density. In this, our concluding section on anharmonic theory, we summarize what is known about two-phonon Raman scattering in the critical region. We shall see that (III.3.21) retains its validity close to criticality, so that thermodynamic fluctuations contribute to the light scattering an intensity with the singular behaviour of the specific heat; however, we shall show that there may exist other contributions (which we associate with non-thermodynamic fluctuations) that are more strongly singular, and actually dominate near T_c. We shall also make passing reference to the intimately related theory of ultrasonic attenuation in the critical region (although we have not attempted to survey the corresponding experiments).

We take as our starting point the expression for the two-phonon Raman-scattering cross section which follows immediately from (III.3.18) and (III.3.19) as

$$I^{(2)}_{\alpha\beta\gamma\delta}(\mathbf{k},\omega)=\sum_{\substack{\mathbf{q}_1 j_1 j_2\\ \mathbf{q}_2 j_3 j_4}} P_{\alpha\beta}\begin{pmatrix}\mathbf{q}_1 & -\mathbf{q}_1-\mathbf{k}\\ j_1 & j_2\end{pmatrix} P^*_{\gamma\delta}\begin{pmatrix}\mathbf{q}_2 & -\mathbf{q}_2+\mathbf{k}\\ j_3 & j_4\end{pmatrix} \tag{III.3.42}$$

$$\times\frac{1}{2\pi}\int_{-\infty}^{\infty}\left\langle Q\begin{pmatrix}\mathbf{q}_1\\ j_1\end{pmatrix},\tau\right) Q\begin{pmatrix}-\mathbf{q}_1-\mathbf{k}\\ j_2\end{pmatrix},\tau\Big) Q\begin{pmatrix}\mathbf{q}_2\\ j_3\end{pmatrix},0\Big) Q\begin{pmatrix}-\mathbf{q}_2+\mathbf{k}\\ j_4\end{pmatrix},0\Big)\right\rangle_{\mathrm{c}} \exp(i\omega\tau)$$

where c denotes connected part, implying that the subtraction of the static average in (III.3.19) is to be understood. The contributions to (III.3.42) that have the strongest temperature-dependence in the vicinity of the critical point are those for which the four participating modes belong to one of the (in general n) soft branches and have wavevectors close to (within radius Λ of) the soft-mode wavevector $\mathbf{q}_S$ which we take to be unique. Writing

$$Q\begin{pmatrix}\mathbf{q}_s+\mathbf{k}\\ j\end{pmatrix}\equiv N^{-1/2}Q_j(\mathbf{k}),$$

the cross section (III.3.42) may thus be approximated by

$$I^{(2)}_{\alpha\beta\gamma\delta}(\mathbf{k},\omega)\approx\sum_{\substack{j_1 j_2\\ j_3 j_4}} P_{\alpha\beta}\begin{pmatrix}\mathbf{q}_s^1 & -\mathbf{q}_s\\ j_1 & j_2\end{pmatrix} P_{\gamma\delta}\begin{pmatrix}\mathbf{q}_s & -\mathbf{q}_s\\ j_3 & j_4\end{pmatrix} \times\Gamma_{j_1 j_2 j_3 j_4}(\mathbf{k},\omega) \tag{III.3.43 a}$$

where

$$\Gamma_{j_1 j_2 j_3 j_4}(\mathbf{k},\omega)=\int_{\mathbf{k}_1}^{\Lambda}\int_{\mathbf{k}_2}^{\Lambda}\frac{1}{2\pi}\int_{-\infty}^{\infty}\langle Q_{j_1}(\mathbf{q}_1,\tau)Q_{j_2}(-\mathbf{k}_1-\mathbf{k},\tau) \times Q_{j_3}(\mathbf{k}_2,0)Q_{j_4}(-\mathbf{k}_2+\mathbf{k},0)\rangle_{\mathrm{c}}\exp(i\omega\tau)d\tau, \tag{III.3.43 b}$$

and we have adopted the notation of eqn. (II.4.5). The j indices in (III.3.43 a) now run over the n soft branches.

As discussed in §I.3.6.2 (and, briefly, in the course of §III.2.1) it has been customary, in interpreting the Raman spectra of systems whose high-temperature-phase soft modes are Raman inactive, to approximate the low-temperature-phase spectrum by the distortion-induced one-phonon term residing in (III.3.43), so that

$$\Gamma_{j_1j_2j_3j_4}(\mathbf{k},\omega)\approx 4N Q_0^2\delta_{j_2j}\,\delta_{j_4j}\cdot\frac{1}{2\pi}\int_{-\infty}^{\infty}\langle\hat{Q}_{j_1}(\mathbf{k},\tau)\hat{Q}_{j_3}(-\mathbf{k},0)\rangle \times\exp(i\omega\tau)d\tau. \tag{III.3.44}$$

Here we suppose that the low-temperature phase is characterized by a distortion

$$\left\langle Q\begin{pmatrix}\mathbf{q}_s\\ j'\end{pmatrix}\right\rangle\equiv N^{1/2}Q_0\delta_{jj'} \tag{III.3.45}$$

and denote by $\hat{Q}_j(\mathbf{k})$ the fluctuating part of the coordinate $Q_j(\mathbf{k})$. This approximation (which has the same mean-field character as the approximation (III.3.16)) is not unreasonable deep in the low-temperature phase. Near the critical point, however, it is certainly inadequate. This inadequacy has been explored quantitatively by Yacoby *et al.* (1978). Analysing a simple model, within quasi-harmonic theory, they make estimates of the one-phonon contribution to the scattering, the two-phonon 'background' (which must ultimately dominate since it provides the only source of scattering above T_c) and the interference terms. Their results suggest that the customary form of analysis of such experiments (where the cross section is fitted to that of a one-phonon process) may lead to very misleading values for the soft-mode frequency and line width, close to the transition temperature.

Indeed, in analysing the $T<T_c$ critical behaviour of the correlation function (III.3.43*b*), it is probably quite inappropriate to make *separate* studies of the induced one-phonon, two-phonon and interference contributions, since subtle cancellations are to be expected. (Such a phenomenon is discussed in §II.5.4.1.). We will make further brief comment on this point below. For the present, however, we focus on the situation above T_c, surveying the results of R.G. analyses of (III.3.43*b*), which reveal once more the importance of the order-parameter dimensionality n.

We consider first the case of a scalar ($n=1$ component) order parameter, when the tensor (III.3.43*b*) has only the single component

$$\Gamma_T(t,\mathbf{k},\omega)=\int_{\mathbf{k}_1}^{\Lambda}\int_{\mathbf{k}_2}^{\Lambda}\frac{1}{2\pi}\langle Q(\mathbf{k}_1,\tau)Q(-\mathbf{k}_1-\mathbf{k},\tau)Q(\mathbf{k}_2,0)Q(-\mathbf{k}_2+\mathbf{k},0)\rangle_c \times\exp(i\omega\tau)d\tau. \tag{III.3.46}$$

R.G. analysis establishes that this spectral function has the dynamic scaling form (Hohenberg and Halperin 1977; cf. eqn. (III.3.28*b*))

$$\Gamma_T(t,\mathbf{k},\omega)\approx t^{-\alpha-z\nu}\tilde{\Gamma}(k/t^{\nu},\omega/t^{z\nu}), \tag{III.3.47}$$

where the dynamical exponent z is given by (III.3.38). It follows immediately that the $\mathbf{k}=0$ scattering has peak height

$$\Gamma_T(t,\mathbf{k}=0,\omega=0)\sim t^{-\alpha-z\nu}, \tag{III.3.48a}$$

a spectral width

$$\omega_T\sim\omega_Q\sim t^{z\nu}, \tag{III.3.48b}$$

and an integrated intensity

$$\int_{-\infty}^{\infty} \Gamma_T(t, \mathbf{k}=0, \omega)\, d\omega \sim \omega_T \Gamma(t, \mathbf{k}=0, \omega=0) \sim t^{-\alpha}. \qquad \text{(III.3.48}\,c\text{)}$$

The last result may also be deduced from the discussion of elastic compliance behaviour in §II.5.1.2. It shows that in this case, subject to the constraints imposed by the symmetry of the polarizability coefficients in (II.3.43 *a*), the components of the Raman tensor carry the singular behaviour of the specific heat (Schwabl 1973 *b*)

$$\int_{-\infty}^{\infty} I^{(2)}_{\alpha\beta\gamma\delta}(\mathbf{k}=0, \omega)\, d\omega \sim t^{-\alpha} \qquad \text{(III.3.49)}$$

in agreement with the prediction of phenomenological theory, eqn. (III.3.21). The accord reflects the fact that (III.3.46) describes the scattering due to fluctuations in the energy density associated with the soft coordinates.

The result (III.3.48 *b*) shows that the width of the associated scattering has the temperature-dependence of the characteristic frequency ω_Q of the ordering coordinate fluctuations. That it does not have the k-dependence one might anticipate reflects, presumably, the fact that the scattering resides in fluctuations of only *part* of the energy density.

The result (III.3.48 *a*), for the low-frequency limit of the spectral function, is of interest since it determines the ultrasonic attenuation (Schwabl 1973) which, therefore, diverges with exponent

$$\rho_1 = \alpha + z\nu. \qquad \text{(III.3.50)}$$

Finally, we remark that the explicit forms cited above are appropriate for an $n=1$ system with *short-range* forces, where the specific heat has the power-law behaviour (cf. eqn. (II.2.11)) displayed in (III.3.49). In uniaxial ferroelectric systems, the specific heat (and hence the integrated intensity (III.3.49)) has a logarithmic anomaly (eqn. (II.5.19)) while the spectral width (III.3.48 *b*) has the temperature-dependence given in (III.3.40). The behaviour of the ultrasonic attentuation has not, to our knowledge, been determined with due account taken of the influence of energy conservation on the critical dynamics. The behaviour of this and other dynamical features of uniaxial ferroelectrics in the absence of (effective) energy conservation has been discussed by Stauffer (1978).

When the order parameter is not a scalar quantity, the situation is considerably more complicated: the tensor (III.3.43 *b*) has a number of independent components carrying, in general, different critical singularities. With the perovskites in mind we shall consider the case of an ($n=d=3$)-component system whose critical behaviour, we shall assume (cf. §II.5.2.1), is dominated by the Heisenberg fixed point. The tensor (III.3.43 *b*) then has two independent components

$$\Gamma_T(t, \mathbf{k}, \omega) \equiv \frac{1}{2\pi} \int_{-\infty}^{\infty} \langle \psi_T(\mathbf{k}, \tau) \psi_T(-\mathbf{k}, 0) \rangle_c \exp(i\omega\tau)\, d\tau \qquad \text{(III.3.51}\,a\text{)}$$

$$\Gamma_{NT}(t, \mathbf{k}, \omega) \equiv \frac{1}{2\pi} \int_{-\infty}^{\infty} \langle \psi_{NT}(\mathbf{k}, \tau) \psi_{NT}(-\mathbf{k}, 0) \rangle_c \exp(i\omega\tau)\, d\tau, \qquad \text{(III.3.51}\,b\text{)}$$

where

$$\psi_T(\mathbf{k},\tau)\equiv\int_{\mathbf{k}_1}^{\Lambda}\sum_{\alpha=1}^{d}Q_\alpha(\mathbf{k}_1,\tau)Q_\alpha(-\mathbf{k}_1-\mathbf{k},\tau) \tag{III.3.52 a}$$

$$\psi_{NT}(\mathbf{k},\tau)\equiv\int_{\mathbf{k}_1}^{\Lambda}\Bigg\{Q_1(\mathbf{k}_1,\tau)Q_1(-\mathbf{k}_1-\mathbf{k},\tau) -\frac{1}{d-1}\sum_{\alpha\neq 1}^{d}Q_\alpha(\mathbf{k}_1,\tau)Q_\alpha(-\mathbf{k}_1-\mathbf{k},\tau)\Bigg\}. \tag{III.3.52 b}$$

The correlation function (III.3.51 a) again has the scaling form (III.3.47) (Hohenberg and Halperin 1977), with z and α taking on their appropriate values. The correlation function (III.3.51 b), however, has a different scaling behaviour (Hohenberg and Halperin 1977)

$$\Gamma_{NT}(t,\mathbf{k},\omega)\approx t^{-\tilde{\gamma}-z\nu}\tilde{\Gamma}_{NT}(k/t^{z\nu},\omega/t^{z\nu}), \tag{III.3.53}$$

where $\tilde{\gamma}$ is given by (cf. eqn. (II.5.32 b))

$$\tilde{\gamma}=2\phi+\alpha-2, \tag{III.3.54}$$

with ϕ the crossover exponent characterizing the instability of the Heisenberg fixed point against a perturbation of uniaxial symmetry. The relations analogous to (III.3.48 a–c) are then, for the peak height of the scattering

$$\Gamma_{NT}(t,\mathbf{k}=0,\omega=0)\sim t^{-\tilde{\gamma}-z\nu}, \tag{III.3.55 a}$$

for the width

$$\omega_{NT}\sim\omega_Q\sim t^{z\nu} \tag{III.3.55 b}$$

and for the integrated intensity

$$\int_{-\infty}^{\infty}\Gamma_{NT}(t,\mathbf{k}=0,\omega)\,d\omega\sim\omega_{NT}\Gamma_{NT}(t,\mathbf{k}=0,\omega=0)\sim t^{-\tilde{\gamma}}. \tag{III.3.55 c}$$

Again the last result mirrors elastic compliance behaviour (eqn. (II.5.32)).

Depending upon the experimental configuration, a light-scattering experiment (or an ultrasonic probe) samples different linear combinations of the tensors (III.3.51 a,b): however, as emphasized by Murata (1976 b) in the context of ultrasonic attenuation, there is no configuration which probes simply the tensor (III.3.51 a) so that, quite generally, we expect the stronger singularities associated with the tensor Γ_{NT} to prevail. Thus, the components of the Raman tensor will diverge as

$$\int_{-\infty}^{\infty}I^{(2)}_{\alpha\beta\gamma\delta}(\mathbf{k}=0,\omega)\,d\omega\sim t^{-\nu}; \tag{III.3.56}$$

the intensity again being distributed over a frequency range with the temperature-dependence of ω_Q. Correspondingly the ultrasonic attenuation will diverge with exponent (Murata 1976 b)

$$\rho_2=\tilde{\gamma}+z\nu. \tag{III.3.57}$$

The situation is somewhat more complicated if the critical behaviour is governed by a fixed point of cubic symmetry: there are then three independent components of the tensor (III.3.43*b*) and different Raman scattering or ultrasonic-attenuation configurations may display different critical singularities, as discussed in detail by Murata (1976*b*).

In conclusion we re-emphasize that the results cited above are known to characterize the critical singularities in the limit at $T \to T_c^+$. The situation below T_c is rather less clear. A naïve inspection of (III.3.44) would suggest that the integrated Raman intensity should typically contain a term diverging with exponent $\gamma - 2\beta \equiv \alpha + 2(\gamma - 1)$ as $T \to T_c^-$. However, we believe that this is incorrect, and that the 'background' contributions to the Raman scattering (ignored in the approximation (III.3.44)) provide a cancelling term so that the leading singularity has the same form as that above T_c–at least in the scalar order-parameter case. It is possible that similar cancellations occur between the background fluctuation contributions and the Landau–Khalatnikov contribution to the ultrasonic attenuation below T_c. However, this problem does not yet appear to have been resolved (Hohenberg and Halperin 1977).

III.3.3. *Discussion*

We conclude this section with an overview of the theory we have developed, interlacing our discussion with comparisons between the theory's predictions, and the experimental results gathered in § III.2. In the spirit of this section we shall consider separately the non-critical and critical regimes.

III.3.3.1. *The non-critical region*

The theory's predictions for a region remote from a critical point (or in the absence of a phase transition at all) are relatively clear cut: quasi-elastic response may occur in the spectral function of any mode whose symmetry permits coupling to phonon-density fluctuations (§ III.3.1.1); the intensity of the quasi-elastic component observed in a scattering experiment may be significantly affected by direct coupling of the scattering probe to the phonon-density fluctuations (§ III.3.1.2).

The theory accounts well fcr the two non-critical quasi-elastic components identified in the diagonal Brillouin spectrum of $SrTiO_3$ (Fleury and Lyons 1978; fig. III.9). The narrower component reflects the coupling of the L.A. mode probed in the experiment to the thermodynamic phonon-density fluctuations: its width displays the characteristic wavevector-dependence implied by (III.3.10); its intensity relative to the Brillouin peak is larger than the simple Landau–Placzek result (III.3.17) would suggest, but is reported to be in accord with the enhanced Landau–Placzek ratio (III.3.22), attesting to the importance of direct coupling between the light and the thermodynamic fluctuations. The broader component may reasonably be attributed to the contributions to the L.A. mode self-energy (together with corresponding contributions arising from direct coupling to the light) of the non-thermodynamic phonon-density fluctuations: the width of this component lies within the range to be expected on the basis of (III.3.4).

While thermodynamic fluctuations in the phonon density can contribute only to the scattering from a mode transforming as the identity representation, the contributions of the non-thermodynamic fluctuations are not thus restricted. In particular, the theory predicts additional low-frequency scattering from the coupled soft ferroelectric and transverse acoustic modes in the *high-temperature* phase of KH_2PO_4 (this scattering being peculiar to piezoelectric systems: fig. III.27); estimates indicate that the effect should be small in this case (Young and Elliott

1974), and it has not been detected experimentally. In the *low-temperature* phase of KH_2PO_4, the coupled soft ferroelectric and (no longer strictly transverse) acoustic modes interact linearly with the thermodynamic phonon-density fluctuations: the theory then again predicts the existence of additional low-frequency scattering (this scattering being common to the spectral functions of *all* low-temperature soft modes: figs. III.27, 28). Such scattering has been observed (if at all) only within the critical region (Mermelstein and Cummins 1977), to be discussed in the following section.

Despite the dearth of experimental data, it should be apparent that the quasi-elastic scattering discussed here should prove a very general phenomenon: it appears to us to merit more extensive studies with the high-resolution Brillouin-scattering techniques best suited for probing the ranges of relatively low-frequency and small wavevector transfer involved.

III.3.3.2. *The critical region*

In some respects the theory's predictions for dynamic behaviour near a critical point are still more specific—at least in the regime above T_c.

The one-phonon spectral function (§III.3.2.1), probed by inelastic neutron-scattering measurements in particular, will invariably be dominated asymptotically as $T \to T_c^+$ by a central component whose integrated intensity diverges with the static susceptibility exponent γ, and whose width obeys the scaling relation (III.3.30), involving the dynamic exponent z: the exponents γ and z are (generally) well-established quantities specific to each universality class.

The two-phonon spectral function (§III.3.2.2) measured in (some) light-scattering experiments will also be dominated as $T \to T_c^+$ by a central component whose integrated intensity diverges in a manner (specified by the theory) which may depend upon the scattering geometry, and whose width obeys the same scaling relationship as that of the one-phonon spectral function.

The dynamic behaviour in the critical region of the ordered phase appears to have attracted less detailed attention. Nevertheless it is to be expected that both the one- and two-phonon spectral functions should be characterized by the same exponents (though not the same scaling functions) as are appropriate above T_c, while the characteristic frequency ω_Q should have a unique (k-dependent) form at T_c. We have also seen that the thermodynamic central peak present in the low-temperature-phase spectral functions of any soft mode (or, more precisely, in one *component* of any soft mode) carries a finite proportion of the spectral intensity, asymptotically as $T \to T_c^-$, if and only if the specific heat diverges at the phase transition.

There are relatively few experimental results against which these predictions can be tested.

None of the scattering experiments we have described supplies a meaningful measure of the dynamic critical exponent z.

As discussed in §I.3.6.2, it is difficult to extract a reliable value for the exponent γ from an integration over the one-phonon neutron-scattering cross section. Light-scattering techniques probe the critical behaviour of the one-phonon spectral function only when the soft mode is Raman active above T_c: measurements of the Brillouin intensity in the high-temperature phase of KH_2PO_4 (Courtens 1978) are consistent with the classical (§II.5.1.4) or tricritical (§II.5.3.1) value $\gamma = 1$.

The weakly critical dynamic central component observed in the off-diagonal spectrum of $SrTiO_3$ (Lyons and Fleury 1977; fig. III.10; note our comments in §III.2.1!) may be a reflection of the predicted singularity (III.3.56) for the appropriate component of the two-phonon Raman tensor. Similarly, the weakly-singular dynamic central component observed in $Pb_5Ge_3O_{11}$ (Lyons and Fleury 1978; fig. III.18) may reflect the weak specific-heat anomaly expected for a uniaxial ferroelectric.

Finally, the observation of a dynamic central component in the spectrum of the soft mode in KH_2PO_4 immediately below the critical point, is consistent with the inferences drawn from (III.3.37): the dynamic central component should grow critically in this case, with exponent $\gamma = 1$, since the specific heat itself 'diverges' (in a manner compatible with the proximity of a tricritical point: Garber and Smolenko 1973). However, this argument does not account for the dielectric relaxation observed in a very similar frequency regime in the ordered phase of KH_2PO_4 (Tornberg and Lowndes 1977).

While the theory thus, clearly, makes predictions which are as yet largely untested by experiment, it is equally clear that existing experimental results raise questions which the theory has not yet satisfactorily answered. Two qualitatively similar and largely unresolved issues are worthy of specific mention.

Firstly, despite the considerable body of literature devoted to the problem, it remains unclear how the one-phonon spectral function evolves from its high-temperature form characterized by a single softening resonance, to its asymptotic form dominated by a central component. More specifically it remains uncertain whether, or to what extent, one should expect to identify *two* time scales in the spectral function. The non-R.G. arguments discussed in the early part of § II.3.2.1 do indicate that two regions of response are to be expected; however, these arguments are not reliable near the critical point and, in any case, do not confront the further question of the universality-class-dependence of the phenomenon. In contrast, R.G. analyses, while elucidating the universality-class-dependence of critical singularities, seem (to us) to leave unanswered the more basic question of the existence of higher-frequency features (a saturated soft phonon) in the spectral function, additional to the dominant central component.

Nevertheless, in the light of all the theoretical evidence we have surveyed we surmise that the combination of a saturating soft phonon and a critical dynamic central component should be a general feature of any near-displacive system exhibiting a phase transition where critical fluctuations play an important role.

We shall discuss further evidence supporting this view in § III.4. For the moment we turn to the experimental situation which, while instrumental in raising this issue, also offers no definitive answers.

Probably the best 'experimental' evidence that a second time scale may emerge near the critical point of a displacive system, characterized by non-classical exponents, is to be found in the molecular-dynamics studies of the $d = 2$ Ising-like model, described briefly in § III.2.5 (Schneider and Stoll 1976; fig. III.24). The results reported are in accord with the conclusions of R.G. analyses to the extent that (except presumably in the asymptotic regime where critical narrowing occurs) the central component observed is not anomalously narrow on the scale of the soft-phonon width.

The significance of the 'two time scale' behaviour exhibited by the phonon neutron-scattering cross section in $SrTiO_3$ (Shapiro *et al.* 1972; fig. III.1) is undermined by the width of the observed central component: with the caveats noted in § III.3.2.1 it seems that a component as narrow as this is incompatible with R.G. estimates which suggest that for a displacive system the width of the critical scattering should not be very different from that expected for an overdamped soft mode. (Bausch and Halperin 1978; note that, although $SrTiO_3$ is in principle an $n=3$ system, it may in certain circumstances be better represented by the Ising-like model studied by these authors: cf. §II.5.3.3) Thus it appears that, to some extent at least, this central component reflects the influence of crystal imperfections (§III.5)—like the unambiguously 'static' features in the Raman spectra of $Pb_5Ge_3O_{11}$ and KH_2PO_4 discussed in §III.2.

Nevertheless, the light-scattering measurements on $SrTiO_3$ (off-diagonal spectrum, Lyons and Fleury 1977) and $Pb_5Ge_3O_{11}$ (Lyons and Fleury 1978), already discussed above, do provide some indication of a soft-phonon saturation as $T \rightarrow T_c^-$ while the weight of the spectral function is growingly transferred to a *dynamic* central component. However, the interpretation of these results is complicated by the fact that, in each case, the light-scattering probe measures the full two-phonon spectral function (III.3.43) (including the distortion-induced one-phonon part, (III.3.44)) rather than a pure one-phonon spectral function. (Distortion-induced interactions between soft and acoustic modes introduce further complications.)

Indeed this brings us to the second of the two major unresolved issues alluded to above: like the first it reflects the difficulties in interfacing theories of critical and non-critical regimes. In the non-critical region of the low-temperature phase, it is clear that the two-phonon spectral function (III.3.43) will be adequately approximated by the one-phonon part (III.3.44); asymptotically the spectrum must be dominated by a central component to which two-phonon and one/two phonon interference terms contribute significantly: again the nature of the intermediate behaviour is obscure.

Nevertheless we surmise that the limiting form of the spectrum should again display traces of a saturated soft phonon (whose coupling to the light, like the mode frequency itself, remains at a fluctuation-induced finite value at T_c) together with a dynamic central component carrying the generally weak singularity expected for the two-phonon intensity.

§III.4 Precursor clusters

In the preceding section we have surveyed one group of theories developed in an attempt to understand the dynamical properties of systems undergoing structural phase transitions, and motivated in a large part by the failure of the simple soft-mode theory to account for the behaviour of the critical one-phonon spectral function in $SrTiO_3$. These theories are linked by a common strategy: the adoption of a *reciprocal-space* basis of harmonic phonon coordinates, and the use of *perturbation theory*, of varying degrees of sophistication. The virtues of this strategy are clear: the techniques (of perturbation theory) it requires are well-established, the essential task it poses (the calculation of the soft-phonon self-energy) is well-defined, and the predictions that ensue (typified by (III.3.38) and (III.3.39)) are explicit and precise. Its deficiencies are also clear: the calculation of the soft-mode self-energy in the critical region is hard to accomplish in detail, and the results obtained are conceptually opaque.

In this section we survey a second class of studies of dynamical behaviour near structural transitions, also largely motivated by the observations of anomalous behaviour in $SrTiO_3$. These studies (Schneider and Stoll 1973, 1976, Krumhansl and Schrieffer 1975, Aubry 1975, 1976 a, b, Varma 1976) have two themes in common: firstly, the conviction that the principal features of the dynamical behaviour are best understood (and perhaps uniquely captured) by methods which treat the inescapably anharmonic character of the problem in a *non-perturbative* way; secondly, their common emphasis on *direct-space* configurations of the atomic displacements—particularly the distinctive clusters of dynamic precursor order that inevitably form with the approach to criticality.

In marked contrast to the anharmonic phonon strategy, this second approach presents the key problems in a form which is not, in general, susceptible to any well-established theoretical technique: correspondingly this section contains no results as explicit as those gathered in § III.3.2.1. Nevertheless if this approach does not, as yet, exactly constitute a 'theory' of structural phase transitions, it does, at the very least, constitute a particularly fruitful way of thinking about the problem: in particular it highlights potentially crucial excitations which may not be captured by (and are certainly not readily identifiable in) phonon perturbation theories; it suggests a conceptually-simple and very general origin for a two time-scale spectral function; and, in emphasizing the underlying unity of order-disorder and displacive systems, it offers an appealing insight into the origins of the universality of the critical region.

It is with this last remark in mind that we begin our discussion (§ III.4.1) by recalling the dynamical behaviour of strongly order–disorder systems, described in some detail in I. We shall set our discussion in the context of the simple model of a structural phase transition extensively described in II, and offering realizations of both order–disorder and displacive systems. Molecular-dynamics studies of the two-dimensional form of this model, and analytic studies of its particularly tractable, but phase-transition-less, one-dimensional variant, constitute the principal sources of the 'cluster picture' of displacive phase transitions to be developed in § III.4.2. Finally, in § III.4.3, we summarize the outstanding open questions prompted by the cluster picture, and assess its plausibility in the light of the experimental results assembled in § III.2.

III.4.1. *Preamble: order–disorder behaviour in a simple model system.*

As we have already remarked, the 'cluster picture' of displacive phase transitions suggests dynamic behaviour with features that are strongly reminiscent of those manifestly present in an order–disorder system. It is appropriate, therefore, to review at this stage the basis for the order–disorder/displacive classification, and to recall the (largely self-evident) characteristics of order–disorder dynamics, discussed in § I.5.2. We shall illustrate our discussion with references to the simple model system introduced in § II.2: for convenience we recall its essential features here.

The model is defined by the classical hamiltonian (eqns. (II.2.1), and fig. III.1):

$$H=\sum_{l}\left\{\frac{1}{2}\dot{u}^2(l)+V_s(u(l))\right\}+\frac{C}{2}\sum_{l,l'}^{n.n}(u(l)-u(l'))^2 \qquad \text{(III.4.1 } a)$$

describing a group of atoms, each of which resides in a single-particle potential

$$V_s(u)=\frac{\bar{A}}{2}u^2+\frac{B}{4}u^4; \quad \bar{A}<0 \qquad \text{(III.4.1 } b)$$

located at the sites of a d-dimensional simple cubic lattice of periodicity a. Each atom interacts with its immediate neighbours with harmonic force constant C. The fundamentally anharmonic character of the model, typical of any system undergoing a structural phase transition, is expressed in the double-well nature of the potential $V_s(u)$. Evidently at zero temperature all atoms reside at the bottom of their left-hand (or all at the bottom of their right-hand) well, yielding a homogeneously ordered phase with order parameter

$$\langle u(l)\rangle\bigg|_{T=0} \equiv Q_0(T=0) = \pm\sqrt{(|A|/B)} \qquad \text{(III.4.2)}$$

Here $Q_0(T=0) \equiv Q_0(0)$ locates the well minima with respect to its maximum, to which we shall refer as the 'high-symmetry site'. In the $d=1$ case, the order parameter Q_0 vanishes at any finite temperature, and a phase transition occurs only in a limited sense, precisely at $T=0$ K (cf. further comments in § III.4.2.2. below); in dimensionalities $d>1$, long-range order persists up to a finite model-parameter-dependent critical temperature T_c, where a continuous phase transition occurs. The universal features of the associated critical behaviour are those of the Ising universality class with conserved energy: the static (§ II.2.2) and dynamic (§ III.3.2; eqn. (III.3.38)) critical exponents are non-classical and dimensionality-dependent for $d<4$, and assume their classical, dimensionality-independent values for $d>4$.

It is the essence of universality (though nonetheless remarkable) that the critical exponents are, in particular, quite independent of the location of the model on the order–disorder/displacive spectrum, whose limits represent, microscopically, very different systems. These differences may be highlighted by introducing the dimensionless ratio g of the local potential well depth $V_s^0 \equiv |V_S[Q_0(0)]|$ and the critical thermal energy

$$g = V_s^0/k_B T_C \qquad \text{(III.4.3)}$$

If one makes use of the (partially) known parameter dependence of the transition temperature ((eqns. (II.2.9) and (II.2.14)) one finds that, for systems with $d>2$, the parameter dependence of g is essentially the same as that of the more explicit, but less transparent, ratio s introduced in (II.2.8). One may, then, in full accord with the classification adopted in II, associate order–disorder and displacive limits with the regimes $g \gg 1$ and $g \ll 1$, respectively.

The essential qualitative characteristics of the order–disorder regime follow immediately: the critical thermal energy is small in comparison with the local potential barrier and, except at temperatures $T \gg T_c$, each atom must largely reside near one or other of its well minima. The local dynamical behaviour is then manifestly characterized by two time scales: a relatively short time scale associated with small-amplitude, essentially harmonic motion about each of the two local potential minima, and a relatively long time scale associated with large-amplitude, strongly anharmonic motion between the two potential wells. To each of these distinct types of local dynamical behaviour there will exist a corresponding collective motion: the collective analogue of the intra-well local motion will have the character of a phonon of the low-temperature structure (a 'low-temperature phonon'); the collective analogue of the inter-well local motion is less immediately transparent without anticipating the discussion of later sections. It is however, intuitively clear that the critical singularities will reside in this collective inter-well motion, rather than in the collective intra-well mode. Consequently, it is to be expected that the dynamic susceptibility or spectral function of the ordering coordinates in such systems will display a non-critical resonant high-frequency feature associated with the intra-well, low-temperature phonon mode and a low-frequency ('central') feature arising from the inter-well motion—the latter carrying critical singularities (diverging intensity and narrowing width) with the universal features discussed in §III.3.2.1.

While this picture appears to offer a qualitatively satisfactory account of order–disorder behaviour (cf. experiments on $NaNO_2$ discussed in §I.3.5), it is important to appreciate that a detailed microscopic theory (even of as simple an order–disorder model system as that used as illustration here) presents formidable challenges. Even at temperatures remote from the critical point, the equations of motion constitute a strongly anharmonic problem whose solutions are clearly inaccessible to any conventional phonon-based perturbation theory. Accordingly these equations are usually themselves modelled by simpler equations which omit the intra-well degrees of freedom altogether, and which describe the strongly anharmonic inter-well motion in a phenomenological fashion (cf. the discussion of the dynamic Ising model in §I.5.2.1).

The physics of the displacive regime, which has been our primary focus of attention in this paper, presents—superficially at least—a total contrast. In this regime the critical thermal energy is *large* compared with the local potential barrier which does not, therefore, localize the atoms near $\pm Q_0(0)$. The local dynamical behaviour does not obviously possess two time scales, nor is it obviously strongly anharmonic. Indeed the harmonic approximation (cf. the coefficient of the quadratic term in eqn. (II.2.3*c*)) gives real ('stable') values for the frequencies of all but a few, long-wavelength modes of vibration about the high-symmetry sites ('high-temperature phonons'). This is the rationale for the use of high-temperature phonon coordinates as a basis for a perturbation treatment of the problem, with the familiar results: the perturbation theory appears at first sight to be analytically tractable (§I.5.1) and yields a dynamic behaviour dominated by a softening quasi-harmonic high-temperature phonon. (Note that the single anharmonic coupling constant B controls both the location of the system on the order–disorder/displacive spectrum, and the life time of the soft mode, which is, therefore, necessarily weakly damped in the displacive limit. This is a feature specific to the simple model (III.4.1). A displacive system with an overdamped soft mode could be realized by introducing further anharmonic terms into the hamiltonian.)

There are a number of reasons to suspect that this picture is *qualitatively* inappropriate near the critical point and, in particular, that its stark contrast with the picture of order–disorder behaviour, sketched above, is misleading. Firstly, the ultimate inadequacy of (finite-order) perturbation theory, within the critical region, suggests that the 'weakly anharmonic' appearance of the displacive system actually disguises a problem that is as 'strongly anharmonic' as the order–disorder system. Secondly, the contrast between the two pictures seems hard to reconcile with the concept of universality. Specifically, we re-emphasize that the central components dominating the asymptotic behaviour in displacive and order–disorder systems (belonging to the same universality class) should exhibit critical behaviour described by precisely the same universal quantities: it thus appears implausible that the displacive central component should be viewed (cf. §III.3.3.2.) simply as the limiting form of a softening quasi-harmonic phonon, while the order–disorder central component is attributed to strongly anharmonic collective inter-well motion.

There is, then, some reason to expect that it should be possible to construct a *unified* picture of order–disorder and displacive phase transitions: indeed this is the essential merit of the cluster picture to which we now turn.

III.4.2. *The cluster picture*

III.4.2.1. *Molecular-dynamics studies*

In § III.2.5 we made brief reference to a sequence of molecular-dynamics studies of a model system (essentially that described in the preceding section) undergoing a structural phase transition (Schneider and Stoll 1973, 1976). We noted, in particular, the results (fig. III.24) of a computation of the spectral function of the ordering coordinates in a model with a displacive character ($g \approx 0{\cdot}19$, in the notation of eqn. (III.4.3)); with the approach to criticality, a softening phonon spectrum gives way to a spectrum dominated by a critically-growing central component together with a phonon whose frequency appears to remain finite at the transition.

These molecular-dynamics results thus afford (highly-controlled) 'experimental' evidence that a two time-scale spectral function is indeed an intrinsic feature of the $T > T_c$ critical region of a displacive system. In addition, however, the molecular-dynamics framework makes it possible to forge clear-cut connections between the features of the spectral function and the time-evolution of specific configurations of the atomic coordinates. It is in this process that the cluster picture of displacive phase transitions has its origins (Schneider and Stoll 1973, 1976). The essential insights of the cluster picture are evident in the molecular-dynamics results displayed in figs. III.31 and III.32: these results characterize the behaviour in the same two-dimensional displacive model as that whose spectral function (fig. III.24) was discussed above.

Figure III.31 depicts the instantaneous configurations of the atoms, typical of a variety of temperatures: notice that only the signs of the local distortions are represented. It is evident that the critical growth in correlation length, with reducing temperature, is expressed in the appearance and growth of clusters—compact groups of adjacent atoms, all of which (at a given instant) are displaced to the same side of the high-symmetry site. Though of conceptual importance this phenomenon is *not* inconsistent with a quasi-harmonic soft-mode picture.

Figure III.32 gives an impression of the manner in which such configurations evolve with time, close to the critical point of the model: it records the displacements $u(l)$ along a typical row of atoms, as a function of time, with the contributions of high-frequency and wavevector Fourier components removed. It appears that the clusters of precursor order, visible in fig. III.31, persist in a relatively well-defined form, for some typical finite lifetime τ, during which their boundaries diffuse (or, perhaps, propagate) through the system. This behaviour—specifically, the persistence of the cluster boundaries—*does* seem qualitatively different from what one might expect for the time evolution of a configuration built from a superposition of quasi-harmonic optic phonons.

In particular these results (more clearly represented in an illuminating motion picture: Stoll and Schneider 1974) make it quite apparent that, in the cluster-dominated region close to the critical point, the time evolution of a typical local coordinate, describing the behaviour of a single typical atom, may reasonably be represented by *two* distinct, easily visualized, components. Firstly, each atom exhibits a relatively fast motion about a quasi-equilibrium position displaced from the high-symmetry site by some typical amount Q_I, determined by the character of the cluster of which the atom is a member. Secondly, the local quasi-equilibrium position itself evolves at a relatively slow rate determined by the time for which the (typical) cluster encompasses the atom in question.

Fig. III.31

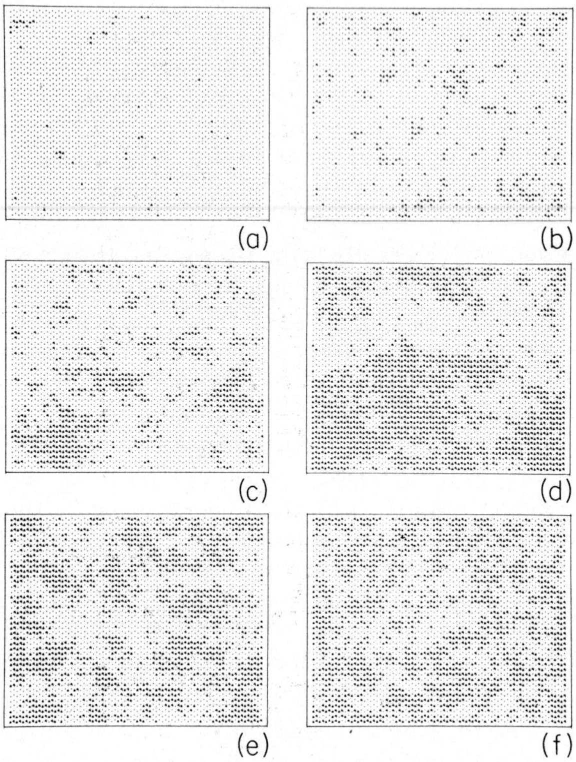

Instantaneous configurations of the atomic displacements in the $d=2$ model defined by eqn. (III.4.1), at a variety of reduced temperatures t: (*a*) $-0{\cdot}59$ (*b*) $-0{\cdot}1$ (*c*) $-0{\cdot}02$ (*d*) $+0{\cdot}02$ (*e*) $+0{\cdot}22$ (*f*) $+2{\cdot}26$. The two possible *signs* of the local displacements are indicated by light and dark points (Schneider and Stoll 1976).

It is then easy to identify the origins of the two time scales in the collective dynamical behaviour implied by the character of the spectral function (fig. III.24) close to the critical point: for the time evolution of the long-wavelength Fourier coordinate, describing the net displacement of many atoms, will also have two components. Firstly, this coordinate will have a relatively fast oscillatory component associated with the collective quasi-harmonic oscillations of the atoms about their cluster-determined distorted quasi-equilibrium positions. This component may be identified with the phonon side band in fig. III.24. Outside the cluster-dominated regime this mode has the softening 'high-temperature' phonon character envisaged in the simple theory. With the critical growth in cluster size, and the accompanying growth in the magnitude and spatial coherence of the precursor order, the mode begins, at the characteristic reduced temperature t_I($\approx 0{\cdot}5$ for this model), to assume the character of a 'low-temperature' phonon, strongly analogous to the collective intra-well mode exhibited by order–disorder systems (§ II.4.1). The saturation of the mode frequency at a finite value ω_∞^c is a natural corollary of this gradual metamorphosis: the phonon begins to be stabilized by the growing cluster-induced precursor order *above* T_c in much the same way as, within simple theory, it begins to be stabilized (only) *at* T_c by the onset of homogeneous order.

Fig. III.32

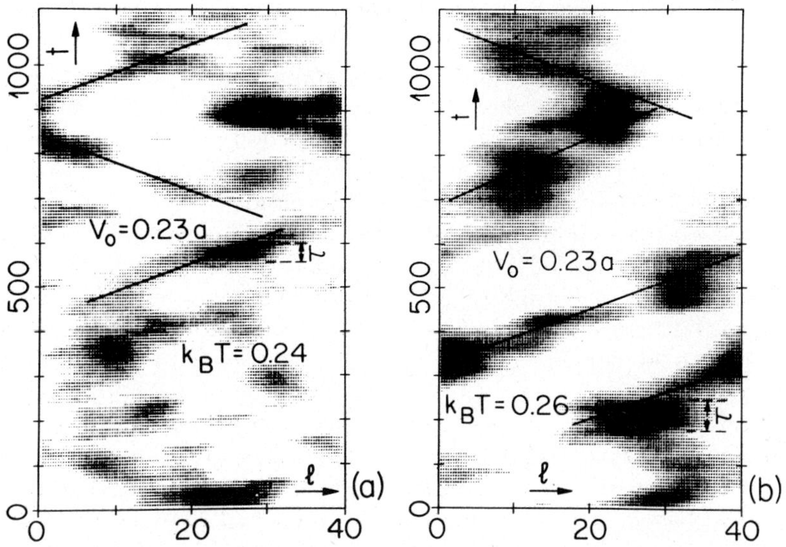

Time-dependent configurations of the atomic displacements in the $d=2$ model defined by eqn. (III. 4.1), (*a*) immediately below and (*b*) immediately above the critical temperature. The darkness of a point measures the distortion $u_l(t)$ of the atom at site l (horizontal axis) and time t (vertical axis). Negative values, and high-frequency and wavevector components of the displacement profile have been eliminated (Schneider and Stoll 1976).

The time evolution of the collective coordinate will, secondly, have a relatively slow component reflecting the relaxation of the precursor order within the clusters. This component may be identified with the central peak appearing in the cluster-dominated critical region below t_I (fig. II.24). This central peak is wholly analogous to that which we associated with the order–disorder system's 'collective inter-well mode', whose physical character we did not discuss. Within the framework we have developed, it is now clear that such a central mode is best regarded as the precursor of the superlattice Bragg peak to appear below T_c. The true Bragg peak has a vanishing k and ω width, reflecting the long-range and static character of the low-temperature phase order; the 'precursor Bragg' central peak must have a non-zero (but critically-narrowing) width in q and ω, reflecting the finite (but growing) range and lifetime of the order associated with the precursor clusters.

When interpreted within this cluster picture, the molecular-dynamics results shown in fig. III.24 thus reveal an aesthetically pleasing unity in the critical behaviour of order–disorder and displacive systems— a unity which it is natural to regard as an expression of universality. However, it is clear that the cluster picture also prompts many questions which the molecular-dynamics technique cannot satisfactorily answer. The evident need for a precisely formulated and quantitatively predictive expression of the cluster picture presents a considerable challenge to analytic theory—the concern of the following section.

III.4.2.2. *Analytic studies in d=1*

It is clear that the principal challenge posed by the cluster picture is to establish formally and explicitly that cluster-wall motion and low-temperature phonons do indeed constitute the essential components of the excitation spectrum immediately above the critical temperature of a displacive system. Perturbation theories employing a high-temperature phonon basis (§ III.3) do little to illuminate this issue. Considerably more insight is to be found in a series of non-perturbative studies of the $d=1$ form of the model (III.4.1) (Krumhansl and Schrieffer 1975, Aubry 1975, 1976 a, b, Varma 1976) to which we devote this section.

Sadly, the very attraction of the $d=1$ model—its unique analytic tractability—reflects a severe limitation of its utility: it displays a phase transition, only in a limited sense, precisely at $T=0$ K (cf. § II.2.2, and below). As a result, studies of the $d=1$ case must be regarded as supplying, at best, circumstantial evidence as to the behaviour of the higher-dimensional ($d=3$, in particular) systems of greatest physical interest.

The grounds for this note of reserve become apparent when one appreciates that, for *any* system of the form (III.4.1), irrespective of its g or d value, the double-well character of the local potential *guarantees* the situation envisaged in the cluster picture, *at low enough temperatures*: that is, there will inevitably exist a low-temperature regime characterized by a high degree of (at least) short-range order, where each atom is predominantly localized on one or other side of its local potential barrier. Equally inevitably, at low enough temperatures the spectral function $\mathscr{S}(\mathbf{k}, \omega)$ will clearly be characterized by a 'low-temperature phonon' resonance and a 'central component': the real issue is whether or not a regime with this character exists to an identifiable extent *above* the temperature at which true long-range order sets in (i.e. T_c itself), so that the guaranteed 'central component' is not merely a Bragg peak. The lack of long-range order at any finite temperature in $d=1$ (and in manifestly 'finite'-sized-systems: cf. Beck 1976) obviously guarantees that such a regime exists 'above T_c'; it is not so obviously guaranteed in systems with finite T_c.

Nevertheless, while this recognition takes the edge from the challenge identified above, it is highly instructive to see the strategy by which the cluster picture is explicitly vindicated—particularly since the procedure highlights the *reason* for the absence of long-range order in $d=1$, which we have not yet discussed, and which, from the perspectives of any quasi-harmonic phonon theory, is wholly non-trivial.

The strategy is simply described: one takes some essentially exact but opaque result of conventional statistical mechanics, and shows that it can be interpreted in terms of an explicit and transparent phenomenology incorporating low-temperature phonons and cluster walls as elementary excitations. This procedure was devised by Krumhansl and Schrieffer (1975), and implemented by them in a semi-quantitative fashion, subsequently refined both quantitatively and conceptually by Currie *et al.* (1980) (as reported by Bishop (1978 a, 1978 b)), and by Mazenko and Sahni (1978).

The evolution of this particular art is a reflection of a general growth in interest in the physics of strongly non-linear systems (cf. Scott *et al.* 1973, Bishop and Schneider 1978). Here, however, we shall simply present, in outline, those results which do most to illuminate the structural phase transition problem, referring the reader to recent reviews by Bishop (1978 a, b) both for further detail, and for a discussion of the wider context in which the results should be viewed.

We will consider, in particular, the properties of the $d=1$ form of the model (III.4.1), in the continuum limit (whose domain of validity we define more explicitly below), in which the hamiltonian is written as a functional of two one-dimensional scalar fields $\dot{u}(x)$ and $u(x)$:

$$H=\int_x \left\{\frac{\dot{u}^2(x)}{2}+\frac{\bar{A}u^2(x)}{2}+\frac{Bu^4(x)}{4}+Ca^2\left(\frac{du(x)}{dx}\right)^2\right\}, \tag{III.4.4}$$

where

$$\int_x \equiv (1/a)\int_0^L dx,$$

and $L=Na$ is the length of the system. The equilibrium properties of the model are specified by the canonical partition function which (for the classical problem in hand) separates into two functional integrals associated, respectively, with kinetic and potential contributions to the configurational energy:

$$Z=Z_{\dot{u}}Z_u \tag{III.4.5a}$$

$$Z_{\dot{u}}=\int_{\{\dot{u}(x)\}}\exp\left[-\beta\int_x [\dot{u}^2(x)/2]\right] \tag{III.4.5.b}$$

$$Z_{\dot{u}}=\int_{\{u(x)\}}\exp\left[-\beta\int_x \left\{\frac{\bar{A}u^2(x)}{2}+Ca^2\left(\frac{du(x)}{dx}\right)^2+\frac{Bu^4(x)}{4}\right\}\right]. \tag{III.4.5c}$$

The kinetic-energy contribution is simply found:

$$Z_{\dot{u}}=[P_0\beta^{-1/2}]^N, \tag{III.4.6}$$

where P_0 is an unimportant phase space constant. In *this* (one-dimensional) case the potential-energy contribution (III.4.5c) can also be determined with the aid of transfer operator techniques (Scalapino *et al.* 1972). In the thermodynamic $N\to\infty$ limit one may effectively write

$$Z_u=[\exp(-\beta\varepsilon_0)]^N, \tag{III.4.7a}$$

where ε_0 is the smallest eigenvalue of the transfer operator. In the continuum limit the determination of ε_0 becomes *formally* equivalent to the calculation of the smallest eigenvalue of a Schrödinger-like equation (no 'real' quantum mechanics is involved!) describing a single particle with parameter- and temperature-dependent effective mass, moving in an isolated double-well potential, whose spatially-varying part is simply $V_s(u)$. At low temperature (large effective masses), the smallest eigenvalue has the form (Krumhansl and Schrieffer 1975):

$$\varepsilon_0=E_0(T)-t_0(T), \tag{III.4.7b}$$

where, in the pseudo-quantum framework, one may identify E_0 with the lowest energy level in one (effectively harmonic) well, and t_0 with the energy-lowering effect of pseudo-quantum tunnelling between the two wells.

To secure a completely successful matching to the phenomenological theory (below) it is particulary important (Bishop 1978a, b, Currie *et al.* 1980) that the tunnelling contribution $t_0(T)$ be evaluated in a low-temperature approximation consistent with that employed in the evaluation of $E_0(T)$. The results of these calculations (which we shall not describe in detail), together with eqns. (III.4.5)–(III.4.7) complete the evaluation of the partition function, and thence of all the associated thermodynamic properties: we shall cite explicitly only the result for the internal energy $U=-\partial/\partial\beta \ln Z$, which proves to be particularly illuminating. The result may be written as (Bishop 1978 b)

$$U=U_{\mathrm{p}}+U_{\mathrm{c}} \tag{III.4.8a}$$

$$U_{\mathrm{p}}=(N-n_m N_{\mathrm{c}})k_{\mathrm{B}}T \tag{III.4.8b}$$

$$U_{\mathrm{c}}=N_{\mathrm{c}}\left(E_{\mathrm{c}}^0+\frac{1}{2}k_{\mathrm{B}}T+(n_m-1)k_{\mathrm{B}}T\right), \tag{III.4.8c}$$

where

$$n_m=2 \tag{III.4.9a}$$

$$E_{\mathrm{c}}^0=\frac{4}{3}\frac{|\bar{A}|^{3/2}C^{1/2}}{B}=\frac{8}{3\sqrt{2}}\frac{V_s^0}{s^{1/2}} \tag{III.4.9b}$$

$$N_{\mathrm{c}}=4\left(\frac{3}{\pi}\right)^{1/2}N\left(\frac{sE_{\mathrm{c}}^0}{k_BT}\right)^{1/2}\exp\left(-\frac{E_{\mathrm{c}}^0}{k_BT}\right), \tag{III.4.9c}$$

while $V_s^0\equiv|\bar{A}^2|/4B$ is the well depth and (cf. eqn. (II.2.8)) s is the dimensionless ratio

$$s\equiv\frac{|\bar{A}|}{8C}. \tag{III.4.10}$$

Apart from the numerical prefactor in (III.4.9 c) (whose accuracy is limited by that of the W.K.B. approximation used to evaluate the tunnelling contribution $t_0(T)$: cf. Bishop 1978 b), these equations are asymptotically exact at low temperatures and in the continuum limit, which we may now identify more explicitly as the regime of small s (eqn. (III.4.10)). Furthermore physical insight into the basis of the continuum approximation is to be found in the phenomenological interpretation of eqns. (III.4.8)–(III.4.10), to which we now turn.

Intuitively, we have argued, it is clear that the low-temperature properties of the $d=1$ system should be dominated by two types of excitation (low-temperature phonons and cluster walls) which should, therefore, be reflected in the results of the calculation sketched above. One finds a formal basis for this assertion—and a key to the development of a convincing phenomenology of eqns. (III.4.8)—in the general observation (Bishop 1978 a) that the partition function (III.4.5), viewed as a 'path integral' (see, for example, Feynman 1972), will be dominated by the stationary phase trajectories defined by the Euler–Lagrange equation of motion for the field $u=u(x,t)$, which follows from eqn. (III.4.4):

$$\left[c_0^2\frac{\partial^2}{\partial x^2}-\frac{\partial^2}{\partial t^2}\right]u(x,t)=\frac{\partial V_s(u(x,t))}{\partial u(x,t)}=\bar{A}u(x,t)+Bu^3(x,t), \tag{III.4.11}$$

where

$$c_0\equiv(2Ca^2)^{1/2} \tag{III.4.12}$$

defines a characteristic velocity.

Among the solutions to this equation one readily identifies two classes with the character we have anticipated.

Firstly, linearizing (III.4.11) about the profile of a homogeneously and completely ordered phase, and assuming periodic boundary conditions, one identifies the approximate solutions

$$u(x,t) = \pm Q_0(0) + \delta u(x,t), \qquad \text{(III.4.13}\,a)$$

with

$$\delta u(x,t) = \delta u \exp\left[i(kx - \omega(k)t)\right] \qquad \text{(III.4.13}\,b)$$

$$\omega^2(k) = 2|\bar{A}| + (c_0 k)^2. \qquad \text{(III.4.13}\,c)$$

These equations define a class of periodic, spatially-extended solutions describing collective small-amplitude oscillations about the potential well minima: they represent the low-temperature phonons.

Secondly, taking the full non-linear equation (III.4.11), and applying 'domain-wall' boundary conditions, gives the exact solutions

$$u(x,t) = Q_0(0) \tanh\left[\frac{x - vt}{d_c(1 - v^2/c_0^2)^{1/2}}\right], \qquad \text{(III.4.14}\,a)$$

with

$$d_c = \frac{a}{(2s)^{1/2}}. \qquad \text{(III.4.14}\,b)$$

These equations define a class of solutions localized (at any time t) on the scale of the system, but with spatial extent d_c (eqn. (III.4.14 b)) large in comparison with the lattice spacing (an explicit expression of the continuum approximation $s \ll 1$ used to replace (III.4.1 a) with (III.4.4)). They represent cluster or microdomain walls (separating regions where $u(x,t) \approx -Q_0(0)$ from regions where $u(x,t) \approx +Q_0(0)$, alternatively known as 'kinks', 'solitary waves' or (somewhat inappropriately) 'solitons'. Their propagation, with velocity v, is accompanied by large-amplitude inter-well atomic motion. The energy $E_c(v)$ required to create a single wall is readily found by substituting the solution (III.4.14) into the hamiltonian (III.4.4): the result is

$$E_c(v) = E_c^0(1 - v^2/c_0^2)^{-1/2}, \qquad \text{(III.4.14}\,c)$$

where E_c^0 is precisely the characteristic energy defined in (III.4.9 b). The functional form of $E_c(v)$ reflects the pseudo-relativistic invariance of the equation of motion (III.4.11), where c_0 plays the role of the light velocity (Aubry 1976 a).

Equations (III.4.13) and (III.4.14) constitute the essential elements of a phenomenological cluster theory of the $d = 1$ system, at low temperatures. The gross features of (III.4.8) and (III.4.9) in particular are now immediately transparent (Bishop 1978 b). Ignoring for the moment the terms involving n_m, one sees that U_c is readily interpreted as the internal energy associated with N_c cluster walls, regarded as quasi-particles each with rest ('potential') energy E_c^0 and mean kinetic energy $k_B T/2$. The latter (equipartition) result is a natural consequence of the 'newtonian' character of the low-velocity cluster walls (cf. eqn. (III.4.14 c)) which will predominate at low temperatures, as the 'activation form' of N_c (eqn. (III.4.9 c)) itself implies. The remaining contribution U_p is then clearly the internal energy associated with the low-temperature phonons.

The additional terms in U_c and U_p are expressions of more subtle effects which have only recently come to be appreciated (Currie *et al.* 1980, Bishop 1978*a, b*).

The remaining contribution $(n_m - 1)\, k_B T \equiv k_B T$ to the cluster-wall term U_c reflects the fact that, in addition to its simple, and very general, translation ('Goldstone') mode, an isolated domain wall has a second purely internal vibrational mode (a feature specific to the particular "ϕ^4" form of our potential $V_s(u)$: see, for example, Goldstone and Jackiw 1975).

The modifying term $-n_m N_c k_B T$ in U_p is a reflection of the fact that phonon and cluster-wall solutions cannot simply be superposed in this essentially non-linear problem: the presence of cluster walls does not change the phonon dispersion (III.4.13*c*) (within the approximations inherent in the functional integral calculation of Z), but it *does* alter the phonon density of states, each of the N_c cluster walls reducing the number of phonon states by $n_m = 2$—the number of modes of a single wall.

With these interpretations it is clear that the *temperature* derivatives of thermodynamic potentials (the specific heat in particular) will be dominated by the phonon contributions at low temperatures, where cluster-wall contributions are exponentially small. However, it is also clear, even without explicit analysis, that the *field* derivatives (specifically the macroscopic order parameter and its susceptibility) will be dominated by the behaviour of the cluster walls. In particular it is evident that the absence of macroscopic order (in zero field) at any finite temperature is an immediate consequence of the presence of cluster walls which, as pointed out by Varma (1976), are the essential microscopic ingredients of the general Landau argument (Landau and Lifshitz 1958: see also Fisher 1972), demonstrating the impossibility of a phase transition at finite temperature in a one-dimensional system. In effect, for any $T > 0$, a homogeneously ordered phase is free-energetically unstable against the formation of cluster walls, which destroy true long-range order, but leave short-range order with a spatial coherence length that is effectively given by the typical cluster size (Krumhansl and Schrieffer 1975)

$$\xi \approx L/N_c \tag{III.4.15}$$

and thus grows exponentially eqn. (III.4.9*c*) at low temperatures.

Thus far we have focused exclusively on the *equilibrium* properties of the $d = 1$ model. The low-temperature dynamical properties are less tractable analytically, and there exist no formally exact results for the spectral function. Nevertheless it has been shown that the cluster picture passes the less stringent tests set by the results of molecular-dynamics studies of the low-temperature dynamics (Aubry 1976 a, b: cf. figs. III.25 (i), (ii); Schneider and Stoll 1975, Koehler *et al.* 1975).

In particular the assumption that the cluster walls constitute an ideal one-dimensional gas of pseudo-relativistic particles (cf. eqn. (III.4.14*c*)) leads to a qualitatively satisfactory account of the behaviour of the low-frequency component observed in the spectral function (figs. III.25 (i) and (ii). The phenomenology can be developed in a particularly simple fashion when $k_B T \gtrsim E_c^0$; the majority of domain walls then have velocities $v \approx \pm \bar{v}$, where $\bar{v}$ is slightly smaller than the characteristic velocity c_0 (eqn. (III.4.12)). A calculation of the spectral function ignoring both the finite thickness of the cluster walls, and the effects of the low-temperature phonons then gives (Aubry 1976 b)

$$\mathscr{S}(\mathbf{k}, \omega) = \frac{S_0 Q_0^2(0) \xi^2/\bar{v}}{\{1 + [(\omega\xi/\bar{v}) + k\xi]^2\}\{1 + [(\omega\xi/\bar{v}) - k\xi]^2\}}, \tag{III.4.16}$$

where S_0 is a constant, and ξ is the correlation length (III.4.15). For $k\xi \gtrsim 1$, (III.4.16) predicts cluster-induced scattering at a finite energy transfer, in accord with figs. III.25 (i) and (ii). For $k\xi \lesssim 1$, the scattering resides in a central component of width

$$\omega_Q \approx \bar{v}/\xi. \tag{III.4.17}$$

The 'critical' narrowing of the central component (as $T \rightarrow 0$) is thus a reflection of the growing time $\tau_Q \equiv 1/\omega_Q$ taken for a typical domain wall (of speed $\bar{v}$) to travel the 'critically' growing distance ξ.

We now turn to consider the ranges of temperature and wavevector over which the one-dimensional system exhibits this cluster-induced central peak.

A number of conceptually-different criteria have been devised to locate the temperature T_I below which a central peak might be expected to appear. Each of these criteria leads to a result of the form

$$k_B T_I = \alpha_T \frac{|\bar{A}|^{3/2} C^{1/2}}{B} \sim \frac{V_s^0}{s^{1/2}} \tag{III.4.18}$$

with α_T a (criterion-dependent) constant of order unity; this result is in rough agreement with molecular-dynamics experiments. However, the accord may be somewhat misleading, since the homogeneity properties of the $d=1$ model in the continuum limit (Aubry 1975) actually *guarantee* the functional form of (III.4.18). Nevertheless, it is worth recording some of the ways in which (III.4.18) has been 'derived', since each gives its own distinctive slant to the physics of the 'crossover' occurring in the vicinity of T_I.

The result (III.4.18) follows most immediately from the development of this section if one regards $k_B T_I$ as the activation energy E_c^0 (cf. eqn. (III.4.9*b*)) for the cluster walls (Bishop and Krumhansl 1975).

A less immediate but perhaps (cf. § III.4.3.1) more widely applicable criterion leading to (III.4.18) identifies T_I with the temperature $T_0^{\text{I.M.}}$ (§ II.3.2) at which, within the independent-mode approximation, the system exhibits a first-order phase transition (S. Aubry, private communication). This identification is immediately plausible when one recalls (§ II.3.2) that $T_0^{\text{I.M.}}$ locates the temperature below which, in a variational sense, the dynamic behaviour is 'better' represented as quasi-harmonic motion about positions displaced from the high-symmetry sites (i.e. by low-temperature phonons) than quasi-harmonic motion about the high-symmetry sites themselves (high-temperature phonons). It then seems reasonable to conclude that the I.M.A. registers the quasi-local ordering precipitated by cluster formation, but identifies the latter with macroscopic ordering since it cannot accommodate the strongly non-linear order-parameter configurations associated with the cluster walls, whose persistence to $T = 0$ K actually precludes true long-range order.

Physically similar motivation underlies the identification of T_I with the temperature at which the probability distribution function for the local displacement begins to exhibit a double-peaked structure (Aubry 1975, Bruce and Schneider 1977) or with the temperature at which the specific heat exhibits a maximum (Aubry 1975): the former is a microscopic, and the latter a macroscopic, signal of the formation of cluster-induced local precursor order.

Finally, we outline one further argument which highlights the displacive/order–disorder character of the crossover occurring at T_I (Bruce 1978). As the temperature is lowered, conventional high-temperature phonon softening leads to a growingly coherent motion characterized by a correlation length which may readily be shown to have the form

$$\xi \sim \frac{C^{2/3}a}{(Bk_{\mathrm{B}}T)^{1/3}}. \qquad \text{(III.4.19)}$$

It is then plausible that the motion of groups of the ξ/a atoms in a cluster of size ξ will begin to assume an order–disorder character when the 'cluster-potential' barrier

$$V_\xi \equiv \frac{\xi}{a} V_s^0 \qquad \text{(III.4.20)}$$

to their collective inter-well motion becomes comparable with the thermal energy (whereafter the low-temperature approximation (III.4.15) begins to be a better approximation than the high-temperature approximation (III.4.19)). Equating V_ξ to $k_{\mathrm{B}}T_I$, and employing (III.4.19), one recovers (III.4.18).

The wavevector k_I bounding the regime within which strongly non-linear effects appear in the spectral function (cf. § III.2.5) also has a functional dependence on the model parameters that is wholly determined by the homogeneity properties of the one-dimensional continuum limit (Aubry 1975):

$$ak_I = \alpha_k \left(\frac{|\bar{A}|}{C} \right)^{1/2} \sim s^{1/2}, \qquad \text{(III.4.21)}$$

with α_k a dimensionless constant. Again this result may be 'derived' (perhaps 'interpreted' would be more appropriate) in an ambiguous variety of ways, of which we cite only two.

Firstly, with the perspectives of the low-temperature (order–disorder) spectrum, k_I may be thought of as the inverse of the cluster-wall thickness d_{c} (eqn. (III.4.14 *b*), Bishop 1978 a).

Secondly, adopting the perspectives of the high-temperature (displacive) spectrum, one sees immediately from (III.4.4) that (III.4.21) simply characterizes the range of wavevectors associated with unstable (bare) high-temperature phonons.

III.4.3. *Discussion*

III.4.3.1. *Open questions*

The insights afforded by the studies surveyed in the two preceding sections may be tentatively summarized in the following 'cluster picture' of a displacive phase transition.

With the onset of criticality the growing coherence of the motion drives a displacive/order–disorder crossover at some characteristic temperature $T_I > T_{\mathrm{c}}$. In the displacive region ($T > T_I$), the atoms oscillate predominantly about their high-symmetry sites and the excitation spectrum is dominated by a softening high-temperature phonon. In the order–disorder regime ($T_I > T > T_{\mathrm{c}}$), the atoms oscillate about quasi-equilibrium positions displaced from the high-symmetry sites to form clusters of dynamic precursor order: under the stabilizing influence of the clusters the high-temperature phonon evolves into its low-temperature counterpart without exhibiting a complete softening, while the critical slowing-down is transferred to a cluster-induced 'quasi-Bragg' central component.

This picture prompts numerous questions—both of principle and of detail—the most significant of which we attempt to identify in this section: to the majority we will be able to offer, at best, speculative answers.

The central question of principle naturally concerns the extent to which the cluster picture does indeed capture the essential physics of displacive phase transitions in the three-dimensional systems of primary physical interest.

The direct supporting evidence is not extensive.

The molecular-dynamics studies reviewed in §III.2.5 and §III.4.2.1 have been extended to the $d=3$ form of the model (III.4.1) (Schneider and Stoll 1978), but with model parameters ($g \approx 0{\cdot}94$ in the notation of eqn. (III.4.3)) which do not give such a good realization of a displacive system. Indeed, the relatively little mode softening displayed by the model, and the large range of reciprocal space over which a central peak is evident, are more reminiscent of the simple order–disorder behaviour characteristic of an atom in an isolated double-well potential, than of the cooperative cluster-driven order–disorder behaviour arising in a regime of high intersite correlation.

The programme of analytic studies outlined in the preceding section has not been extended to systems of higher dimensionality. Two key (and presumably related) obstacles may be identified. Firstly, with the exception of certain highly-anisotropic 'quasi-one-dimensional' systems (Bishop and Krumhansl 1975), the equilibrium statistical mechanics of systems with $d > 1$ is not susceptible to the techniques which, in the $d=1$ case, establish the exact results used to assess the predictions of a cluster phenomenology. Secondly, the development of a cluster phenomenology is hampered by the fact that the higher-dimensional analogues of the cluster-wall solutions (III.4.14) to the Euler–Lagrange equation lack the formal stability properties characteristic of the one-dimensional solution (see Jackiw 1977 for a discussion in a field-theoretic context). While this observation implies that the $d=1$ cluster phenomenology is not (trivially) generalizable to higher dimensions, it does not imply that cluster walls are unimportant: excitations which are formally unstable may contribute significantly to finite-temperature properties (witness high-temperature phonons!) provided only that their lifetime is sufficiently long (Bishop 1978 a). This observation does, however, highlight the need to take careful account of those features of real systems which will influence cluster-wall stability. Of these the most obvious (and universal) feature is the underlying lattice structure, whose essential role is witnessed by the pathologies of the continuum limit in $d \geqslant 2$. The stabilizing influence of symmetry-breaking defects (§III.5) may prove to be still more important in the interpretation of experiments on real systems.

The circumstantial evidence for the picture is not much more impressive. Studies of the single-particle distribution function of the $d=3$ model (III.4.1) show that (cf. the $d=1$ case, §III.4.2.2) it does not develop a double-peaked structure as $T \to T_c^+$ if the model is sufficiently displacive (Bruce and Schneider 1977). Again this null result does not signify the absence of precursor cluster order—merely that the local equilibrium properties of the displacive system are overwhelmingly dominated by essentially harmonic fluctuations (cf. further comments below, and in the following section).

Of the other criteria used to locate (or interpret) the crossover temperature of the $d=1$ model, only that which identifies T_I with $T_0^{\text{I.M.}}$ (the transition temperature in the independent-mode approximation) seems immediately generalizable. It seems likely (but has not, to our knowledge, been formally demonstrated) that the true transition temperature T_c lies below $T_0^{\text{I.M.}}$. Since the ratio $T_c/T_0^{\text{I.M.}}$ tends to unity in the displacive limit (II.3.37)), the argument of the preceding section then suggests a cluster-dominated region with a width t_I which, like the conventional critical region itself (cf. eqn. (II.3.50)), vanishes with the parameter s (eqn. (II.2.8)) or g (eqn. (III.4.3)) characterizing the order–disorder 'content' of the model.

Indeed—and here we turn to the strongest evidence for the picture, its intuitive appeal—it seems plausible that the critical region *is* a 'cluster-dominated' regime, with 'clusters' (suitably defined) the direct space expressions of 'critical fluctuation'; t_I and k_I would then simply characterize the ranges of reduced temperature and wavevector over which non-classical critical behaviour is evident. With these perspectives, it is tempting to view the universality of the critical region as the universality of cluster behaviour, with each universality class hallmarked by its cluster topology, and the extent of the deviations from classical critical behaviour determined by (or reflected in) the degree to which cluster walls are important constituents of the critical excitation spectrum.

To give some substance to these conjectures it is helpful to juxtapose the dimensionality dependence of the conventional critical properties of the model (III.4.1) with the dimensionality dependence of its cluster topology.

In one dimension, deviations from classical critical behaviour are at their largest, as witnessed by the complete suppression of the critical temperature. In progressively higher dimensions the departures from classical behaviour, as reflected in critical exponents, become progressively smaller, the classical predictions becoming exact for $d>4$.

In one dimension, the notion of a cluster (and a cluster wall) is unambiguous and, indeed, the cluster walls are explicitly identifiable as the constituents of the excitation spectrum which promote the suppression of the critical temperature. In higher dimensions, the notion of a cluster is less clear cut: in $d=2$, pockets of ordering of one polarity can be embedded in large regions of the other polarity; in $d>2$, networks of opposite polarity can interpenetrate one another and the concept of a cluster is no longer unambiguous. (See Kretschmer *et al.* 1976, for a discussion.)

These trends may be interrelated through the dimensionality dependence of the R.G. fixed-point potential for the model, $V_s^*(u)$ (cf. §II.4.5.2 *c*) which one might reasonably think of as a 'cluster potential' since it characterizes the limiting critical behaviour of groups of atoms in a volume of linear dimension ξ (cf. eqn. (III.4.20)). This potential itself has a definite location on the displacive/order–disorder spectrum (Bruce 1978). In one dimension, $V_s^*(u)$ lies at the order–disorder end of the spectrum, and is characterized by an infinite value of the ratio g^* (cf. eqn. (III.4.3)) of well-to-critical thermal energy. In higher dimensions, g^* becomes progressively smaller (fig. II.4): in $d=3$, Wilson's approximate recursion formula gives $g^*\approx 0{\cdot}55$; for $d=4-\varepsilon$, the result is $g^*=\varepsilon/4+0(\varepsilon^2)$, so that, as $d\to 4$, the fixed-point potential tends to the displacive end of the spectrum.

While g^* is not strictly universal (D. J. Wallace, private communication) so that the quantitative significance of these results is dubious, their qualitative implications are clear: the stronger the critical fluctuations, and the greater the associated deviations from classical critical behaviour, the more pronounced is the order–disorder character of the fixed-point potential, and the more plausible is the role of the cluster wall.†

These considerations bring us naturally to the second major question of principle—the extent to which the essentially non-linear effects highlighted by the cluster picture are actually captured by (static and dynamic) phonon-based renormalization-group theories of the critical region. In fact this question is inextricably bound up with the reliability with which 'small' parameters—

† For a recent development of these ideas see Bruce *et al.* 1979 b.

principally $\varepsilon = (4-d)$—may be used to truncate an otherwise exact but intractable set of R.G. equations.

The manifest inappropriateness of this technique in 'low' dimensions is a clear reflection of the fact that the fixed point lies too far to the order–disorder end of the spectrum to be located through studies of the R.G. flow emanating from a displacive ('physical': §II.4.3.3) hamiltonian. By contrast, an exact R.G. treatment of the equilibrium properties of the $d=1$ model is quite straightforward when the 'physical' hamiltonian has an order–disorder (Ising) character (Nelson and Fisher 1975).

The reliability of the technique in $d=3$ was discussed in §II.4.5.1. Two further comments are appropriate here.

First of all it is important to appreciate that the consistent solution of a set of R.G. equations even to *leading order* in ε implicitly sums an *infinite* series of perturbation contributions in the phonon-coupling constant (cf., for example, eqn. (II.4.93)).

On the other hand it is now known that the ε-expansion does not produce a convergent series, and it appears that the *large-order* behaviour reflects 'tunnelling phenomena' which it is natural to associate with cluster-wall motion (Wallace 1978; see also Mazenko and Sahni 1978).

A third issue of importance is the extent to which the cluster picture is conceptually compatible with R.G. predictions. Specifically, eqn. (III.3.39)—whose validity transcends that of the ε-expansion (Halperin *et al.* 1976)—implies that the critical spectral function of the model (III.4.1) is dominated by a central component residing in the dressing of a high-temperature soft phonon through its non-linear coupling to a heat-diffusion mode; the development of this section suggests a critical spectral function dominated by a cluster-induced 'quasi-Bragg' peak.

Though clearly very different, the two modes of thought are, we believe, not incompatible. The phonon-based theory asserts that the dominant critical order-parameter fluctuations decay at a rate set by thermal diffusion; it does not expose the qualitative form in which the order-parameter fluctuations are realized. The cluster picture asserts that the dominant critical order-parameter fluctuations are realized in the motion of cluster walls; it does not characterize this motion quantitatively. The two points of view may be synthesized in the single conclusion that the dominant fluctuations in the order parameter are to be associated with the diffusion of cluster walls with a diffusion constant $D_c \sim D_T$ (Halperin and Varma 1976). The characteristic width of the $T > T_c$ central peak, eqn. (III.3.39), can then be thought of (cf. eqn. (III.4.17) *et seq.*) as the inverse of the time taken for a typical cluster wall to diffuse the critically-growing distance $\xi = 1/\kappa$. The conventional heat-diffusion central peak appearing below T_c as a 'halo' on the superlattice Bragg peak (cf. the discussion following eqn. (III.3.15)) may be interpreted in a similar fashion.

Nor, we believe, is there any conflict between cluster and phonon-based theories over the existence of additional (non-critical) structure in the spectral function: the cluster picture (with the backing of universality arguments) strongly suggests the existence of such structure, associated with fluctuations *within* the clusters; current R.G. theory does not capture such structure (cf. §III.3.3.2) but does not disprove its existence.

Of the numerous questions of detail raised by the cluster picture, we make brief mention of those with a bearing on some of the experimental results to be discussed in the following section.

Firstly, it is evidently of interest to establish more precisely the character of precursor clusters—their shapes and the distribution of their sizes. The precise definition of the term 'cluster' is clearly all-important here (Kretschmer *et al*. 1976); however one or two very general remarks can be made. Specifically, while in $d=1$ the most probable cluster size coincides with the correlation length, in higher dimensions this will not be the case. One expects (cf. Kadanoff 1971) a spectrum of cluster sizes with a maximum linear dimension set (roughly) by the correlation length, and a minimum set (presumably) by the value of the correlation length at the temperature t_I where clusters first become viable.

A second question to be addressed is the extent of the cluster-induced precursor order—i.e. the magnitude of the typical local cluster-induced distortion Q_I. In the strict order–disorder limit where (low-temperature) phonons carry vanishing weight, Q_I clearly coincides with the root-mean-square local displacement $\langle u^2\rangle^{1/2}$—both quantities being determined by the location $Q_0(0)$ of the minimum in the single-particle potential (the zero-temperature order parameter). In the order–disorder critical regime of a displacive system, we no longer expect such a correspondence: Q_I represents a collective (temperature-dependent) property not determined simply by the characteristics of the local potential: we anticipate that its value (at T_c) will be small on the scale of the zero-temperature order parameter, and small relative to the r.m.s. local displacement at T_c—the latter inequality reflecting the fact that the majority of modes in a displacive system have an essentially harmonic phonon character: the fluctuations about the distorted quasi-equilibrium positions will thus be large in comparison with the local quasi-equilibrium distortions themselves.

In characterizing the extent of the cluster-induced precursor order, one must therefore discriminate against the quasi-harmonic fluctuations by which it will otherwise be masked. This aim is realized in the choice

$$Q_I^2 \equiv \langle u^2\rangle^{\mathrm{C.P.}} \equiv \alpha_u \int_{\mathrm{C.P.}} \mathscr{S}(\mathbf{k}, \omega), \tag{III.4.22}$$

which identifies the precursor order with that part of the local mean-squared fluctuation characterized by frequencies in the central-peak range: the integral represents the net ($\mathbf{k}$- and ω-integrated) central-peak intensity, with α_u the appropriate normalization constant. The conceptual merit of this choice is clear; its practical merit will be evident in the following section. Notice that, in the order–disorder limit, the result $Q_I^2 = Q_0^2(0)$ is recovered; it is, of course, in the displacive regime that the existence of a non-zero value of Q_I (i.e. the existence of a central peak) represents a non-trivial cooperative phenomenon.

Thirdly, we remark that the limiting frequency $\omega_\infty(T_c)=\omega_\infty^c$ of the soft phonon also provides an interesting quantity characteristic of the cluster picture and (presumably) closely related to the precursor order parameter $Q_I(T_c)$. Indeed, given the empirical result that deep in the low-temperature phase, and into the fringes of the critical region, the soft-mode frequency varies linearly with the macroscopic order parameter $Q_0(T)$ (cf. fig. III.7 (*a*))

$$\omega_\infty(T \ll T_c) \approx \alpha_\omega Q_0(T), \tag{III.4.23 a}$$

one might expect (cf. §III.4.2.1) that the saturation frequency would be well approximated by

$$\omega_\infty^c \approx \alpha_\omega Q_I(T_c). \tag{III.4.23 b}$$

The relationship (III.4.23*a*) has itself yet to be put onto a firm theoretical basis: presumably it may be justified as a high-frequency approximation to the low-temperature phonon self-energy. The relationship (III.4.23*b*) is undiluted surmise.

We conclude this section with a brief reference to a number of further studies relevant to the issues we have surveyed.

Cook (1977) has extended Frenkel's (1946) phenomenological droplet theory of heterophase fluctuations near a first-order transition to the problem of precursor-order-induced quasi-elastic scattering at a continuous phase transition. His arguments display many of the features implied by the cluster picture: in particular the crossover temperature T_I is identified with a 'mean-field' transition temperature; however, the analysis does not accommodate critical fluctuations correctly.

Stamenkovic *et al*. (1976) have attempted to develop a unified microscopic theory of the order–disorder/displacive spectrum of the hamiltonian (III.4.1). In the spirit of the cluster picture, they suppose that the ordering process should be characterized by *two* parameters—one reflecting the degree of local precursor order, and the other controlling the degree of long-range order. Their arguments, though qualitatively illuminating, rely on a combination of independent-site and independent-mode approximation schemes, and are thus invalid in detail.

Binder *et al*. (1974, 1975) and Kretschmer *et al*. (1976) have developed a general cluster phenomenology of critical dynamics offering, in particular, a basis for dynamic scaling. Their studies do not otherwise illuminate the problems we have discussed.

III.4.3.2. *Precursor effects in real systems*

The cluster picture has two essential elements. It embodies, first of all, the key realization that the approach to criticality, in any system, is *necessarily* accompanied by the formation of precursor order with growing spatial and temporal coherence. It suggests, secondly, that the growing precursor order will be manifested in the growth of clusters inducing a two time-scale order–disorder component into the dynamical behaviour of a displacive system.

The first element is largely self-evident, and is implicit in the simplest soft-phonon theory. Its explicit and central role in the cluster picture is, however, well merited: for it is the key to an immediate understanding of a whole range of experiments demonstrating, in effect, that short-range precursor order induces, above T_c, very similar effects to those which long-range (macroscopic) order induces in the low-temperature phase. We discuss a number of illustrative examples below.

The second element is wholly non-trivial. The prediction of a two time-scale dynamics—not an obvious result of phonon-based theory—we have already effectively assessed against experiment in the course of §III.3.3.1: we shall have little to add on this issue. However, the notion of an order–disorder component in the dynamics of a displacive system focuses attention on certain quantities—the precursor order parameter Q_I in particular—which do not figure explicitly in the phonon-based theory: these we shall discuss quantitatively in the light of experiments on $SrTiO_3$.

In fig. III.33 we gather together a number of experimental results which have already appeared in the course of these articles and which may now be profitably reconsidered as expressions of precursor order. We will discuss each of the results in qualitative terms before returning to examine those relating to $SrTiO_3$ in more quantitative detail.

Fig. III. 33.

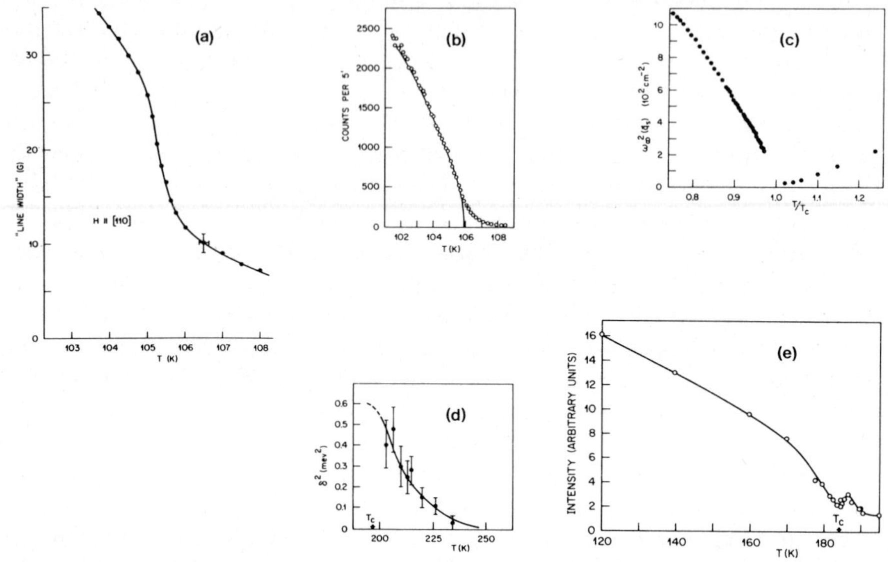

Precursor effects near structural phase transitions. (*a*) shows the effective width of an E.P.R. line, split by macroscopic order in $SrTiO_3$ (W. Berlinger, private communication), (*b*) the sum of superlattice Bragg and central-peak neutron-scattering intensities in $SrTiO_3$ (after Töpler *et al.* 1977), (*c*) the square of the soft-mode frequency in $SrTiO_3$ (after Steigmeier and Auderset 1973, and Shapiro *et al.* 1972), (*d*) the difference between high- and low-frequency soft-mode self-energies in $RbCaF_3$ (after Almairac *et al.* 1977, and (*e*) the intensity of the hard $227\,\mathrm{cm}^{-1}$ E_g mode in $KMnF_3$ (after Bruce *et al.* 1980).

Figure III.33 (*a*) effectively combines the E.P.R. results shown in figs. II.17 and II.25: it shows the temperature dependence of the width of an E.P.R. line in a sample of $SrTiO_3$ transforming to a monodomain low-temperature phase. For the line in question, the shift in Larmor frequency of the (Fe^{3+} V_0) E.P.R. centre varies linearly with the rotation angle of the surrounding oxygen octahedron about the monodomain axis—the local ordering coordinate for an Ising universality-class phase transition (cf. §II.5.3.3). In such a case the shape of the E.P.R. line effectively mirrors the shape of the distribution function of the local distortions set-up by motion slow on the time scale of the E.P.R. probe; the latter time scale is itself determined by the E.P.R. line width (see, for example, Abragam 1961). The width of the E.P.R. line thus gives a measure of the contributions to the local root mean-square distortion made by Fourier components of the displacement field with frequencies small compared with the inverse of this time scale—the 'resolution' of the E.P.R. probe. Deep in the low-temperature phase, the octahedra oscillate predominantly about positions distorted by $\pm Q_0(T)$ from the high-symmetry site, and the E.P.R. 'line' actually consists of two components (symmetrized through the existence of macroscopic domains of opposite polarity); the 'width' of the line—the separation of its two components—then gives a measure of the true equilibrium order parameter. As $T \to T_c^-$, however, the order parameter tends towards zero, ultimately becoming small in comparison with the (slow) fluctuations about the equilibrium sites: the two lines merge into a single component, and the influence of the truly static equilibrium order is lost in a background of precursor order that is effectively static on the time scale of the E.P.R. probe. The precursor order itself disappears (the E.P.R. line narrows to a weakly-temperature-dependent minimum) as the upper boundary of the critical region is approached. Evidently this phenomenon sets a limit on the accuracy with which an E.P.R. experiment can locate the transition temperature and measure the order-parameter exponent β.

Figure III.33 (*b*) reproduces the results (fig. III.3) for the temperature dependence of the quasi-elastic neutron-scattering cross-section in the vicinity of the Brillouin-zone corner in $SrTiO_3$. The intensity may be regarded as a measure of the contributions to the local mean-square value of the ordering coordinate, made by Fourier components with wavevectors lying within the instrumental resolution. (The contributions at phonon frequencies are also filtered out.) Deep in the low temperature phase, the macroscopically occupied 'mode' of wavevector $\mathbf{q}_s$ dominates, and the intensity varies as the square of the order parameter. Again, however, deviations from the simple power-law behaviour occur when the order parameter is no longer large compared with the ($\mathbf{q} \approx \mathbf{q}_s$, slow) critically-enhanced fluctuation contributions, which produce order that is 'effectively' macroscopic, and static, on the scale set by the resolution of the experiment. Again this precursor phenomenon hampers accurate determination of T_c and β.

Figure III.33 (*c*) synthesizes the results for the soft-mode frequency, ω_∞, in $SrTiO_3$ obtained by neutron-scattering techniques above T_c (fig. III.4) and light-scattering measurements (on the A_1 symmetry mode) in the ordered phase (fig. III.7 (*a*)). Little comment is necessary: we simply reiterate that the growing precursor order has a stabilizing influence on the soft phonon above T_c qualitatively similar to that exerted by the macroscopic order below the phase transition (where, again, a t^β behaviour is exhibited in an intermediate range of temperatures—the basis for equation (III.4.23 (*a*)).

Figure III.33 (*d*) presents the same kind of result in a different guise: it recalls (cf. fig. III.12) the behaviour of the difference δ^2 between the high- and low-frequency soft-mode self-energies in the case of $RbCaF_3$. (Our choice of the $RbCaF_3$ data at this point is prejudiced: it will be recalled that, possibly as a result of deconvolution difficulties, neutron-scattering experiments on the perovksites seem to suggest a variety of forms for the temperature dependence of the parameter δ^2. The form obtained in the case of $RbCaF_3$ seems to us the most plausible.) In the low-temperature phase, for which no data are available, simple theory (ignoring critical fluctuations) again predicts a behaviour following the square of the macroscopic order parameter (cf. eqn. (III.3.16) *et seq.*); again it is natural to view the non-zero value at and immediately above T_c as an expression of the existence of precursor order which, on the time scale of the soft phonon, is effectively static.

Lastly, in fig. III.33 (*e*), we recall the measurements (fig. III.11) of the intensity of the 227 cm^{-1} hard E_g mode in $KMnF_3$, whose Raman activity is induced by the ordering of the soft-mode coordinates. Again, with the approach to T_c, the intensity deviates from the simple $|t|^{2\beta}$ behaviour appropriate deep in the ordered phase, and persists for the now familiar reasons into the high-temperature phase.

It is to be emphasized that, with the exception of the behaviour depicted in figs. III.33 (*c*) and (*d*), none of these results carries with it an immediate inference that the precursor order in which it originates has a non-trivial character: *qualitatively* each result could be understood within a simple softening-phonon theory. To establish the appropriateness of the cluster picture of the precursor order, and to characterize it quantitatively, further considerations are necessary: with these intentions we examine the results for $SrTiO_3$ (figs. III.33 (*a*)–(*c*)) in more detail.

That the precursor order detected by the E.P.R. study is not soft-phonon induced is clear from its time scale: one may estimate from the measurements that the local precursor order in question persists for a time longer than 10^{-8} s—much too long to be associated with soft-phonon contributions. The contributing fluctuations can only be those which give rise to the central component: indeed, an upper limit to the width of the central peak may be inferred from such measurements with the aid of further model-based calculation (Müller *et al.* 1974, Reiter *et al.* 1980). It follows that the E.P.R. line width of T_c affords an immediate measure of the precursor order parameter as defined in (III.4.22): the E.P.R. probe constitutes an explicit realization of the low-pass filter necessary to discriminate against the fast quasi-harmonic fluctuations which otherwise dominate the local behaviour (Bruce *et al.* 1979 a). With this identification one finds (cf. von Waldkirch *et al.* 1973)

$$Q_I(T_c) \approx 0{\cdot}4°, \qquad \text{(III.4.24 } a\text{)}$$

which is to be compared with a zero-temperature order parameter $Q_0(0) \approx 2°$ (von

Fig. III.34

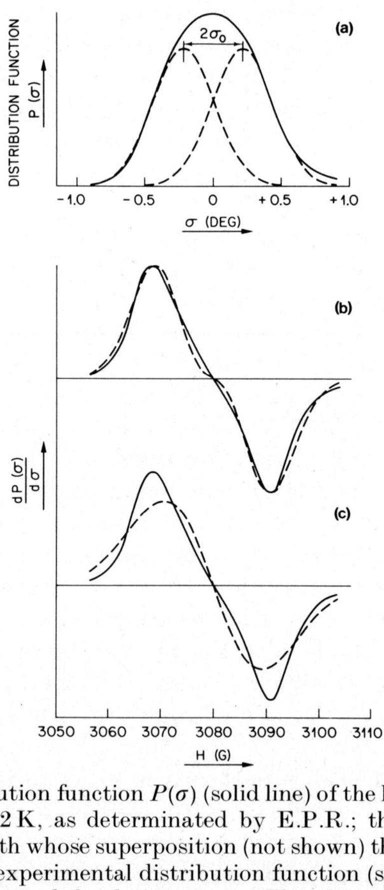

(*a*) The experimental distribution function $P(\sigma)$ (solid line) of the local ordering coordinate σ, in $SrTiO_3$, at T_c +0·2 K, as determinated by E.P.R.; the dashed lines define two displaced gaussians with whose superposition (not shown) the spectrum is modelled. (*b*) The derivative of the experimental distribution function (solid line) and of its double-gaussian representation, defined in (*a*). (*c*) The derivative of the experimental distribution function (solid line), and of its optimal single-gaussian representation.

Waldkirch *et al.* 1973), and a local root mean-square fluctuation near T_c of $\langle u^2 \rangle^{1/2} \approx 2°$ (Meyer *et al.* 1978).

Further, more direct evidence for the appropriateness of the cluster picture of the precursor order—and a modified estimate of its extent—is to be found in the *shape* of the E.P.R. line immediately above the critical point (Feder 1974, Bruce *et al.* 1979 a). Its perceptibly non-Gaussian character (figs. III.34 (*a*) and (*c*)) in a small range of temperature above T_c (Müller and von Waldkirch 1976) suggests that the distribution of local quasi-static distortions which it mirrors might be better modelled by a superposition of two gaussians (figs. III.34 (*a*) and (*b*)). This concrete, if crude, expression of the expectations of the cluster picture suggests that, close to but above T_c, the oxygen octahedra oscillate predominantly about quasi-equilibrium positions displaced by an amount (σ_0 in fig. III.34 (*a*))

$$Q_I(T_c) \approx 0{\cdot}22° \tag{III.4.24 b}$$

from the high-symmetry site.

The precursor order measured by the neutron-scattering experiment (fig. III.33 (*b*)) is manifestly cluster- rather than phonon-induced—the soft-phonon contributions lying outside the frequency window of the experiment. With the assumption that the wavevector window of the experiment is sufficiently wide to capture the bulk of the central -peak intensity, one may apply (III.4.22) (calibrating α_u by the Bragg intensity below T_c) to give the estimate

$$Q_I(T_c) \approx 0{\cdot}28°. \tag{III.4.24 c}$$

From the wavevector resolution of the experiment, one may deduce that the degree of precursor order implied by this result has a spatial coherence of at least 25 lattice spacings.

Finally, we turn to consider the behaviour of the soft-mode frequency. In the low-temperature phase, over a wide temperature range, the frequency is well described by (III.4.23 *a*) with a proportionality constant $\alpha_\omega \approx 0{\cdot}69$ THz deg^{-1}. The conjecture (III.4.23 *b*) combined with the estimates (III.4.24 *b*) and (III.4.24 *c*)—we take their mean—suggests a saturation frequency $\omega_\infty(T_c) \approx 0{\cdot}17$ Thz, in reasonable accord with the experimental value of $0{\cdot}13 \pm 0{\cdot}02$ THz.

This success, together with the close correspondence of the two very differently-based estimates (III.4.24 *b*, *c*) of the precursor order parameter, lends considerable plausibility to the cluster picture. However, it must be emphasized that the time scale of the precursor order—i.e. the central-peak width—is no less problematic for the cluster picture than it is for the phonon-based R.G. calculations (the two theories being, we believe, practically if not conceptually equivalent). Thus, with the same caveats as noted in § III.3.3.1., it seems unlikely that the *time scale* of the precursor phenomena (in effect, the *motion* of the clusters) in $SrTiO_3$ can be explained without invoking extrinsic mechanisms. One must then enquire to what extent the *other* characteristics of the cluster behaviour (the saturation frequency ω_∞^c, the precursor order parameter Q_I, indeed the very existence of clusters) might also reflect extrinsic influences. This is an issue which we can properly address only in the light of the results surveyed in the following section.

We must emphasize that our preoccupation, throughout this section, with the perovskites in general, and $SrTiO_3$ in particular, reflects the fact that, with their relatively well-defined soft modes, they constitute the best realizations of displacive systems where the existence of a non-zero precursor order parameter and of a central peak are, we believe, of greatest significance (cf. eqn. (III.4.22) *et seq.*). By comparison, the experimental evidence for a non-zero precursor order parameter in the hydrogen-bonded ferroelectrics (specifically KH_2PO_4 and its isomorphs) is actually much better documented; its significance, however, is rather less clear. The (magnetic-resonance) experiments in question have recently been surveyed by Blinc (1978 and by Müller (1979); consequently we shall merely record the key references, emphasizing the results of N.M.R. rather than E.P.R. techniques, since the interpretation of the latter is complicated by the possible influence of the E.P.R. centre upon the local dynamics which it probes (cf. Müller *et al.* 1976, Müller and Berlinger 1978).

The pioneering work is that of Blinc and Bjořkstam (1969), who found that the ^{75}As nuclear magnetic resonance in KH_2AsO_4 displays a line spectrum with the symmetry of the low-temperature phase up to 60 K above the phase transition. Adriaenssens (1975) has since shown that it is possible to account fully for these observations with a model presupposing the existence of fully polarized mobile clusters, having a lifetime of at least 10^{-3} s, and moving at a rate such that a typical As nucleus remains within a particular cluster for between 10^{-5} and 10^{-8} s.

The ambiguous character of the dynamical behaviour observed in the KH_2PO_4 family (cf. comments in § III.1) makes it difficult to evaluate the significance of these results. We remark only that the time scale of the phenomena in question (compatible with the very-low-frequency dielectric relaxation observed in the paraelectric phase of KH_2AsO_4: cf. fig. III.22) again seems too long to be associated with intrinsic cluster dynamics. Similar reservations are appropriate in the interpretation of the cluster effects seen in magnetic resonance experiments on K_2OsCl_6 (Armstrong and Martin 1975) and squaric acid (Mehring and Suwelack 1979).

Finally, it is appropriate to recall the results of the neutron-scattering investigations of our other prototype phase transition—that exhibited by the moderately displacive system $Pb_5Ge_3O_{11}$ (§ III.2.2). In this case the $\mathbf{q} \approx \mathbf{q}_s = 0$ portion of the quasi-elastic peak is masked by Bragg scattering, so that the precursor order parameter cannot readily be estimated. Nevertheless the wavevector dependence of the scattering at larger momentum transfers provides (as we saw in § III.2.2.) an explicit demonstration of the existence above T_c of precursor clusters (not associated with the soft phonon) with a topology similar to that of the ferroelectric domains found in the low-temperature phase (cf. fig. III.15). Whether such structure reflects, at least partially, intrinsic weak (cf. § II.5.1.3.) critical fluctuations or whether it should be entirely attributed to the influence of crystal imperfections is again an open issue.

§ III.5. Crystal defects

III. 5.1. *Types of defects*

In this the final portion of our review, we turn to consider the influence which crystal imperfections exert on systems undergoing structural phase transitions. We shall be concerned primarily with the possibility—noted and developed to varying degrees by a number of authors (Axe *et al.* 1974, Levanyuk *et al.* 1976, Halperin and Varma 1976)—that the quasi-elastic scattering observed near structural phase transitions arises from the coupling of cystal defects with the soft mode. However, we shall also discuss some of the wider issues raised by the existence of defects, including their influence on static critical phenomena, to which we alluded briefly in

§II.5.4.4. The only defects we shall consider explicitly will be impurities: we expect that structural imperfections, such as dislocations or grain boundaries, will give rise to qualitatively similar effects.

We begin by considering the classification of defects—an issue discussed, with a variety of perspectives, by several authors (Halperin and Varma 1976, Schmidt and Schwabl 1978, Höck and Thomas 1977). It is natural, though not altogether satisfactory, to base such a classification scheme upon the symmetry of the site 'occupied' by the defect. Consider, then, a single defect localized at some specific site S. Let ΔV denote the change in the crystal potential energy resulting from the introduction of the defect. In the spirit of §I.2.1, ΔV may be expanded as a power series in the displacements of the atoms from their high-symmetry sites. The essential displacements are, naturally, those whose Fourier components are associated with the soft branch. Thus, assuming for simplicity that the soft mode $(\mathbf{q}_s j)$ is non-degenerate, and writing

$$Q\begin{pmatrix} \mathbf{q}_s+\mathbf{k} \\ j \end{pmatrix} \equiv Q(\mathbf{k}),$$

we have an expansion of the form

$$\Delta V = U_0 + \frac{1}{\sqrt{N}} \sum_{\mathbf{k}} U_1(\mathbf{k})\, Q(\mathbf{k}) + \ldots . \qquad \text{(III.5.1)}$$

The coefficients in this expansion do not, in general, conserve wavevector because the defect, by its very nature, breaks the translational symmetry. More importantly, for our present purposes, the coefficients in the expansion (III.5.1) embody the essential symmetry properties of the defect site S. Specifically, for one class of defect site, symmetry permits a non-zero value for $U_1(\mathbf{k}=0)$: we shall call a defect 'occupying' such a site 'symmetry-breaking', since it sets up a symmetry-breaking field conjugate to the local ordering coordinate. Defects 'occupying' other sites—those whose symmetry imposes the constraint $U_1(\mathbf{k}=0)=0$—we shall term 'symmetry-conserving'.

As an illustration consider the case of $SrTiO_3$. It is readily seen that any substitutional defect is necessarily symmetry-conserving; the same is true of an interstitial defect located at a point such as A (fig. III. 35). In contrast, a defect located at an interstitial site such as B (fig. III. 35) belongs to the symmetry-breaking class: the coefficient $U_1(\mathbf{k}=0)$ (we consider one specific component of the three-fold degenerate soft mode) is, in general, non-vanishing for a defect occupying such a site. It is of considerable importance to note, however, that if B constitutes a possible site for a specific defect, there must exist a further symmetry-related site (B′ in fig. III. 35) which could, equally well, be occupied by that defect: the coefficient $U_1(\mathbf{k}=0)$ for a defect occupying this site is equal in magnitude but opposite in sign to the coefficient characterizing a defect at site B. It should be clear that these remarks highlight a general feature of symmetry-breaking defects.

The inadequacies of this classification scheme should also be apparent. A defect will not be frozen at a particular site: its position will fluctuate with time. The extent to which a defect should be thought of as symmetry-breaking or symmetry-conserving then involves a careful consideration of time scales. More specificially, a defect occupying a symmetry-breaking site may be free to hop to the symmetry-related site (it may also be free to hop from one unit cell to another, but we shall not

Fig. III.35

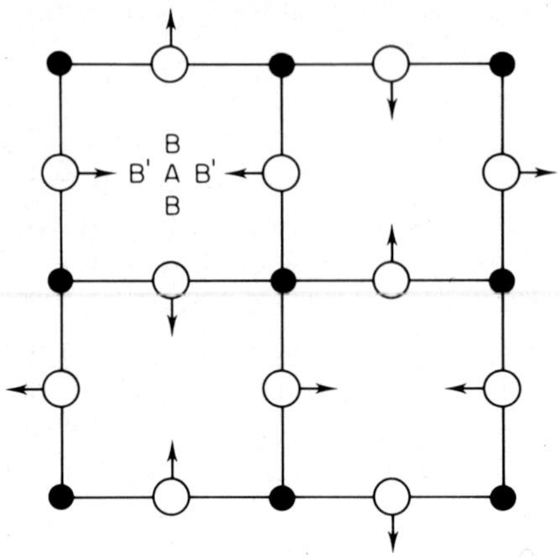

Atomic positions in a (100) plane of $SrTiO_3$, above T_c: the filled and open circles denote, respectively, titanium and oxygen ions. The arrows indicate the pattern of displacements of the oxygen ions in one possible low-temperature phase-configuration. An impurity at site A would be 'symmetry conserving'; an impurity at any of the sites B, B′ would be 'symmetry-breaking' (Halperin and Varma 1976).

discuss this case: cf. Halperin and Varma 1976). For an experiment whose time scale is long compared with the typical hopping interval, such a defect effectively belongs to the symmetry-conserving class. On the other hand we shall see (cf § III.5.3) that a defect which at high temperatures 'occupies' a symmetry-conserving site (i.e. a defect whose fluctuations break the local symmetry only on a typical phonon time scale) may so affect its environment that, below some characteristic temperature $T_l > T_c$, it may favour a non-zero value for the local ordering coordinate for times which become arbitrarily long for $T \ll T_l$. The ensuing behaviour is effectively that of a system containing 'symmetry-breaking' defects.

III.5.2. *Symmetry-breaking defects*

III.5.2.1. *The isolated defect regime*

We consider a crystal containing $N_D = pN$ symmetry-breaking defects. We suppose that the local potential experienced by a single defect is sufficiently deep that the possibility of intersite hopping may be neglected; we also neglect its oscillation about its position of local equilibrium. The essential effect of such a (frozen, Ising-like) symmetry-breaking defect is similar to that of a static local field conjugate to the ordering coordinate: it produces a static local distortion with Fourier components

$$\langle Q(\mathbf{k})\rangle_D = -\chi(\mathbf{k})\frac{U_1(\mathbf{k})}{\sqrt{N}}, \tag{III.5.2}$$

where $\chi(\mathbf{k}) \equiv \chi(t, \mathbf{k}, \omega = 0)$ is the static wavevector-dependent susceptibility of the host, $U_1(\mathbf{k})/\sqrt{N}$ is the corresponding Fourier component of the effective field set up by the defect, and we have neglected the higher-order terms which, in principle, are present in the expansion (III.5.1). Provided the coefficient $U_1(\mathbf{k})$ is regular at $\mathbf{k} = 0$ (cf. below), the direct space displacement profile of the distortion (III.5.2) decays exponentially in space, at a rate determined by the correlation length ξ of the host crystal. It follows that, provided the defects are separated, typically, by distances large compared with ξ, they may be regarded as being isolated from one another. The distortion around each defect then acts as an independent centre of elastic scattering.

For a scattering probe which couples linearly to the ordering coordinate, the cross-section at wavevector trans $\mathbf{q}_s+\mathbf{k}$ is then readily found to be of the form (Axe *et al.* 1974, Halperin and Varma 1976)

$$\mathscr{S}_D(\mathbf{q}_s+\mathbf{k},\omega)=pN\chi^2(\mathbf{k})|U_1(\mathbf{k})|^2\delta(\omega). \qquad \text{(III.5.3)}$$

In the case of a scattering probe which couples bilinearly to the ordering coordinate (cf. light scattering when the soft mode is nominally Raman inactive), the cross-section in the limit of vanishing wavevector transfer is proportional to (Levanyuk *et al.* 1976)

$$\Gamma_D(\mathbf{q}=0,\ \omega)=pN\left(\frac{1}{N}\sum_{\mathbf{k}}\chi^2(\mathbf{k})|U_1(\mathbf{k})|^2\right)^2\delta(\omega). \qquad \text{(III.5.4)}$$

The process giving rise to (III.5.3) is known as Huang scattering. Provided $U_1(\mathbf{k})$ is regular, this scattering has (essentially) a lorentzian form as a function of wavevector transfer, is linear in the concentration p, and (for $\mathbf{k}=0$) diverges as $t^{-2\gamma}$, where γ is the susceptibility exponent of the host. The defect-induced ('one-phonon') scattering thus grows more strongly (as $T\rightarrow T_c$) than the scattering associated with the host crystal.

Equation (III.5.4) suggests that the same is true of the 'two-phonon' contribution to the scattering. Assuming once again the regularity of $U_1(\mathbf{k})$, making use of appropriate scaling relations (eqn. III.3.28a); table II.1), and assuming that no untoward cancellations take place (cf. § II.5.4.1), one finds that (III.5.4) diverges as $t^{2(2\beta-\gamma)}$. This result must be viewed cautiously, however, since its derivation employs an approximation (the neglect of thermal fluctuations within the distorted regions) similar to that which we have already suggested (§ II.3.2.2) leads to an incorrect prediction for the intrinsic low-temperature phase scattering.

Some further remarks on the range of validity of these results are now appropriate.

Firstly, the assumption that the coefficients U_1 $(\mathbf{k})$ are regular at $\mathbf{k}=0$ will not be valid if the soft mode is ferroelectric and the defect is charged. We are unaware of any studies of this situation.

Secondly, we recall that eqn. (III.5.2) incorporates the effects only of the leading-order terms in the expansion (III.5.1). We do not expect that the inclusion of higher-order terms will generate qualitatively different results.

Thirdly, eqns. (III.5.3) and (III.5.4) presuppose that, as far as its scattering behaviour is concerned, the impure crystal may be regarded as a pure host subject to random symmetry-breaking fields: this approximation neglects the fact that the defect atoms will, in general, have scattering properties different from those of the host atoms. We expect that this approximation should be reasonable except in the high-temperature (low-correlation length) limit.

Fourthly, we re-emphasize that the arguments presented in this section presuppose that the defects are frozen in their symmetry-breaking positions. If the defects can, in fact, hop between symmetry-related sites, we expect that (III.5.3) and (III.5.4) will be modified only to the extent that the spectra involved assume a finite frequency width set by the inverse of the typical time interval τ_D between intersite hops.

Finally, we note that all of the foregoing results will certainly fail close enough to criticality—the regime which we discuss next.

III.5.2.2. *The critical regime*

It is clear that the arguments presented in the preceding section must break down close enough to criticality. Each symmetry-breaking defect polarizes the surrounding medium to a distance determined (in the simplest picture) by the correlation length ξ of the host; as the host-crystal transition temperature T_c is approached, these pre-ordered clusters will therefore grow in size. It is clear that they can no longer be regarded as isolated, independent, scattering centres when the correlation length becomes sufficiently large that the clusters overlap: i.e. when $\xi=\xi_p\sim p^{-1/3}a$, where a is the lattice spacing. Alternatively, and perhaps more appropriately, one might argue that the isolated defect approximation will fail when the static distortion (cf. eqn. (III.5.2)) induced mid-way between two typical defects is comparable with the root mean-square value of the thermally-fluctuating local-ordering coordinate; this criterion is fulfilled when $\xi=\xi_p\sim\alpha_p p^{-1/3}a$, where α_p depends weakly upon T and p.

The nature of the behaviour in this critical, 'interacting-defect', regime has not been fully elucidated. It is apparent, however, that it will depend crucially upon whether or not the defects are mobile (Halperin and Varma 1976). We shall discuss first of all the situation where the defects are frozen on the time scale of any experiment. In the language of critical phenomena, such defects constitute 'quenched impurities coupling linearly to the order parameter'; the problem they pose is known (for self-evident reasons) as the 'random-field' problem.

We subdivide our discussion according to the symmetry and dimensionality of the order parameter, considering first of all the simplest 'Ising-like' case where the order parameter has only the one component. Each defect will then favour local order with one of the two possible ('positive' or 'negative') senses. The distortion fields of two neighbouring defects favouring order in the same sense will merge when $\xi\approx\xi_p$ to give a larger region of local order. The distortion fields of two neighbouring defects favouring order of opposite senses will remain separated by a domain wall. Since the percolation concentration for 3-d lattices is typically much less than 50% (Essam 1972), we anticipate the formation of macroscopic interpenetrating clusters of the two senses of order.

It is not clear whether the temperature T_{tr} at which this 'phase transition' occurs will lie above or below the transition temperature T_c of the unperturbed host. The foregoing arguments indicate that it should lie close to the temperature at which $\xi=\xi_p$, suggesting $T_{tr}>T_c$. However, as emphasized by Halperin and Varma (1976), the very existence of the regions of local order acts to inhibit phonon softening, and thence to inhibit the growth in correlation length, thus depressing the temperature at which the correlation length ξ attains its 'critical' value ξ_p.

Still more fundamentally it is not clear even that a well-defined transition temperature T_{tr} (associated with some fixed point of the renormalization group) actually exists. Thus the limiting behaviour of (for example) the scattering cross sections (III.5.3) and (III.5.4) is not known.

The characteristics of the phase favoured below T_{tr} are also somewhat obscure. The state consisting of interpenetrating clusters does not have true long-range order in the conventional sense: it seems to us, however, that the differences (from a conventionally-ordered phase) would not be readily discernible, at least within a scattering experiment. It is unclear whether such a state will remain favoured with further reduction in temperature. An argument proposed by Imry and Ma (1975) seems to suggest that it will not. They consider the stability of a homogeneously-ordered Ising-like phase against a break-up into domains, given the presence of random symmetry-breaking frozen defects. The formation of a compact domain containing $\mathcal{N}$ sites 'costs' a wall energy proportional to $\mathcal{N}^{2/3}$ in three dimensions. The total number of defective sites among such a group will be $p\mathcal{N}$. According to the central limit theorem, the excess of defects favouring one orientation of the order parameter over defects favouring the other, in this group, will then be proportional to $\mathcal{N}^{1/2}$. The field energy saved by domain formation will thus also be proportional to $\mathcal{N}^{1/2}$, and will therefore be less than the wall energy.

This argument thus certainly establishes that a homogeneously-ordered phase is energetically favoured over a phase in which long-range order is destroyed by the existence of *compact* clusters of 'positive' and 'negative' ordering. However, the picture developed above suggests that one should actually enquire whether the homogeneously-ordered phase is energetically favoured over a phase in which the long-range order is broken by percolating clusters, interpenetrating to maximise the energy saving associated with the random fields. The field energy is then proportional to $\mathcal{N}$; the wall ('bond') energy cost is less obvious, but it seems likely that it will also be proportional to $\mathcal{N}$. The homogeneously-ordered phase does not therefore necessarily represent the ground state. (It is clear that it will not if the local fields are sufficiently strong: cf Aharony 1978a.) Furthermore, even if the conventionally-ordered phase were energetically favoured, it is by no means clear that it would be accessible to a system having the inhomogeneous order we envisage, on any reasonable time scale.

We turn now, briefly, to consider those cases where the order parameter has more than one component. When the order parameter has continuous symmetry, it is well-established that a phase with true long-range order does not represent the ground state in the presence of frozen symmetry-breaking impurities (Imry and Ma 1975): in this case an argument similar to that sketched above shows that a homogeneously-ordered state is *un*-stable against a break-up into compact domains within which the sense of the local order is determined by the local arrangement of the defects. Such a phase is reminiscent of a spin glass (Edwards and Anderson 1975). The only structural phase transitions to which these arguments apply directly are transformations to incommensurate phases, where the arbitrariness of the phase of the distortion profile implies an order parameter with continuous symmetry (cf. §I.4.4; §II.5.2.1). It is expected, then, that random symmetry-breaking defects will suppress long-range order in such systems, by pinning the phase of the distortion profile (Sham and Patton 1976, Imry and Ma 1975).

In general, however, the continuous symmetry of an n-component order parameter is broken by crystal anisotropy (cf. §5.2.1). We are unaware of any studies of the influence of random fields in such instances.

Thus far we have considered only static properties. In a system with frozen impurities, the dynamic effects are reflected entirely in the soft-phonon response. The limiting behaviour of the soft-mode frequency, close to the transition temperature T_{tr} (to the extent that is defined!), is not known: a speculative discussion of this point is given by Halperin and Varma (1976).

If the symmetry-breaking defects are not frozen on the time scale of experiment, the detailed critical behaviour may differ considerably from that suggested above. We defer discussion of this situation to § III.5.3.2.

III.5.3. *Symmetry-conserving defects*

In this section we consider the second of the two classes of defect described in § III.5.1, namely, defects which 'occupy' sites of sufficiently high symmetry that they do not *a priori* break the symmetry seen by the local ordering coordinate.

Defects of this type may themselves be subdivided into two groups. First there exist 'soft' defects, which *favour* the formation of the ordered phase. Such defects may be thought of as raising the 'local transition temperature' from its bulk value T_c to some characteristic temperature T_l, below which the defect preferentially occupies a symmetry-breaking position. In fact, the mobile symmetry-breaking defects discussed in the preceding section may be thought of as soft symmetry-conserving defects well below their local instability temperature T_l. Secondly there exist 'hard' defects which continue to favour their high-symmetry site even into the ordered phase: such defects act to lower the 'local transition temperature'.

The (universal) features of the critical region do not reflect these distinctions: the relevant theoretical predictions are summarized in § III.5.3.4. The problems posed by the approach to criticality, however, vary according to the relative values of T_l and T_c. We shall discuss three limiting cases: a very soft defect, having $T_l \gg T_c$ (§ III.5.3.1), a not-so-soft defect having $T_l \gtrsim T_c$ (§ III.5.3.2), and a hard defect, having $T_l < T_c$ (§ III.5.3.3).

III.5.3.1. *The soft-defect limit:* $T_l \gg T_c$

We consider a crystal containing defects which, at high (but still physically interesting) temperatures 'occupy' symmetry-conserving sites. To expose the essential behaviour as clearly as possible, we make a number of simplifying assumptions. We suppose, in the first instance, that the host crystal can be treated with the framework of quasi-harmonic lattice dynamics. We assume that the defects are substitutional and that the introduction of a particular defect changes only the (harmonic part of) the local single-particle potential at the defect site. The expansion (III.5.1) then assumes the form

$$\Delta V = U_0 + \frac{1}{N} \sum_{\mathbf{k}_1 \mathbf{k}_2} U_2 \, Q(\mathbf{k}_1) Q(\mathbf{k}_2), \qquad \text{(III.5.5)}$$

where the coefficient U_2 is independent of wavevector, and we have, once again, assumed a scalar order parameter.

We consider the local susceptibility $\chi_D(\omega)$ measuring the response of a defect to a local field: Elliott and Taylor (1967) calculated this quantity within the quasi-harmonic approximation. When the influence of all other defects is ignored, the result is

$$\chi_D(\omega) = \frac{\chi_L(\omega)}{1 + U_2 \chi_L(\omega)}, \qquad \text{(III.5.6)}$$

where $\chi_L(\omega)$ is the local susceptibility of the unperturbed host

$$\chi_L(\omega) = \frac{1}{N} \sum_{\mathbf{k}} \frac{1}{\omega(\mathbf{k})^2 - \omega^2} \qquad \text{(III.5.7)}$$

The vanishing of the denominator in (III.5.6) locates the frequency of any local mode associated with the defect. If the coefficient U_2 is negative (the defining characteristic of a 'soft' defect), the local-mode frequency will soften as the bulk-phonon mode frequency $\omega(k=0)$ decreases and the (local) host susceptibility grows. According to (III.5.6), the local-mode frequency will 'vanish' and the static local defect susceptibility will 'diverge' if a characteristic temperature T_l is reached such that

$$1+U_2\chi_L(0)\Big|_{T_l}=0. \tag{III.5.8}$$

Taken at face value, (III.5.6)–(III.5.8) suggest that T_l locates the temperature of a 'local phase transition' at which the crystal becomes unstable against the local mode associated with each defect (Höck and Thomas 1977, Schmidt and Schwabl 1977). Indeed an extended analysis along similar lines (Höck and Thomas 1977) explicitly suggests that, below T_l, there will exist a static distortion field around each defect site.

It is clear, however, that this picture must be incorrect in detail. In the first instance a static *local* susceptibility cannot exhibit a true divergence: local susceptibilities reflect the strength of local fluctuations and are necessarily finite. The 'divergence' of $\chi_D(0)$ suggested by (II.5.6) and (III.5.8) reflects the inadequacies of the (quasi-harmonic or mean-field) approximations implicit in its derivation. By the same token a *sharp* phase transition, characterized by a well-defined instability temperature, is an essentially cooperative many-body phenomenon: the temperature T_l cannot therefore locate a sharp instability, as long as the local mode is of finite spatial extent; correspondingly, one cannot expect a truly static distortion field around each defect below T_l.

To give a wholly satisfactory account of the behaviour near T_l, it would be necessary to treat the non-linear dynamics of both host and defect: we are unaware of any attempt to confront this problem. Nevertheless, the analysis outlined above does serve to suggest the likely behaviour (Höck and Thomas 1977, Schmidt and Schwabl 1977, Höck *et al*. 1979). Specifically, we anticipate that there will exist a characteristic temperature T_l (*not* sharply-defined) separating two regimes in which the local dynamical behaviour, near a defect, is qualitatively different. In the regime $T\gtrsim T_l$ the defect sees an effective single-particle potential which has a single minimum; there is a fairly well-defined local mode associated with its motion. On cooling, the increasing fluctuations of the soft-phonon coordinates change this effective potential, and the local mode softens. At $T\approx T_l$ the effective potential develops a double-minimum form, the depth of the wells growing as the temperature is further reduced. Thus in the regime $T\lesssim T_l$ it is reasonable to expect that the defect response will exhibit the by now familiar two components—a relatively fast oscillatory motion in each well of the effective potential, and a relatively slow hopping between the minima, manifesting itself in quasi-elastic scattering. As the temperature is lowered and the correlation length of the host increases, the effective mass of the defect (reflecting the size of the locally polarized region which follows its dynamics) will grow, and the width of the quasi-elastic scattering (and, possibly, the frequency of the oscillatory mode: cf. Höck *et al*. 1979) will decrease. The system will then display effectively the behaviour described in §III.5.2.1. Its behaviour close to criticality will be either that discussed in §III.5.2.2, or that discussed below in §III.5.3.4, according to whether the time scale which the central-mode motion has acquired when the critical region is reached is, or is not, long on the scale set by experiment.

III.5.3.2. *The intermediate case*

In developing the picture described in the preceding section we assumed, implicitly, that the local instability temperature was large compared with the transition temperature of the host, so that the host correlation length near T_l would be small compared with the inter-defect distance, and the defects could, correspondingly, be regarded as independent of one another at this temperature. This situation should be realized in the limit of particularly soft defects (a large negative coupling coefficient U_2) of low concentrations. If these conditions are not fulfilled—i.e. if the magnitude of U_2 is near the critical value below which (III.5.8) cannot be satisfied, or if the defect concentration is high—the isolated defect approximation will break down before the defect-mode instability temperature is reached. The approach to criticality then presents somewhat different problems—most notably the need to consider, from the outset, the nature of the excitations in a system with a high concentration of impurities.

There have been numerous attempts to develop theories of crystals containing many defects. These theories vary in sophistication from the average crystal approximation (employed by Halperin and Varma 1976) which treats the crystal as an assembly of identical 'average' atoms, to the many variants of the coherent potential approximation (CPA) as reviewed by Elliott *et al.* (1974).

Qualitatively one finds that systems containing an admixture of, say, two types of atoms often give rise to two bands of excitations: each band corresponds to the propagation of the excitation on, predominantly, one type of atom. Behaviour of this type has been demonstrated experimentally in several magnetic systems (Cowley 1976) although, in detail, the excitation spectrum may be considerably more complicated than is suggested by this simple picture (Cowley *et al.* 1977).

In the problem of interest here it is thus tempting to suppose that there will be two (important) bands of excitations—the one set propagating on the host atoms, the other set propagating on the defects. If the defects are soft, the latter band of excitations will be built from the local modes discussed in the preceding section. The approach to criticality will then be signalled by a softening of one mode of this band. This picture is implicit in the 'average-crystal' theory developed by Halperin and Varma (1976). It should be emphasized, however, that this picture *presupposes* that the excitations built from the defect modes have a spatially extended form. Whether, or when (at what temperature), this will in fact, be the case cannot itself be determined within the usual average crystal or coherent potential approximations because they assume *ab initio* that the excitations are spatially extended. Indeed, in $Rb_2Mn_{0\cdot5}Ni_{0\cdot5}F_4$ one finds experimentally a band of excitations whose spectral distribution is described well within the CPA but which (according to computer simulations) actually have a localized character (Kirkpatrick and Harris 1975).

The 'crossover' from excitations that are localized to excitations that are spatially extended seems to merit further attention. It may be possible to exploit the analogy with the much-studied problem posed by the nature of electronic states in random alloys and amorphous semiconductors, where it is now well established that there is a sharp energy, the mobility edge (Mott 1966), separating localized and extended states.

If the picture described above is correct, it is reasonable to expect that the actual transition temperature (which is expected to be sharp at least in some cases: cf. § III.5.3.4) will lie close to—perhaps be identical with—the instability temperature of the extended defect mode (note that the latter temperature cannot be determined by a relationship of the form (III.5.8), which presupposes a localized defect mode). Thus we suspect that the range of temperatues, above the phase transition, over which the defect mode is extended (making the average-crystal approximation reasonable) *and* below its condensation temperature (giving quasi-elastic scattering) will in fact be very small.

III.5.3.3. *Hard defects*

Lastly, we consider the case of 'hard' defects which tend to stabilize the high-temperature phase. In the notation developed in § III.5.3.1, such defects are characterized by a positive value of the coefficient U_2, so that (cf. eqn. (III.5.6)) the local static susceptibility of the defect is less than that of the host. In such instances we anticipate that (except within the critical region) the usual CPA methods should be quite reliable. Qualitatively we expect that defects of this type will increase the soft-phonon self-energy, resulting in a lowering of the true transition temperature, and in an enhancement of the soft-mode damping. Defect-induced quasi-elastic scattering is not predicted by any simple theory.

III.5.3.4. *The critical regime*

In the language of critical phenomena, symmetry-conserving defects (or symmetry-breaking defects that move between symmetry-related sites on the time scale of experiment) constitute 'quenched impurities coupling quadratically to the order parameter'. The problem they pose is known as the 'random bond' or 'random exchange-constant' problem, and has been the subject of considerable theoretical activity in recent years: for a review see Aharony (1978 b). Two issues of particular importance may be identified: it is of interest to establish, firstly, under what conditions a system containing such defects will, nevertheless, exhibit a sharp phase transition and, secondly, the nature of the critical behaviour in such instances.

The first of these issues is readily resolved with the aid of a pleasingly simple argument due to Harris (1974). Consider a system with a concentration p of symmetry-conserving defects; suppose, for the sake of argument, that the system does exhibit a sharp phase transition characterized by a well-defined transition temperature, and well-defined critical exponents. In the light of the discussion at the beginning of § III.5.3, we expect that the transition temperature will be a function of the defect concentration, $T_c = T_c(p)$. Now the *actual* concentration of impurities within a volume V containing $\mathcal{N}_v$ sites will be in the range $p \pm \Delta p_v$ where, according to the central limit theorem,

$$\Delta p_v \approx \left(\frac{p}{\mathcal{N}_v} \right)^{1/2} \tag{III.5.9}$$

Thus the transition temperature controlling the behaviour of the atoms within the volume V will itself be subject to a spread

$$\frac{\Delta T_{c,v}}{T_c(p)} \approx \frac{\Delta p_v}{T_c(p)} \frac{dT_c(p)}{dp}. \tag{III.5.10}$$

If the phase transition of the crystal as a whole is to *appear* sharp (as supposed), the spread in the transition temperatures of the groups of $\mathcal{N}_v \equiv (\xi/a)^d$ atoms contained within each coherence volume $V \equiv \xi^d$ must be small on the scale of the reduced temperature t. Thus, combining (III.5.9) and (III.5.10) one concludes that

$$\frac{1}{T_c(p)} \frac{dT_c(p)}{dp} p^{1/2} \left(\frac{\xi}{a}\right)^{-d/2} \ll t, \qquad \text{(III.5.11)}$$

a condition which it will be possible to satisfy arbitrarily close to the critical point only if $d\nu > 2$, or (making use of the scaling relationships listed in Table II.1) provided the specific-heat exponent satisfies

$$\alpha < 0. \qquad \text{(III.5.12)}$$

Thus the hypothesis of a sharp phase transition in the impure system is consistent only with a finite (non-divergent) specific heat.

This plausible argument finds more formal justification in R.G. analyses (Grinstein and Luther 1976, Aharony 1975 a) which also go some way towards answering the second of the two questions raised above. In these analyses the 'replica' technique devised by Emery (1975) is employed to replace the statistical problem posed by the impure, inhomogeneous system with a formally equivalent problem described by a tractable (translationally-invariant) effective hamiltonian. One finds that the presence of impurities is reflected in the existence of a term in this effective hamiltonian, whose scaling exponent (cf. § II.4.3.5) with respect to the pure-system fixed point is given by

$$\lambda_p = \alpha/\nu, \qquad \text{(III.5.13)}$$

where α and ν are the pure-system exponents. If the exponent α is negative (i.e. if the specific heat of the pure system has a cusp), the impurity-induced perturbations are 'irrelevant' and a sharp phase transition, with pure-system exponents, may occur in accord with the Harris argument. If the exponent α is positive, the impurities are 'relevant' and the system cannot simply exhibit pure-crystal critical behaviour.

Before turning to the nature of the critical behaviour in such cases, we must note that the foregoing arguments are actually incomplete: it transpires (Aharony 1975 b) that in systems with multi-component order parameters (for which α may be negative), symmetry will often permit *additional* impurity-induced perturbations with scaling exponents other than (III.5.13): these perturbations may be quite strongly relevant with respect to the host-crystal fixed point. It follows that a negative value of the pure-system specific-heat exponent constitutes a necessary but not sufficient condition that the critical behaviour will be unchanged by the presence of impurities. (The situation is reminiscent of that arising in the case of order-parameter–strain coupling: cf. § II.5.1.2 and § II.5.2.2. Indeed, the deformations set up by strain modes may be thought of as 'annealed' symmetry-conserving defects: cf. Schuster 1977.) In fact, in most of the physical systems we have encountered in the course of these articles, it would appear that the presence of symmetry-conserving defects *will* affect the critical behaviour, over a range of temperature which, in customary crossover style (cf. § II.4.3.7), will reflect the concentration of impurities and the size of the crossover exponent ϕ_p($\phi_p = \alpha$ in the simple example cited above). In practice this range of temperature may therefore be very small.

Within the asymptotic regime a number of forms of behaviour are possible. The defect-induced crossover may take the system to a new fixed point, yielding new critical behaviour (including, necessarily, a non-divergent specific heat). This appears to be the situation in the case of the short-range Ising ($n=1$) system, where an impurity-dominated fixed point has been located within a novel perturbation expansion in $\sqrt{\varepsilon}$ (Khmelnitskii 1975), and in the case of the uniaxial ferroelectric (§ II.5.1.3) where the defect-dominated regime is characterized by singularities of a bizarre form (Aharony 1976). The behaviour of systems with multi-component order parameters is less straightforward: the pure-system fixed point, stable against defects which preserve the local symmetry fully (Grinstein and Luther 1976), is unstable against defects which induce a local anisotropy favouring one component of the order parameter over the others (Aharony 1975 b). The ensuing critical behaviour is not known: the phase transition might be smeared, or of first order.

These results are, in any case, undermined somewhat by doubts as to the veracity of the replica technique (cf. Aharony 1978). Their immediate relevance to real, dirty crystals is also doubtful: *macroscopic* gradients in impurity concentrations typically produce an unavoidable smearing of the transition temperature observed, making it hard to determine critical exponents, and harder still to establish that they have the necessary foundation of an intrinsically-sharp transition.

Finally, we turn to the dynamical behaviour. The dynamic critical properties of this universality class have been examined by Grinstein *et al*. (1977) with the aid of the dynamic R.G. methods discussed in § III.3.2.1. They find that (when a 'defect-dominated', random fixed point is stable) the leading corrections to the dynamic critical exponent z are $0(\varepsilon)$, in contrast with the behaviour typical of pure systems, where the corrections are $0(\varepsilon^2)$. Calculations of the dynamic response function are also reported: the impurities result in a shape function $f_x(y)$, eqn. (III.3.31), that is qualitatively different from that expected for the pure system, being sharply enhanced at low frequencies.

The behaviour of the soft phonon itself has not been studied with a comparable degree of sophistication. Available calculations suggest however (Halperin and Varma 1976, Schmidt and Schwabl 1977, 1978) that (except possibly in the case of hard defects) the phonon frequency will tend towards a finite limiting value $\omega_\infty^c \equiv \delta(T_c)$, where δ^2 is linear in the defect concentration p,

$$\delta^2 \propto p, \tag{III.5.14}$$

and reflects the stabilizing influence of the (defect-induced) clusters of precursor order. We anticipate, however, that a more refined theory might show a limiting frequency ω_∞^c independent of p at sufficiently large defect concentrations.

III.5.4. *Discussion*

It will be clear from the preceding sections that the current theory of the role of defects near structural phase transitions is far from satisfactory: we shall review the outstanding open questions in the course of our final summary (§ III.6). In this section we survey the results of relevant experiments (to many of which we alluded in § III.2) and attempt a comparison with the theoretical predictions, such as they are.

Any such comparison must necessarily be both qualitative and tentative, not only because of the deficiencies of the theory itself, but for a number of other reasons, of which two merit special mention.

Firstly, it is a key presupposition of all the theory surveyed above that the defects in question are distributed through the crystal in a truly random fashion. In real crystals, however, there may be clustering effects or, as we have already remarked, macroscopic concentration gradients which invalidate this assumption.

Secondly, it is clear that the application of the foregoing theory necessarily requires a knowledge of the changes in the interatomic forces induced by the presence of a defect—or, at the very least, some knowledge of the qualitative character of the defect. Such information is seldom available: indeed there are relatively few instances of defect-induced scattering phenomena where the specific defects involved have been identified.

In keeping with our original survey in §III.2 we shall focus, in what follows, on scattering investigations, referring the reader to recent articles by Blinc (1978) and by Müller (1979) for discussion of the local defect-induced phenomena probed by N.M.R. and E.P.R. techniques. The experiments to be discussed may conveniently be subdivided into two categories: experiments on nominally 'pure' crystals which, nevertheless, display the influence of defects; and experiments performed on crystals containing controlled quantities of specific impurities. We shall discuss the former group first of all.

The most unambiguous demonstration of the influence of defects on scattering properties near structural phase transitions is to be found in light-scattering experiments on $KH_3(SeO_3)_2$ (Yagi *et al.* 1977) and KH_2PO_4 (Courtens 1978). The observation that the strength of the quasi-elastic scattering displayed by these systems could be strongly reduced, in the one case by repeated cycling through the transition, in the other by annealing, provides clear evidence that these central peaks do not represent ideal crystal behaviour. The specific defects responsible for the scattering have not, however, been identified. The suggestion (Courtens 1977) that the quasi-elastic peak in KH_2PO_4 might reflect the presence of deuterium ions (which would constitute 'soft defects', of the type envisaged in §III.5.3.1 or §III.5.3.2) has not been substantiated by further experiment (Courtens 1978). More encouragingly, however, the temperature-dependence of the residual central-peak intensity in KH_2PO_4, after annealing, is consistent with the strong divergence predicted by (III.5.3), with a classical susceptibility exponent $\gamma=1$.

Among the other experiments surveyed in §III.2 there are several which offer somewhat more circumstantial evidence for the influence of impurities.

The critical growth of the quasi-elastic light scattering in $SrTiO_3$ reported by Steigmeier *et al.* (1973, 1974) is certainly more consistent with the type of strong singularity suggested by the defect-dominated result (III.5.4), than with the relatively weak singularity predicted by the theory of intrinsic critical behaviour (eqn. (III.3.55.*c*)).

The quasi-elastic light scattering observed in $Pb_5Ge_3O_{11}$ (§III.2.2) also appears to grow considerably more rapidly than intrinsic theory would suggest (eqn. (III.5.4) does not, however, apply to the case of a uniaxial ferroelectric). In any event, the essentially static character of this central peak, clearly established in this case, leaves no doubt as to the involvement of defects.

As we have indicated on earlier occasions the upper limit established for the width of the quasi-elastic neutron scattering observed in $SrTiO_3$ also suggests the involvement of impurities. Halperin and Varma (1976) estimate that a defect concentration $p\approx 10^{-4}$–10^{-5} is sufficient to account for the observed central-peak intensity. However, the apparent sample-independence of the saturation frequency of the soft phonon (fig. III.4) is hard to reconcile with the prediction of the simple defect theory (eqn. (III.5.14)). The critical growth in the intensity of the quasi-elastic component, characterized by an exponent in the range 1·2–1·5 (Töpler *et al.* 1977), is also quite consistent with ideal crystal behaviour, and markedly different from the srongly singular behaviour suggested by the defect theory (eqn. (III.5.3)). We defer further comment to our concluding section.

We turn now to discuss experiments on systems doped with specific impurities. The number of such experiments is small, but on the increase.

Neutron-scattering experiments on $SrTiO_3$ reduced with hydrogen (Hastings *et al.* 1978) showed a relatively weak dependence of central-peak intensity upon defect concentration, although the transition temperature was strongly suppressed by the doping (§III.2.1, fig. III.5). The most plausible interpretation of these results is that the reduction process introduces predominantly hard symmetry-conserving defects which (cf. §III.5.3.3) should, indeed, lower the transition temperature without contributing significant quasi-elastic scattering.

Recent neutron-scattering experiments on Ba-doped $Pb_5Ge_3O_{11}$ have revealed similar behaviour (Cowley *et al.* 1980). The transition temperature is lowered and the quasi-elastic scattering enhanced somewhat—but not in proportion to the defect concentration.

The light-scattering properties of $Pb_5Ge_3O_{11}$ doped with Si (Gebhardt and Müller-Lierheim 1978, Hisano and Toda 1978) or Ba (Taylor *et al.* 1978, Hosea *et al.* 1978) have also been studied. The low-temperature phase soft-mode frequency is reduced (in accord with the lowering of the transition temperature) and its damping increased. It was also found that the Ba-doped samples exhibit a considerably stronger quasi-elastic peak, persisting over a wider range of temperatures than was observed in the case of a nominally pure system.

Finally, we consider perhaps the most extensively investigated—and in some respects the most interesting—doped systems studied to date: $KTaO_3$ containing various concentrations of Li, Na or Nb. The particular fascination of these systems lies in the fact that (nominally) pure $KTaO_3$ is an incipient ferroelectric: the ferroelectric soft mode is, apparently, stabilized by quantum fluctuations at low temperatures (cf. §II.5.4.3) so that no phase transition actually occurs in the pure system. When doped with a few percent of the aforementioned impurities, however, $KTaO_3$ exhibits dielectric behaviour indicative (cf. further comment below) of a phase transition at a finite temperature T_{tr} (Davis 1972, Höchli *et al.* 1977, 1978).

We discuss first the case of $KTaO_3$: Li. Light-scattering studies on this system (Yacoby and Just 1974, Yacoby 1978a) have indicated the existence of a defect mode (distinct from the host ferroelectric mode) showing considerable softening as the temperature is lowered toward T_{tr} (approximately 80 K for a Li concentration $p=0{\cdot}05$), and with an intensity roughly proportional to the Li concentration. Moreover, sharp ('first-order') Raman features of the host crystal, symmetry-forbidden in the pure system, were observed well above this transition temperature. Subsequent measurements (Höchli *et al.* 1978) have revealed, in addition, a dynamic central peak in the dielectric response whose (small) width decreases exponentially fast as the temperature is lowered, in accord with an earlier result reported by Yacoby (1978a). Remarkably, further experiments (Höchli *et al.* 1979) seem to indicate that below the transition temperature T_{tr} the Li-doped samples do not, in fact, exhibit (significant) 'spontaneous' polarization (i.e. the residual polarization left after application and removal of the largest coercive field that can be applied without sample breakdown). The samples do, however, exhibit a polarization if they are cooled through T_{tr} in the presence of a biasing field.

These observations prompt the following plausible, though tentative, picture. The Li impurities, substitutional for K ions, are *a priori* symmetry-conserving defects: the light-scattering measurements suggest, however, that they occupy predominantly off-centre (symmetry-breaking) positions. They should thus be regarded as soft symmetry-conserving defects below their local instability temperature T_l (§III.5.3.1). The soft defect mode observed in the light-scattering experiments is then presumably to be associated with the oscillations of the Li ions about these displaced positions, while the low-frequency response observed in the dielectric measurements reflects the slow hopping of the defect with its accompanying cloud of polarized host. The absence of a spontaneous polarization at low temperatures may be an indication of the situation envisaged in §III.5.2.2: the reorientation time for the defects is so long when the defect-centred clusters begin to overlap that the system enters a 'glassy' phase which does not possess true long-range order (cf. Noolandi and Varma 1975) unless the coherence of the different (effectively frozen) polarized clusters is enforced by the presence of a biasing field.

The behaviour of the Nb- and Na-doped systems is similar to that of $KTaO_3$: Li, at least in some respects, the Nb and Na impurities being substitutional for Ta and K ions, respectively. Light-scattering studies (Yacoby and Just 1974, Yacoby 1978 b) again suggest that these defects are soft, and already occupy predominantly off-centre positions well above the respective transition temperatures. In contrast to the Li-doped system however, the system containing Nb does apparently exhibit spontaneous polarization at low temperatures (Boatner *et al.* 1977). It is natural to correlate this difference with the fact that, according to dielectric measurements (Höchli, private communication), the Nb defect hopping time is considerably shorter than that observed in the Li-doped system, suggesting that in this case the defect-centred clusters may reorient themselves (rather than freeze incoherently) to give a phase with true long-range order.

§III.6. Conclusions

While we have touched on a number of subsidiary questions in the course of this article, it has been our primary purpose to describe and assess our current understanding of the critical dynamical behaviour and the quasi-elastic scattering phenomena displayed near displacive phase transitions. It is, therefore, on this issue that we shall focus in this our concluding summary. We shall consider successively the problems posed by model systems, and the problems posed by real crystals.

The bulk of the theoretical activity we have reviewed has been devoted to the study of model systems with an ideal, defect-free, crystal lattice. Indeed it seems to us that it is in this area (as distinct from the others touched on below) that the most significant progress has been made. The theory of non-critical quasi-elastic scattering phenomena (§III.3.1) is well established, if largely untested. The predictions of the dynamic R.G. method (§III.3.2) are explicit and precise, if conceptually opaque and entirely untested. The cluster picture described in §III.4 constitutes a conceptually compelling refinement of the simple soft-mode picture. We have argued that it is probably not incompatible with R.G. predictions, but it is clearly essential that a more concrete link be forged between them if the cluster picture is to be given a firm analytic foundation. In particular, it might prove illuminating to study with dynamic R.G. methods (cf. Freedman and Mazenko 1975, Grinstein *et al.* 1977) and the techniques of crossover theory (cf. §§II.4.3.7., II.4.5.2) the evolution in the behaviour of the spectral function of a system exhibiting a soft underdamped phonon outside the critical region. R.G. studies of this problem, performed to date, have invariably presupposed, from the outset, a soft mode that is overdamped even in the absence of critical fluctuations; they have focused, moreover, on the critical exponents and the amplitudes of critical singularities rather than on the dynamic scaling functions themselves. Calculations along the lines proposed would serve to clarify the conditions under which the distinct central peak and softening phonon observed in molecular-dynamics studies may actually be expected, and would establish the extent to which the phenomenon is indeed an expression of non-classical fluctuations. In the longer term it will also be necessary to assess the influence of dipolar forces and order-parameter symmetries and dimensionalities more complex than we have supposed (cf. the model hamiltonian III.4.1) throughout much of this paper.

In this area, then, much of the basic physics is understood; the questions which remain to be addressed (cf. §§III.3.3, III.4.3.1) are fairly clear; some of them, at least, appear tractable with available techniques.

The state of the theory of model systems with an imperfect lattice structure is, by contrast, much less satisfactory: the open questions here are of a vary basic nature. We know the likely dynamical behaviour near an isolated defect only qualitatively: a detailed study of a single soft defect in a growingly critical host would be illuminating. We know still less about the cooperative behaviour occurring in the defect-dominated critical region: the nature of the phase transition (sharp or smeared?) in the presence of effectively frozen symmetry-breaking impurities is unclear; the character of the low-temperature phase in such instances (a 'glass'?) is intriguing; and the factors which define the borderline between this case (cf. §III.5.2.2) and the rather different situation arising (cf. §III.5.3.4) when relaxation of the defect-centred clusters is possible, are clearly deserving of detailed study. It is also of particular importance to establish a better theory of the limiting behaviour of the soft phonon. In what circumstances does it saturate? Does its saturation frequency obey the 'average crystal' result (III.5.14)? The defect concentration dependence of ω_{∞}^{c} may prove the feature most capable of distinguishing between extrinsic and intrinsic central-peak phenomena in real systems.

This brings us naturally, and finally, to the problems posed by quasi-elastic phenomena in real systems: a three-fold subdivision is useful here.

Firstly, we have seen (cf. §III.3.3.1) that there exists a class of central component (most accessible to light-scattering techniques) that is unambiguously dynamic and non-critical: this phenomenon is well understood, but is irrelevant to the phase-transition problem.

Secondly, there is sparse evidence (cf. §§III.2.1, III.2.2, III.3.3.2) for quasi-elastic scattering that is critical (if only weakly) and which is characterized by time scales that are not incompatible with ideal crystal theories. Detailed interpretation of these observations is hampered, however, by the fact that they probe the two-phonon rather than the one-phonon spectral function.

The third, and by far the largest group, of quasi-elastic phenomena is that involving time scales that are very probably incompatible with defect-free crystal behaviour; the neutron scattering (§III.2.1) and E.P.R. (§III.4.3.2) measurements on $SrTiO_3$ fall into this category. It is ironic that our understanding of these observations, which prompted much of the theoretical development we have surveyed, should have progressed so little. Nevertheless, the fact remains that we do not know the extent to which such phenomena are representative of ideal crystal behaviour. It *is* clear that the phenomena observed imply the existence of clusters of critically growing precursor order. It is (fairly) clear that the time scale for this precursor order is set by crystal imperfections. However it is *not* clear what factors are predominant in controlling the *extent* of the precursor order. It is possible that the 'precursor order parameter' is set by the intrinsic order–disorder component that appears to be characteristic of the critical dynamics of a displacive system: in this view, the defects, passive as far as the static behaviour is concerned, simply control the cluster life time through some, as yet uncharacterized, pinning effect. It is also possible that the degree of precursor order is itself controlled by the local field exerted by the defect, together with the defect concentration, so that the intrinsic cluster phenomenon is totally masked. As we remarked in the preceding section, there is some, far from conclusive, evidence favouring the former view.

It is apparent, however, that this issue can be resolved (if at all) only by further careful experiments on crystals characterized with greater thoroughness than has been customary in this area (cf. Elwell and Scheel 1975). The defects primarily responsible for the cluster pinning must be identified and the extent of their role established by the systematic investigation of crystals containing the defects in various specified concentrations. The high-quality flux-grown crystals of $SrTiO_3$ that have recently become available (Scheel 1976, Scheel *et al.* 1976) merit particular attention.

Of the other experiments one might envisage, further investigations of the wavevector-dependence of quasi-elastic scattering phenomena would be of particular value. The wavevector width of quasi-elastic peaks has, to date, received far less attention than the frequency width: thus it remains unclear whether or not the q-width (in $SrTiO_3$, for example) is well described, as it has been customary to suppose, by the phenomenological form (III.1.2), where all wavevector-dependence is assumed to reside in the soft-mode dispersion. It might well be that crystal defects significantly modify not only the time scale of the precursor order (i.e. the characteristic frequency ω_Q) but also its length scale (i.e. the correlation length ξ)—at least when the correlation length becomes comparable with the inter-defect distance.

In short, we believe that the experimental observation of quasi-elastic scattering phenomena, and the theoretical activity which it has motivated, have together contributed much to the understanding of structural phase transitions. Additional experiments are needed, both to test existing theories of extrinsic and intrinsic central-peak behaviour, and to guide the further theoretical development that will certainly be necessary if this diverse phenomenon is to be satisfactorily understood.

Acknowledgments

We are happy to acknowledge the contributions made to our understanding of structural phase transitions by collaboration and discussion with (among many others) S. Aubry, J. D. Axe, H. Beck, A. R. Bishop, W. Cochran, G. J. Coombs, E. Courtens, U. T. Höchli, T. J. Hosea, M. Iizumii, J. A. Krumhansl, K. A. Müller, A. F. Murray, R. J. Nelmes, T. Schneider, E. Stoll, W. Taylor and H. Thomas. The financial support of the Science Research Council is also acknowledged gratefully.

References

Abragam, A., 1961, *The Principles of Nuclear Magnetism* (Oxford: Clarendon).
Adriaenssens, G. J., 1975, *Phys. Rev.* B, **12,** 5116.

AHARONY, A., 1975a, *Phys. Rev.* B, **12,** 138; 1975b, *Ibid.*, **12,** 1049; 1976, *Ibid.*, **13,** 2092; 1978a, *Ibid.*, **18,** 3318; 1978b, *J. Magn. mag. Mater.*, **7,** 198.

AHARONY, A., and HOHENBERG, P. C., 1976, *Phys. Rev.* B, **13,** 3081.

ALMAIRAC, R., ROUSSEAU, M., GESLAND, J. Y., NOUET, J., and HENNION, B., 1977, *J. Phys., Paris*, **38,** 1429.

ANDERSON, P. W., 1960, *Fizika Dielectrikov*, edited by G. I. Skanavi (Moscow: Akad. Nauk. SSR).

ANDRADE, P. R., RAO, A. D., KATIYAR, R. S., and PORTO, S. P. S., 1973, *Solid St. Commun.*, **12,** 847.

ARMSTRONG, R. L., and MARTIN, C. A., 1975, *Phys. Rev. Lett.*, **35,** 294.

AUBRY, S., 1975, *J. chem. Phys.*, **62,** 3217; 1976a, *Ibid.*, **64,** 3392; 1976b, *Ferroelectrics*, **12,** 263; 1977, *Ibid.*, **16,** 313.

AXE, J. D., KEATING, D. T., and MOSS, S. C., 1975, *Phys. Rev. Lett.*, **35,** 530.

AXE, J. D., SHAPIRO, S. M., and SHIRANE, G., 1974, *Anharmonic Lattices, Structural Transitions and Melting*, edited by T. Riste (Leiden: Noordhoff), p. 23.

AXE, J. D., and SHIRANE, G., 1974, *Phys. Rev.* B, **8,** 1965.

BARTOLOME, M. L. P., and KERR, W. C., 1977, *Solid St. Commun.*, **21,** 253.

BAUSCH, R., and HALPERIN, B. I., 1978, *Phys. Rev.* B, **18,** 190.

BECK, H., 1976, *J. Phys.* C, **9,** 33.

BECK, H., and MEIER, P. F., 1973, *Helv. phys. Acta*, **46,** 480.

BINDER, K., STAUFFER, D., and MUELLER-KRUMBHAAR, H., 1974, *Phys. Rev.* B, **10,** 3853; 1975, *Ibid.*, **12,** 5261.

BIRGENEAU, R. J., KJEMS, J. K., SHIRANE, G., and VAN UITERT, L. G., 1974, *Phys. Rev.* B, **10,** 2512.

BISHOP, A. R., 1978a, *Proceedings of the International Conference on Lattice Dynamics*, edited by M. Balkanski (Paris: Flammarion), p. 144; 1978b, see Bishop and Schneider (1978), p. 85.

BISHOP, A. R., and KRUMHANSL, J. A., 1975, *Phys. Rev.* B, **12,** 2824.

BISHOP, A. R., and SCHNEIDER, T. (editors), 1978, *Solitons and Condensed Matter Physics* (Berlin, Heidelberg: Springer-Verlag).

BLINC, R., 1978, *Ferroelectrics*, **20,** 121.

BLINC, R., and BJORKSTAM, J. L., 1969, *Phys. Rev. Lett.*, **23,** 788.

BLINC, R., and ZEKS, B., 1974, *Soft Modes in Ferroelectrics and Antiferroelectrics* (Amsterdam: North-Holland).

BOATNER, L. A., HÖCHLI, U. T., WEIBEL, H. E., 1977, *Helv. phys. Acta*, **50,** 620.

BRUCE, A. D., 1978, see Bishop and Schneider (1978), p. 116; 1980, *Phys. Rev. Lett.*, **44,** 1.

BRUCE, A. D., and COWLEY, R. A., 1973, *J. Phys.* C, **6,** 2422.

BRUCE, A. D., MÜLLER, K. A., and BERLINGER, W., 1979a, *Phys. Rev. Lett.*, **42,** 185.

BRUCE, A. D., SCHNEIDER, T., and STOLL, E., 1979b, *Phys. Rev. Lett.*, **43,** 1284.

BRUCE, A. D., and SCHNEIDER, T., 1977, *Phys. Rev.* B, **16,** 3991.

BRUCE, A. D., TAYLOR, W., and MURRAY, A. F., 1980, *J. Phys. C*, **13,** 483.

BURNS, G., and SCOTT, B. A., 1972, *Phys. Lett.* A, **39,** 177.

COCHRAN, W., 1960, *Adv. Phys.*, **9,** 387.

COOK, H. E., 1977, *Phys. Rev.* B, **15,** 1477.

COOMBS, G. J., and COWLEY, R. A., 1973, *J. Phys.* C, **6,** 121.

COURTENS, E., 1977, *Phys. Rev. Lett.*, **39,** 561; 1978, *Ibid.*, **41,** 1171.

COURTENS, E., GAMMON, R. W., and ALEXANDER, S., 1979, *Phys. Rev. Lett.*, **43,** 1026.

COWLEY, R. A., 1967, *Proc. phys. Soc.*, **90,** 1127; 1970, *J. phys. Soc. Japan*, Suppl., **28,** 239; 1974, *Ferroelectrics*, **6,** 163; 1976, *AIP Conf. Proc.*, **29,** 243.

COWLEY, R. A., AXE, J. D., and IIZUMI, M., 1976, *Phys. Rev. Lett.*, **36,** 806.

COWLEY, R. A., and COOMBS, G. J., 1973, *J. Phys.* C, **6,** 143.

COWLEY, R. A., COOMBS, G. J., KATIYAR, R. S., RYAN, J. F., and SCOTT, J. F., 1971, *J. Phys.* C, **4,** L203.

COWLEY, R. A., LUCAS, B., and HOSEA, T. J., 1980, (unpublished).

COWLEY, R. A., and SHIRANE, G., 1978, *J. Phys.* C, **11,** L939.

COWLEY, R. A., SHIRANE, G., BIRGENEAU, R. J., and GUGENHEIM, H. J., 1977, *Phys. Rev.* B, **15,** 4292.

COX, D. E., SHAPIRO, S. M., COWLEY, R. A., EIBSCHUTZ, M., and GUGGENHEIM, H. J., 1979, *Phys. Rev.* B, **19,** 5754.

Currat, R., Müller, K. A., Berlinger, W., Denoyer, F., 1980, *Phys. Rev.* B, **17,** 2937.

Currie, J. F., Krumhansl, J. A., Bishop, A. R., and Trullinger, S. E., 1980 (to be published).

Darlington, C. N. W., Fitzgerald, W. J., and O'Connor, D. A., 1975, *Phys. Lett.* A, **54,** 35.

Darlington, C. N. W., and O'Connor, D. A., 1976, *J. Phys.* C, **9,** 3561; 1978, *Proceedings of the International Conference on Lattice Dynamics*, edited by M. Balkanski (Paris: Flammarion), p. 750.

Davis, T. G., 1972, *Phys. Rev.* B, **5,** 2530.

Denoyer, F., Lambert, M., Comes, R., and Currat, R., 1976, *Solid St. Commun.*, **18,** 441.

Durvasula, L. N., and Gammon, R. W., 1977, *Phys. Rev. Lett.*, **38,** 1081.

Edwards, S. F., and Anderson, P. W., 1975, *J. Phys.* F, **5,** 965.

Elliott, R. J., Krumhansl, J. A., and Leath, P. L., 1974, *Rev. mod. Phys.*, **46,** 465.

Elliott, R. J., and Taylor, W., 1967, *Proc. R. Soc.*, **296,** 161.

Elwell, D., and Scheel, H. J., 1975, *Crystal Growth from High-Temperature Solutions* (London: Academic Press), Chap. 9.

Emery, V. J., 1975, *Phys. Rev.* B, **11,** 239.

Enz, C. P., 1972, *Phys. Rev.* B, **6,** 4695; 1974, *Solid St. Commun.*, **15,** 459.

Essam, J. W., 1972, *Phase Transitions and Critical Phenomena*, Vol. II, edited by C. Domb and M. S. Green (New York: Academic Press), p. 197.

Feder, J., 1971, *Solid St. Commun.*, **9,** 2021; 1974, *Anharmonic Lattices, Structural Transitions and Melting*, edited by T. Riste (Leiden: Noordhoff), p. 113; 1976, *Local Properties at Phase Transitions*, edited by K. A. Müller and A. Rigamonti (Amsterdam: North-Holland), p. 312.

Ferrell, R. A., Menyhard, N., Schmidt, H., Schwabl, F., and Szepfalusy, P., 1967, *Phys. Rev. Lett.*, **18,** 891.

Feynman, R. P., 1972, *Statistical Mechanics* (Reading, Massachusetts: Benjamin).

Fisher, M. E., 1972, *Essays in Physics*, Vol. 4 (London: Academic Press), p. 43.

Fleury, P. A., and Lyons, K. B., 1976, *Phys. Rev. Lett.*, **37,** 1088; 1978, *Proceedings of the International Conference on Lattice Dynamics*, edited by M. Balkanski (Paris: Flammarion) p. 731.

Fleury, P. A., Scott, J. F., and Worlock, J. M., 1968, *Phys. Rev. Lett.*, **21,** 16.

Folk, R., Iro, H., and Schwabl, F., 1978, *Proceedings of the International Conference on Statistical Physics, Annals of Israel Physical Society*, Vol. 2 (Bristol: Adam Hilger), p. 412.

Freedman, R., and Mazenko, G. F., 1975, *Phys. Rev. Lett.*, **34,** 1575.

Frenkel, J., 1946, *Kinetic Theory of Liquids* (Oxford: Clarendon), p. 374.

Garber, S. R., and Smolenko, L. A., 1973, *Zh. éksp. teor. Fiz.*, **64,** 181: *Soviet Phys. JETP*, **37,** 94.

Gebhardt, W., and Müller-Lierheim, W., 1978, *Proceedings of the International Conference on Lattice Dynamics*, edited by M. Balkanski (Paris: Flammarion), p. 689.

Goldstone, J., and Jackiw, R., 1975, *Phys. Rev.* D, **11,** 1486.

Grinstein, G., and Luther, A. H., 1976, *Phys. Rev.* B, **13,** 1329.

Grinstein, G., Ma, S. K., and Mazenko, G. F., 1977, *Phys. Rev.* B, **15,** 258.

Halperin, B. I., and Hohenberg, P. C., 1967, *Phys. Rev. Lett.*, **19,** 700.

Halperin, B. I., Hohenberg, P. C., and Ma, S. K., 1972, *Phys. Rev. Lett.*, **29,** 1548; 1974, *Phys. Rev.* B, **10,** 139; 1976, *Ibid.*, **13,** 4119.

Halperin, B. I., and Varma, C. M., 1976, *Phys. Rev.* B, **14,** 4030.

Harris, A. B., 1974, *J. Phys.* C, **7,** 1671.

Hastings, J. B., Shapiro, S. M., and Frazer, B. C., 1978, *Phys. Rev. Lett.*, **40,** 237.

Heller, P., 1970, *Int. J. Magn.*, **1,** 53.

Hisano, K., and Toda, K., 1978, *Solid St. Commun.*, **27,** 915.

Höchli, U. T., Weibel, H. E., and Boatner, L. A., 1977, *Phys. Rev. Lett.*, **39,** 1158; 1978, *Ibid.*, **41,** 1410; 1979, *J. Phys.* C, **12,** L563.

Höck, K. H., Schaffer, R., and Thomas, H., 1979, *Z. Phys.* B, **36,** 151.

Höck, K. H., and Thomas, H., 1977, *Z. Phys.* B, **27,** 267.

Hohenberg, P. C., and Halperin, B. I., 1977, *Rev. mod. Phys.*, **49,** 435.

Hosea, T. J., Lockwood, D. J., and Taylor, W., 1979, *J. Phys.* C, **12,** 387.

Hosea, T. J., Taylor, W., and Lockwood, D. J., 1978, *Ferroelectrics*, **21,** 321.

Imry, Y., and Ma, S. K., 1975, *Phys. Rev. Lett.*, **35,** 1399.

IWASAKI, H., MIYAZAWA, S., KOIZUMI, H., SUGII, K., and NIIZEKI, N., 1972, *J. appl. Phys.*, **43,** 4907.

JACKIW, R., 1977, *Rev. mod. Phys.*, **49,** 681.

KADANOFF, L. P., 1971, *Critical Phenomena*, edited by M. S. Green (New York: Academic Press), p. 118.

KAMINOW, I. P., and DAMEN, T. C., 1968, *Phys. Rev. Lett.*, **20,** 1105.

KATIYAR, R. S., RYAN, J. F., and SCOTT, J. F., 1971, *Light Scattering in Solids*, edited by M. Balkanski (Paris: Flammarion), p. 436.

KERR, W. C., see Bishop and Schneider (1978) p. 150.

KHMELNITSKII, D. E., 1975, *Zh. ésp. teor. Fiz.*, **68,** 1960; *Soviet Phys. JETP*, **41,** 981.

KIRKPATRICK, S., and HARRIS, A. B., 1975, *Phys. Rev.* B, **12,** 4980.

KJEMS, J. K., SHIRANE, G., MÜLLER, K. A., and SCHEEL, H. J., 1973, *Phys. Rev.* B, **8,** 1119.

KLEIN, R., and WEHNER, R. K., 1968, *Phys. kondens. Mater.*, **8,** 141; 1969, *Ibid.*, **10,** 1.

KOEHLER, T. R., BISHOP, A. R., KRUMHANSL, J. A., and SCHRIEFFER, J. R., 1975, *Solid St. Commun.*, **17,** 1515.

KRETSCHMER, R., BINDER, K., and STAUFFER, D., 1976, *J. statist. Phys.*, **15,** 267.

KRUMHANSL, J. A., and SCHRIEFFER, J. R., 1975, *Phys. Rev.* B, **11,** 3535.

LAGAKOS, N., and CUMMINS, H. Z., 1974, *Phys. Rev.* B, **10,** 1063; 1975, *Phys. Rev. Lett.*, **34,** 883.

LANDAU, L. D., and LIFSHITZ, E. M., 1958, *Statistical Physics* (Oxford: Pergamon Press).

LEVANYUK, A. P., 1976, *Zh. ésp. toer. Fiz.*, **70,** 1253; *Soviet Phys. JETP*, **43,** 652.

LEVANYUK, A. P., OSIPOV, V. V., and SOBYANIN, A. A., 1976, *Theory of Light Scattering in Condensed Matter*, edited by B. Bendow, J. L. Birman and V. M. Agranovich (New York: Plenum), p. 517.

LOCKWOOD, D. J., ARTHUR, J. W., TAYLOR, W., and HOSEA, T. J., 1976, *Solid St. Commun.*, **20,** 703.

LOCKWOOD, D. J., and TORRIE, B. H., 1974 a, *Anharmonic Lattices, Structural Transitions and Melting*, edited by T. Riste (Leiden: Noordhoff), p. 147; 1974 b, *J. Phys*, C, **7,** 2729.

LOWNDES, R. P., TORNBERG, N. E., and LEUNG, R. C., 1974, *Phys. Rev.* B, **10,** 911.

LYONS, K. B., and FLEURY, P. A., 1976, *Phys. Rev. Lett.*, **37,** 161; 1977, *Solid St. Commun.*, **23,** 477; 1978, *Phys. Rev.* B, **17,** 2403.

MAZENKO, G. F., and SAHNI, P. S., 1978, *Phys. Rev.* B, **18,** 6139.

MEHRING, M., and SUWELACK, D., 1979, *Phys. Rev. Lett.*, **42,** 317.

MERMELSTEIN, M. D., and CUMMINS, H. Z., *Phys. Rev.* B, **16,** 2177.

MEYER, G. M., NELMES, R. J., and HUTTON, J., 1978, *Proceedings of the International Conference on Lattice Dynamics*, edited by M. Balkanski (Paris: Flammarion), p. 652.

MEZEI, F., and HAYTER, J., 1974, EPS Conference, Budapest (unpublished).

MOSS, S. C., KEATING, D. T., and AXE, J. D., 1973, *Phase Transitions*, edited by C. E. Cross (New York: Pergamon), p. 179.

MOTT, N. F., 1966, *Phil. Mag.*, **13,** 989.

MOUNTAIN, R. D., 1966 a, *Rev. mod. Phys.*, **38,** 205; 1966 b, *J. Res. natn. Bur. Stand* A, **70,** 207.

MÜLLER, K. A., 1979, *Lecture Notes in Physics*, Vol. 104, edited by C. P. Enz (Berlin: Springer-Verlag), p. 210.

MÜLLER, K. A., and BERLINGER, W., 1978, *Z. Phys.* B, **31,** 151.

MÜLLER, K. A., BERLINGER, W., WEST, C. H., and HELLER, P., 1974, *Phys. Rev. Lett.*, **32,** 160.

MÜLLER, K. A., DALAL, N. S., and BERLINGER, W., 1976, *Phys. Rev. Lett.*, **36,** 1504.

MÜLLER, K. A., and VON WALDKIRCH, TH., 1976, *Local Properties at Phase Transitions*, edited by K. A. Müller and A. Rigamonti (Amsterdam: North-Holland), p. 134.

MURATA, K. K., 1975, *Phys. Rev.* B, **11,** 462; 1976 a, *Ibid.*, **13,** 2028; 1976 b, *Ibid.*, **13,** 4015.

NELSON, D. R., and FISHER, M. E., 1975, *Ann. Phys.*, **91,** 226.

NIKALASSON, G., 1970, *Ann. Phys.*, **59,** 263.

NOOLANDI, J., and VARMA, C. M., 1975, *Phys. Rev.* B, **11,** 4743.

OHNARI, I., and TAKADA, S., 1979, *Prog. theor. Phys.*, **61,** 11.

PEERCY, P. S., 1975, *Phys. Rev.* B, **12,** 2725.

POLLINA, R. J., and GARLAND, C. W., 1975, *Phys. Rev.* B, **12,** 362.

PYTTE, E., 1973, *Phys. Rev.* B, **8,** 3954.

REITER, G. F., BERLINGER, W., MÜLLER, K. A., and HELLER, P., 1980, *Phys. Rev.* B, **21,** 1.

RISTE, T., SAMUELSEN, E. J., OTNES, K., and FEDER, J., 1971, *Solid St. Commun.*, **9,** 1455.

Ryan, J. F., and Hisano, K., 1973, *J. Phys.* C, **6,** 566.

Scalapino, D. J., Sears, M., and Ferrell, R. A., 1972, *Phys. Rev.* B, **6,** 3409.

Scarparo, M. A. F., Katiyar, R. S., Srivastava, R., and Porto, S. P. S., 1978, *Phys. Stat. Sol.* (b), **90,** 543.

Scheel, H. J., 1976, *Z. Kristallog.*, **143,** 417.

Scheel, H. J., Bednorz, J. G., and Dill, P., 1976, *Ferroelectrics*, **13,** 507.

Schmidt, H., and Schwabl, F., 1977, *Phys. Lett.* A, **61,** 476; 1978, *Proceedings of the International Conference on Lattice Dynamics*, edited by M. Balkanski (Paris: Flammarion), p. 748.

Schneider, T., and Stoll, E., 1973, *Phys. Rev. Lett.*, **31,** 1254; 1975, *Ibid.*, **35,** 296; 1976, *Phys. Rev.* B, **13,** 1216; 1978, *Ibid.*, **17,** 1302; 1980 (to be published).

Schulhof, M. P., Nathans, R., Heller, P., and Linz, A., 1971, *Phys. Rev.* B, **4,** 2254.

Schuster, H. G., 1977, *Z. Phys.* B, **27,** 251.

Schwabl, F., 1972, *Phys. Rev. Lett.*, **28,** 500; *Z. Phys.*, **254,** 57; 1973 a, *Solid St. Commun.*, **13,** 181; 1973 b, *Phys. Rev.* B, **7,** 2038; 1974, *Anharmonic Lattices, Structural Transitions and Melting*, edited by T. Riste (Leiden: Noordhoff), p. 87.

Scott, A. C., Chu, F. Y. F., and McLaughlin, D. W., 1973, *Proc. Inst. elect. electron. Engrs.*, **61,** 1443.

Sham, L. J., and Patton, B. R., 1976, *Phys. Rev.* B, **13,** 3151.

Shapiro, S. M., Axe, J. D., Shirane, G., and Riste, T., 1972, *Phys. Rev.* B, **6,** 4332; 1974; *Anharmonic Lattices Structural Transitions and Melting*, edited by T. Riste (Leiden: Noordhoff), p. 135.

She, C. Y., Broberg, T. W., Wall, L. S., and Edwards, D. F., 1972, *Phys. Rev.* B, **6,** 1847.

Siggia, E. D., 1975, *Phys. Rev.* B, **11,** 4736.

Silberglitt, R., 1972, *Solid St. Commun.*, **11,** 247.

Silvermann, B. D., 1970, *Phys. Rev. Lett.*, **25,** 107.

Simanek, E., and Ward, R., 1976, *Solid St. Commun.*, **18,** 841.

Sinha, S. K., 1978, *Proceedings of the International Conference on Lattice Dynamics*, edited by M. Balkanski (Paris: Flammarion), p. 1.

Stamenkovic, S., Plakida, N. M., Aksienov, V. L., and Siklos, T., 1976, *Phys. Rev.* B, **14,** 5080.

Stanley, H. E., 1971, *Introduction to Phase Transitions and Critical Phenomena* (Oxford: Clarendon Press).

Stauffer, D., 1978, *Ferroelectrics*, **18,** 199.

Steigmeier, E. F., and Auderset, H., 1973, *Solid St. Commun.*, **12,** 565; 1975, *Phys. Lett.* A, **51,** 285.

Steigmeier, E. F., Auderset, H., and Harbeke, G., 1973, *Solid St. Commun.*, **12,** 1077; 1974, *Anharmonic Lattices, Structural Transitions and Melting*, edited by T. Riste (Leiden: Noordhoff), p. 153; 1975, *Phys. Stat. Sol.* (b), **70,** 705.

Taylor, W., Lockwood, D. J., and Vass, H., 1978, *Solid St. Commun.*, **27,** 547.

Teitelbaum, G. B., 1975, *JETP Lett.*, **21,** 154.

Thomas, H., 1974, *Anharmonic Lattices, Structural Transitions and Melting*, edited by T. Riste (Leiden: Noordhoff), p. 231.

Töpler, J., Alefeld, B., and Kollmar, A., 1975, *Phys. Lett.* A, **51,** 297.

Töpler, J., Alefeld, B., and Heidemann, A., 1977, *J. Phys.* C, **10,** 635.

Tornberg, N. E., and Lowndes, R. P., 1977, *J. Phys.* C, **10,** L549.

Varma, C. M., 1976, *Phys. Rev.* B, **14,** 244.

von Waldkirch, Th., Müller, K. A., and Berlinger, W., 1973, *Phys. Rev.* B, **7,** 1052.

Wallace, D. J., 1978, see Bishop and Schneider (1978), p. 104.

Wehner, R. K., and Klein, R., 1971, *Physica*, **52,** 92; 1972, *Ibid.*, **62,** 161.

Widom, A., Tornberg, N. E., and Lowndes, R. P., 1978, *J. Phys.* C, **11,** L433.

Wilson, C. M., and Cummins, H. Z., 1971, *Light Scattering in Solids*, edited by M. Balkanski (Paris: Flammarion), p. 420.

Wilson, K. G., 1971, *Phys. Rev.* B, **4,** 3174; 3184.

Yacoby, Y., 1978 a, *Proceedings of the International Conference on Lattice Dynamics*, edited by M. Balkanski (Paris: Flammarion), p. 453; 1978 b, *Z. Phys.* B, **31,** 275.

Yacoby, Y., Cowley, R. A., Hosea, T. J., Lockwood, D. J., and Taylor, W., 1978, *J. Phys.* C, **11,** 5065.

YACOBY, Y., and JUST, S., 1974, *Solid St. Commun.*, **15,** 715.
YACOBY, Y., KRUHLER, W. W., and JUST, S., 1976, *Phys. Rev.* B, **13,** 4132.
YAGI, T., TANAKA, H., and TATSUZAKI, I., 1977, *Phys. Rev. Lett.*, **38,** 609.
YAMAZAKI, Y., 1976, *Prog. theor. Phys.*, **55,** 1733.
YOUNG, A. P., and ELLIOTT, R. J., 1974, *J. Phys.* C, **7,** 2721.

Index